Scientific Research in the Field of Asian Art:

Proceedings of the First Forbes Symposium at the Freer Gallery of Art

Scientific Research in the Field of Asian Art:

Proceedings of the First Forbes Symposium at the Freer Gallery of Art

Edited by Paul Jett

with Janet G. Douglas, Blythe McCarthy, and John Winter

Archetype Publications

in association with the Freer Gallery of Art, Smithsonian Institution

First published 2003 by Archetype Publications Ltd.

Archetype Publications Ltd.
6 Fitzroy Square
London W1T 5HJ

www.archetype.co.uk

Tel: 44(207) 380 0800
Fax: 44(207) 380 0500

Smithsonian
Freer Gallery of Art

ISBN 1-873132-38-7

British Library Cataloguing in Publication Data
A catalogue record for this book is available from the British Library.

Edited by Jennifer Alt
Cover design by Kimberly Glyder
Cover illustration: Ring. China, Eastern Zhou period, 5th–4th century B.C.E. Jade, 7.0 cm. Purchase, F1939.33

Typeset by Kate Williams, Abergavenny
Printed and bound in Great Britain by Antony Rowe Ltd., Chippenham, Wilts

Contents

Metalworking Traditions

Paintings

Glass, Jade, and Ceramics

Foreword

In late September 2001, the Freer Gallery of Art and the Arthur M. Sackler Gallery held their first symposium on scientific research in the field of Asian art. The symposium was supported by the Forbes Fund, a bequest given to the Freer Gallery by John Thacher, former director of the Dumbarton Oaks Research Library and Collection. Mr. Thacher studied at Harvard University, and, under the tutelage of Edward Waldo Forbes, director at the Fogg Art Museum, came to believe in the importance of using technical study to understand works of art. In honor of his mentor, Mr. Thacher left his bequest to the Freer with the intent that the Forbes Fund be used to advance the scientific study of the care, conservation, and preservation of works of art. To that end, the fund has been used for a number of purposes.

In addition to the symposium, activities under the aegis of the Forbes Fund have included many pursuits. First, it has made possible the creation of a fellowship for scientific research at the Freer. Fellows from Korea, China, Sri Lanka, and the United States have come to the gallery to pursue conservation research on such topics as ancient Chinese bronze casting, binding media for Central Asian manuscripts, and the physical nature, deterioration, and conservation of Sri Lankan mural paintings. Fellowship support also enabled two of the fellows to complete their Ph.D. dissertations. The Forbes Fund has also enabled the Freer to bring in short-term guest workers and guest lecturers in order to pursue or give presentations on their current research. The research topics covered included the study of the megalithic copper alloys of India and the use of amber in China. Moreover, the fund gave support for the study and conservation of Mughal paintings and an ancient Chinese manuscript, and for the overall scientific research effort of the Freer and Sackler's staff.

In addition, in conjunction with the Middle Eastern Culture Center in Japan, financial support was provided for the publication of Field Notes, two series of pamphlets in Turkish and English that cover a variety of practical conservation issues for archaeological conservators working in Turkey.

We feel a tremendous sense of both gratitude and responsibility to Mr. Thacher for his bequest. In both his memory and that of Mr. Forbes, the Freer and Sackler galleries will continue to advance the scientific study and technical understanding of works of art. The first Forbes Symposium on Scientific Research in Asian Art is, we hope, a worthy contribution to this effort, and the first symposium of many.

Julian Raby
Director
Freer Gallery of Art and Arthur M. Sackler Gallery

Acknowledgments

I would like to thank the members of the Department of Conservation and Scientific Research for all of their efforts in organizing and carrying out the symposium. Conservation scientists Blythe McCarthy and Janet Douglas and conservation technician LaHoma Lee, who were instrumental in the successful result of this effort, deserve particular mention. Audiovisual specialist Andy Finch ably dealt with the variety of presentation media that the speakers had and with many last-minute arrangements. Special events coordinator Caroline Bedinger, as usual, was helpful above and beyond the call of duty. Milo Beach, former director of the Freer Gallery of Art and Arthur M. Sackler Gallery; Vidya Dehejia, former deputy director of the Freer and Sackler galleries; and Tom Lentz, director of the International Art Museums, have always been supportive and encouraging of the Department of Conservation and Scientific Research and its efforts, and they were particularly so for the symposium.

Senior conservation scientist John Winter contributed to the selection of the papers and contributed to the editing of the papers. Lynne Shaner, head of publications, and Jennifer Alt, assistant editor, provided both guidance and a tremendous amount of assistance in making these proceedings a reality. Also, the support of Dr. Julian Raby, director of the Freer and Sackler galleries, for the publication of the proceedings is greatly appreciated. Last, I would like to express heartfelt thanks to the speakers at the symposium and the authors of the papers published here. It was the speakers who made the meeting the success that it was and who braved more trying circumstances than usual in coming to Washington, D.C., in late September of 2001.

Paul Jett
Head, Department of Conservation and Scientific Research
Freer Gallery of Art and Arthur M. Sackler Gallery

List of Contributors

Pierre BAPTISTE
Musée Guimet
Paris, France

Thierry BOREL
Centre for Research and Restoration of the French Museums
Paris, France

David BOURGARIT
Centre for Research and Restoration of the French Museums
Paris, France

Ellen Salzman CHASE
Freer Gallery of Art/Arthur M. Sackler Gallery
Smithsonian Institution
Washington, D.C., USA

W. T. CHASE
Chase Art Services
Bethesda, Md., USA

CHEN Jianli
Institute of the History of Metallurgy and Materials
University of Science and Technology
Beijing, China

Michael COWELL
Department of Scientific Research
The British Museum
London, UK

Paul CRADDOCK
Department of Scientific Research
The British Museum
London, UK

Janet G. DOUGLAS
Freer Gallery of Art/Arthur M. Sackler Gallery
Smithsonian Institution
Washington, D.C., USA

Kate DUFFY
Williamstown Art Conservation Center
Williamstown, Mass., USA

Soodabeh DURALI
Institut für Mineralogie
Universität Frankfurt
Frankfurt, Germany

Jacki ELGAR
Museum of Fine Arts
Boston, Mass., USA

Elisabeth West FITZHUGH
Freer Gallery of Art/Arthur M. Sackler Gallery
Smithsonian Institution
Washington, D.C., USA

Jennifer GIACCAI
Freer Gallery of Art/Arthur M. Sackler Gallery
Smithsonian Institution
Washington, D.C., USA

HAN Rubin
Institute of the History of Metallurgy and Materials
University of Science and Technology
Beijing, China

Paul JETT
Freer Gallery of Art/Arthur M. Sackler Gallery
Smithsonian Institution
Washington, D.C., USA

Hyung Tae KANG
Department of Conservation Science
National Museum of Korea
Seoul, Korea

Takayasu KOEZUKA
Center for Archaeological Operation
Nara National Research Institute for Cultural Properties
Nara, Japan

Mohammad LAMEHI
Atomic Energy Organization of Iran
Tehran, Iran

Susan LA NIECE
Department of Scientific Research
The British Museum
London, UK

Marco LEONA
Los Angles County Museum of Art
Los Angeles, Calif., USA

LIAN Haiping
Research Laboratory for Conservation and Archaeology
Shanghai Museum
Shanghai, China

MA Qinglin
Analytical and Conservation Laboratory
Gansu Provincial Museum
Lanzhou, China

Blythe MCCARTHY
Freer Gallery of Art/Arthur M. Sackler Gallery
Smithsonian Institution
Washington, D.C., USA

MEI Jianjun
The Needham Research Institute
Cambridge, UK

C. MICHAELSON
Department of Oriental Antiquities
The British Museum
London, UK

Benoit MILLE
Centre for Research and Restoration of the French Museums
Paris, France

Mitsuru MIMURA
Nara National Research Institute for Cultural Properties
Nara, Japan

Ryu MURAKAMI
Nara National Research Institute for Cultural Properties
Nara, Japan

B. D. NANDADEVA
Department of Fine Arts
University of Kelaniya
Kelaniya, Sri Lanka

Jack OGDEN
Osmiridium Consulting
Hildesheim, Germany

Parvin OLIAIY
Atomic Energy Organization of Iran
Tehran, Iran

Javad RAHIGHI
Atomic Energy Organization of Iran
Tehran, Iran

Jessica RAWSON
Merton College
Oxford University
Oxford, UK

Chandra L. REEDY
University of Delaware
Newark, Del., USA

Josef RIEDERER
Rathgen Research Laboratory
Museums of Berlin
Berlin, Germany

Masanori SATO
Nara National Research Institute for Cultural Properties
Nara, Japan

Masaaki SAWADA
Nara National Research Institute for Cultural Properties
Nara, Japan

D. A. SCOTT
Getty Conservation Institute
Los Angeles, Calif., USA

Farah SHOKOUHI
Atomic Energy Organization of Iran
Tehran, Iran

SU Rongyu
Institute for the History of Natural Sciences
Chinese Academy of Sciences
Beijing, China

TAN Derui
Research Laboratory for Conservation and Archaeology
Shanghai Museum
Shanghai, China

John TWILLEY
Conservation Scientist
Hawthorne, N.Y., USA

WAN Xin
Institute of Cultural Relics and Archaeology of Liaoning Province
Shenyang, China

H. WANG
Department of Coins and Medals
The British Museum
London, UK

M. L. WAYMAN
Department of Chemical and Materials Engineering
University of Alberta
Edmonton, Alberta, Canada

John WINTER
Freer Gallery of Art/Arthur M. Sackler Gallery
Smithsonian Institution
Washington, D.C., USA

Kazuo YAMASAKI
Nagoya University
Nagoya, Japan

Heisun YU
Department of Conservation Science
National Museum of Korea
Seoul, Korea

Thierry ZÉPHIR
Musée Guimet
Paris, France

Introduction

The vast realm comprised by the subject of Asian art is in many ways beyond comprehension. In time and space alone it stretches from the Neolithic period to the present, from the edge of Europe east to the tip of North America, and from the Arctic Circle to the equator. The part of the story of Asian art that scientific research may reveal is a very small part of the whole, yet even the telling of a small part of the story helps to bring to life the artists of the past and their worlds. For those who believe art is important, this is no mean accomplishment. And for those who strive to understand the world today, the world of the past is the bedrock for that understanding. These are the fundamental reasons why scientific research in the field of Asian art is undertaken, why it is important, and why we here present proceedings from a symposium on the topic.

These proceedings are the papers chosen for a symposium held at the Freer Gallery of Art in the fall of 2001. The timing of the symposium was intended to celebrate the fiftieth anniversary of the pursuit of scientific research at the Freer. A tradition of using scientific methods of research as part of the study of works of art began at the Freer in 1951, when R. J. Gettens left his position as research chemist at the Fogg Art Museum at Harvard University and came to Washington. Mr. Gettens founded the Technical Laboratory at the Freer, and for the next seventeen years until his retirement, he built upon an already distinguished body of work. Mr. Gettens's contributions to the advancement of conservation and the use of scientific methods in the study of works of art are well known. Perhaps less recognized but also of great importance was how Mr. Gettens combined the use of scientific methods of research with the efforts of a team of experts—art historians, archaeologists, and epigraphers—to more fully understand the nature of works of art. The two-volume publication on the Freer's collection of ancient Chinese bronzes is a stellar example of such a collaboration. This approach to the study of collections has served as a model for many later projects at the Freer and at other institutions as well.

It may be that Mr. Gettens's inspiration for his collaborative approach was his association with Edward Waldo Forbes, director of the Fogg Art Museum from 1909 until 1944. As one of the earliest and strongest supporters of the use of scientific methods in the study of works of art, Mr. Forbes founded the Laboratory for Conservation and Technical Research at the Fogg in 1928. Mr. Forbes seems to have instilled in those around him his belief that scientific methods of study can make important contributions to our understanding of works of art. Another one of his associates at the Fogg was John S. Thacher, who went on to become director of the Dumbarton Oaks Research Library and Collection in Washington, D.C. When Mr. Thacher passed away in 1982, the Freer Gallery received a generous bequest from his estate for the establishment of a fund for the scientific study of works of art. Mr. Thacher directed that the fund be named after Mr. Forbes. The Forbes Fund has supported a range of research activities that include fellowships, publications, and lectures, and it was this fund that made the symposium and the publication of the papers possible.

While the occasion for the symposium was the fiftieth anniversary of scientific research at the Freer, the underlying and far more important purpose of the symposium was to survey the current state of scientific research in the field and to bring together researchers from Asia, Europe, and America. It was gratifying to learn of the amount of research that was being undertaken, but it was also frustrating to learn that the symposium could only offer a small sampling of that research. We hope that future symposia will provide forums for continued discussion and presentation of work in the field.

A few of the papers included here were not presented at the meeting due to the inability of some of the speakers to travel at the time of the symposium because of circumstances beyond their control. Also, while some were unable to attend the symposium, many speakers and attendees overcame unusual difficulties and concerns with regard to travel in order to make the trip to Washington and join in the meeting. The efforts were made both by those who came to the symposium and those who tried but could not give testimony to their commitment to the study of Asian art. That commitment is inspiring, and we hope that these proceedings, in at least some small measure, pay tribute to those persons.

Paul Jett

Ornament

1 Connections between Islam, Europe, and the Far East in the Medieval Period: The Evidence of Jewelry Technology

Jack Ogden

ABSTRACT A preliminary microscopic examination of gold jewelry from the Far East, including China and Java, suggests three main waves of influence. The wires made by the strip-twist technique, loop-in-loop chains, and granulation were all in use in China before the beginning of the Christian era, and an introduction from the West seems more likely than a wholly independent development. The use of beaded wires in Javanese gold jewelry dating from the Early Classic period (ca. seventh to tenth century) probably reflects Hindu/Buddhist influence, and strip-twist wire was possibly introduced in Java at the same time. No evidence has been found for Islamic influence on Javanese or other Far Eastern goldworking techniques before the second millennium C.E., although it might then be manifested in the adoption of the wire-drawing technique. However, the earliest certain use of wire drawing identified to date is in Korea, and the technique may have ultimately originated and been disseminated from there.

Discussion

Gold jewelry that has survived from the past is seldom blighted by the corrosion or decay that affects most other material remains, and so it will often retain near-original surfaces and tool marks. From the microscopic examination of these tool marks we can deduce much about the methods of manufacture used in the construction of a gold object and combine this knowledge with other art-historical and technical evidence to discuss which factors influenced how, when, and why those techniques were used. Then, in turn, we can make proposals regarding chronology and cultural links.

Today there are few ancient goldsmithing techniques that remain a mystery to us, and researchers are now considering how and when these techniques were combined and what characteristics can be used to differentiate regions, workshops, and even craftsmen. Indeed, we are beginning to be able to identify individual workshops in some instances, and there is undoubtedly enormous potential here (Williams 1988; Williams and Ogden 1994). However, work is hampered by the relatively low number of researchers and the view, still too frequently encountered, that gold is simply "treasure" and so is denied the detailed study applied to most other categories of artifacts in excavation reports and catalogues.

The present study developed out of my work on jewelry produced in the ancient Old World and in medieval European and Islamic lands (for example, Ogden 1982, 1987, 1994). It was prompted by the examination of a Korean earring that raised a question as to the nature, if any, of links between the Far East and the Islamic and European worlds between about 650 and 950. These three centuries mark the initial rise and expansion of Islam, and, indeed, there has often been an assumption of some influence on or connection with the jewelry then being produced in southeast Asia. For example, a recent scholarly catalogue of Islamic rings saw fit to include several gold rings of well-known pre-tenth century southeast Asian or Javanese types (Wenzel 1993, 38).[1] The purpose of the present study was to see if technical examination of the jewelry of these three centuries could shed any light on such connections.

A problem that immediately arises that makes the study both more pertinent and more difficult is the dearth of relevant material to examine from Europe or the Near East. Minimal jewelry survives from Byzantium or Islamic lands that can be securely dated to the eighth to tenth century. This is not the place to discuss the reasons for this lack of gold jewelry—including, perhaps, the rise of Islam with its associated redirection of gold to military expansion, the not-unrelated move to iconoclasm in the Byzantine church, and so on—but it is true that if a link could be found between East and West, then the East might usefully help fill in some of the gaps in our knowledge concerning the development of jewelry technology in the West.

There was a wide variety of goldsmithing techniques used in the past that might help us to unravel connections and chronologies. However, here we will look at just one: the production of gold wire. The benefit of considering wire is that it is present in one form or another in a large proportion of ancient goldwork and that there was a limited variety of production methods. Within reason, these production methods are usually relatively easy to distinguish with a microscope or even a hand lens and so can be noted in catalogues and excavation reports. There are many other aspects of jewelry manufacture that can provide useful

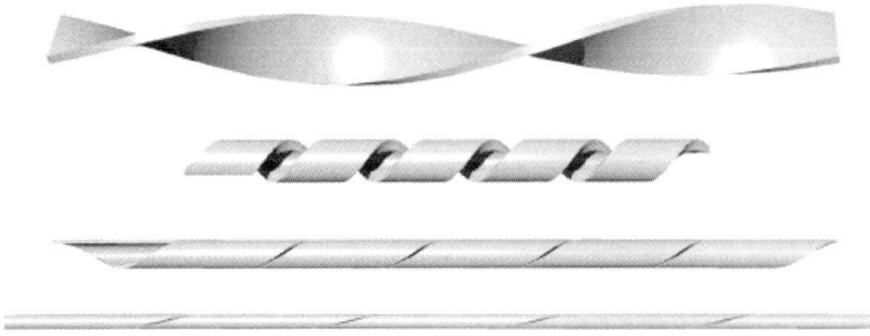

Figure 1. *Drawing showing the manufacture of strip-twist wire. A narrow strip of wire is twisted and then finally rolled to obtain an even, circular section wire, though typically retaining spiral seam lines. Copyright 2001 Jack Ogden*

information, such as the composition of solder alloys, but these typically require specialist resources.

The methods of production of gold wire in antiquity have been the subject of considerable study since Williams and Hein's pioneering work in the 1920s (Williams 1924). To sum up briefly, the gold wire used in most jewelry in the Old World between about 2000 B.C.E. and around 800 C.E. was made by twisting and then rolling smooth a strip of gold. The basic process is shown in figure 1. This technique is usually called the strip-twist technique, and although subdivisions have been proposed, it is my view that variations relate more to the sheet-thickness-to-width proportions of the initial strip used than to any conscious production alternatives (Ogden 1991).

Strip-twist wires are characterized by spiral seam lines that wind along the wires. The extent to which these seam lines remain visible will depend on various factors, including the degree of compaction; the heating processes, such as soldering during manufacture; and the amount of wear during use. Nevertheless, most ancient gold wires produced by the strip-twist technique will reveal at least some vestiges of the original seam lines that can be spotted under modest magnification. The appearance of a typical example, from a Roman gold chain, is shown in figure 2 (similar wire can also be seen in figure 6, from an Early Classic Javanese ornament). Wires over about 2 mm in diameter, such as earring hooks, were more usually hammered to shape.

In Europe, wire drawing superseded strip-twist wire manufacture around 800 (Oddy 1977; Whitfield 1990; Ogden 1991). There is not enough extant Islamic gold jewelry that can be securely dated to this period to prove a similar date of introduction, but strip-twist wires were normal in Byzantine jewelry well into the seventh century, and wire drawing was universal in the Byzantine and Islamic worlds by about 1000.

In the wire-drawing process, a strip of metal is pulled through a series of diminishing-sized holes in an iron or steel plate. This passage through the "drawplate" draws the metal out into an even-sectioned wire of the desired diameter. The wire has to be heated (annealed) at regular intervals to keep it soft and ductile. In practice, wires much over about 2 mm in diameter are tricky to draw by hand, and, as with strip-twist wires, wires of greater diameter were generally hammered to shape until after the introduction of mechanical improvements, such as the draw bench. Some such improved drawing processes came into use in Europe during the Middle Ages, possibly as early as the twelfth century.

It would seem that the drawn wires produced in the medieval European and Islamic worlds were almost invariably circular section, and I have seen little evidence that other section wires were drawn much before the fourteenth century, and even then these were rare in jewelry. For example, the rectangular-section gold strips that play such a major role in medieval Islamic gold filigree work were typically made by hammering flat a drawn circular-section wire.

Drawn wires are characterized by parallel striations that lie along the axis of the wire (fig. 3). As with the seam lines on strip-twisted wires, these striations are usually discernable under magnification, but they are typically fainter and can be all but obliterated by subsequent soldering operations or wear. It must also be remembered that various processes, such as burnishing, can leave marks along the axis of the wire. Care must be taken to distinguish drawn wire with confidence, especially if its presence is being considered as a criterion for forgery.

From what has been said, it is clear that the major change from the strip-twist to the drawing technique for wire manufacture appears to have occurred in the Old World during the very centuries from which we have minimal surviving jewelry. The abundant Dark Age European and early Byzantine jewelry up to the early seventh century C.E. contains just strip-twist wire. The wire in jewelry of two centuries or so later, right across the Old World from Viking Scandinavia to Fatamid Egypt, is all drawn.[2] We thus have

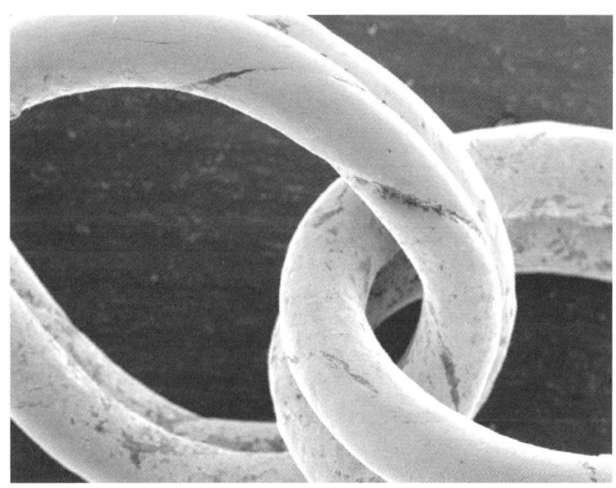

Figure 2. *Example of strip-twist gold wire showing the spiral seam lines. Detail from a Roman-period chain link*

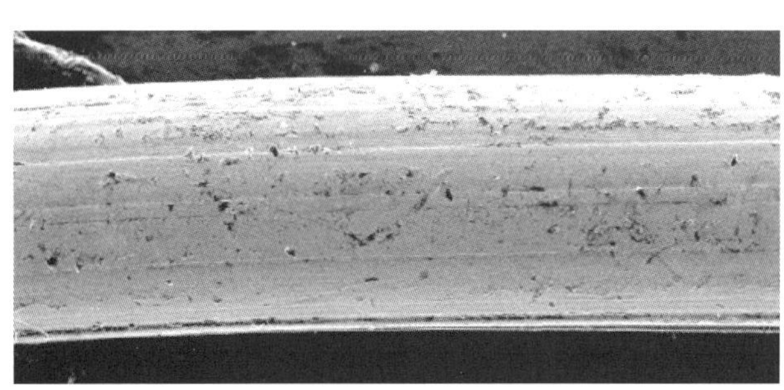

Figure 3. *Example of modern wire, showing the parallel striations along the wire left by the drawing process*

very little evidence to confirm just when and where wire drawing first appeared or how rapidly it superseded the older strip-twist method.

I was understandably intrigued when I had the opportunity to examine an example of Korean goldwork of the Silla kingdom (fifth to sixth century) in 1978 and found the wires to be drawn (Christie, Manson, and Woods 1978). However, it would have been presumptuous to redefine the wire-making chronology on the basis of a single, unprovenanced example.[3] However, my subsequent examination of other Korean goldwork of the Silla period, including a pair of earrings in the British Museum, acquired in 1938 (British Museum OA 1938.5-24.243), confirmed that their wires were indeed drawn (Ogden 1991). If, as it seemed, wire drawing was in use in at least parts of the Far East three centuries prior to clear evidence for its use in the West, might the technique of wire drawing have been introduced to the Islamic and European world from the Far East? Over the last ten years, other researchers have also observed drawn wire on Korean goldwork of the Silla period (Meyers pers. comm.; Douglas pers. comm.).

The next stage was obviously to look at other Far Eastern goldwork contemporary with the Korean Silla kingdom. Until a few years ago, the amount of Chinese jewelry available for study in the West was limited, and much was difficult to date with accuracy. Recently, far more Chinese gold jewelry has become accessible for study, and it is hoped that it can be the subject of a fuller research program in due course. However, my research to date suggests that the strip-twist technique was in use in China well into the Tang period (618–907). For example, the strip-twist technique was used on a pair of elaborate gold hairpins in the British Museum (British Museum OA 1938.5-24.252) and a very similar single pin in the collections of the Metropolitan Museum of Art, New York (Department of Asian Art, 31.54.1). These pins are elaborately decorated with settings for inlays, delicate filigree, and openwork granulation. An eighth-century date has been proposed. In China, the strip-twist technique seemingly outlived the introduction of wire drawing in Korea by at least a century.

I have not had the opportunity to examine later Tang goldwork, but on the basis of photographic evidence only, an adoption of wire drawing in China by the end of the Tang period seems possible but not proven. Wire drawing was certainly widespread in China by the thirteenth to fourteenth century.

Incidentally, the direction of twist in the wires on the pins in the British Museum is, like that in much Chinese goldwork examined so far, opposite to that seen in the vast majority of ancient wire from further West (Douglas and Chase 2001). The single Metropolitan Museum of Art pin, however, is an exception.

When we look further south to the jewelry produced in Java contemporary with the Tang period, we see a similar development in gold-wire technology pattern. The strip-twist process was used for wires in the gold jewelry of the Early Classic period—that is, for at least part of the period in the seventh to tenth century. For example, the wires on the three small gold earrings, or "hawk tags" (their original use is still debated), in plate 1 are all made by the strip-twist process.

So far I have not had the opportunity to examine any of the infrequent Javanese gold jewelry that can be dated to the eleventh to twelfth century, but once we get to the more abundant Javanese jewelry of the Late Classic period, after about 1200, we find that gold wire was drawn (for example, on the ornament in plate 2). Gold jewelry fragments excavated at Kota Cina in northeast Sumatra and dating to the twelfth century included drawn wire (Swan and Scott 1986).

The date of the earliest Islamic influence in Indonesia is still debated. There were Indian trading links with Java by the end of the second century C.E. and seemingly at least an awareness of its existence at this period further west, as shown by the writings of the geographer Ptolemy. The artistic influences most likely to be relevant to Early Classic jewelry were Hindu-Buddhist, presumably predominantly from India, and the same might well be true for technical influences, since the two tend to go hand in hand. Few authorities would now suggest any significant Islamic influence in Indonesia much before the beginning of the twelfth century. The twelfth-century Kota Cina find referred to above included ceramics and coins of Indian, Arab, and Chinese origin (Swan and Scott 1986).

When we turn to the gold jewelry produced around the tenth to twelfth century in other parts of southeast Asia, such as Khmer gold jewelry from Cambodia, the study is hindered by the lack of accurate chronology and the embarrassing abundance of fakes that will confuse all but the most impartial and diligent of future researchers. However, once again, on the basis of the objects I have been able to examine, the same pattern seems to emerge: wire drawing was probably established by around 1000.

On the basis of this limited research, there appears to be little evidence for early Islamic influence on gold-wire manufacture in the Far East, but there is perhaps an argument for a correlation between the apparent date of the widespread adoption of wire drawing in the Far East and the expansion of Islamic influence and trade in the century or so after 1000. However, we cannot ignore the far earlier Korean drawn wires. Either we accept that this was a purely local and isolated development or we propose that wire-drawing technology may have begun in the Far East and spread westward with the growing trade links in the days of the early Byzantine and Sasanian empires.

If this study of wire technology poses as many questions as it answers regarding East–West links, can any other jewelry techniques help us? If we remain with wire and look at various types of decorative wires, much the same picture becomes apparent. For example, the detail of Javanese wire shown in figure 4 is a type of decorative wire called beaded wire. The spiral seam left by the original strip-twist process used to make the wire can still be seen even though the wire was subsequently rolled under an edge to deform it into a series of grooves to impart a beaded appearance. The process is shown in figure 5. In an alternative version, spiral-beaded wire is rolled under an oblique edge to produce a groove almost like a screw thread along the wire (figs. 6 and 7).

Beaded and spiral-beaded wires appeared in the Mediterranean world in about the seventh century (possibly helped by the introduction of iron tools) and rapidly became a common decorative motif in jewelry in the Mediterranean

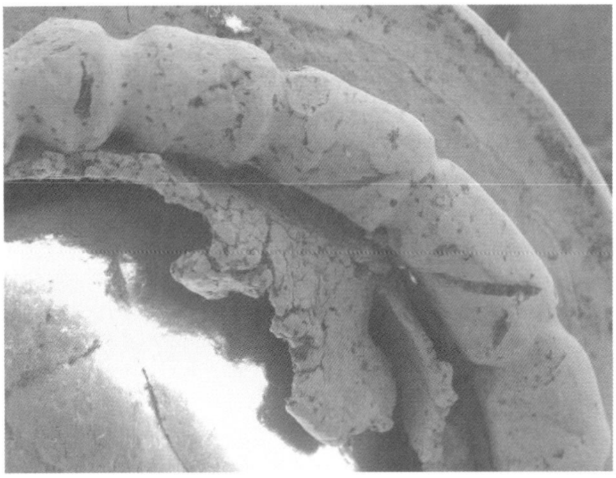

Figure 4. *Detail of beaded wire from one of the ornaments in plate 1, Java, Early Classic period, private collection*

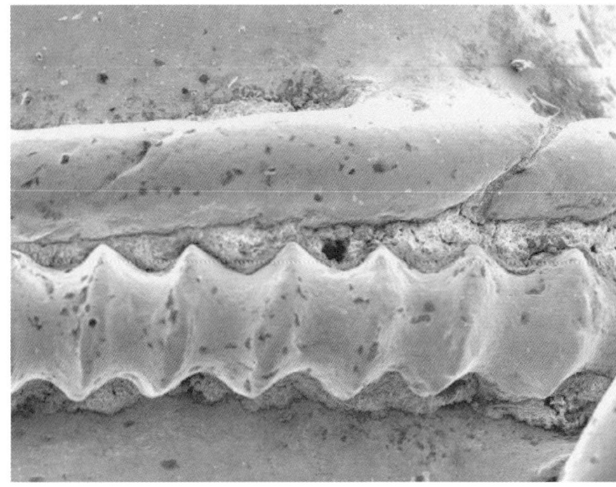

Figure 6. *Detail of spiral-beaded wire from one of the ornaments in plate 1, Java, Early Classic period, private collection*

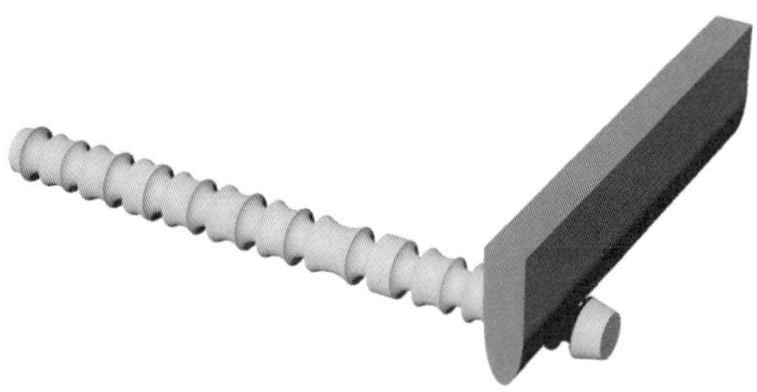

Figure 5. *Drawing showing one method of manufacture for beaded wire. The wire is rolled under a single edge to produce a series of grooves. Copyright 2001 Jack Ogden*

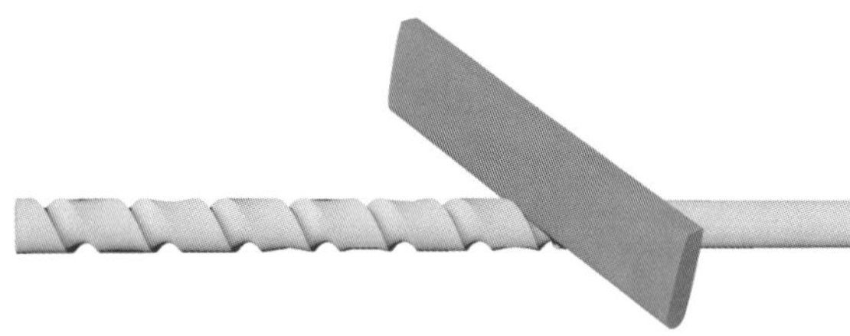

Figure 7. *Drawing showing the manufacture of spiral-beaded wire. The wire is rolled under a single edge held at an angle to produce a groove spiraling along the wire. Copyright 2001 Jack Ogden*

region (Williams and Ogden 1994), continuing to be used in medieval European and Byzantine jewelry both before and after the eighth-to-tenth-century "jewelry hiatus." Beaded wire and its variants were not common in the Hellenistic or Roman East or in contemporary jewelry from India or Gandara but had spread east at least as far as India by or during Sasanian times. The Javanese jewelry that can be dated to the Early Classic period often incorporates beaded and spiral-beaded wires. However, the introduction of this decorative wire into Java from the Islamic world seems highly improbable—beaded wire is conspicuous by its absence in early Islamic gold jewelry. So again, here is evidence that might count against any Islamic influence on Early Classic Javanese jewelry manufacture. We might propose that beaded and spiral-beaded wires reached Java with direct or indirect Buddhist influence from India. In a future study, the implications of the apparent dearth of beaded wires in Tang Chinese goldwork must be considered.

Jewelry historians divide technology into two convenient although not always wholly distinct categories. There are the basic technologies (wire drawing and sheet-gold manufacture, for example) and the applications of those technologies (that is, how the wire and sheet were shaped and decorated). This division can be useful when using jewelry technology as evidence for cultural contact.

As an example, we can take the basic technology of wire drawing and consider two jewelry applications that depended on the long, regular lengths of wire that wire drawing made available. The first is a springlike coil of wire—helicoid wire—that can resemble and may have been an alternative to beaded wire. An example in figure 8 is from the Later Classic Javanese ornament in plate 2, and indeed, this type of decorative wire is quite common in Late Classic Javanese goldwork and in some other categories of southeast Asian jewelry—for example, on the terminals of Cambodian gold necklaces dated to the tenth to twelfth century. However, helicoid wire is almost unknown in medieval Islamic goldwork of this period.

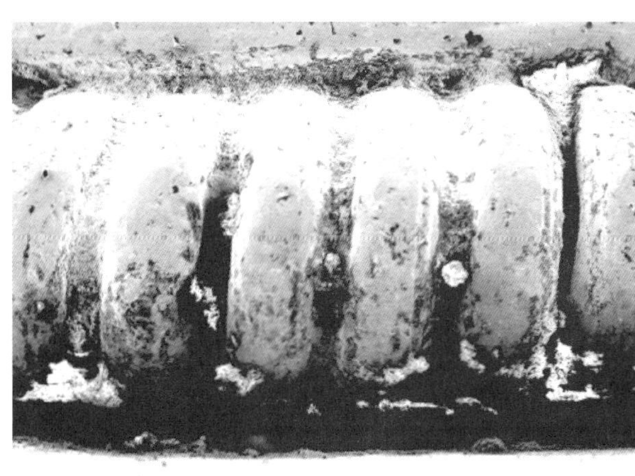

Figure 8. *Detail of helicoid-wire decoration from the ornament in plate 2, Java, Late Classic period, private collection*

This brings us to a second application of drawn wire. In southeast Asia this helicoidal-wire decoration is often seen on the terminals of chains that resemble complex loop-in-loop chains; these chains are one of the most common forms in Khmer jewelry. In fact, these chains are typically a sort of woven wire tube, rather like the French or spool knitting produced by children with a length of wool and a tubular spool (figs. 9 and 10). This type of braided wire was popular in Western Europe in the ninth and tenth centuries and was seemingly described by Theophilus in the early twelfth century in the sixty-second chapter of his third book (Hawthorne and Smith 1979).

Braided chains of this type are not unknown in Islamic jewelry of about the eleventh to twelfth century. An Eastern example is from the Chimkent hoard that probably dates to between the eleventh and thirteenth centuries,[4] but that is uncommon in Islamic jewelry. Therefore, although the basic technique of wire drawing may have spread with Islamic influence in the East, the evidence of the two most obvious decorative applications of drawn wire implies that there was little medieval Islamic influence on Far Eastern jewelry design or within Far Eastern jewelry-making centers. The braided chains and helicoid wire in medieval European jewelry suggest we might need to consider alternative cultural links.

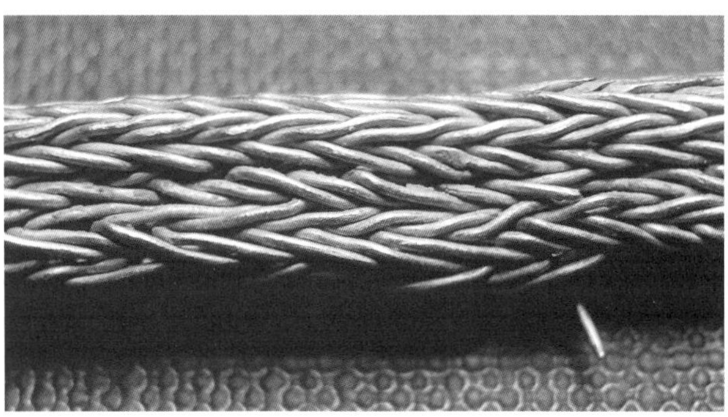

Figure 9. *Detail of braided chain of gold wire, southeast Asia, probably Cambodia, ca. tenth century* C.E., *private collection. Photo by Jack Ogden*

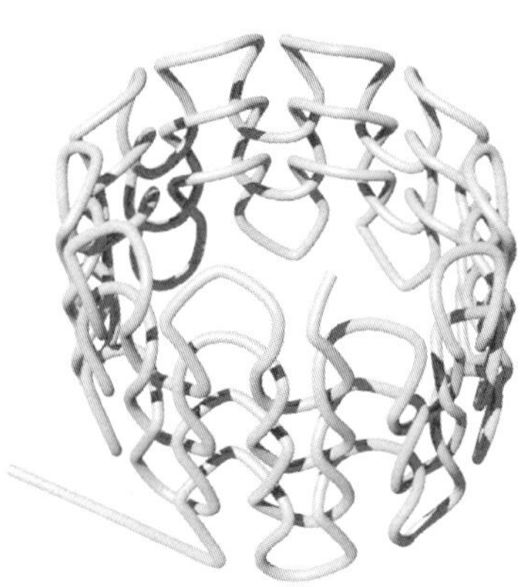

Figure 10. *Drawing showing the construction of a braided chain of the type in figure 9 (drawn in a far looser weave to show construction). Copyright 2001 Jack Ogden*

Acknowledgments

I am grateful to many people who have allowed me to examine Islamic, Oriental, and other medieval jewelry over the years and to those who have facilitated such examination. In particular, I would like to thank David Khalili and the Nour Foundation, Sue La Niece, Nigel Meeks, Carol Michaelson, and Benjamin Zucker.

Notes

1. I would be wary of assuming an Islamic connection solely on the basis of the inscribed gemstones now set in some of these rings.
2. The continued use of the strip-twist method for some eleventh-century Fatamid gold jewelry has been suggested and is not impossible, but we might suspect confusion due to the visual similarity between strip-twist wires and a decoration formed by twisting two wires together (so-called rope) or by twisting a single strip into a looser decorative length. Such is seemingly the case with the example of strip-twist wire noted by Keene (Jenkins and Keene 1982, 143), a pendant in the Metropolitan Museum of Art, New York, that I had the opportunity to examine microscopically. The comprehensive Nour collection of Fatamid goldwork I have examined has provided no instances of true strip-twist wires.
3. Similar considerations led to the assumption that the wires on a pair of Korean earrings on the Swiss market in 1990 were recent restorations (Habsburg, Feldman 1990).
4. The construction is seen in some of the better drawings, such as that reproduced in Arne (1914, fig. 52).

References

Arne, T. J. 1914. *La Suède et l'Orient: études archéologiques sur les relations de a Suède et de l'Orient pendant l'âge des Vikings*. Archives d'Études Orientales 8. Uppsala: Appelbergs.

Christie, Manson, and Woods. 1978. *Important Chinese and Korean ceramics, bronzes, enamels, paintings, and works of art.* Auction sale, London, 10 July, lot 232.

Douglas, J. G., and W. T. Chase. 2001. Examination of two Eastern Zhou jade and gold assemblages from Jincun, near Luoyang, Henan Province, China, and implications for the original configuration of the Freer pectoral. *Studies in conservation* 46(1):35–48.

Habsburg, Feldman. 1990. *Gold: Important ancient and ethnic jewellery and works of art in precious metal.* Auction sale, Geneva, 14 May, lot 63.

Hawthorne, J. N., and C. Stanley Smith. 1979. *Theophilus on divers arts*. New York: Dover.

Jenkins, M., and M. Keene. 1982. *Islamic jewelry in the Metropolitan Museum of Art.* New York: Metropolitan Museum of Art.

Miksic, J. N. 1988. *Small finds: Ancient Javanese gold.* Singapore: National Museum.

Oddy, W. A. 1977. The production of wire in antiquity. *Masca journal* 1(2):44–45.

Ogden, J. 1982. *Jewellery of the ancient world*. London: Trefoil Books.

——. 1987. Islamic goldsmithing techniques in the early medieval period. In *Islamic rings and gems,* ed. D. J. Content, 408–27. London: Philip Wilson.

——. 1991. Classical gold wire: Some aspects of its manufacture and use. *Jewellery studies* 5:95–105.

——. 1994. The technology of medieval jewellery. In *Ancient and historic metals: Conservation and scientific research,* ed. D. A. Scott, J. Podany, and B. B. Considine, 153–82. Malibu, Calif.: Getty Conservation Institute.

Les ors de l'archipel Indonésien. 1995. Musée national des Arts Asiatiques-Guimet. 5 May–31 July. Paris: Réunion des Musées Nationaux.

Swan, R., and D. Scott 1986. Gold work from Kota Cina: A technical study. *Southeast Asian archaeology.* Oxford: British Archaeological Reports International Series 561. 261–73.

Wenzel, M. 1993. *Ornament and amulet: Rings of the Islamic lands.* Oxford: Nour Foundation, in association with Azimuth Press and Oxford University Press.

Whitfield, N. 1990. Round wire in the early Middle Ages. *Jewellery studies* 4:13–28.

Williams, C. R. 1924. *Catalogue of Egyptian antiquities, numbers 1–60, gold and silver jewelry and related objects.* New York: New York Historical Society.

Williams, D. 1998. *Three groups of fourth-century south Italian jewellery in the British Museum.* Volume 95 of *Mitteilungen des Deutschen Archaeologischen Instituts Roemische Abteilung.* Mainz: Phillip von Zabern.

Williams, D., and J. Ogden. 1994. *Greek gold: Jewellery of the classical world.* London: British Museum Press.

2 Studies on Archaeological Ambers in Japan

Masanori Sato, Mitsuru Mimura, and Kazuo Yamasaki

ABSTRACT Although amber articles are commonly excavated at archaeological sites in Japan, scientific research on their characteristics has not yet been fully performed. In these studies, the chemical structure of amber was investigated by using microscopic Fourier transform infrared spectroscopy (FTIR) and fluorescence spectroscopy. From microscopic observations on thin film samples, amber is composed of a random mixture of components in either a powdery state or a malleable state (translucent). The infrared spectrum showed a different peak pattern according to the difference in apparent state. Fluorescence spectroscopy also contributed to confirm the difference in states due to the difference in the fluorescence intensity. The fluorescence intensity of the malleable state is stronger compared with that in the powdery state, the former state being richer in the amount of unsaturated structures.

To study further the classification of amber articles in terms of the place of provenance, the infrared spectral peaks for each sample were decomposed by assuming a Gaussian peak function. The correlation among many sample spectra was explained by principal component analysis (PCA). A few groups of different provenance could be distinguished from each other. The classification of sample characteristics by cluster analysis was also useful to distinguish samples of different provenance.

The potential of FTIR microscopy for the investigation of excavated organic materials is great because the method is versatile and requires the least amount of sample.

Introduction

Since the early stage of human history, mankind has utilized a lot of natural materials for various kinds of decoration, not only in daily life but also in expressing their artistic sensibility or for religious purposes. Articles made of amber, a fossil plant resin, are found in various archaeological sites throughout the world even though amber is made of organic components and is hence very fragile and apt to heavily degrade over the long term.

In Europe, ambers found in the Baltic Sea area have already been studied using several kinds of instrumental analysis: thin-layer chromatography, infrared absorption spectroscopy (nondispersive as well as wavelength dispersive), x-ray diffraction, gas chromatography/mass spectroscopy, differential thermal analysis, emission spectroscopy, neutron activation analysis, and C-13 nuclear magnetic resonance (Fraquet 1987; Lambert, Frye, and Poinar 1985).

In Japan one place famous for discovering ambers is Kuji City on the northeast coast of Honshu Island. Ancient amber articles have been found at various sites dating as far back as the Jomon period of ten thousand years ago. Most excavated ambers are usually small beads a few millimeters in diameter. Although the surfaces are usually degraded, the interior of most samples is still in very good condition. In Japan scientific studies on excavated amber articles are still few, except for some studies of provenance using infrared spectroscopy on a dispersive instrument (Muroga 1985). A few years ago, we began to investigate excavated amber articles found at the Heijō-kyō palace site of the Nara period (circa eighth century) in order to estimate their provenance (Sato 1999). Since with cultural properties the removal of even small amounts of materials for analysis should be avoided, we decided to use microscopic FTIR, a versatile procedure for the analysis, because only a small amount of sample (< 1 mg) is required. This is worth mentioning, as the method is practically nondestructive, which is very important when analyzing precious samples from cultural properties.

Fluorescence spectroscopy was also used to further differentiate the samples. Throughout the comparative studies on excavated ambers, we focused our effort to first understand the difference in the microstructure of amber.

Experimental

FTIR microscopy was performed using a Horiba FTIR 520 instrument. The small amount of sample, less than 1 mg, was removed from an amber object using an optical microscope of low magnification. Then it was pressed in a metallic die to make a very thin film. FTIR measurement was then performed in transmittance mode, using sample films laid on the smooth metallic surface.

Peak deconvolution was performed using Igor Pro ver. 3.16 (Wavemetrics); principal component and cluster analyses used Statistica (Statsoft).

Fluorescence spectroscopy was also applied to samples to understand the chemical characteristics in detail. The instrument (Hitachi fluorescence spectrophotometer F-4500) is capable of presenting a three-dimensional plot with fluorescence intensity as contours on a coordinate system of excitation wavelength and emission wavelength.

Over one hundred samples of diverse provenance and different archaeological sites in Japan were measured using the above procedures.

Results

FTIR Spectroscopy

Generally speaking, the infrared spectrum is similar to that of related compounds, such as copal or dammar resins, and rosin (IRUG 2000). Typical infrared spectra of amber found at Kuji district are shown in figure 1 to illustrate briefly the assignment of main absorption peaks, as follows (Mills and White 1987).

In figures 1a and 1b, the small difference of spectral patterns is shown in samples taken from the two different place in Kuji district. Figures 2a and 2b show the difference in spectra of two different states of the same amber. Figure 2a is a spectrum of the powdery state, and figure 2b is a spectrum of the malleable state. The enlarged photograph of figure 3 shows that the amber is not a homogeneous substance but is rather a random mixture of two states. The powdery state is seen in the upper half of the photo, whereas the malleable state is shown in the lower half. The distinct difference in peak patterns of figures 2a and 2b is remarkable in the low wavenumber (1300–1000 cm^{-1}) region.

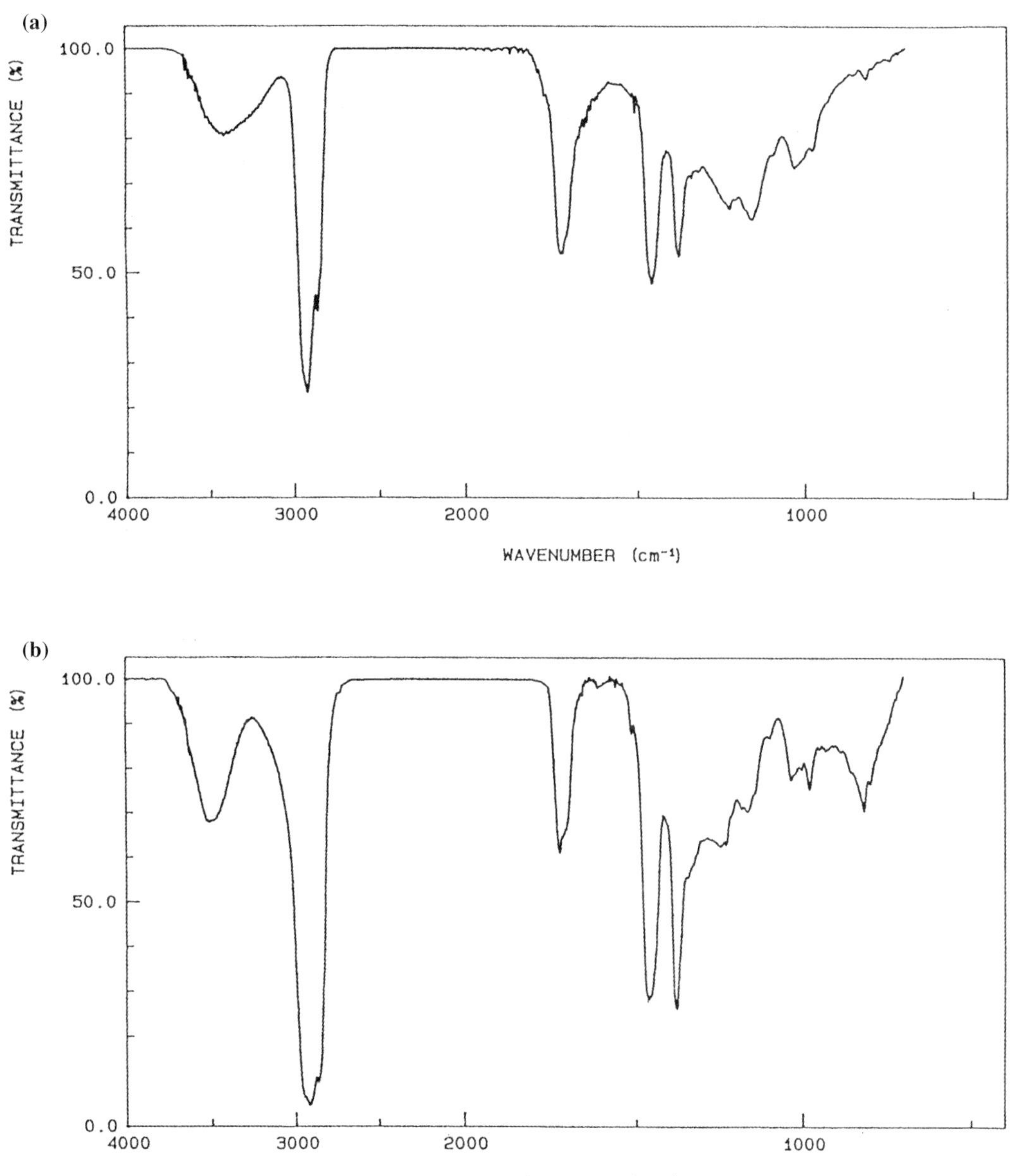

Figure 1. *Infrared spectra of amber (Kuji city)*

(a) Miyauchi district:
circa 1160 cm^{-1}: ester (succinate)
circa 3040, 1640, 885 cm^{-1}: exo-methylene
circa 2900–2800 cm^{-1}: methylene

(b) Ube-kitunebata district
circa 1710–1740 cm^{-1} (broad): ester and carboxyl
circa 1025 cm^{-1}: axial hydroxymethylene
circa 3400 cm^{-1}: hydroxyl

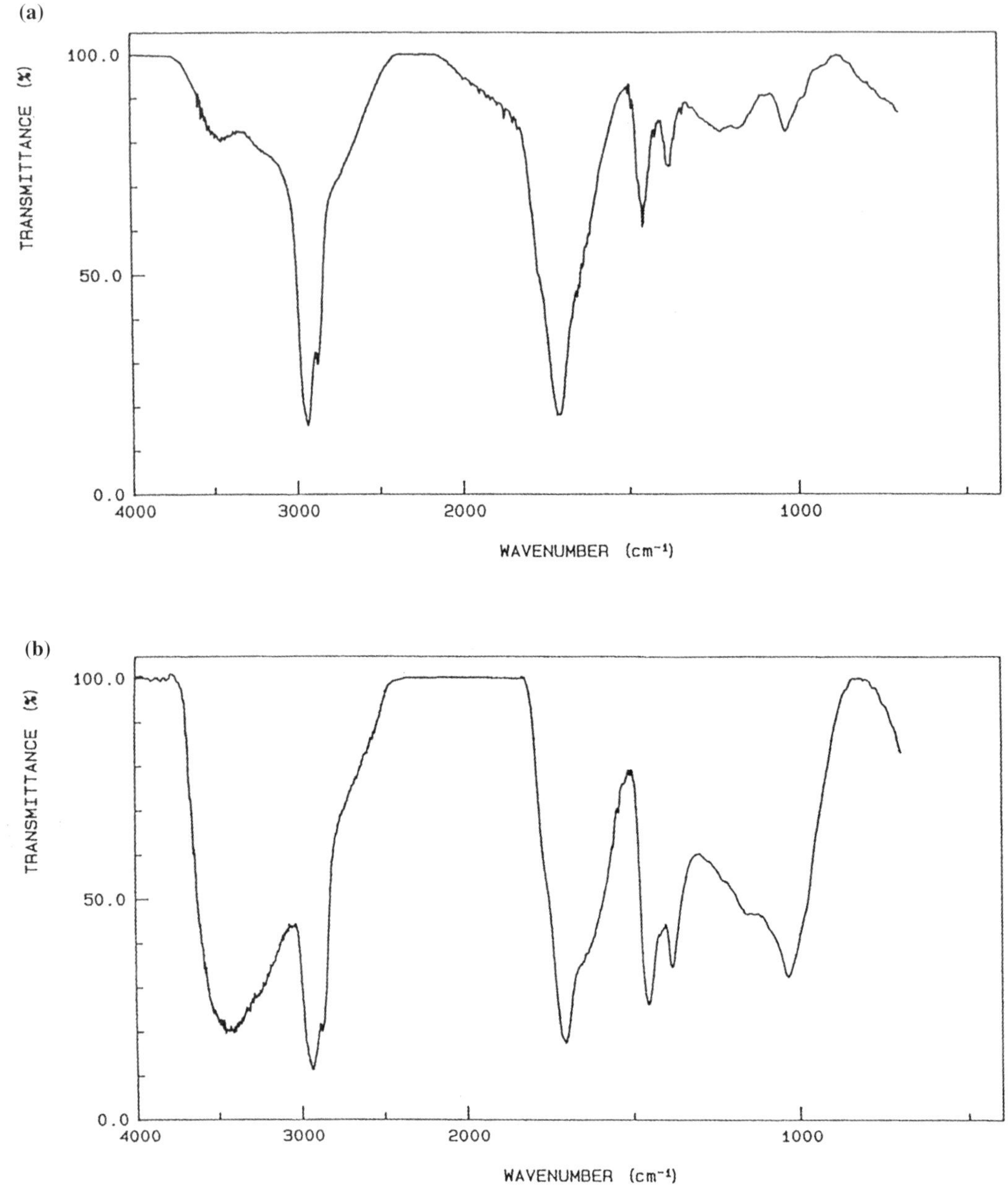

Figure 2. *Infrared spectra of amber beads (Heijō-kyō archaeological site): (a) Powdery state; (b) Malleable state*

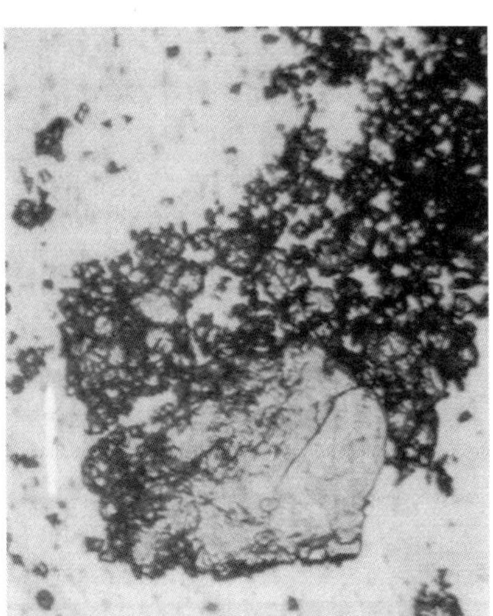

Figure 3. *Photograph of pressed thin layer of amber (upper half: powdery state; lower half: malleable state)*

By comparing spectra of several amber samples, the spectra show a generally similar pattern, irrespective of the provenance. It is therefore rather difficult to differentiate the samples from each other by simply comparing the infrared spectral pattern.

Mathematical Analysis of Infrared Spectra

In the infrared spectrum, most of the peaks show peak broadening due to the superposition of multicomponent systems and the presence of some kind of intermolecular interactions. We proceeded to do peak decomposition into component peaks, assuming the peak shape to be a Gaussian function ($f_{peak}(\nu)$). Here:

$$f_{peak}(\nu) = k_0 \exp\left[-\left(\frac{\nu - k_1}{k_2}\right)^2\right] \quad (1)$$

where k_0 = peak height, k_1 = peak position, and k_2 = peak width. This equation is based on the assumptions that molecules have an isotropic spatial distribution and that specific interactions among molecules are not too strong.

Each broad peak observed in an infrared spectrum could then be decomposed to a few simple peaks ($f_{obs}(\nu)$) by the following equation:

$$f_{obs}(\nu) = \sum_i f_{peak,i}(\nu) \qquad (2)$$

where Σ is the summation for each component *i*.

In figure 4, the decomposed spectrum of an ancient sample is shown. Figure 5 shows the enlarged spectrum of the preceding figure for the region 3100–2500 cm^{-1}. The experimental spectrum is clearly decomposed into three peaks.

Further samples showed the variety of decomposed peak patterns, and this diversity enabled us to distinguish several samples by multivariate analysis.

To reveal correlations of factors among many sample spectra, PCA and cluster analysis were performed. An example of PCA is shown in figure 6, presented in a two-dimensional coordinate system as a function of factors 1 and 2. The factors were extracted by the main component analysis. In this figure, three isolated groups are clearly seen. The one group is attributable to the Kuji district, the second to the Heijō-kyō in Nara district, and the third to Shizuoka district.

An example of cluster analysis is shown in figure 7. The samples excavated from the Heijō-kyō site comprise one unique assembly rather isolated from samples of other districts.

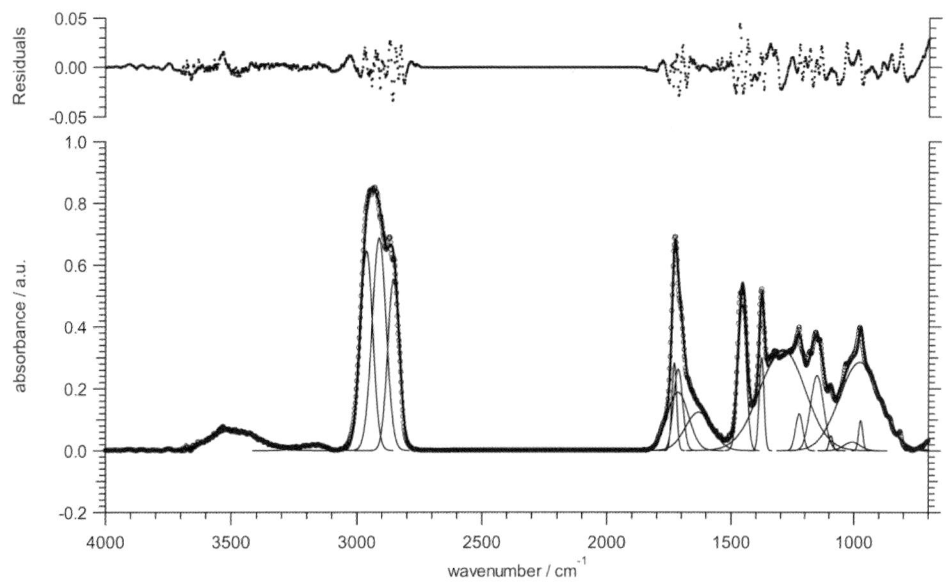

Figure 4. *Decomposed infrared spectrum (sample: Heijō-kyō 801)*

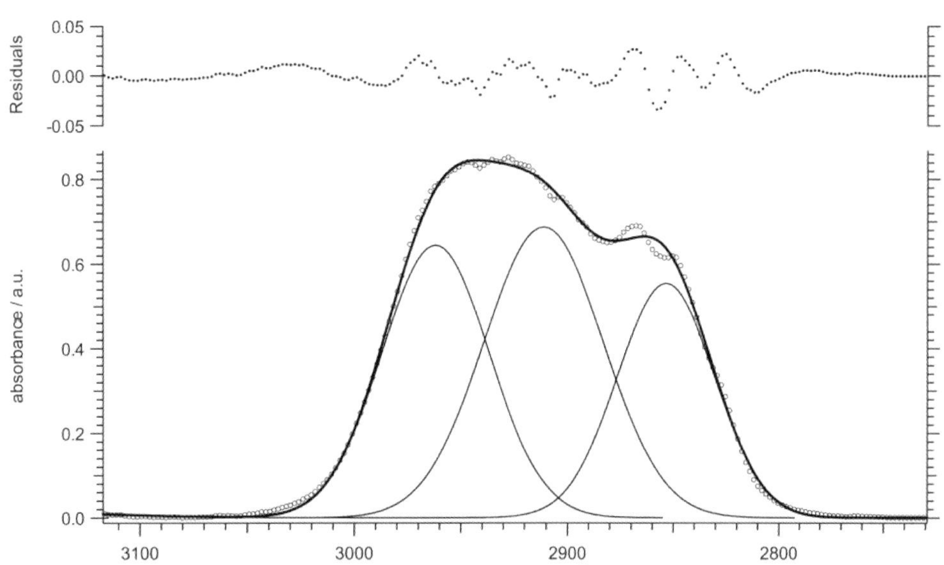

Figure 5. *Decomposed infrared spectrum (detail of figure 4)*

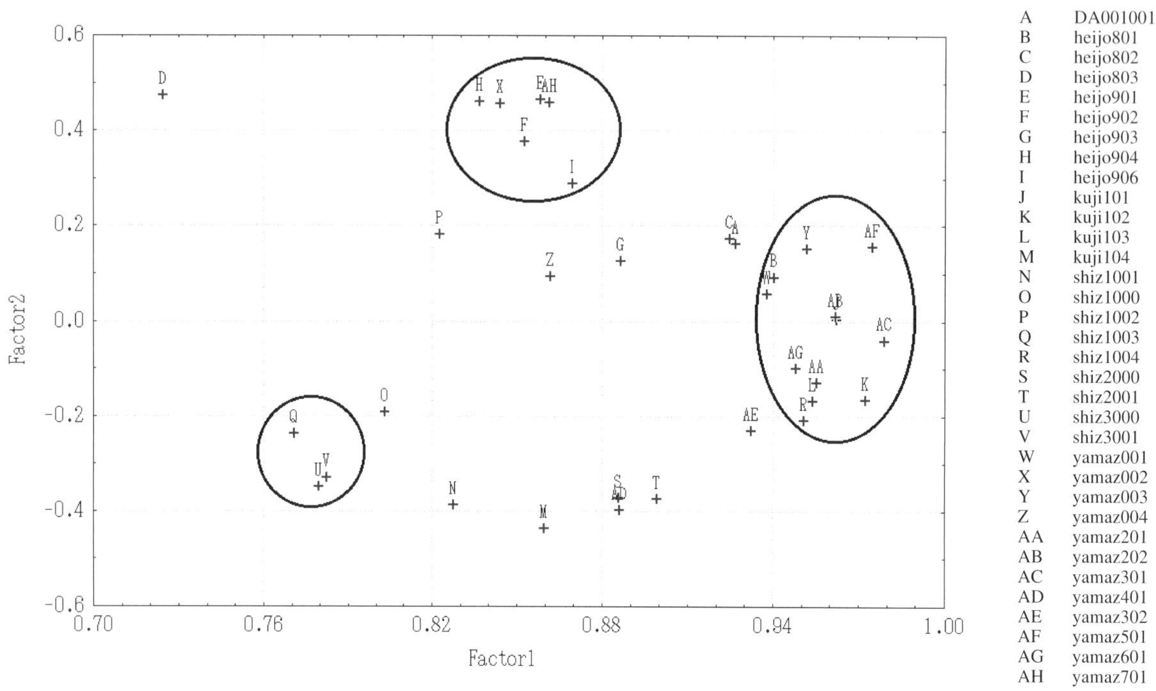

Figure 6. *Principal component analysis of amber*

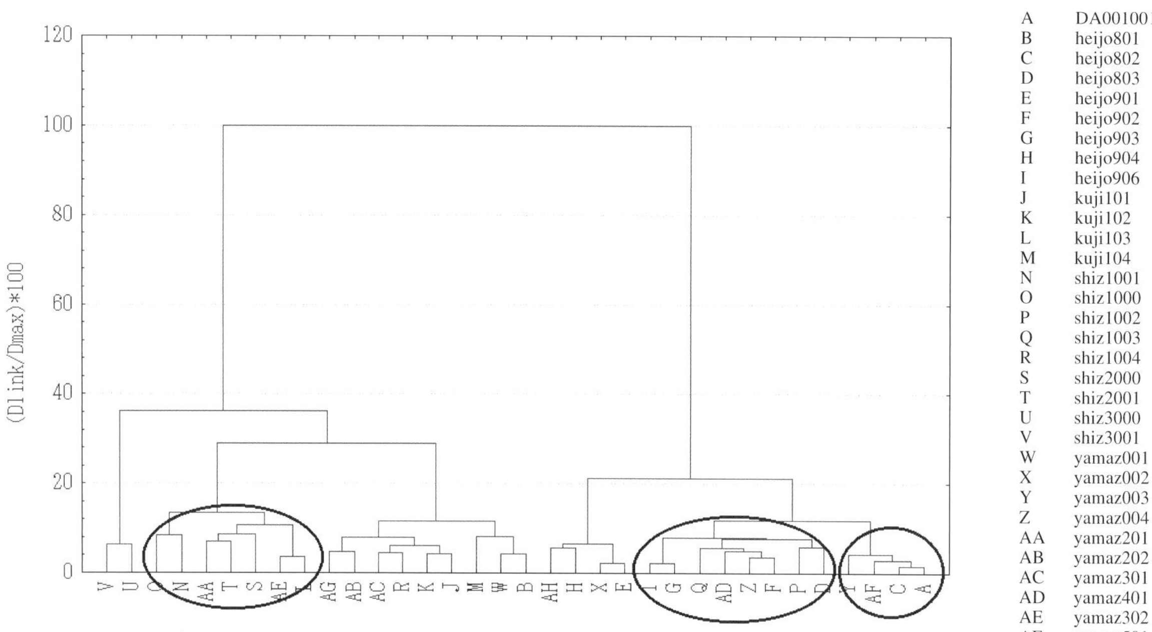

Figure 7. *Cluster analysis of amber*

Fluorescence Spectroscopy

Figure 8 explains the principle of three-dimensional representation of fluorescence. It shows typical fluorescence spectra represented as an assembly of contour lines having the same fluorescence intensity on a two-dimensional coordinate system in which the ordinate is the excitation wavelength and the abscissa is the emission wavelength. The fluorescence contours show the position of a maximum in fluorescence intensity as a function of excitation wavelength and emission wavelength.

In figure 8a, the sample shows a maximum in fluorescence intensity situated at 425 nm in excitation and at 485 nm in emission. Its relative fluorescence intensity is 54. On the other hand, the spectrum of the translucent part of the same sample (fig. 8b) shows the fluorescence peak at 350 nm in excitation and 415 nm in emission, and the relative fluorescence intensity is 162. Comparing both figures, the molecular structure in a translucent part contains presumably more unsaturated components compared with that of a nontranslucent part, because the maximum emission wavelength is situated at a shorter wavelength compared with that of the translucent part. The results made clear that the chemical composition of amber is not homogeneous on a microscopic scale but rather that the amber is composed of a powdery component and a malleable component (translu-

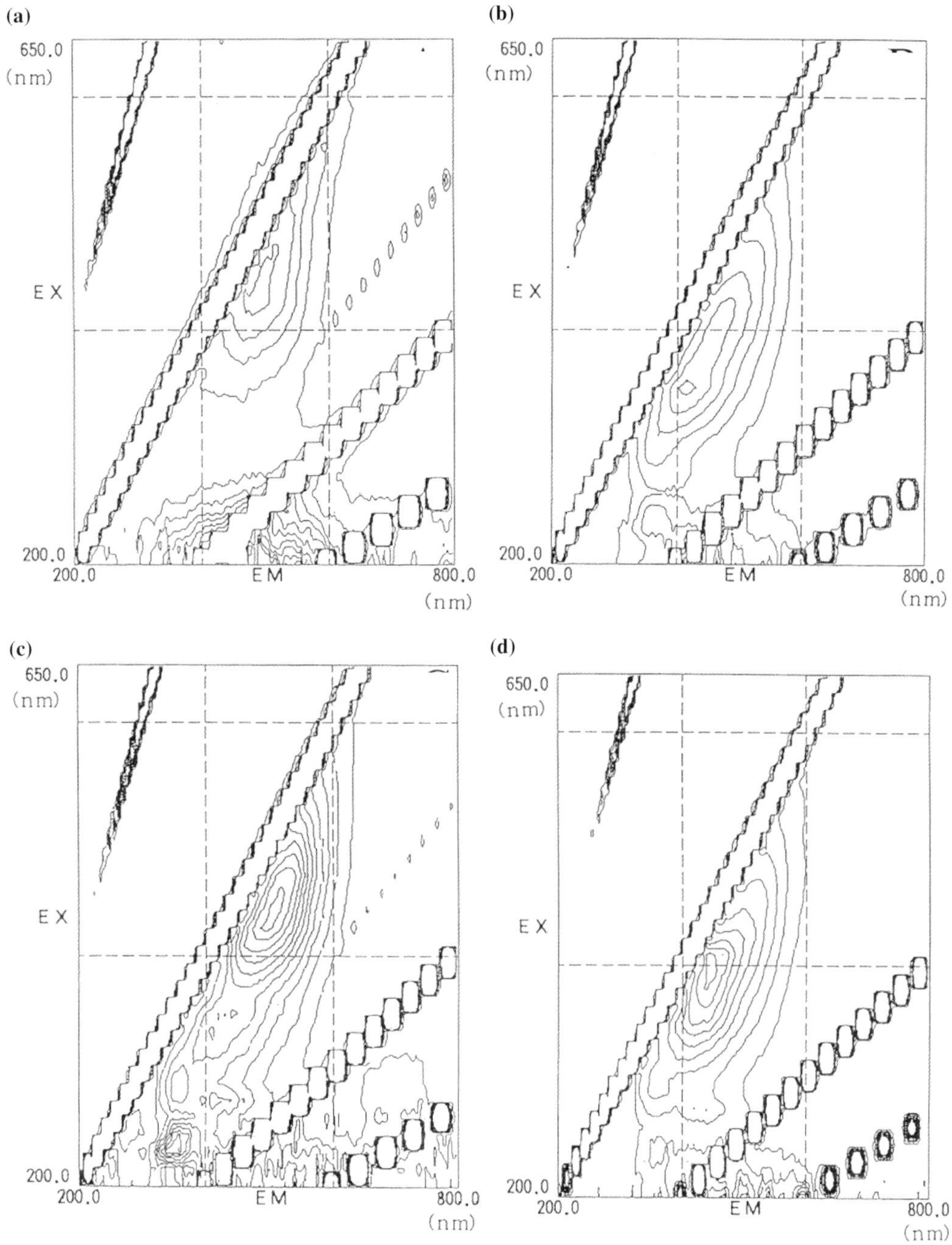

Figure 8. *Fluorescence of amber*
(a) Sample from Kuji-ube district (powdery state)
(b) As for (a) (malleable state)
(c) Sample from Kuji-goishi district
(d) Sample from Heijō-kyō archaeological site (red-colored)

cent). In other words, the amber is a random mixture of both components.

The red-colored amber (fig. 8c) found at the Heijō-kyō palace site has a fluorescence intensity peak at 395 nm in excitation and at 445 nm in emission. The maximum relative intensity is 95—more intense than the values of other samples in the Kuji district. This kind of amber might be produced in another district. Although the spectrum is not shown, black-colored amber found at Heijō-kyō palace site has a maximum fluorescence intensity of 470 nm in excitation wavelength and 530 nm in emission wavelength. The relative fluorescence intensity is about 7–10. As a first approximation, the characteristics of this sample are apparently analogous with that of Kuji district. Since the Nara district was the ancient capital at that period, various kinds of amber articles must have been transported there from all parts of Japan.

The fluorescence intensity depends on both the kind and the amount of fluorescent components included in amber. Since each amber sample showed variable fluorescence intensity, chemical components are also diverse among them. The greater the proportion of unsaturated compound, the stronger the fluorescence expected.

Conclusions

The characteristics of ancient amber articles excavated at archaeological sites in Japan were investigated by using FTIR microscopy and fluorescence spectroscopy. Both methods of analysis made clear that the amber is composed of a random mixture of powdery components and malleable components. It is therefore not sufficient to use the peak pattern of infrared spectra as a method of distinguishing the place of provenance of amber articles. To study further the details of infrared spectrum, the composite broad peaks were decomposed into constituent peaks by using Gaussian peak functions. By applying this mathematical treatment to over one hundred spectra of ancient amber articles, the correlation of each sample was analyzed by PCA to give an assembly of experimental points distributed on a two-dimensional coordinate system as a function of factor 1 and factor 2. As a first approximation, the presence of three different assemblies, Kuji, Heijō-kyō, and Shizuoka, was observable, making possible a rough classification of provenance for amber articles. The classification of data by using cluster analysis also showed that the Heijō-kyō samples have unique characteristics compared with those of another provenance.

FTIR microscopy is suitable for the analysis of cultural properties because of its versatility for measurement and the very small sample requirement—less than one milligram.

Acknowledgments

We appreciate the donation of amber samples by Dr. K. Sasaki, director of the Amber Museum in Kuji City, Japan; the late Dr. K. Masutomi, former director of the Masutomi Geology Museum in Kyoto, Japan; and Mr. T. Fujiwara, curator of the Masutomi Geology Museum in Kyoto, Japan.

References

Fraquet, H. 1987. *Amber*. London: Butterworths.

IRUG. 2000. The Infrared and Raman Users Group Spectral Database.

Lambert, J. B., J. S. Frye, and G. O. Poinar Jr. 1985. Amber from the Dominican Republic: Analysis by nuclear magnetic resonance spectroscopy. *Archaeometry* 27(1):43–51.

Mills, J. S., and R. White. 1987. *The organic chemistry of museum objects*. London: Butterworths. 96–99.

Muroga, T. 1985. Search on transported route of ancient amber articles. In *Commemorated collections, Kon, for Professor Suenaga. Nara Prefecture, Japan,* 1861–1869. Nara Prefecture: Kashihara Archaeological Institute.

Sato, M. 1999. Scientific analysis of excavated organic materials. In *Proceeding of the research seminar on conservation science,* chapter 5, 1–16. Nara: Nara National Research Institute for Cultural Properties.

3 Feathers of Blue on a Field of Gold: Chinese Ornament with Kingfisher Feather Cloisonné

Blythe McCarthy and Ellen Salzman Chase

ABSTRACT Kingfisher feather cloisonné was used from the Han dynasty to the early twentieth century as a means of decorating objects in Chinese court circles, in theatrical productions, and for personal adornment. This paper discusses the range of objects decorated with kingfisher feathers and the various materials and methods employed in their manufacture. It focuses on a collection of jewelry in the Dr. Paul Singer Collection of Chinese Art of the Arthur M. Sackler Gallery as well as pieces from the National Museum of Natural History, Smithsonian Institution. Several colors and types of feathers were found on the objects, including blue feathers from the White-breasted Kingfisher and purple feathers from the Indian Roller and the Fairy Bluebird. Cut sections of feather would have been applied to a variety of substrate materials, including metal and paper, with a protein-based glue.

Introduction

Kingfisher feathers have been used in Chinese culture to decorate headdresses, hairpins, hair combs, earrings, and other types of personal ornament as well as larger items, such as bed hangings, screens, and lanterns. Kingfisher feathers were also used to decorate costumes in Chinese opera. Part of a tradition of luxury and beauty, some even argue that the first telling of a Cinderella story includes the kingfisher feather. Originating from the ninth century, the version includes not only the wicked stepmother and a fairy godmother (in this case, a magical fish), but the girl has shoes of gold and a cloak of blue halcyon (kingfisher) feathers (Schafer 1967).

The iridescent colors of the feathers were highly prized in the court and later among the larger population. The use of kingfisher feathers seems to have occurred during the Han dynasty (Chambers, Rawlings, and Ritchie 1981; Gyllensvard 2001; Hulsewé 2001; Liu 1991), but in general, references did not mention their specific use in jewelry prior to the Tang dynasty (Garrett 1994; Schafer 1963). First used by the court alone, by the late nineteenth century jewelry decorated with kingfisher feathers was available more widely in China as well as for export to Europe. Along with other types of jewelry, some later kingfisher pieces appear to have been adapted specifically for Western tastes (Liu 1984). Guangzhou (Canton) is the most frequently cited place of production and is where the last major consignment of kingfisher jewelry was supposed to have been manufactured during the 1930s (Garrett 1994; Hartman 1980). Chang'an (Xian) was listed as a source of manufacture but only in the context of royal artisans in the Tang dynasty (Schafer 1963). Beijing is most frequently noted as a modern-day source, although in one instance it also was mentioned as a place of manufacture in the eighteenth to nineteenth century (Untracht 1982). Although newly made kingfisher jewelry is still available, it is not found as readily as in the past.

Thirty-three pieces of Chinese jewelry in the Singer collection of the Arthur M. Sackler Gallery (AMSG) that are decorated with kingfisher feather or that might have formerly had kingfisher on the surface were examined in terms of their materials and techniques.[1] Several of these objects are shown in plate 3. In addition, approximately sixty pieces with kingfisher feather, also from China, that are in the collection of the National Museum of Natural History (NMNH), Smithsonian Institution, were examined.[2]

There are two main forms of blue feather found on the objects, one very fine and the other a more coarse type with visible black barbules. In addition to blue feathers, purple, magenta, and green feathers were also observed on the objects examined, although always in comparatively small amounts and always in conjunction with blue feathers. Feathers are applied to a number of materials for the jewelry and headdresses, including metal and paper. They are cut to shape and set into cells bound by cloisons in a manner similar to that used in enamels. On objects with metal substrates, the cloisons are virtually identical to those seen on enamels, frequently formed of twisted wire. In the case of nonmetal substrates, the delineation of the cell is maintained but with materials such as string or gold-colored paper.

Methods

The objects from both the Singer collection and from the NMNH were surveyed for their physical characteristics of substrate type, feather texture, cloisonné form, and support structure. More detailed examinations were carried out on the Singer collection objects, including identification of the feathers used, examination under a binocular microscope to

Table 1. *Characteristics of Kingfisher cloisonné and related objects*

Object number	Object name	Feather types	Substrate	Cloison wire	Diameter of wire (mm)	Twists/mm on wire	Other decoration
RLS1997.48.1	Crown	NV	Ag, Cu	T	.16–.21	3.8–4.7	E, PBL
RLS1997.48.843	Bird/Phoenix	NV	Ag, Cu	T	.13	2.7, 4.0–4.2	NV
RLS1997.48.2915	Hairpin	NV	Ag (Au, Cu)-G	T	.16	3.8	GB
RLS1997.48.2916	Hairpin	SWT	Cu, Zn (Ag)-G	T	.08–.11	3.8–1.5	GB
RLS1997.48.2917	Hairpin	SWT	Cu (Ag, Zn)	T	.11–.16	3.4	S
RLS1997.48.3043	Hairpin	SB	Cu, Ag	T	.13–.16	5.2	PBL
RLS1997.48.4260	Hairpin	SB (remains)	Cu, Ag, Au (As)-G	T	.21	3.0–3.4	NV
RLS1997.48.4261	Hairpin	SB, SWT, C	Ag (Au, Cu)-G	T	.11–.16	5.3	G, PBL, S
RLS1997.48.4262.1/2	Hairpin	SB (remains)	Cu (Ag, Au, Zn)-G	T, P	.16	4.2	PBL
RLS1997.48.4262.2/2	Hairpin	SB	Cu (Ag, As, Zn)-G	T, P	.18, .21	2.9–3.0	G, PBL
RLS1997.48.4265	Ornament	SB (remains)	Ag, Cu (Au)-G	T	.18	3.4	NV
RLS1997.48.4474	Hairpin	NV	Cu, Ag, Au-G	T	.16	4.2	P
RLS1997.48.4625	Pin	SB, C	Ag (Au, Cu)-G	T	.05–.08	3.0	F, RPG
RLS1997.48.4627	Ornament	SB	Au, Ag (Cu)	T	.21	3.0	PBL
RLS1997.48.4631	Ornament	SB (remains)	Au (Ag, Cu)	T	.16–.21	4.2	NV
RLS1997.48.4633	Pin	NV	Au, Ag (Cu)	T	.16–.21	2.7–3.0	PBL
RLS1997.48.4634	Ornament	SWT	Cu, Ag (Au)-G	T	.11	5.7	GB, S
RLS1997.48.4635	Hairpin	SWT	Ag, Cu	T	.11–.16	4.2	GB
RLS1997.48.4636	Hairpin	SWT	Cu, Ag (Zn)	T	.11–.16	2.3–2.7	PBL
RLS1997.48.4637	Hairpin	SWT	Cu, Ag (Zn)-G	T	.11–.16	3.0	GB, PBL
RLS1997.48.4639	Hairpin	SB	Cu, Ag (Au, Zn)-G	T	.16	2.3	GB
RLS1997.48.4640	Hairpin	SWT	Ag (Au, Cu)-G	T	.13	2.3	GB, PBL
RLS1997.48.4641	Hairpin	SWT	Cu, Ag	T	.16	2.3	GB
RLS1997.48.4642	Hairpin	SWT	Cu, Ag	T	.16	3.8	GB
RLS1997.48.4643	Hairpin	SB	Ag, Cu	T	.16	3.4	NV
RLS1997.48.4644	Hairpin	SB, C, I, CD	Paper	N/A	N/A	N/A	P, PR, T, TM, RPG
RLS1997.48.4645	Hairpin	SB, P	Ag (Cu)	T	.16	4.6	GB, P
RLS1997.48.4771	Bird	SB (remains)	Cu, Ag (Zn)-G	P	.21	N/A	P
S1989.13	Crown	SWT	Cu, Zn	P	.11	N/A	GB, M, T
S2001.3	Hairpin	SWT	Ag, Cu	T	.16	4.6	GB
S2001.4	Hair ornament	SB	Cu (Au, Ag)	T	.16	3.4	G, GB
LTS1985.1.272	Crown	NV	Au (Cu)	T	.24–.26	3.4	E, GB, P
LRN 4883	Ornament	SWT	Ag, Cu, Au-G	T	.13, .21	1.9, 2.6	GB, PBL

Key: N/A = not applicable; NV = none visible

Feather types: C = *Coracias garrulous benghalensis*; CD = chicken or duck; I = *Irena puella*; P = purple, unidentified; SB = *Halcyon smyrnensis*, back; SWT = *Halcyon smyrnensis*, wing or tail

Substrate: For metals, elements listed are > 1% of composition and are in order of amount. If < 10%, the elements are listed in parentheses. G = gilded

Cloison wire: P = plain; T = twisted

Other decoration: E = enamel; F = silk floss; G = glass; GB = glass beads; M = mirror; P = pearls; PBL = pearls or beads, lost; PR = paper; RPG = red-pigmented protein glue; S = stone; T = thread; TM = thread, metal wrapped

determine working methods (methods of feather application and metal cuts), and measurement of the wire diameters and number of twists per millimeter. A summary is given in table 1. Identification of the metals used in the substrates of objects in the Singer collection was carried out with x-ray fluorescence spectroscopy (XRF). Areas were chosen for analysis, when possible on the reverse of the piece, which exhibited minimal corrosion and appeared ungilded. Although the surface technique is only semiquantitative, it allowed for a comparison of the objects in the collection. Analysis of adhesives and other organic materials was accomplished using Fourier transform infrared spectroscopy (FTIR).[3]

Feather Type and Source

The blue feathers used in Chinese kingfisher jewelry have been attributed to several different types of kingfishers, including both *Alcedinidae* (water-dwelling) and *Halcyonidae* (forest-dwelling) types. Although there is a distinction in Chinese between the alcedinine kingfishers *(cui niao)* and the halcyon kingfishers *(fei cui),* it is apparently a relatively recent linguistic differentiation (cjvlang.com/Birds 2001), and this could have contributed to confusion about their source. Of the forest kingfisher group, *Halcyon chloris* (Jenyns 1967), the White-collared Kingfisher, and *Halcyon smyrnensis* (Untracht 1982), the White-breasted Kingfisher, were mentioned frequently. Plate 4 shows specimens of both of these bird types with a piece of jewelry from the Singer collection. After comparing the specimens with each piece in the Singer collection, none of the feathers on the objects seemed to correspond to the *Halcyon chloris;* however, all of the pieces with the coarse blue feather type were comparable to the wing or tail feathers of the *Halcyon smyrnensis.* The finer type of feather was identified as the back feathers of the *Halcyon smyrnensis.* The blue feathers on the pieces in the NMNH collection also seem to be of these two types. There is a certain amount of variation seen in the color of *Halcyon smyrnensis*, depending on the subspecies or geographic area of the particular specimen. However, both the male and female seem to exhibit similar pattern and coloration. An occasional discoloration of the feathers, where they appear darker blue due to compaction of their structure caused by damage, was observed both on the actual birds and on the jewelry.

The color of certain feathers on one type of water-dwelling kingfisher *(Alcedo meninting proxima)* is similar to that seen on the jewelry, but the texture and structure of these feathers do not correspond. The blue feathers of the Indian Roller were also mentioned in one text as a substitute for the kingfisher when it became scarce (Untracht 1982); however, none of the objects examined exhibited the blue feathers from this bird. It is possible that other types of birds were used, but they were not observed among the ninety or so objects examined in this study.

In addition to blue feathers, five pieces in the Singer collection were found to have purple feathers applied to their surfaces. Two feather types were identified from bird types other than kingfisher: *Corracias garrulous benghalensis* (Indian Roller) and *Irena puella* (Fairy Bluebird). One piece also had magenta and green feathers. The magenta-colored areas appear to be modified purple Indian Roller feathers, as will be discussed below. The green feathers were not easily identified but may be from a chicken or duck.

Several places are mentioned by literary sources as areas of origin for the kingfisher feathers on the jewelry, including Cambodia, India, and Vietnam as well as places in southern China (Chou 1992; Jenyns 1967; Schafer 1963; Untracht 1982), as shown in the map in figure 1. The source of the feathers on the Singer objects, however, is not known. Numerous references mentioned that the birds were hunted to extinction in China due to the popularity of the jewelry in the late nineteenth to early twentieth century (Chambers, Rawlings, and Ritchie 1981; Hartman 1980; Liu 1991). It seems possible, however, that the birds may have become scarce, rather than extinct, in China during that time. The areas of China mentioned as sources for the bird are still included in the range for the *Halcyon smyrnensis* kingfisher in modern field guides (DeSchauensee 1994; Fry and Fry 1992). Additionally, the Animals of the World Database on the Convention of International Trade in Endangered Species (CITES) website lists *Halcyon smyrnensis* as not only being present in China, but the bird is also listed as breeding in the area rather than just being present as a vagrant (CITES.org 2001).

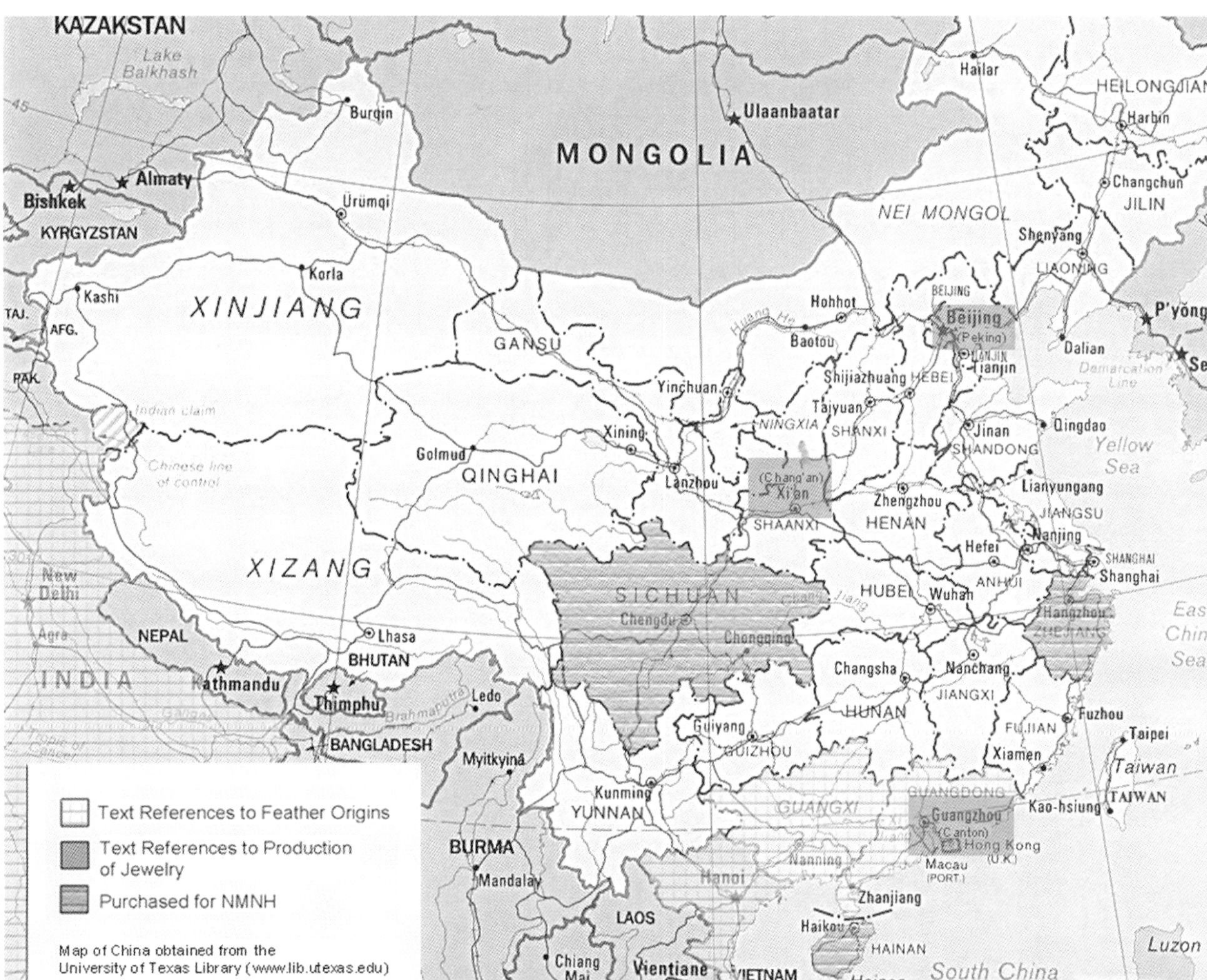

Figure 1. *Map of China showing kingfisher feather origins and jewelry sources*

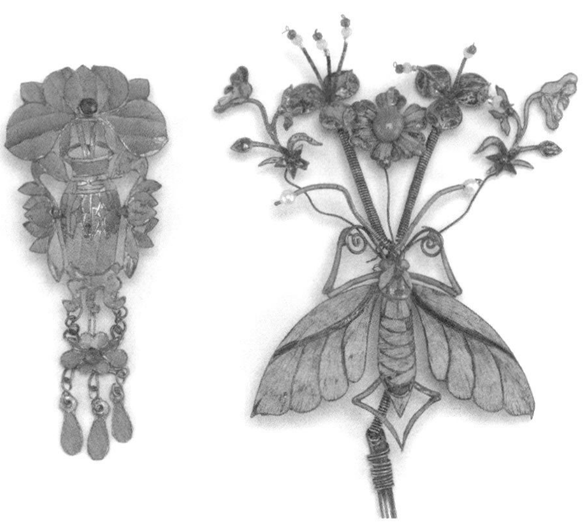

Figure 2. Example of coarse-textured (S2001.3) and fine-textured (RLS1997.48.4645) Halcyon smyrnensis

Feather Usage

The more coarsely textured feathers seen on some of the objects are from the tail and wing of the *Halcyon smyrnensis.* These feathers are an iridescent turquoise on the upper side and black on the underside. More of the tail feather is usable than the wing, as the wing feather has a portion that is black both on the front and back. Figure 2 shows an object inlaid with this type of feather as well as an object inlaid with a finer-textured feather. The rachis (see figure 3 for definition of the term) on the tail and wing feathers is black with a blue rachilla and black barbules on the barbs of the feather. The blue color and the iridescence are caused by structural properties of the feather and will disappear temporarily when a sample is immersed for a few minutes in ethanol.

The second, finer type of feather seen on the jewelry is from the back of the *Halcyon smyrnensis.* The back feathers on the bird are finer and somewhat less green in color than the wing and tail. The overall appearance of the back feathers is quite different from that of the wing and tail feathers, with only the tip of the feather colored blue and the remainder brown. The tip of the back feather is blue on both sides, not black on the underside like the wing and tail feathers. The barbules, small and sparse, are a light brown in color.

The barbules from the wing feather are larger and denser than those of the back feather. They can more easily act on their own to hold the barbs of the cut feather pieces together and are a more visible component on the jewelry. The barbules of both feather types are shown in figure 4.

Application of Feathers

The barbs were cut off of the black rachis of the wing and tail feathers for use in the jewelry, while only the tip of the back feather was used. In general, the back feather was cut so that the very end of the rachis was included in the cut, to avoid separation of the barbs. The black of the rachis is visible in many areas on the jewelry when examined closely. Plate 5 demonstrates the overlapping feather tips both on the bird and then on an object. On the jewelry the tips point

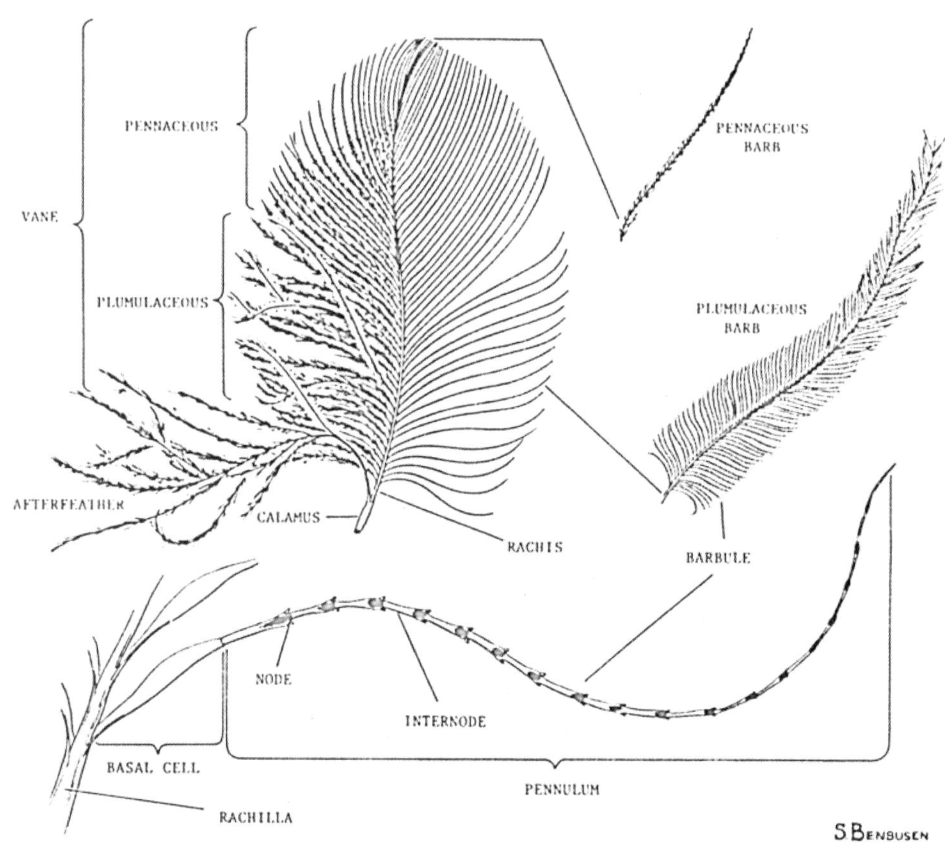

Figure 3. Topography of a contour feather (Laybourne, Deedrick, and Hueber 1994)

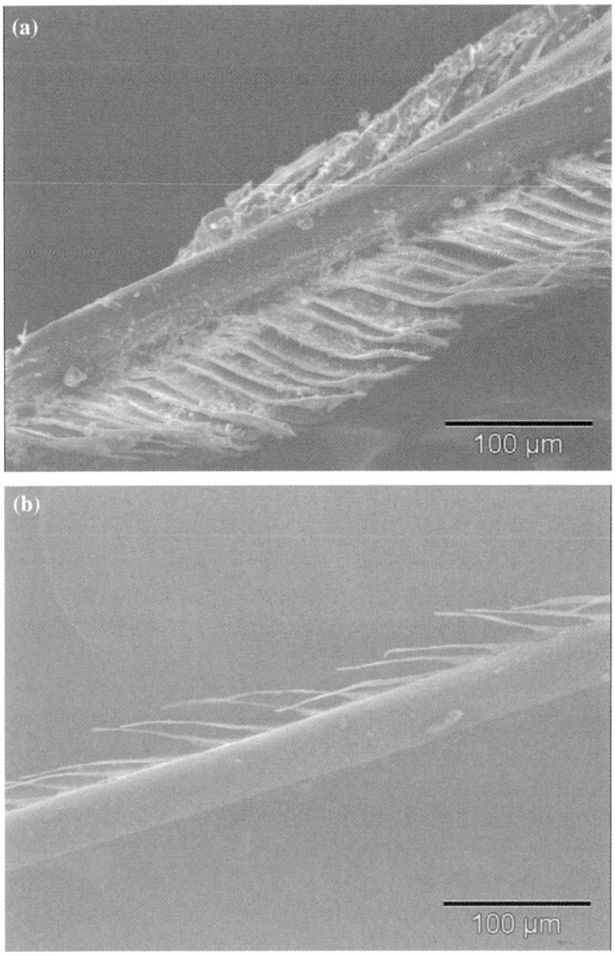

Figure 4. SEM images of coarse- and fine-textured Halcyon smyrnensis

toward the interior of the cell with the back edge cut to fit the contour. Where possible, they are overlapped to cover the small part of the rachis that is holding them together. This type of application seems consistent for the back feather, although it is not always carefully done.

There is a range of quality to the feather work that was examined. In general, the feather pieces within a cell tend to run in the same direction as each other. The majority of cloisons have two or three larger pieces with slight overlap covering the surface of the metal. In some objects the feather is carefully cut to follow the curves of the cloison, leaving little bare metal visible except where damaged. In other cases the application of feather is more haphazard, with standard rectangular pieces of feather applied rather than being cut to a specific shape. The direction of the barbs on the applied sections of feather is not always consistent.

FTIR analysis in several instances indicates that the adhesive used to set down the feathers was a type of protein glue. It seems possible that adhesive was also used to hold together the feathers in order to cut them. In addition, the feather may have been attached to a facing, such as a thin sheet of paper, before cutting. Once in place on the object, the facing could have been removed. Many of the objects examined show evidence of adhesive on the top surface of the feather, but many may also have been coated with some type of material as part of a later treatment, making confirmation of the use of a facing problematic.

In a few examples, purple feathers were applied to a layer of red-pigmented protein glue. A separate, clear layer of a protein adhesive was found underneath the red, suggesting that the red was used to alter the color and not simply as an adhesive. It was never found underneath blue feathers, and the only purple feather type identified that has this red underlayer is the Indian Roller. This red underlayer is not always present when purple Indian Roller feathers are used. The red material does not extend beyond the edges of the feather pieces, which could indicate that the feathers were placed on the red background material prior to transfer to the metal and that the red material had the secondary function of holding the barbs together. A painted red underlayer, different from the aforementioned red material, was observed on some of the NMNH pieces with paper and wood substrates, both underneath blue feathers and under areas of gilding or silvering.

Evidence of feather-application methods was observed in only one instance on the Singer objects (RLS1997.48.4625). Circular depressions were noted in an area of purple feather with underlying red-pigmented glue. These marks seem to be where the feathers were pushed onto the surface. Since the marks were not found elsewhere, it is unclear if they were made when a tool pushed the feather into the underlying red material or when pushed onto the surface of the substrate to attach it. Forceps or another type of fine grabbing instrument have been mentioned as being used to place the feather in the cells (Chin 1991); however, no visual indications of this method were observed.

Metal Substrates

The majority of the objects analyzed contained mainly copper or silver or an alloy of the two metals as the substrate (fig. 5). Zinc was found as an alloying element in four of the copper alloy samples, including examples with both feather types. 1–2% arsenic was identified in three of the examples, all of which were heavily corroded. The arsenic present in the analysis could be a component of the solder. Two of the objects were primarily gold; feather remnants on both of these objects were of the fine textured type. For all objects analyzed, lead, tin, and iron contents were less than 1%. No conclusive correlation was seen between feather type, style of piece, and alloy type. This was not unexpected for this collection, presumably manufactured in different workshops or even geographical areas over a lengthy time span.

Wires

Two wires were twisted together to form the majority of the cloisons, most easily seen in areas where the wire has partially unraveled, as in figure 6. Wire diameters measured from 0.05 to 0.25 mm and the number of twists of the wire per millimeter ranged from 2 to 6. The wires are uniform in thickness and were most likely drawn, but in many cases the surface is obscured by gilding and/or corrosion. An exam-

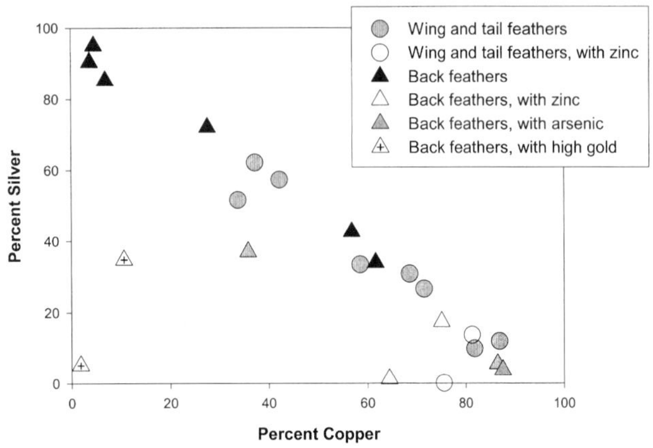

Figure 5. XRF analyses of metal substrates

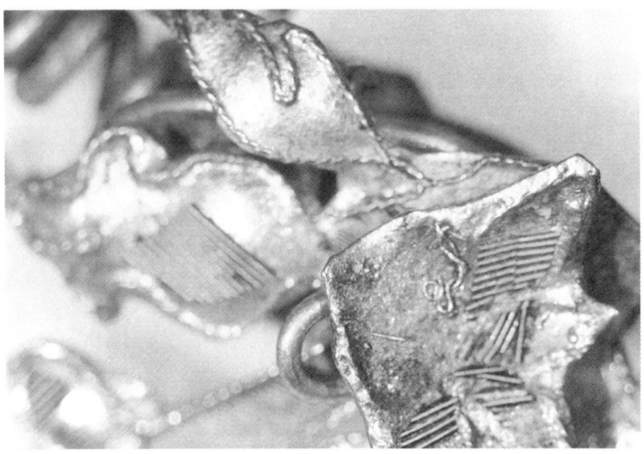

Figure 6. Close-up of area with partially unraveled twisted wire (note also that feather is applied without preferential direction)

ple of a headband with kingfisher cloisonné in the Department of Conservation and Scientific Research's reference collection (LRN 4883) was sampled and analyzed with scanning electron microscopy (SEM) and energy dispersive x-ray spectroscopy (EDXRF). The entire surface of the metal is gilded, and areas of silver extended outward from the wire, indicating that a silver solder was used. The undermetal of the substrate is a copper alloy, while the twisted wires are almost pure silver. Only this one piece has been sampled for metal at this time, and it is not known if such an alloy difference between wire and substrate is a common occurrence.

In addition to the wires forming the cloisons, the majority of the objects had some form of support structure of plain wires attached to the reverse of the metal substrates. This varied from a single wire soldered to the center of the metal piece to a complete outline of the shape.

Formation of Substrate and Cloisonné

A series of small cuts was used to shape the outline of the substrate. The number of cuts and the smoothness of the outline varied from piece to piece. The majority of the objects examined had excess metal on the edge of the cut on the reverse of the piece, suggesting that the cuts were made from front to rear. There was no evidence to distinguish between knife and scissor cuts in most instances, but ghost marks characteristic of scissor cuts were detected (RLS1997.48.4633, RLS1997.48.2916). Most objects in the Singer collection appear to have been gilded. On some, both the obverse and reverse were gilded, but on others gilding was limited to the obverse surface. When gilding is present, it extends over the wires and across the entire cloisonné surface. In some examples, the metal substrate was cut prior to forming the cloisons, while in others the substrate was cut or at least trimmed afterwards.

Studio Marks

Marks in both roman letters and Chinese characters are found on Chinese silver, following the practice of European hallmarking. In the jewelry studied, several examples of Chinese characters were found. Marks used by Chinese artisans include workshops' and makers' names and those indicating silver or gold content. A few publications have listed several of these marks, correlating them to workshop and place of production where possible (Forbes et al. 1975). The majority of marks found on the objects studied do not seem to be among the published marks. Table 2 lists marks found on both Singer collection and NMNH objects. One pin from the Singer collection (RLS1997.48.4625) has stamped marks on the reverse that indicate a workshop name and that the piece is made of pure silver. A piece in NMNH's collection (cat. # E405436, acc. # 268187) closely related in size, shape, and technique has a similar mark for pure silver stamped on the reverse. The donor of the NMNH piece stated that her father purchased it new on a trip to China in 1910, providing an approximate date for the Singer pin (Department of the Registrar, NMNH).

Paper Substrates

A hairpin examined from the Singer collection (RLS1997.48.4644) and shown in figure 7 has a support of several layers of paper built up in a manner like papier-mâché rather than a metal substrate. Despite being on a paper backing, this intricate piece exhibits a high level of craftsmanship. In addition to blue feathers, the hairpin has feathers that are purple, magenta, and green in color. The techniques used are similar to those seen with the metal pieces but in some ways are very different. The feathers are still adhered with a protein glue. There are still support wires, but here they are placed in between the layers of paper. Interestingly, the black surface visible on the reverse of the piece appears to be tarnished silver leaf. This is particularly of note, as so many of the examples with metal substrates are composed of silver alloys or silver-colored alloys.

On some portions of the piece, gilded paper outlines sections of blue feather, forming "pseudo-cloisonné," as shown in figure 8. The gilded paper does not extend below the entire surface of these areas but only continues a few

Table 2. Workshop marks and their translations

Chinese characters	Romanization	Rough translation	Object
華興	Huaxing	China workshop	AMSG hairpin RLS2001.1997.48.4261
□□，足紋	□□, Zuwen	unreadable, 100% silver	AMSG ornament RLS2001.1997.48.4265
怡安紋銀	Yi'an wenyin	pure silver from Yi'an workshop	AMSG pin RLS2001.1997.48.4625
德興	Dexing	Workshop	AMSG hairpin RLS2001.1997.48.4640
泰豊	Taifeng	Peaceful harvest (workshop)	AMSG hairpin RLS001.1997.48.4645
義記, □	Yi Ji, □	Mark of righteousness, unreadable	NMNH hairpins 379217, 379227
董萬-□	Dong wan-□	A proper name (workshop)	NMNH hairpin 379219
景福	Jingfu*	Wealthy prospect	NMNH hairpin 379223
紋銀	Wenyin	Pure silver	NMNH pendant 405436

*Active in Shanghai, 1800–1920 (Chin 1991)
□ = illegible character

millimeters underneath the feathers. In these areas the feathers are applied using a method similar to that used in the metal cloisonné. This object also has kingfisher feathers applied in a manner reminiscent of paper-cut designs. Here the feather is laid down on a black background, in one area a piece of black paper saturated with a protein glue and in others a black-colored wax. Although no traces of residual paper or other materials were seen on the feathers, a method utilizing a facing material or at least a consolidating adhesive prior to cutting the shapes is highly probable in this instance. Cut pieces of feather in this part of the design do not seem to have a portion of the rachis remaining, or it is not easily visible.

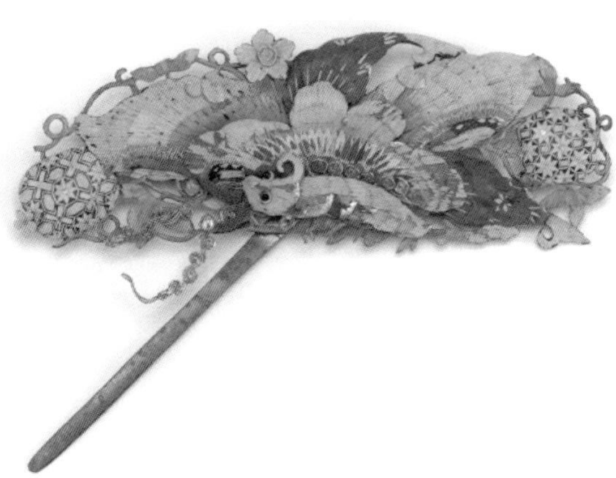

Figure 7. *Hairpin with paper substrate (RLS1997.48.4644)*

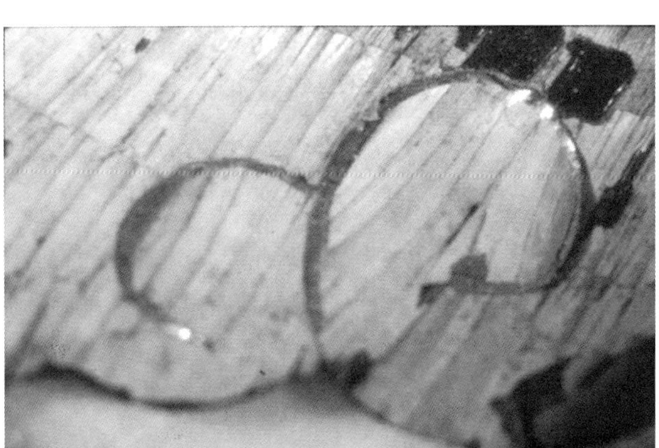

Figure 8. *Close-up of paper cloisons (RLS1997.48.4644)*

The "magenta" feathers are purple Indian Roller feathers, but a larger amount of red-pigmented glue is present in between the barbs than on other pieces examined. This shifts the perceived color of the barbs toward magenta rather than purple. Since there is also a purple feather present on the same piece, this greater application of the red material would seem to be an intentional manipulation of the technique to vary color. As on the other objects examined that have this red material, it is a pigmented protein glue. XRF analysis confirms the presence of mercury in the red material, making it likely that the pigment is cinnabar (HgS) or the synthetic version, vermilion. The purple seen on this hairpin is from the Fairy Bluebird. As noted previously, small amounts of an iridescent green feather on the piece are most likely from a chicken or duck. Although there were numerous examples at NMNH with paper substrates, they were on a larger scale than this piece and more crude in construction.

Other Inlay Materials

In addition to other colored feathers, nonfeather materials were used on a number of objects, further increasing the palette of colors beyond the kingfisher blue. Silk floss sheet, a feltlike material made of silk, was found inlaid between the cloisons of RLS1997.48.4625 (plates 3 and 6). Red, yellow, and green sections of floss were observed. Floss sheet has been used at least since the Tang dynasty in clothing and was used in the Ming and Qing dynasties for a variety of handicrafts (Weiji 1992). The structure of the floss used here, a linear arrangement of fibers, mimics that of the fine-textured feather, and in some cases where its color has faded, it is difficult to tell the floss from feather.

Enameling was used on cloisonné surfaces that seem virtually identical to those used for the feathered jewelry. One enameled crown at the Sackler Gallery (LTS1985.1.272) is an example of this. The crown is constructed from sterling silver with twisted wire cloisons. There are a number of techniques that were used to produce "kingfisher" blue-colored decoration in China other than the feather itself, including enamel. These seem to have been in use at the same time as the feather, and it has been suggested that they are less expensive versions or that the use of alternative materials increased with the decreased availability of kingfisher at the end of the nineteenth century. In many cases, however, enamel would certainly have been used for its own merit rather than as a substitute. There are some suggestions that both the enamel and feather techniques may have been developed as substitutes for gemstones (Hartman 1980).

When no fragments of feather or enamel were found on an object, it was difficult to determine if an object would have been enameled, feathered, or left unadorned. In a number of cases, however, small amounts of feather, remains of feather, or possibly feather pseudomorphs were found on pieces that initially were thought either to have lost feather or to not have had feather. Indirect evidence of feather was also found in one instance where the impressions of the feather remained in the glue layer.

Initially it was surmised that the twisted wires were used on both enamel and feather but that straight wire cloisons were used only with enamel. Pieces with straight wire, however, were found to have remnants of feather, disproving this theory. We have speculated that gilding the entire surface of the metal substrates was used as a time-saving device but also may have effectively prepared them for the application of either enamel or feather, allowing pieces to be fashioned without exact knowledge of their decorative scheme.

Modern

Modern examples of jewelry were made available for analysis by the Freer and Sackler shops. As with some of the older pieces, the feather type of these pieces appears to be *Halcyon smyrnensis* wing or tail feathers.[4] The modern pieces examined use a much larger overlap of feather than the majority of the older Smithsonian objects. The overlap on the modern pieces mimics the visual appearance of the older pieces, where feather pieces were applied with light and dark edges adjacent, utilizing the gradation in color across the wing feathers. The metal substrate of the new jewelry, a copper-nickel alloy, is unlike any of the others analyzed. The entire surface, front and back, is gilded. No support wires were observed on any of the new pieces examined. Their metalwork seems slightly thicker or heavier in appearance than the older jewelry, and excess adhesive from manufacture is much more common on the surface of the feathers than on the historic pieces.

Conclusions

This study has identified a broad range of materials and techniques used in kingfisher ornaments. Such a range is only to be expected for these objects, popular for a long time and produced in many different workshops and for various purposes, including those of personal adornment, theatrical use, as decorative arts, and for export goods. At the same time, the diversity of the objects and the scarcity of published sources make provenancing and dating the objects difficult.

Acknowledgments

Several people were instrumental to us in completing this project. Included among those are the staff of the Freer and Sackler's Department of Conservation and Scientific Research, Mary Becker, Martin Bernstein, W. T. Chase, James Dean, Carla Dove, Natalie Firnhaber, David Furth, Gary Graves, Neil Greentree, Christine Giuntini, Don Harvey, Marcy Heacker-Skeans, Walter Hopwood, Johanna Humphrey, Jan Stuart, Su Rongyu, Susan Wilkerson, and Jodi Zeppelin. We are grateful to Barbara Chu for her donation of a kingfisher headpiece for use in the study and to the Freer and Sackler shops for their loan of the modern jewelry.

Notes

1. The objects were a joint gift of the Arthur M. Sackler Foundation; Paul Singer; the Arthur M. Sackler Foundation for the Arts, Sciences, and Humanities; and the children of Arthur M. Sackler to the Arthur M. Sackler Gallery.
2. The objects, which included not only jewelry but also pieces used in opera and even architectural elements, such as a chandelier, were acquired by the museum from various sources over the last hundred and twenty years.
3. The analyses were performed using an Omega Five XRF and a Mattson Polaris FTIR.
4. Sale of "non-North American" birds and parts from birds that are imported is legal, provided they were properly imported to the United States, they were properly exported from the country of origin, they are not considered endangered by the United States, and they are not listed by CITES as endangered.

References

Chambers, Marlene, Carol Rawlings, and Margaret Ritchie, eds. 1981. *Secret splendors of the Chinese court: Qing dynasty costume from the Charlotte Hill Grant collection*. Denver: Denver Art Museum.

Chin, Edmond. 1991. *Gilding the phoenix: The Straits Chinese and their jewellery*. Singapore: National Museum.

Chou, Ta Kuan. 1992. *The customs of Cambodia*. Trans. Paul Pelliot. 2nd. ed. Bangkok: Siam Society.

CITES.org. 2001. Animals of the World Database.

cjvlang.com/Birds/alcedin.html. 2001. *Alcedinidae*.

Department of the Registrar, NMNH. Accession Records. Washington, D.C.: Smithsonian Institution.

DeSchauensee, Rodolph Meyer. 1994. *The birds of China*. Washington, D.C.: Smithsonian Institution Press.

Forbes, H., A. Crosby, John Devereux Kernan, and Ruth S. Wilkins. 1975. *Chinese export silver, 1785–1885*. Milton: Museum of the American China Trade.

Fry, C. Hilary, and Kathie Fry. 1992. *Kingfishers, bee-eaters, and rollers*. Princeton: Princeton University Press.

Garrett, Valery M. 1994. *Chinese clothing: An illustrated guide*. Hong Kong: Oxford University Press.

Gyllensvard, B. V. 2001. China: Jewelry. www.groveart.com.

Hartman, Roland. 1980. Kingfisher feather jewelry. *Arts of Asia* 75–81.

Hulsewé, A. F. P. 1979. *China in central Asia: The early stage, 125 B.C.–A.D. 23: An annotated translation of chapters 61 and 96 of* The history of the former Han dynasty. Leiden: Brill.

Iroens, Neville John. 1983. *Silver and carving of the old China trade*. Oriental Art Series. London and Hong Kong: House of Fans, Ltd., and Kaiserreich Kunst, Ltd.

Jenyns, Soame. 1967. Feather jacket of the Momoyama period supposed to have belonged to Hideyoshi. *The British Museum quarterly* XXXII(1–2):48–52.

Laybourne, R. C., D. Deedrick, and F. Hueber. 1994. Feather in amber is earliest new world fossil of Picidae. *Wilson bulletin* 106:18–25.

Liu, Robert K. 1984. Imported Chinese jewelry. *Ornament* 7(4):56–61.

——. 1991. Kingfisher feather ornaments. *Ornament* 15(2):16–17.

Schafer, Edward H. 1963. *The golden peaches of Samarkand: A study of T'ang exotics*. Berkeley, Calif.: University of California Press.

——. 1967. *The vermilion bird: T'ang images of the south*. Berkeley, Calif.: University of California Press.

Untracht, Oppi. 1982. *Jewelry concepts and technology*. Garden City, N.Y.: Doubleday and Company.

Weiji, Cheng. 1992. *History of textile technology of ancient China*. New York: Science Press.

Bronze and Iron in China

4 A History of the Study of Bronze Casting in Ancient China

Su Rongyu

ABSTRACT This paper focuses on the brief history of technical studies on bronze casting in ancient China. Western scholars began the scientific study on Chinese bronzes in the early twentieth century, first in Europe and then in China. Japanese scholars subsequently became involved in the study of Chinese bronzes. The main thrust of bronze study by Chinese scholars started in the 1930s because of the Shang-dynasty field excavations at Anyang, Henan Province.

From that time to the present, studies in this field can be divided into two stages. During the first stage, from the 1930s to the mid-1970s, archaeologists mainly addressed issues related to archaeology. During the second stage, from the late 1970s to the present, metal historians began to address issues related to the history of bronze-casting technologies. The scholars learned from the West and Japan and used technical traces left on the bronzes along with chemical composition and physical characteristics of bronze objects and clay molds to determine the casting processes, including the methods of model and mold making, molds assemblage, and runner systems. Metal historians taught their ideas to graduate students in China who later helped to develop these studies. Although their results on bronze-casting technologies are distinguished and pioneering, they rarely adopted archaeological and art-historical theory and method.

The present generation is covering new ground and is able to study bronze casting from a variety of perspectives. Some of these researchers are from the West and have a familiarity with bronze-casting technologies from around the world, which gives them keen knowledge and insight.

Introduction

Although the collection and study of Chinese bronzes started in the Song dynasty in the eleventh century, the traditional research through the Qing period focused basically on the epigraphy and cataloguing of bronze objects. Very few scholars were interested in the manufacturing and inscription process.[1]

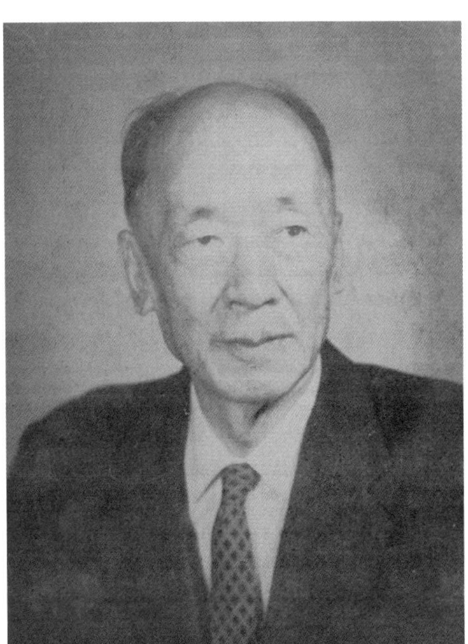

Figure 1. *Dr. Li Ji (Li 1979)*

The modern archaeological approach has been transmitted to China since the beginning of the twentieth century. For instance, such scholars as S. W. Bushell and P. V. Yetts thought about Chinese casting methods in comparison to those for Egyptian and European bronzes. They concluded that the manufacturing method was lost-wax casting.[2] Subsequently, the understanding of Chinese bronzes grew as Chinese students studying abroad returned to China to do fieldwork.[3] Between 1926 and 1929, with the support of the Freer Gallery of Art, Li Ji (fig. 1) carried out three excavations. The first was undertaken at the Neolithic site of Xiyin Cun, Xiaxian, Shaanxi Province, and in 1929 two excavations were made at Anyang, the late capital of the Shang dynasty, where a large number of oracle bones had been discovered.[4] Scientific excavation at Anyang was a milestone in Chinese archaeology.

First Stage: Early Technical Studies by Archaeologists, 1930s–1970s

Scientific research on Chinese bronzes started at the same time as modern archaeology in China, with the important excavation at Anyang. The new excavations showed that many of the bronzes probably originated in the Shang period. At that time many questions remained, such as whether the bronzes were cast at Anyang or imported from elsewhere, who made them, and what materials were used.

Metallurgical remains like molds, charcoal, furnaces, fragments of crucibles, slag, and malachite were excavated at Anyang in 1928. Liu Yuxia, one of the excavators at Anyang, was the first to study these objects with the aim of understanding metallurgy in the Shang period.[5] Li Ji, the excavation organizer, sought to have some of these materials analyzed. H. C. H. Carpenter analyzed four objects in 1931, and his metallographic results indicated that the bronzes were cast from copper-tin alloys.[6]

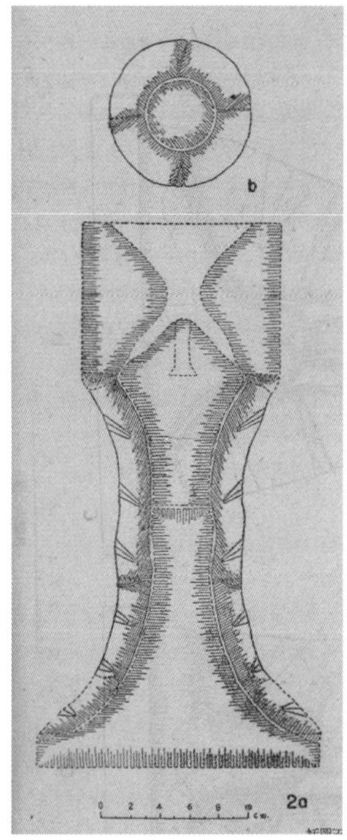

Figure 2. Gao's assemblage of molds for gu *vessel (Gao 1955)*

In Taiwan

After the civil war in China (1946–1949), some Chinese archaeologists moved to Taiwan with the Institute of History and Philology of the Academia Sinica. During this time they could not continue field excavations in mainland China because the political opposition between the mainland and Taiwan interrupted any communication. Instead, they chose to study individual bronze objects that had been brought to Taiwan.[7] Gao Quxun, an excavator at Anyang, also went to Taiwan with the institute and was the first Chinese scholar to describe the clay models and molds unearthed at Anyang as well as the assemblage of molds and the process of decoration. An example of Gao's study of a *gu* vessel is shown in figure 2. In this study he reproduced the model and molds of such objects as the *zun* vessel (Gao 1955).[8]

Li Ji felt strongly that technical study was as important as other aspects of study, such as type, function, decoration, and inscription (Li 1965). In 1962, Li Ji invited Wan Chia-pao, a metallurgical engineer, to join him in the technical study of bronzes unearthed at Anyang at the Institute of History and Philology in Taipei, where they studied 176 bronze vessels from Anyang. The first type they studied was forty-two *jue* vessels along with sixty-three piece molds that may have been used to make them. Wan carefully observed the joining marks of molds on *jue* vessels and analyzed the frequency of their positions as well as the grinding and repairs. In this first project, he also addressed the general process of bronze casting, like the clay material and runner system. He found the rabbet and mortise are not often used for mold assemblage on these bronzes (Li and Wan 1966).

For a second project on *gu* vessels, forty-two *gu* vessels were examined along with twenty-six piece-mold fragments thought to be from these vessels (fig. 3) (Li and Wan 1964). They aimed to understand the casting processes of these vessels through comparison of both vessels and mold fragments. Wan referred to the works by Western scholars as well as Chinese classics.[9] He found that there were two different assemblages of molds for *gu* vessels: four-piece molds and two-piece molds. He also suggested that on the early vessel types, mold joins usually did not align at the flange. Based on Wan's work, Li Ji differentiated the mold patterns into twelve types and contributed to his chronology. These studies showed that the Anyang bronzes were cast with piece molds rather than by the lost-wax process. They also concluded that bronzes of similar shape were made by similar techniques. They described the processes as following: (1) model making with settling clay; (2) painting patterns on model; (3) carving patterns and linking individual decorations; (4) baking model; (5) making molds from model and linking separate patterns on molds; (6) baking molds; (7) cutting model as core(s); (8) assembling molds and cores

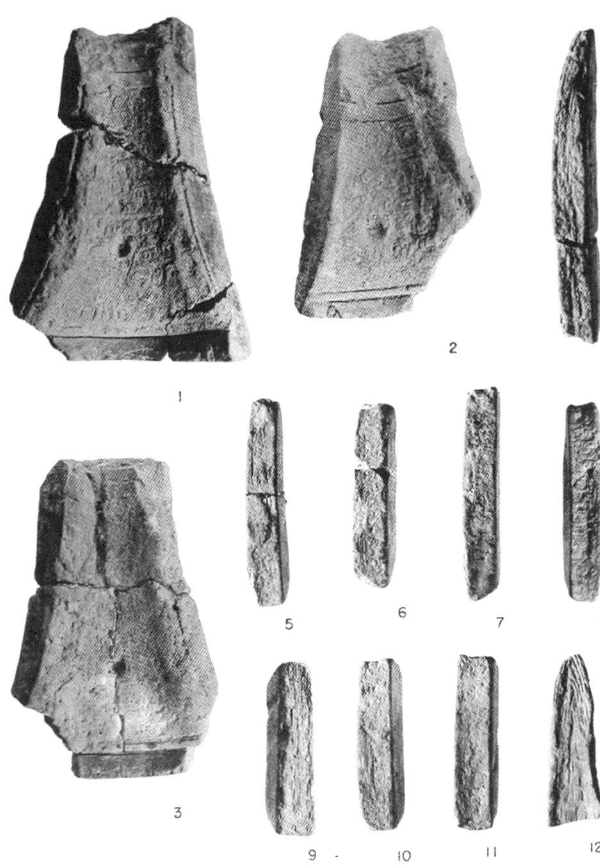

Figure 3. The clay molds for gu *vessels unearthed at Anyang (Li and Wan 1964)*

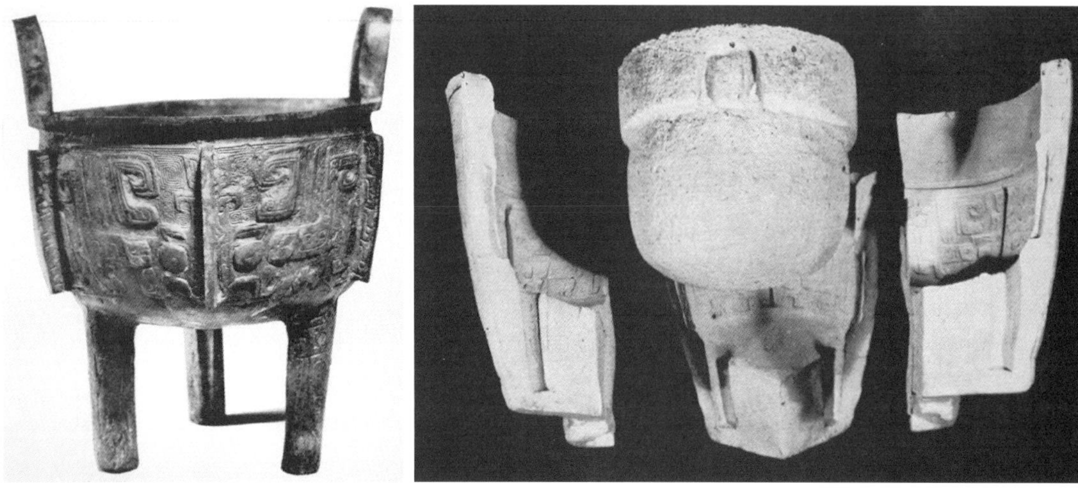

Figure 4. *Reproduction of a* ding *vessel and its molds by Wan Chia-pao (from Li and Wan 1970)*

with spacers; and (9) melting and pouring bronze (normally the mold is positioned upside down). At the conclusion of this research, Wan built a laboratory to do reproduction studies with a grant from the Harvard-Yenching Institute. This work remains unpublished but greatly aided in the understanding of the ancient processes of bronze casting.

The third bronze type studied by Wan was the *jia* vessel, and this work included sixteen-piece *jia* vessels. He also studied the leg structure of *jia* vessels that had been earlier misunderstood by Li Ji (Li and Wan 1968). Subsequently, Wan worked on twenty-three *ding* vessels and fifty-three other vessels. Wan was successful in reproducing bronzes according to his studies (Li and Wan 1970, 1972), as shown with the reproduction *ding* vessel and its molds in figure 4. Bronze-casting studies of 176 vessels unearthed at Anyang from 1928 to 1937 were completed in 1972, and Wan made the following three major conclusions about these bronzes: (1) There is no evidence for the use of the lost-wax process to cast bronze in the Shang dynasty; (2) Objects of the same vessel type were cast by similar casting methods; and (3) The pattern decorations on the bronzes were influenced strongly by technology, and an advanced piece-mold casting technology limited the development of other metal processes (Li and Wan 1970, 1972).

Wan learned from the work at the Freer Gallery (Gettens 1965, 1969), which emphasized the study of casting features on the vessel and molds and the use of radiography for identifying these features. Wan started using some of these methods to aid his research and radiographed a bronze *ding* vessel in 1970 (fig. 5) (Li and Wan 1968, 1970).[10] The studies of Anyang bronze vessels by Li Ji and Wan Chia-pao are shown in figure 6. Subsequently, Wan started studying bronzes unearthed in northern Henan in the Spring and Autumn period using similar methods (Wan 1975). Unfortunately, his work did not offer any solid conclusions on the manufacturing techniques of these bronzes.

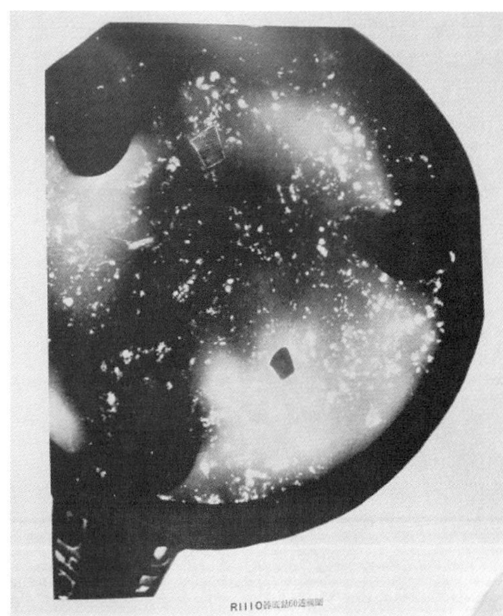

Figure 5. *Radiography of the bottom of a* ding *vessel (Li and Wan 1970)*

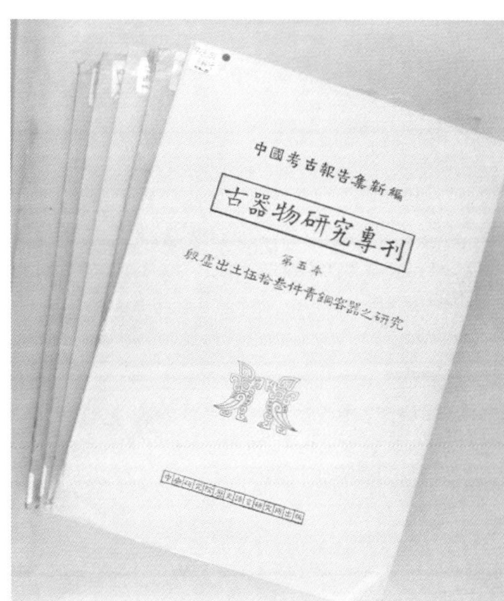

Figure 6. *Studies on Anyang bronze vessels by Li Ji and Wan Chia-pao*

In Mainland China

Little work on the study of bronze-casting methods was done in the 1950s and1960s, and what was done focused on individual objects. The results of these studies were often used to prove the great achievements in early Chinese history. For example, the Simuwu *ding* was studied as a unique bronze with a weight of 875 kg. Yang Gen and Ding Jiaying roughly studied its foundry techniques and chemical composition. They pointed out that the legs and handles are hollow and suggested the assemblage pattern of the molds.[11]

In the 1950s, a great foundry was discovered at Houma, Shaanxi Province, the state foundry of Jin, and at this site thousands of models and molds were unearthed (fig. 7). This discovery encouraged archaeologist Zhang Wanzhong to study the bronze-casting methods used to make these bronzes. His results, however, were problematic.[12] At the same time, Guo Baojun, an archaeologist and also an excavator at Anyang, started a comprehensive project studying Chinese bronzes and the foundry technology involved. He investigated 170 groups of two thousand objects. In his conclusion he divided Chinese bronze casting into six phases: beginning, progressing, developed, assembling, separated, and specialized.[13]

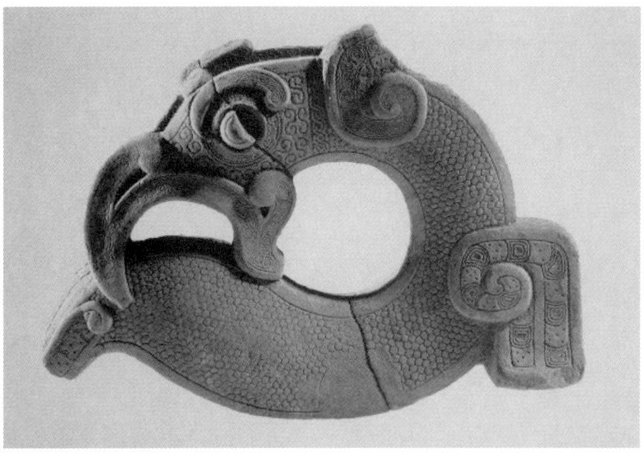

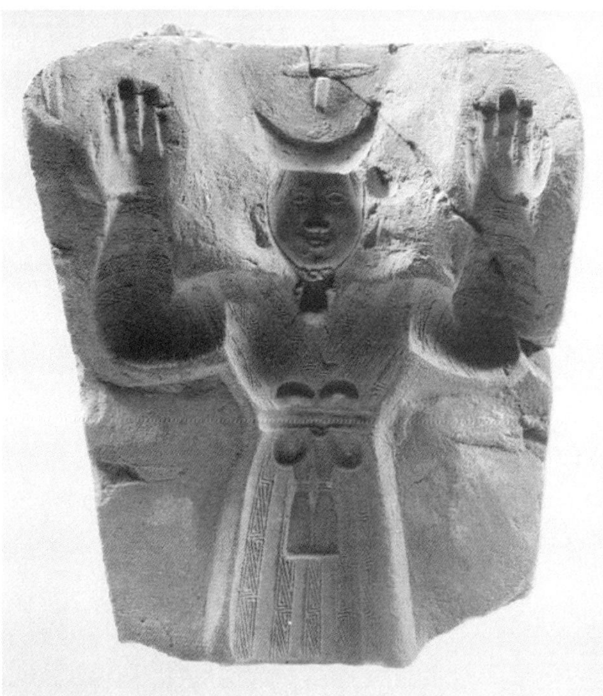

Figure 7. *Clay model and mold from Houma (Beijing 1993)*

Before the end of the 1970s, most of the study of bronze casting was done by archaeologists. Li Ji summarized their basic ideas:

> Technology and art can be viewed as two different human innovations. The highest technological achievement represents the highest level of art. The bronze foundry in the Shang dynasty along with the development and improvements of technology were shown with the designs of decorations. In other words, the accomplishment and formation of most of the designs seen on bronzes were first limited by old tradition or later evolved with new technological processes. The designs changed synchronously with the change of shapes; nevertheless the shape of objects depends, at a given place and time, on the level of development of the manufacturing methods. (Li and Wan 1964)

Most of the bronzes they studied were unearthed at Anyang, and their research focused on how the Shang people cast the bronze vessels at Anyang. Issues such as why those technologies evolved and how technologies varied with place and time were typically not addressed. In the 1960s, archaeologists in both mainland China and Taiwan recognized the need for technical research on bronzes, but there were basic differences in their approach. Archaeologists in Taiwan sought cooperation between archaeologists and metallurgists, whereas the archaeologists in the mainland did not have this inclination. New research from abroad was familiar to the Taiwan scholars, but not the newly excavated materials in the mainland. Mainland scholars, however, were familiar with the excavations but knew little about the manufacturing processes of the excavated artifacts.

Second Stage: The New Approach of Metal Historians, 1970s–1990s

Because the Chinese government and people believed that many great contributions had been made to the history of science and technology in China, the Institute for History of Natural Science was founded in Chinese Academy of Science in Beijing in 1957. Its mission was to find and highlight those contributions. Metal-history studies, including the scientific and technical research on bronzes and iron, became a major research direction of the institute.

In the beginning years, research focused on the chemical compositions of bronze alloys and, later on, bronze casting. During the Great Cultural Revolution in the 1970s, some metallurgists were not able to do research in their fields of specialty and turned to the metal history of ancient China. Ke Jun, a well-known physical metallurgist in the Beijing Iron and Steel College, formed the History of Metallurgy in China Group, composed of scholars in Beijing from the Institute for History of Natural Science in the Chinese Academy of Sciences, the Research Institute of Mechanics, the Research Institute of Non-Ferrous Metals, and the Capital Steel and Iron Company. This action greatly promoted research in the field of metal history and prompted the beginning of the Research Division of History of Metallurgy in the Beijing Steel and Iron College.[14]

At that time, some important sites were excavated and yielded finds that greatly affected scholarly interest in bronzes in China. First was the discovery of the Fuhao tomb at Anyang, which was unique, one of the most complete royal tombs found that dated to the Shang dynasty (fig. 8). Fuhao was a concubine of King Wuding, and more than seven hundred bronzes were recovered from her tomb (Beijing 1980). This excavation attracted Hua Jueming, a metal historian from the Institute for History of Natural Science, who was familiar with the work of N. Barnard (1961) and Gettens (1969) as well as that of Li Ji and Wan Chia-pao. He persuaded Xia Nai (fig. 9), a well-known archaeologist and director of the Institute of Archaeology, to support his scientific research on the Fuhao bronzes in collaboration with other researchers, including Feng Fugen, a foundry engineer, and Wang Zhenjiang and Bai Rongjin, both conservators at the Institute of Archaeology. Hua followed the methodology of Wan and Gettens by carefully examining the objects, comparing them with the clay molds unearthed at Anyang, and putting them in the context of earlier studies. The basic features and casting processes of the Fuhao bronzes were thoroughly studied, and this work provides a standard of bronze casting in the early Anyang period. His work also corrects some problems with the earlier work of Wan Chia-pao (Hua et al. 1981). Hua and his colleagues also carefully compared some of the typical bronzes with some reproductions of ancient Chinese bronzes (Feng et al. 1980, 1982).

At about the same time, news spread through China about the discovery of the tomb of Marquis Yi of State Zeng in the Warring States period, located at Leigudun, Suixian, Hubei Province. More than one thousand bronzes were found, and among these fine objects are a set of sixty-five bells (fig. 10) and other remarkable objects, such as a set of intact *zun* and *pan* objects (fig. 11) (Beijing 1989a). Hua organized a research team of a dozen scholars, including archaeologists, music historians, and engineers, to study the bell set comprehensively.[15] For the first time, bell tone, including its frequency and variation mode, was investigated using x-ray laser techniques (Jia et al. 1981; Lin et al. 1981) The mechanics of two-tone bells were thoroughly explored, and finally a complete set of bells was reproduced.

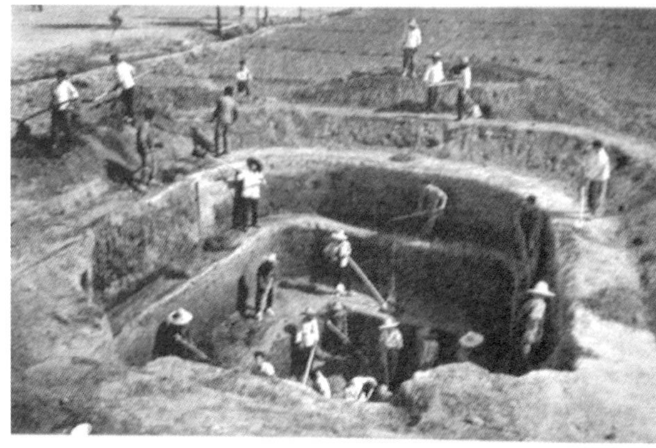

Figure 8. *Excavation of Fuhao tomb (Beijing 1980)*

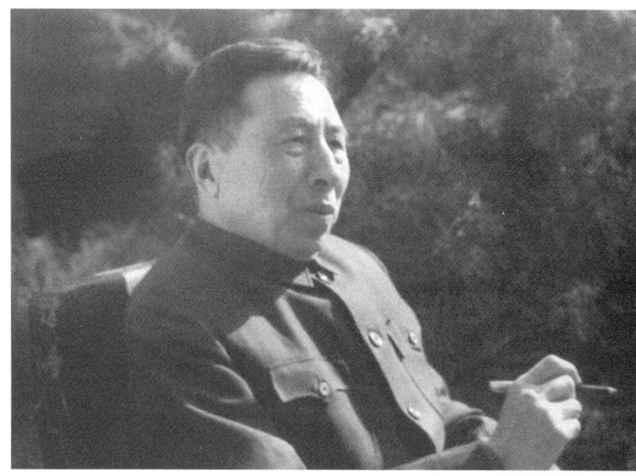

Figure 9. *Dr. Xia Nai (1910–1985)*

Together with the discovery of bronze *zun* and *pan* from Hubei, large cemeteries of the Chu state have been excavated at Xiasi in Xichuan, Henan Province, that date to the late Spring and Autumn period. Hua suggested that the earliest object created by lost-wax casting in China was the bronze *jin* table unearthed at no. 2 tomb in Xiasi (fig. 12). He also stated that other excellent bronzes cast by the lost-wax process are the set of *zun* and *pan* from the tomb of Marquis Yi of State Zeng (Hua 1985). Hua has been keenly involved with the investigation of the origin and development of lost-wax casting in China through the study of traditional processes (Hua et al. 1985a, 1978). Additional researchers, such as Zhao Shigang, Tang Wenxing, and Li Jinghua, have studied the casting for the bronzes unearthed

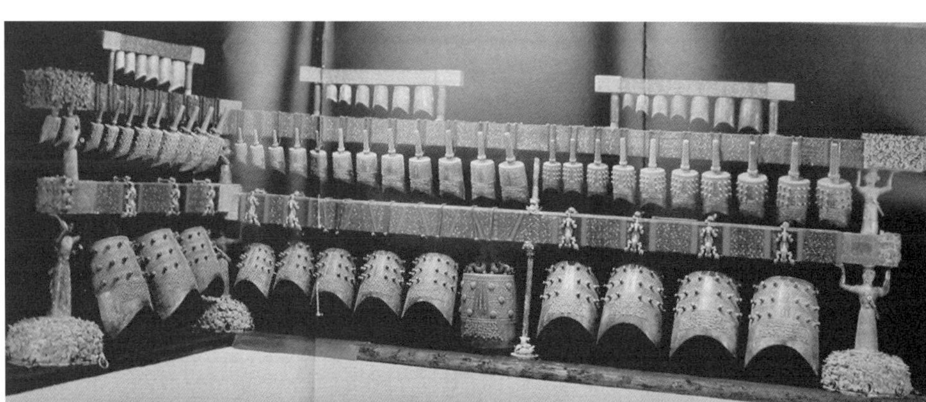

Figure 10. *Chain bells from the tomb of Marquis Yi of State Zeng (Beijing 1989a)*

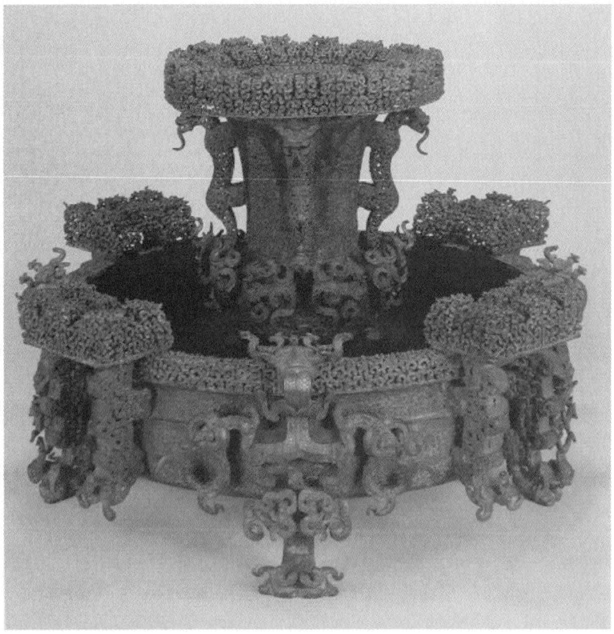

Figure 11. *Bronze* zun *and* pan *from the tomb of Marquis Yi of State Zeng (Beijing 1989a)*

at Xiasi as well (Zhao 1991; Tang 1981). Preliminary studies have also been done on another group of bronzes unearthed at Chu tombs in Baoshan (Li 1991).

Since the 1970s, field excavations have yielded a vast number of bronzes. During this time, unfortunately, many archaeologists did not support technical studies of these materials. Lu Yudong, an editor in the Cultural Relics Publishing House, was instrumental in the promotion of technical studies on excavated objects. Most excavation reports, some of which included technical information, were made available through this publisher in the 1980s.

In 1983, Zhou Jianxun and I were graduate students who started studying ancient bronze casting technology with Professor Hua Jueming. Zhou went to the Division of History of Technology, Institute for History of Natural Science, in Beijing, and I went to the Division of History of Natural Science, Chinese University of Science and Technology, in Hefei. In 1984 we started the technical studies of bronzes in the Western Zhou period. The bronzes Zhou studied were unearthed at Liulihe in Beijing from the cemetery of State Yan. The objects I studied were unearthed at the cemeteries of State of Yu in Baoji, Shaanxi Province. Zhou addressed aspects of bronzes production, such as the placement of spacers, which can be seen through radiography. However, the number of bronzes from the State of Yu is huge; the vessels number more than two hundred pieces as well as hundreds of weapons and fittings for chariots and ornaments. This research investigated the casting details of every worthy object, with the more typical objects studied more carefully. Such details include investigation of the clay model, mold, spacers, runner system, casting defects and repairs, and separated casting, which led to a comprehensive understanding of bronze casting starting in the middle Western Zhou periods. A comparative study of bronzes in early Western Zhou and Shang dynasties showed that they do not have large differences in their manufacture (Su et al. 1986a).

In the 1990s, a great discovery was made at Sanmenxia of Henan Province. Hundreds of bronzes from this site have been included in a preliminary study of casting methods. Most of these bronzes were constructed through multiple castings, and in many cases, the castings are poorly joined (Wu et al. 1999). Bronzes unearthed at Chu tombs in Zhaojiahu in Dangyang, Hubei Province, were also constructed through separate castings, but often solder or metal pegs were used to connect them (Wu 1992).

In 1978, another great discovery was found during the excavation of the tomb of King Cuo for State Zhongshan at Sanji in Pingshan, Hebei Province, of the middle Warring States period, which compares favorably with the tomb of Marquis Yi of State Zeng. Thousands of objects have been recovered, and many of the bronzes are of unusual types and include fine decorations of gold and silver inlay. These bronzes typify the cultural factors from the grasslands in northern China (Beijing 1996). An outstanding bronze seat from this tomb is shown in figure 13. Technical examination showed that these objects were not cast by the lost-wax process (Su et al. 1996) as some researchers had originally thought (Tan 1985, 1989).

Two unbelievable Shang discoveries were made in both the upper and lower Yangzi Valley in 1986 and 1989, respectively. The former is the excavation at Sanxingdui in Guanghan, Sichuan Province, which is famous for its extremely unusual bronze figures and masks. Study of their

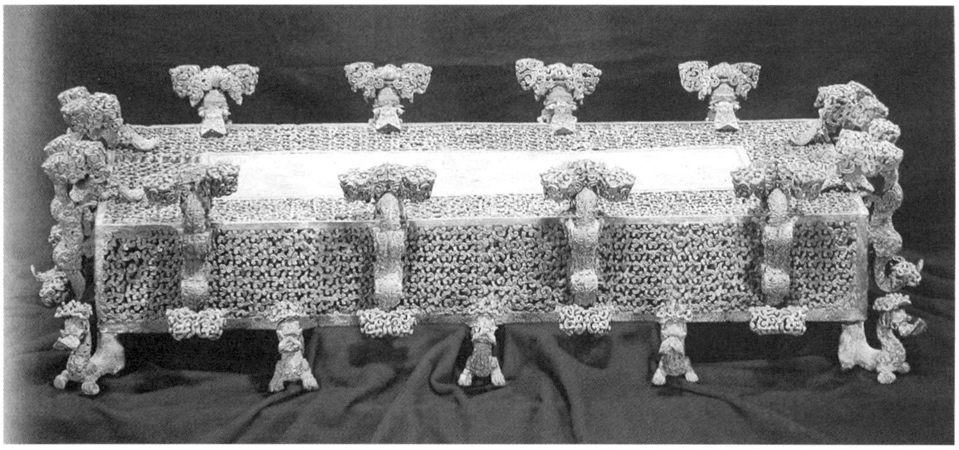

Figure 12. *Bronze* jin *from the no. 2 tomb at Xiasi (Beijing 1991a)*

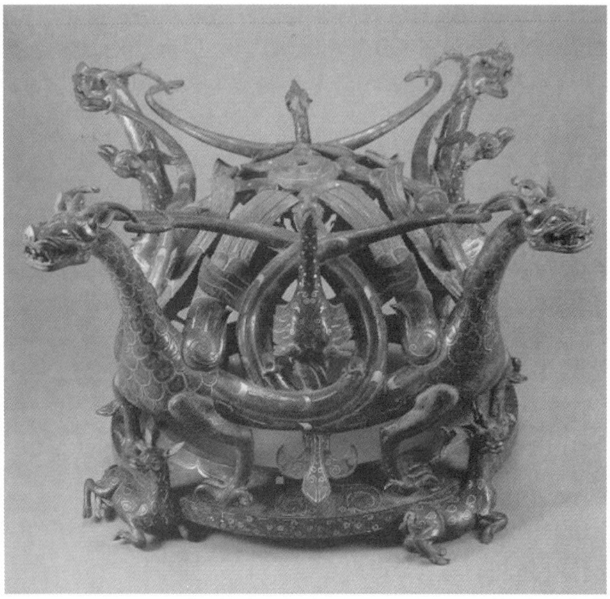

Figure 13. *Bronze seat from the tomb of King Cuo of State Zhongshan (Beijing 1996)*

method and manufacture is currently underway.[16] The later find occurred at Dayangzhou in Xin'gan, Jiangxi Province. Its bronzes are typically decorated with fluid animal decorations. Technical study of these bronzes suggests that the bronze industry in south China developed very early and is similar to the Shang style (Su et al. 1997). Some of the casting details, such as the use of spacers, probably began in the south and spread north.

Tan Derui, a scientist at the Shanghai Museum, has done some detailed work on methods of bronze production by investigating the properties of the clay mold and testing his ideas using reproduction castings (Tan 1999). He found that plant residue plays a vital role in improving the fluidity of molten bronze, which allows fine and complete casting with few flaws. Unfortunately, Tan was unable to find evidence of plant residue in the molds unearthed at Zhengzhou, Henan Province, an early Shang site, so his ideas remain unproven. Other researchers, Wan and Hua, also explored the use of reproduction castings, but they do not support the use of plant-material additives for mold preparation (Li and Wan, 1968, 1970; Hua 1999; Feng et al. 1980, 1982). Tan studied a *you* vessel from the Shanghai Museum and suggested that its rope-shaped handle was cast using plant materials in the clay mold. He supported his work by using reproduction casting. His ideas might have influenced the suggestion that a lost fabric technique might have been used for similar purposes (Bunker 1988). Some bronzes with rope-shaped handles that have no joining marks for molds are described in the study, with the conclusion that they were made by burning out plant materials (i.e., rope) in the mold to create the casting. This work, however, does not address why some vessels with rope-shaped handles do show joining marks and were therefore made by more traditional methods. Some of the clay models for these more traditional bronzes have been found at Anyang (Beijing 1987).

Too few metal historians were available to study the huge number of newly excavated bronzes in the 1970s and 1980s. During this period, Hua was extremely active in this area along with many of his colleagues, and he was able to involve many graduate students and form research teams to address the issues of bronze casting in early China.[17] In 1983, the first graduate student went for a master's degree at the Chinese University of Science and Technology in Hefei. Two years later, two more students went into the Institute for History of Natural Science and Chinese University of Science and Technology. One year later, graduate students started at the Beijing Steel and Iron College. Today, these institutions train Ph.D. students, and the first Department of Archaeometry has been founded at the Chinese University of Science and Technology at Hefei.

Essential work by Hua in the 1970s and 1980s has created a uniform terminology of casting features and related items (Hua 1986); however, more work needs to be done. Two comprehensive books have been published on ancient metals technology in the 1990s. The first focuses on the metal technologies in early ancient China, with a focus on bronze casting (fig. 14) (Su et al. 1995). This book discusses the development of metal technology before the Han dynasty and is based on earlier studies. Included are discussions on the use of clay piece-molds casting, the use of Cu-Sn-Pb alloys, and the lost-wax process, which developed well after clay piece-mold casting. The second book is a general history of metal technologies in ancient China that also has as its focus bronze casting but with a wider relationship between technology and society (fig. 15) (Hua 1999).

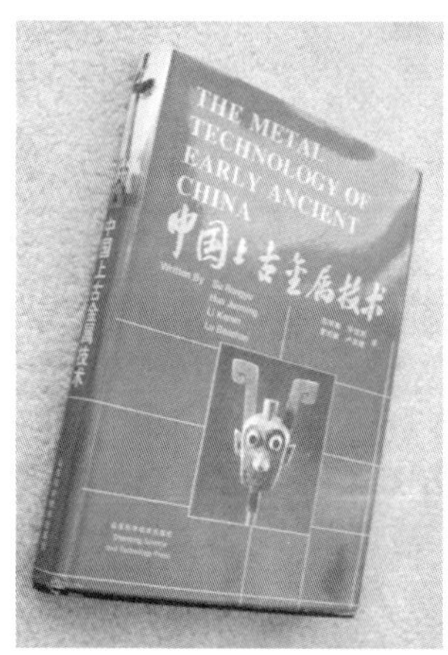

Figure 14. *The metal technology of early ancient China*

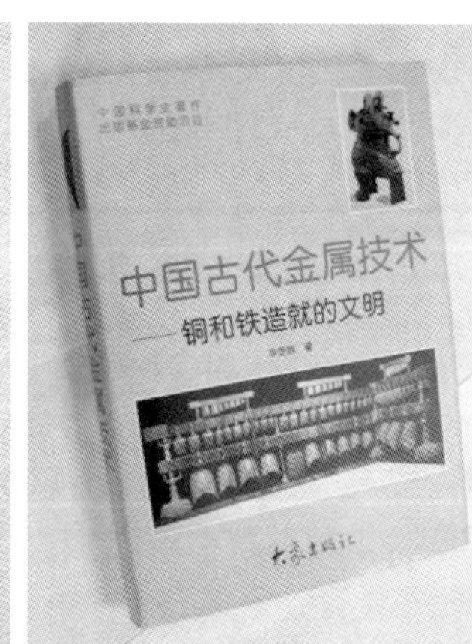

Figure 15. *Metals technology in ancient China*

Conclusion: Remaining Issues and Direction of Future Research

Early archaeologists had extraordinary knowledge and insight and strove to develop the scientific and technological studies of ancient Chinese bronzes. Li Ji initiated the studies on bronze casting for the objects from Anyang in Taiwan. Xia Nai, an academic leader for Chinese archaeology, followed Li Ji in the mainland and conducted and supported several studies on bronze technology (Xia 1979). Many of their successors, however, were less active in technical studies for various reasons involving differences in education and circumstance. This situation has improved in recent times. Beijing University, for instance, is offering an entire course on the technical studies of ancient Chinese bronzes. A new generation of archaeologists may well create a new chapter in the study of these materials. More interaction with Western scholars from disciplines such as archaeology, archaometallury, and art history is also creating a richer environment for studying these materials.

Funding for research on ancient Chinese bronzes continues to be a problem in China. Very few grant programs are able to attract first-class students and engineers to join these studies. Larger coordinated and comprehensive projects are particularly difficult to organize, although some current projects of this type are underway.

The following areas need to be further explored to understand bronze-casting technologies: (1) the technical and cultural development of the piece-molds casting process; (2) the routes and manner of technology transmission of bronze-casting technology; (3) the origin of the lost-wax process in China; (4) the relationship between the clay piece-mold process and the quality of casting, particularly the cause of casting defects; (5) the organization of the foundry process for high-quality casting; (6) identification of forgery castings and patinas; (7) the relationship between the casting, alloys, structure, and corrosion; and (8) the methods of conservation.

Acknowledgments

I have a plan to systematically review the scientific studies on Chinese bronzes, including the origin of metallurgy, copper mining and smelting, bronze casting, bronze alloys, bronze metallography and lead isotope studies, finishing crafts, and other aspects. I hope that this work will facilitate academic communication between Chinese scholars and non-Chinese scholars. This paper is a first step and focuses solely on the Chinese studies on bronze casting. Its sister paper on the research of Chinese bronze-casting technologies by Western scholars will be presented at a later time. This paper was produced during my one-year Forbes fellowship at the Department of Conservation and Scientific Research in the Freer Gallery of Art and Arthur M. Sackler Gallery, Smithsonian Institution, Washington, D.C. Many thanks go to the institution and my colleagues there for their fellowship and help. Special thanks go to Ms. Janet Douglas, conservation scientist at the Freer and Sackler, who gave me much help to make this fellowship successful. She also greatly facilitated the writing of this paper. I appreciate any future discussion about the topic of bronze-casting technologies.

Notes

1. Ruan Yuan is probably the first to study the process of inscriptions, but because he was lacking in technical knowledge, many of his ideas were incorrect. Luo Zhenyu thought the inscriptions for a Western Zhou *gui* vessel were made with clay types. Because Guo Baojun was familiar with the clay molds unearthed at Anyang, he suggested the bronzes unearthed at Junxian were made by piece-mold processing.
2. Stephan W. Bushell is probably the first Western scholar mentioned in the research of bronze casting in China. He suggested that the Shang bronzes were cast by the lost-wax process (Bushell 1904). Later, W. Perceval Yetts thought some Shang bronzes were cast using the second lost-wax method (Yetts 1929). Although their ideas were incorrect, many scholars have cited them (Simpson 1948).
3. Archaeology began in China through two means. First, scholars introduced and translated some discoveries and books from Japanese, and second, foreign scholars conducted field excavations in China. Archaeology as a science was founded in China by students such as Li Ji, who had been trained abroad (Su 1999).
4. The Freer Gallery of Art, Smithsonian Institution, was interested in fieldwork in China. According to Li Ji's writing, in 1924 he met Carl Whiting Bishop, who represented the Freer Gallery, in Beijing to investigate possibilities for scientific excavation. Li Ji was chosen as the field staff when Bishop agreed with his conditions. The first excavation was done in 1926 at Xiyin Village in Xiaxian, Shaanxi Province. Li was invited by the Freer director John E. Lodge to visit the Freer Gallery to discuss future and continuing work. In 1928, Li accepted the new position of the head of the Archaeology Department at the Institute of History and Philology, Beijing. The Freer Gallery continued its support of his excavation at Anyang until 1930 (Li 1977, 56–68). The annual reports of the Smithsonian Institution from 1925 to 1928 outline the collaboration between the Freer Gallery and archaeologists in China (Smithsonian 1925–28).
5. Liu joined the fourth through sixth excavations at Anyang, and his research addressed such issues as chemical composition of the bronzes and casting remains, such as malachite, crucibles, shaft furnace, charcoal, and clay molds. He pointed out that the bronzes had been cast with the piece-mold method. Unfortunately, he had only limited materials to study, and some of his ideas about bronze smelting were incorrect (Li 1933).
6. Two Chinese chemists were asked to do the analysis: Professor Wang Jin of the Institute of Chemistry and Professor Liang Guanyu of the Institute of Geology in Academia Sinica (Li 1933). Carpenter was the mining professor at the Royal College of Mines in London who studied bronzes unearthed in Egypt. He also contributed some research on Anyang bronzes (Carpenter 1933). Japanese scholars like S. Komatsu, Y. Yamanouchi, and S. Umehera analyzed some ancient bronze mirrors (Komatsu and Yamanouchi 1933; Umehera 1933). S. J. F. Jensen analyzed seven bronze mirror samples in Shanghai (Swallow 1937).
7. In the 1950s, Taiwan archaeologists did not have access to information about field excavations in the mainland. Li Ji, for example, did not learn about the new excavations at Erligang in Zhengzhou in the 1950s and Erlitou in Yanshi, Henan Province, until the late 1950s to 1960s from his Japanese and American colleagues (Li and Wan 1966).
8. Gao investigated the traditional lost-wax casting for figures of Buddha statues in Kunming in 1938. He also asked the antiquarians about forgery casting. This work lacks information on the mold pieces and positions so that some of the assemblages for the bronze vessels he suggests are problematic (Gao 1955).
9. He cites the work of H. H. Coghlan (1956) and H. Garland and C. O. Bannister (1927) on Egyptian and European bronze technologies as well as O. Karlbeck (1935), W. Perceval Yetts (1929), and Noel Barnard (1961) on Chinese bronzes.
10. Wan asked a hospital to radiograph a *jia* vessel using a radioactive 60 Co source. A previously unknown inscription

was found on the bottom of the *jia* that appears light on the radiograph, indicating it is in relief rather than incised (Li and Wan 1968). A *ding* was also radiographed using a 60 Co source. The radiograph shows three spacers located near the three legs that include patterns indicating that they are fragments of a bronze vessel (Li and Wan 1970).

11. The idea of the model of raised patterns formed by piling up mud probably came from Gao (1955); and the use of crucibles for pouring from Liu (1932). Yang and Ding suggested that the mold consisted of eight-piece molds, including one for the bottom, four for the inside of the body, and three for each leg. The handles are precast and put into place in the mold. Forty crucibles were used in the pouring, which was done through two legs, and one leg was used as a riser (Yang and Ding 1959).
12. Zhang, one of the excavators at Houma, expresses the limitations of his familiarity with bronze-casting technology (Zhang 1962).
13. Guo was very interested in bronze casting, and he probably learned about piece-mold casting from Huang Jun's *Ye Zhong pian yu,* a catalogue for the objects unearthed at Anyang, in which some clay molds unearthed at Anyang were found. Guo's preface in the 1926 catalogue of the bronzes recovered at Lijialou in Xinzheng mentions casting with the piece-mold method. Guo studied four aspects of Chinese bronze: type and assemblage, shape and casting, patterns and inscriptions, and function and usage (Guo 1980).
14. This college is now called the University of Science and Technology of Beijing, and the division name is the Institute of Archaeometallurgy and History of Materials.
15. The team consisted of a dozen scholars from the Institute of History of Natural Science, the Provincial Museum of Hubei, the Institute of Musicology, the Institute of Mechanic Process in Wuhan, Wuhan College of Engineering, the Scientific and Technological University in Harbin, and Foshan Institute of Spheroidal Iron.
16. I am currently studying this topic with the excavator, Mr. Chen De'an.
17. Hua organized many projects involving a wide range of bronze studies. In addition to the major projects on the bronzes unearthed at the Fuhao tomb and the tomb of Marquis Yi for State Zeng, he also initiated a project on the bronze objects unearthed at the Chu tombs in Xiasi. He encouraged Chinese archaeologists and scientists to join the study of these materials. Several papers on bronzes unearthed both at Jiangxi and Guangdong are the results (Peng 1982; Peng et al. 1985; Xu et al. 1985).

References and Selected Reading

Barnard, N. 1961. *Bronze casting and bronze alloys in ancient China.* Canberra and Tokyo: Australia National University and Mounumenta Serica.

Beijing. 1980. *Yinxu Fuhao mu.* Beijing: Cultural Relics Publishing House.

——. 1985. *Zhongguokaoguxuehui di si ci nianhui lunwenji, 1983.* Beijing: Cultural Relics Publishing House.

——. 1987. *Yinxu fajue baogao, 1958–1961.* Beijing: Cultural Relics Publishing House.

——. 1988, *Baoji Yuguo mudi.* Beijing: Cultural Relics Publishing House.

——. 1989a. *Zenghouyi mu.* Beijing: Cultural Relics Publishing House.

——. 1989b. *Zhengzhou Erligang qi zhu tong jizhi, Kaoguxue Jikan,* no. 6. Beijing: Cultural Relics Publishing House.

——. 1991a. *Xiasi Chunqiu Chu mu.* Beijing: Cultural Relics Publishing House.

——. 1991b. *Baoshan Chu mu.* Beijing: Cultural Relics Publishing House.

——. 1992. *Dangyang Zhaojiahu Chu mu.* Beijing: Cultural Relics Publishing House.

——. 1993. *Houma zhong tong Yizhi.* Beijing: Cultural Relics Publishing House.

——. 1995. *Liulihe Yanguo mudi.* Beijing: Cultural Relics Publishing House.

——. 1996. *Cuo mu—Zhanguo Zhongshang'guo Guowang zhi mu.* Beijing: Cultural Relics Publishing House.

——. 1997. *Xia'gan Shangdai da mu.* Beijing: Cultural Relics Publishing House.

——. 1999a. *Sanxingdui jisi keng.* Beijing: Cultural Relics Publishing House.

——. 1999b. *Sanmenxia guoguo mudi.* Beijing: Cultural Relics Publishing House.

Bunker, E. 1988. Lost wax and lost textile: A universal ancient technique for casting gold belt plaques. In *The beginning of the use of metals and alloys,* ed. R. Maddin, 222–27. Cambridge: MIT Press.

Bushell, S. 1904. *Chinese art.* London: Wyman and Sons, Ltd.

Carpenter, H. 1933. Preliminary report on Chinese bronzes, Shanghai. *Anyang fajue baogao* 4:677–80.

Coghlan, H. 1956. *Notes on the prehistoric metallurgy of copper and bronze in the Old World.* Oxford: Pitt River Museum.

Feng, F., et al. 1980. Shangdai qingtongqi shi zhu jianbao, Beijing. *Kaogu* 1.

——. 1982. Yinxu chutu shangdai qingtong uu zhuzaogongyi fuyuan yanjiu. *Kaogu* 5.

Gao, Q. 1955. Yindai de zhu tong gongyi, Taipei. *Academia Sinica* 26.

Garland, H. 1927. *Ancient Egypt metallurgy.* London: Charles Griffin and Co.

Gettens, R. 1965. Joining method in the fabrication of ancient Chinese bronze ceremonial vessels. In *Application of science in examination of works of art,* 205–17. Boston: Museum of Fine Arts.

——. 1969. *The Freer Chinese bronzes.* Vol. 2. Washington, D.C.: Smithsonian Institution.

Guo, B. 1980. *Shangzhou qingtongqi qun zonghe yanjiu.* Beijing: Cultural Relics Publishing House.

Hua, J. 1981. Zenghouyi bianzhong ji zhuanxu gouji de yezhu jishu. *Jianghan. Kaogu* 1:11–13.

——. 1985. *Shilafa zai zhongguo de qiyuan he fazhan,* no. 13, Shanghai.

——. 1986. *Zhongguo Yezhu shi lunji.* Beijing: Cultural Relics Publishing House.

——. 1999. *Zhongguo gudai jinshu jishu—Tong he tie zaojiu de wenming.* Zhengzhou: Elephant Press.

Hua, J., et al. 1978. Yinheyuan tongting gaojian he bolafa. *Wenwu* 5:67–69.

——. 1979. Zenghouyi mu qingtongqi qun zhu han jishu he shilafa. *Wenwu* 7.

——. 1981. Fuhaomu Qingtongq qun zhuzao jishu de yanjiu. *Kaoguxue jikan* 1:244–72.

——. 1983. Zenghouyu zunpan he shilafa de qiyuan. *Studies of natural science* (2):4.

——. 1985a. *The history of metallurgy.* Beijing: Press of Document of Science and Technology.

——. 1985b. Bolafa de diancha he fuyuan shiyan. Shanghai.

Jia, L., et al. 1981. Yuong Jiguang Chaanxi Jishe Yanjio Zenghouyi Bianzhong de Zhendong Moshi, *Jianghan Kaogu* 1.

Karlbeck, O. 1935. Anyang mould. *The bulletin of Museum of Far Eastern Antiques* 7:39–60, plates I–VII.

Komatsu, K., and Y. Yamanouchi. 1933. A chemical study of ancient Chinese bronzes. *Tôhô Gakunô (Journal of Oriental studies)* 3(1).

Li, J. 1927. Xiyincun shiqian yicun. *The series of the Research Institute at Qinghua School* 3.

——. 1933. *Yinxu tongqi wu zhong ji xiangguan wenti.* Beijing.

——. 1963. *Yin Shang shiqi zhuangshi yishu yanjiu zhiyi, bijiao gu xing qi de huawen suo yinqi de jige wenti*, 699–739.

——. 1965. Ruhe yanjiu Zhongguo qingtongqi, Taipei. *The bulletin for the National Palace Museum* 1.

——. 1977. *Anyang.* Seattle: University of Washington Press.

——. 1991. *Baoshan Chumu qingtognqi zhizao jishu de chubu kaocha.* Beijing.

Li, J., and C. Wan. 1964. Yinxu chutu qingtong gu xing qi zhi yanjiu. *Archaeologia Sinica.*

——. 1966. Yinxu chutu qingtong jue xing qi zhi yanjiu. *Archaeologia Sinica.*

——. 1968. Yinxu chutu qingtong jia xing qi zhi yanjiu. *Archaeologia Sinica.*

——. 1970. Yinxu chutu qingtong ding xing qi zhi yanjiu. *Archaeologia Sinica.*

——. 1972. Yinxu chutu wushisan jian qingtong rong qi zhi yanjiu. *Archaeologia Sinica* 5.

Lin, R., et al. 1981. Zenghouyi bianzhong jiegou de tantao. *Jianghan kaogu* 1:20–25.

Liu, Y. 1932. Yindai yetongshu yanjiu. *Excavation report at Anyang* 4.

Maryon, H., and H. Plenderleith. 1962. *Fine metal work.* In C. Singer, 1962.

Peng, S. 1982. *Jiangxi Shangzhou qingtong zhuzao jishu.* Shanghai.

Peng, S., et al. 1985. *Jiangxi diqu zaoqi tongqi yezhu jishu de jige wenti.* Beijing.

Shi, Z. 1954. Yindai de zhu tong gongyi. *Academia Sinica* 26.

Simpson, B. 1948. *Development of the metal casting industry.* Chicago: American Foundrymen's Association.

Singer, C., et al., eds. 1954. *A history of technology.* Oxford: Oxford University Press.

Smithsonian Institution. 1925–28. *Annual report of the Board of Regents of the Smithsonian Institution, 1925–1928.* Wash-sington, D.C.: Smithsonian Institution.

Su, R. 1999. *The transmitting and founding of archaeology in China.* Symposium of terminology transmitting to China, Gottingen.

Su, Rongyu, et al. 1986a. *Cong Yuguo mudi qingtongqi kan xizhou zaoqi qingtong jishu dui Yinshang de jicheng he fazhan.* Master's paper. Chinese University of Science and Technology, Hefei.

——. 1986b. *Baoji Yuguo mudi qingtongqi zhuzao gongyi kaocha he jinshu wenwu jiance.* Beijing.

——. 1995. *Zhongguo shanggu jinshu jishu.* Ji'nan: Shandong Science and Technology Press.

——. 1996. *Zhongshanwang Cuo mu qingtongqi qun zhuzao gongyi yanjiu.* Beijing.

——. 1997. *Xin'gan Shangdai qingtongqi qun zhuzao gongyi yanjiu.* Beijing.

Swallow, R. W., ed. 1937. *Ancient Chinese bronze mirrors.* Beijing: Henri Vetch.

Tan, D. 1985. Zhongguo gudai shila zhuzao zuyi. *Wenwu* 12:66–69.

——. 1989. *Canlan de Zhongguo gudai shila zhuzao.* Shanghai: Press of Documents of Science and Technology.

——. 1999. Zhongguo Qingtongshidai Taofan zhuzao jishu yanjiu. *Kaogu xuebao* 2:211–50, plates 14–24.

Tang, W. 1981. Xichuan Xiasi yihao mu qingtongqi de zhuzao jishu. *Kaogu* 2:174–76.

Umehera, S. 1927. L'analyse Chimique des bronzes anciens de la Chine. *Artibus Asiae* IV.

——. 1933. An appreciation of the chemical study of ancient Chinese bronzes by Professor Komatsu and Mr. Y. Yamanouchi. *Tôhô Gakunô* 3.

Wan C. 1968. Notes on the casting of the handle of some bronze vessels in Yin dynasty. *The bulletin of the Institute of History and Philology, Academia Sinica* 46.

——. 1975. Huixian ji Jixian chutu Dongzhoushiqi qingtong ding xing qi de zhuzao he hejin yanjiu (Studies on the foundry and chemical compositions of *ding* vessels unearthed at the tombs in Huixian and Jixian in the Eastern Zhou period). *Dalu zaizhi* 60(6).

Wang, Y., et al. 1991. *Xichuan Xiasi Chunqiu Chum u chutu bianzhong de zhandong moshi yu jiegou fenxi.* Beijing.

Wu, K. 1992. *Dangyang Zhaojiahu Chum u tongqi zhuzao gongyi (Foundry of bronzes unearthed at Chu tombs in Zhaojiahu, Dangyang).* Beijing: Wen Wu Chu Ban She.

Wu, K., et al. 1999. *Guoguo mu chutu qingtongqi zhuzao gongyi fenxi yu yanjiu (Studies on the bronzes unearthed at the tombs of State Guo).* Beijing.

Xia, N. 1979. *Kaoguxue he kejishi.* Beijing: Science Press.

Xu, H., et al. 1985. Guangdongsheng chutu qingtongqi yezhu jishu yanjiu. *The essays for the history of science and technology* 13.

Yang, G., and Ding, J. 1959. Simuwu dading de hejin chengfen jiqi zhuzao jishu de chubu yanjiu. *Wenwu* 12:27–29.

Yetts, W. 1929. *The Eumorfopoulos collection.* Vol. 1. London: E. Bean, Ltd.

Zhang, W. 1962. Houma Dongzhou taofan de zangxing gongyi (The making process for the clay mold from Houma). *Wenwu* 37–42.

Zhang, Z. 1964. Zhongguo huaxue shigao (gudai zhi bu) (The brief history of chemistry in China [ancient part]). Beijing: Science Press.

Zhao, S. 1991. *Xichuan Xiasi Chunqiu Chumu qingtongqi zhuzao gongyi.* Beijing: Wen Wu Chu Ban She.

Zhou, J. 1995. *Shangzhou qingtongqi zhuzao gongyi ruogan tantao.* Beijing.

5 Chinese Cast Iron through Twenty-Five Hundred Years

P. T. Craddock, M. L. Wayman, H. Wang, and C. Michaelson

ABSTRACT Studies on Chinese iron production have tended to concentrate on the inception of iron making through the first millennium B.C.E., up to the end of the Han period. This paper reports in outline the metallographic examination of some objects of cast iron dating from the fifth to the nineteenth century C.E., and a selection of cast-iron coins. The latter date from the eleventh and twelfth centuries, during the brief period when iron coinage was prevalent as part of the tremendous general increase in iron production in China. This increase was bound up with the introduction of coal and coke as smelting fuels, and this is reflected in the composition and metallographic structure of the metal reported here.

With this new knowledge the development of Chinese iron smelting from the Han period through the last two thousand years can begin to be more accurately assessed.

Introduction

There has been tremendous progress in recent years in our understanding of the early development of Chinese iron technology, further emphasizing the precocious and fundamental contribution that China made to this all-important area of production and economic activity. However, much still remains to be discovered, and many areas are still little researched. In particular, previous research has tended to concentrate on the earlier periods, especially in the first millennium B.C.E., covering the period from the first use of iron in China up to the Han period, which came to an end in 220 C.E., by which time China was incontestably the world leader in iron technology. By contrast, there has been relatively little research on Chinese iron technology through the last two thousand years.

This is partly the result of a curious mindset that has developed in the West toward Chinese technology generally (Craddock 1994). The Chinese are popularly credited with having invented just about everything, but at the same time, the China of the last few centuries is perceived as being innately inward-looking, conservative, and noninnovating. How these curious and contradictory notions arose is in itself interesting and informative. Through the eighteenth and nineteenth centuries, China was faced with an ever-growing flood of imports from an increasingly industrialized Europe. Large sections of Chinese industry could not compete, and particularly mining and metal production went into sharp decline. This has recently been well documented by Wagner (1984, 1997), who has shown that only the most simple processes producing bulk materials that required little capital investment or training of the operatives could survive, and that the previous sophisticated processes by and large went under. The primitive processes, often of no very great age or tradition, were observed by Western mining engineers and economic geologists in the nineteenth and early twentieth centuries. They all too readily believed them to be the primordial unchanging and unchangeable state of Chinese technology rather than the death throes of a once thriving and innovative technical culture.

In the mid-twentieth century, Joseph Needham entered the lists to redress the balance, documenting in the magisterial volumes of *Science and civilisation in China* (*SCC*) the hitherto neglected extent of China's innovation and primacy in many technologies. Needham paid special attention to iron and steel, dedicating a special monograph (1958) to the subject ahead of the main *SCC* volume that is still forthcoming. In his enthusiasm Needham on occasion could use evidence, especially literary sources, quite uncritically, citing sources, sometimes copied at several removes over hundreds of years, to show that particular techniques or materials were in use much earlier than had been previously believed. For example, Needham cites secondary sources using a particular word that latterly meant zinc to claim that the metal was produced in China over two thousand years ago (Needham 1974, 212–19). The reality is that the production of zinc in China began about five hundred years ago, albeit in a process of very considerable technical ingenuity (Craddock and Zhou 2003).

Pushing the dates of major technical advances ever further back into the distant past had the effect of creating an apparent innovation vacuum for the last two millennia in which little was discovered. This perception, coupled with the pejorative descriptions of Chinese industry and technology in the early twentieth century mentioned above, created a quite false picture, and thus Needham's efforts to dispel the misconceptions over Chinese innovations rather backfired and wound up reinforcing them.

The reality is of a much steadier development, in a China that was much more innovative and open to new ideas and foreign business opportunities, as documented by industries as diverse as porcelain (Medley 1989) and cane sugar (Mazundar 1998). To return briefly to zinc, within fifty years of the independent discovery of a viable distillation process in the sixteenth century, Chinese zinc dominated world markets. Not bad going for a noninnovative and inward-looking society!

With iron technology also it seems that technical development continued long after the Han, although it is still difficult to judge how continuous or sustained these improvements are likely to have been. Especially important developments seem to have taken place in north China during the Tang (618–907) and Song (960–1279) dynasties. Through this period of the later first millennium C.E. into the early second millennium, there was a huge and well-documented increase in iron production (Hartwell 1966, 1967). Associated with this expansion was the introduction of fossil fuel to augment charcoal as fuel and reductant. This is also the period in which some justly famous and massive individual iron castings were made, many of which are still extant and bear testimony both to the scale of iron usage and to the skill of the artisans, handling huge quantities of molten metal with experienced precision (Till and Swart 1993; Paludan 1994; Wagner 2000). There has hitherto been little metallographic study of the surviving artifacts of cast iron from this period to complement the historical evidence.

This paper summarizes research being carried on at the British Museum in part to redress this problem. Here we will describe some of the work carried out on iron castings, principally of Buddhist statuary, dating between the eighth and nineteenth centuries and on the iron coinage that was produced on a very large scale between the eleventh and thirteenth centuries. Some metallographic and analytical work was published many years ago by Read (1913, 1936) and Pinel, Read, and Wright (1938) on Chinese iron castings of the last two thousand years, but otherwise there is little published work. The evidence contained in the structure and composition of the metal gives important clues to the technology by which the metal was smelted and cast.

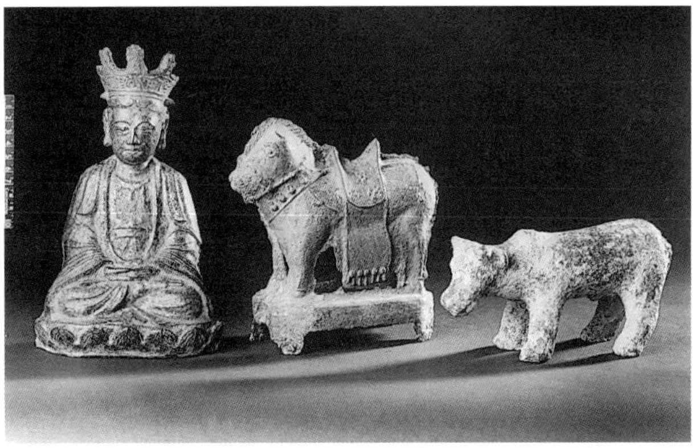

Figure 1. Group of three iron castings; from left, Buddha, BM OA Reg. 1993, 11-11,1; standing horse, BM OA Reg. 1994, 1-2,4; and standing ox, BM OA Reg. 1994, 1-29, 5 (A. Milton, British Museum)

Nineteen cast-iron statues and other artifacts (figs. 1 and 2), together with thirty-seven cast-iron coins (figs. 3 and 4) were examined metallographically by optical and scanning electron microscopy. The majority of the statues were of white cast iron, but a substantial minority were of gray cast iron, and all of the coins were of white cast iron. The full metallographic descriptions are being published elsewhere (Wayman et al. forthcoming; Wayman and Wang forthcoming). The metallographic structures encountered are summarized below, concentrating on features of major technical and historical significance, but first it is necessary to explain a little of the metallurgy of cast iron.

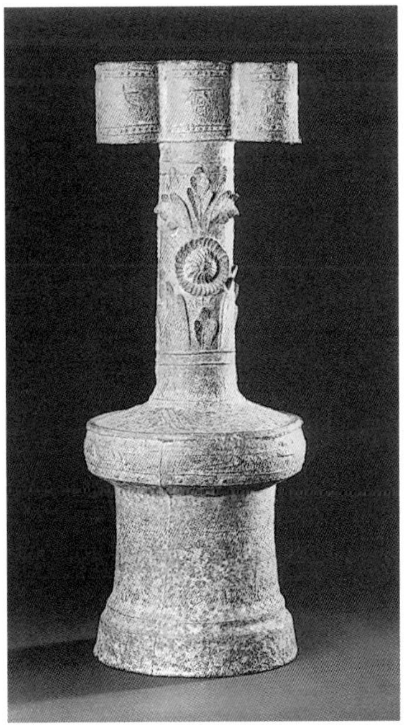

Figure 2. Altar vase, dated Wanli 7th year (1579); BM OA Reg. 1982, 12-13, 1. Note the flash lines on this white cast iron. (A. Milton, British Museum)

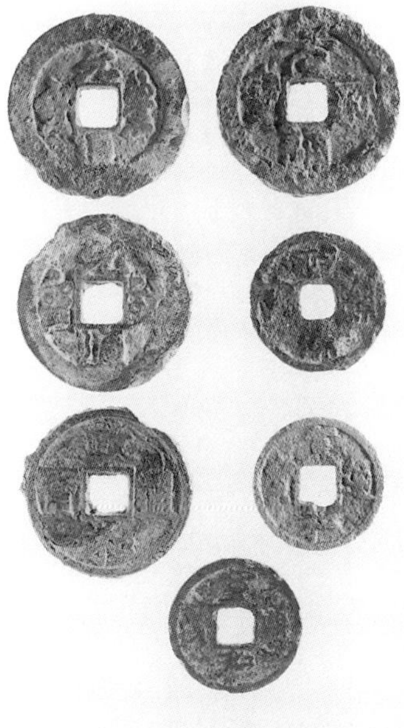

Figure 3. A group of Chinese cast-iron coins of the Song dynasty (obverse face) (A. Milton, British Museum)

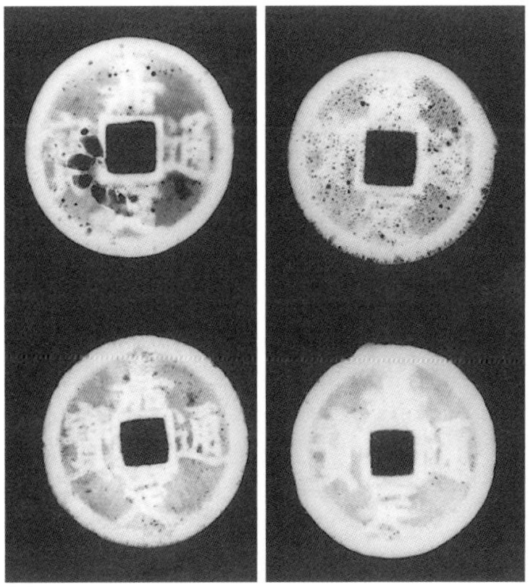

Figure 4. Radiography of a group of Song coins of white cast iron. Note the quite macroscopic porosity in some of the coins. (J. Lang, British Museum)

A Little Metallurgy

Metal ores were, and still are, smelted under reducing conditions with carbon as the fuel and reductant to release the metal from the mineral. The carbon was traditionally in the form of charcoal but is now almost universally in the form of fossil fuel, coal, or coke, a technology pioneered by the Chinese well over a thousand years ago, as documented both historically and by the composition of the metalwork reported here. Most metal ores were smelted to form a liquid metal, although zinc was smelted as a vapor (Craddock 1995, 309–17; Craddock and Zhou 2002). However, outside China and Korea, iron was usually produced as a solid in antiquity, and this continued until much later in many parts of the world. If the iron was smelted at temperatures somewhat higher than those used in the solid state process and under rather more reducing conditions, then some of the carbon from the charcoal actually dissolved in the forming iron and lowered the melting point by several hundred degrees such that it could be run from the furnace as a liquid just like other metals (Craddock 1995, 234–53; Rostoker and Bronson 1990). This is known in English as cast iron, irrespective of whether or not it is intended to be used for castings, and the smelting operation is known as the blast-furnace process. For more on the metallurgy of iron and especially of iron castings, see Wagner (1993, 335–404), Rostoker and Bronson (1990), or any traditional textbook on ferrous metallurgy and foundry work—for example, Howard (1958).

The liquid iron contained between about 3 and 5% of carbon in solution, but as the temperature drops, the solubility of the carbon in the iron drops sharply and the carbon could reappear in two forms, depending on the conditions. If the iron was relatively pure, as it usually was if smelted with charcoal, then the form would depend mainly on the rate at which the iron solidified and cooled. If the iron solidified and cooled down quickly, in a matter of minutes for metal poured into a cold mold, then the bulk of the carbon would react with the iron to form a compound, iron carbide (Fe_3C), known as cementite. Although 25% of the atoms in cementite are carbon atoms, they are much lighter than the iron atoms, which means that cementite actually contains only 6.7% of carbon *by weight*. Hence in a cast iron containing 4% of carbon, with all the carbon in the form of cementite, no less than about 60% of the total weight of the metal will be cementite (fig. 5).

This is known as white cast iron on account of the color of a fractured surface. Cementite is one of the hardest substances known—harder than quartz, for example. The white cast iron castings are virtually impossible to work, and any attempt to shape by hammering would merely shatter them. It is not possible to drill, cut, or even file them with any conventional tool other than by very laborious grinding. This can be very bad news for the smith, as white cast iron is a poor material for castings, tending to shrink and move away from the mold as it cools and to form large cavities within the casting, requiring remedial work that is generally impossible to effect (fig. 4).

If the iron cooled down and solidified slowly, by keeping the molds hot over a period of hours, for example, then most of the carbon would precipitate out of the iron as flakes of graphite (fig. 6). This technology of heat treatment of the cast iron lies at the heart of the great Chinese achievement of making a type of cast iron that produced much better castings. This iron is also much more tractable, such that it is possible to drill, cut, file, and even hammer it a little. This material is known as gray cast iron, and for the past couple of centuries has probably been the most common metal in the world. The familiar iron castings, such as manhole covers, engine blocks, and door handles of the everyday modern world are most likely to be of gray cast iron.

If the iron is relatively pure, as charcoal-smelted iron usually is, then the type of cast iron is dependent on the cooling rate alone. Thus, in some large castings it is possible to have variations in the structure of the iron in different parts of the casting just because of the differing rates of cooling. For example, the main casting of the huge Cangzhou lion (Kunyi, Jinghua, and Minzhi 1984; see below) is of gray cast iron, but the smaller, yet still substantial, lotus must have cooled more quickly and is of white cast iron with some areas of gray cast iron. Such mixtures of white and gray cast iron are known as mottled cast iron.

Iron from the Tang and Song dynasties sometimes contains over 1% of silicon and traces of manganese (tables 1 and 2), indicative of high smelting temperatures. It is

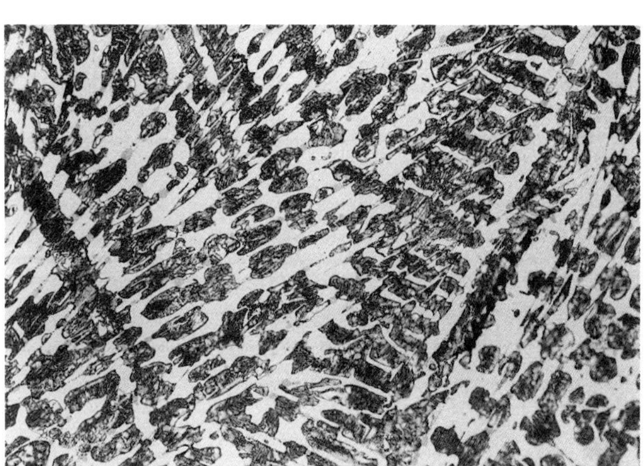

Figure 5. *Metallographic structure of a Chinese cast-iron coin. The dendritic structure is typical of white cast iron. (M. Wayman, British Museum)*

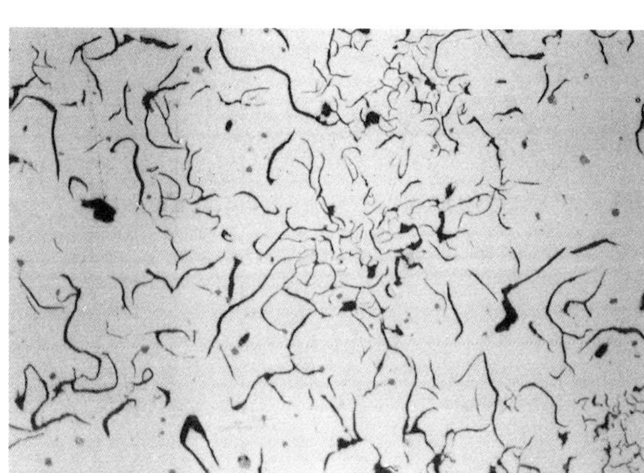

Figure 6. *Metallographic structure of a Chinese cast-iron tool of the Han period. The dark flakes of graphite are typical of gray cast iron. (M. Wayman, British Museum)*

Table 1. *Summary of the metallurgical data on the castings (Wayman et al. forthcoming, table 4)*

Reg. No.	Object	Date	%Si	%P	%S	other	%C (est.)	Type
1992.6–13.1	Plaque	718	1.98	nd	nd	0.62% Mn; 0.35% Cu		Gray
1994.1–29.5	Standing ox	9–10th cent.	na	0.46	0.20		3.9	White
1922.11–17.1	Head of Bodhisattva	10–14th cent.	nd	0.71	nd		5.2	White
1943.2–15.1	Head of Bodhisattva	10–14th cent.	1.83	0.17	0.20	0.48% Mn		Mottled
1993.8–9.1	Warrior	1098	0.51	0.44	1.00		2.8	White
1990.4–13.1	Seated Lohan	1494	0.53	0.41	0.49		3.1	White
1993.7–14.2	Cat	15–16th cent.	1.98	0.24	0.09	0.71% Mn; 0.12% Ti		Gray
1993.7–14.1	Horse	15–16th cent.	1.70	nd	0.14	0.28% Mn		Gray
1982.2–13.1	Altar vase	1579	0.43	0.42	0.88		2.7	White
1990.5–26.1	Standing figure	16th cent.	0.20	1.16	0.52		2.4	White
1989.10–10.1	Jade maiden	16–17th cent.	0.42	0.89	0.88		2.3	White
1993.11–11.1	Buddha	16–17th cent.	0.38	0.77	0.65		2.3	White
1994.1–29.4	Standing horse	16–17th cent.	0.81	0.71	0.23	0.33% Mn	4.5	White
1994.1–29.6	Figure of Bodhisattva	16–17th cent.	0.31	0.83	0.39		2.6	White
1993.8–4.1	Ox on frame	17th cent.	0.51	0.77	0.76		2.2	White
1988.5–19.3	Figure with pot belly	17–18th cent.	1.24	1.02	nd	0.50% Mn		Gray
1993.7–30.1	Judge	17th cent.	3.01	0.26	0.11	0.29% Mn; 0.16% Ti		Gray
1990.11–22.1	Incense burner	18–19th cent.	1.12	0.21	1.36		2.0	White
1988.5–19.2	Bell	1848	1.10	0.37	0.86		1.8	White

(Wayman et al. forthcoming, table 4)

Table 2. *Summary of the metallurgical data on the coins (Wayman and Wang forthcoming, table 2)*

Coin	Reign Period	Date Range	Mint Mark	Coin Date	% Si	% P	% S	% C (est.)	Mn
cm7	Yuanfeng	1078–1085			0.32	0.6	0.9	3.4	nd
cm8	Yuanyou	1086–1093			0.53	0.9	0.73	3.4	nd
cm9	Yuanyou	1086–1093			0.7	1.38	1.06	3.1	nd
cm10	Zhenghe	1111–1117			0.71	0.79	1.71	2.6	nd
cm11	Zhenghe	1111–1117			0.20	0.30	0.15	3.1	0.31
cm12	Xuanhe	1119–1125	Shaan		0.25	0.58	0.39	4.0	nd
cm13	Xuanhe	1119–1125	Shaan		0.57	0.69	1.65	2.3	nd
cm17	Xuanhe	1119–1125	Shaan		0.32	0.68	0.37	4.0	nd
cm14	Xuanhe	1119–1125	Shaan		0.78	0.71	2.05	2.3	nd
cm15	Xuanhe	1119–1125	Shaan		0.55	0.71	1.33	2.9	nd
cm16	Xuanhe	1119–1125	Shaan		0.57	0.73	1.14	2.9	nd
cm18	Qiandao	1165–1173			0.18	1.35	0.41	4.2	nd
cm2	Chunxi	1174–1189	dot in crescent		0.25	1.46	1.97	2.6	nd
cm3	Chunxi	1174–1189			0.50	1.25	2.00	2.6	nd
cm19	Chunxi	1174–1189	dot in crescent		0.36	1.36	1.35	3.7	nd
cm20	Chunxi	1174–1189			0.25	1.65	1.52	3.4	nd
cm21	Chunxi	1174–1189			0.26	1.37	1.08	4.0	nd
cm30	Chunxi	1174–1189	chun		0.18	0.18	nd	4.3	nd
cm32	Chunxi	1174–1189	tong		nd	0.14	nd	4.9	nd
cm33	Chunxi	1174–1189	tong 12	1185	nd	nd	nd	4.0	nd
cm31	Chunxi	1174–1189	chun 15	1188	nd	0.12	nd	4.3	nd
cm34	Shaoxi	1190–1194	chun 1	1190	0.16	nd	nd	3.9	nd
cm35	Shaoxi	1190–1194	chun 2	1191	nd	0.15	nd	4.2	nd
cm36	Shaoxi	1190–1194	tong 2	1191	0.16	0.10	nd	4.0	nd
cm37	Shaoxi	1190–1194	tong 5	1194	nd	nd	nd	4.3	nd
cm6	Qiangyuan	1195–1200			nd	0.31	nd	4.0	nd
cm1	Qiangyuan	1195–1200			0.13	0.13	nd	4.3	nd
cm28	Jiading	1208–1224	han 1	1208	nd	0.22	nd	4.5	nd
cm29	Jiading	1208–1224	han 2	1209	nd	0.16	nd	3.6	nd
cm22	Jiading	1208–1224	chun 2	1209	0.10	0.18	nd	4.8	nd
cm26	Jiading	1208–1224	tong 2	1209	nd	nd	nd	4.3	nd
cm5	Jiading	1208–1224	3	1210	nd	0.25	nd	3.9	nd
cm27	Jiading	1208–1224	tong 3	1210	nd	nd	nd	4.3	nd
cm4	Jiading	1208–1224	chun 3	1210	0.38	0.18	0.65	3.4	nd
cm23	Jiading	1208–1224	chun 3	1210	0.14	0.12	nd	4.5	nd
cm24	Jiading	1208–1224	chun 4	1211	0.13	0.16	nd	4.8	nd
cm25	Jiading	1208–1224	chun 8	1215	0.13	nd	nd	3.9	nd

(Wayman and Wang forthcoming, table 2)

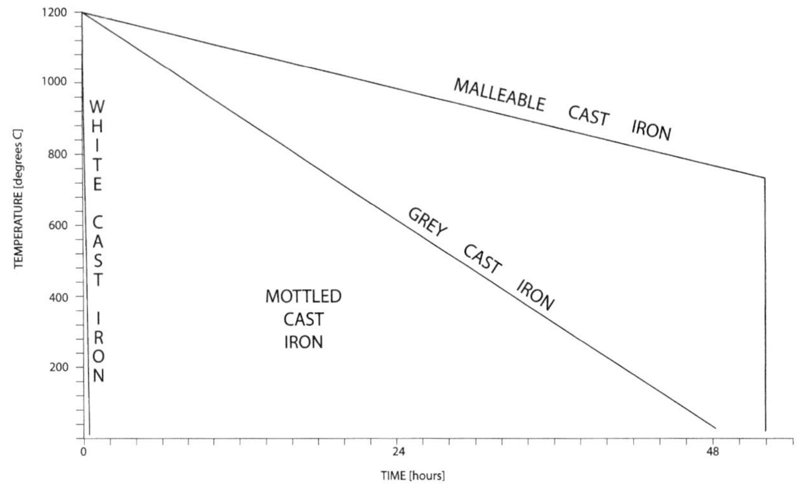

Figure 7. The cooling of low-impurity charcoal-smelted cast iron and the resulting structure of the iron

believed that at this period, fossil fuel, coal, and its derivative coke, began to be used in iron smelting (see below), which introduced substantial traces of sulfur into the iron. Each of these elements influences whether white or gray cast iron forms. Smelting with fossil fuel is generally believed to result in higher temperatures, and modern coke-smelted irons typically contain between 1 and 2% of silicon. Silicon in the iron strongly favors the formation of gray cast iron. However, coal and coke also contain quantities of sulfur, which can enter the iron, and sulfur promotes the formation of white cast iron. Indeed, if there is more than about 0.5% of sulfur dissolved in the iron, then the metal will almost certainly be white cast iron, irrespective of the cooling rate. The hotter smelting conditions can also result in small quantities of manganese being reduced from the ore and dissolved in the iron. By itself manganese favors the formation of white cast iron. However, things are a little more complicated, because the manganese will react with any sulfur in the iron to form inclusions of manganese sulfide, which remove the harmful influence of the sulfur on the structure of the iron. Thus, in most situations of iron smelted with fossil fuel, small quantities of manganese will actually favor the formation of gray cast iron. Furthermore, the higher smelting temperatures associated with fossil fuel tend to drive more of the sulfur from the iron.

Clearly there are a lot of conflicting factors at work here, especially once fossil fuel began to be used regularly from the mid-first millennium onward, with both composition and cooling rates affecting the outcome. As the ancient Chinese castings were usually very low in silicon, long cooling times may have been necessary. There are records, for example, of smiths in nineteenth-century China cooling the clay molds containing cast iron woks slowly over periods of several days to ensure that the product was a gray cast iron (Wagner 1989).

The Chinese also made other types of heat-modified castings (fig. 7). If the iron was kept at red heat only a little below the melting point for several days, enveloped in charcoal dust to prevent oxidization and the loss of the carbon in the iron, and then allowed to cool, the graphite in the iron would form into nodules. These are much less likely to initiate cracking than are the long, thin flakes in ordinary gray cast iron. Alternatively, the casting could be kept red-hot but exposed to the air such that much of the carbon was burnt out of the iron. These last two heat treatments are known now as malleabilizing because of the greater degree of working that can subsequently be carried out on the castings (Howard 1958, 67–69; Wagner 1989, 351–61).

Through most of its history, both in China and the rest of the world, the fate of the majority of the cast iron was to be converted into wrought iron, latterly mild steel. This was achieved by melting the cast iron and working it to expose it to air until almost all of the carbon had been burnt out and wrought iron was produced, in the processes known in English as fining or puddling.

A Short Technical History of Iron in China Up to the End of the Han Period

One of the questions this paper seeks to address is the form of iron in the iron castings from the periods after iron smelting with fossil fuels had become widespread and what control, if any, the Chinese smiths were able or felt they should be able to exert over it. Before discussing the results of our investigations of the Chinese cast iron of the last fifteen hundred years, it is necessary to set the scene by describing the precocious developments in iron technology that took place in China through the first millennium B.C.E. that advanced bulk iron production to levels not attained elsewhere for at least another thousand years.

Our understanding of the inception and early development of Chinese iron technology has increased dramatically over the last few years (Wagner 1993, 1999; Bronson 1999), but major questions still remain. It is, for example, still very uncertain whether the technology for the production of iron in China was introduced from the West—after all, iron had been in use in the Middle East for approximately half a millennium by the time of the first Chinese production—or whether there was an independent discovery in China drawing on the local traditions of copper smelting. Part of the problem in trying to answer these questions is the overreliance on the study of iron artifacts to glean information on how the iron was smelted and the objects made. No iron-smelting sites have yet been discovered that are contemporary with the first few hundred years of iron production in China. The earliest iron-smelting furnaces presently known are those associated with large foundries dated to the last centuries B.C.E. These are easy to recognize due to the presence of copious amounts of distinctive mold fragments. Slag heaps by themselves, without the mold fragments, do not seem to have attracted the attention of the archaeologists, yet they must exist.

The earliest evidence for smelted iron in China is currently iron artifacts from burials in the north of China dated to the eighth and even to the ninth century B.C.E. (Han 1996; Wagner 1999). The presence of long, thin stringers of slag in the iron show beyond doubt that it is smelted rather than meteoritic iron, and it has been claimed that the presence of slag shows that the iron was smelted by the solid-state process. Variants of this process were in use all over the iron-using world during the first millennium B.C.E. and for a long time afterward, and it is an understandable assumption to make. In this process the solid lump of iron was pulled from the smelting furnace and was hammered to remove the majority of the slag that it inevitably contained, so as to produce wrought iron with just a few slag stringers such as are found in the earliest Chinese iron. However, another possibility exists: the wrought iron might have been produced by fining cast iron—that is, by a process whereby cast iron was remelted on a hearth and worked in air to burn out the carbon that it contained, as outlined in the previous section. While this was going on, some of the iron would oxidize and react with silica in the clay of the hearth linings to produce a slag identical to that found in the wrought iron produced by the solid-state process. Thus, the resulting wrought iron would be very similar, made by either process, despite the processes themselves being so different. In fact, differentiating between the irons made by the two processes is one of the major problems worldwide facing archaeometallurgists at present.

The question is of great importance for the history of Chinese iron because if the first iron had been made by fining blast-furnace cast iron, then this would be very strong evidence for an independent discovery of iron smelting in China. This is not as improbable a scenario as it might once have seemed. In Europe it is now known from the study of early blast-furnace smelting sites, dating from about 1000 C.E., that the first cast iron (as opposed to the first iron castings) was fined to produce wrought iron, and that iron castings did not begin to be produced for several centuries after the development of the blast-furnace process (Yalçin and Lychatz 1995; Gassmann, Yalçin, and Hauptmann 1995; Craddock 2003). Possibly a similar developmental sequence is applicable to China. Incidentally, the slow evolution of the process in Europe shows that the blast furnace must have been an independent development rather than being a technology that had been introduced from China. The discovery and excavation of early iron-smelting sites in China would conclusively answer these questions that are presently left open by the study of artifacts alone.

Around the middle of the first millennium B.C.E., the first iron castings began to be produced in China. They were tools, weapons, and general fittings, and fitments and were of both white and gray cast iron, followed by malleable cast iron (Rostoker et al. 1985). To be able to transform cast iron into a viable material for everyday artifacts, either by the production of gray cast iron or by malleabilizing the castings, was the great achievement of Chinese iron technology (Wagner 1989; 1993, 351–61). Malleable iron castings were certainly produced by the Chinese in the last centuries B.C.E., as evidenced by the structures of the surviving iron artifacts. From the limited number of early Chinese tools and weapons of cast iron so far examined, there does not seem to be a consistent pattern of structure emerging. In post-medieval Europe there was a clear delineation between slow cooling of the metal in the mold to produce castings of gray cast iron and the heat treatment of castings of white cast iron to malleabilize them. It is possible that in antiquity the Chinese did not differentiate very carefully between the heat treatments for producing more tractable iron in their castings. Clearly the castings must have been deliberately heat-treated for long periods, but the dividing line between very slow cooling and maintaining a high temperature for some time and then cooling may not have been very rigorously applied (fig. 7). Similar treatments in Europe only began in the post-medieval period and are more normally regarded as a technology of the Industrial Revolution.

Fining hearths have been recognized at smelting sites alongside the iron-smelting furnaces from the Han period but may go back much earlier, even to the inception of the blast-furnace process, as was certainly the case in Europe (see above).

There is also evidence for the partial decarburization of cast iron to produce steel in the Han period (Wagner 1993), but that is beyond the scope of this paper.

Certainly by the end of the first millennium B.C.E., the Chinese were able to produce a variety of different irons from their cast iron and to produce them on a prodigious scale compared to anything that had gone before. The stack molds found at the foundry sites in which hundreds of standard chariot fittings were cast at a time, for example (Hua Jueming 1983), have their closest parallels in the Industrial Revolution of Europe almost two thousand years later. There was one big difference: the furnaces of the Industrial Revolution were fired with fossil fuel, coal, and its derivative coke, but the Chinese iron of the Han period and earlier was smelted with charcoal. Coal was reported from some of the early excavations of smelting foundry sites, and it was initially believed that the coal had been used as the smelting fuel. However, subsequent studies have shown that although coal was certainly present, it was only used to fire the molds and to fuel the heat treatments; it was not in contact with the iron ore in the smelting operation for which charcoal alone was used (Li Jinghua 1996).

In the recent past there was a highly distinctive crucible process in Shaanxi for producing cast iron in direct contact with coal (Read 1913; Needham 1958, 14–15, plates 11–13; Needham 1980, 520; Rostoker and Bronson 1990, 113–14).

Needham asserted that this process was operational "going back to the fourth century C.E. at least," but without providing any evidence for the early history of what would otherwise be the earliest use of coal as a smelting fuel, and it seems likely that this is a more recent technology, possibly coeval with the zinc-smelting process in origin.

In reality the solution to the seemingly intractable problem of smelting iron with coal or coke took place during the Tang and Song dynasties and represents another major advance for Chinese iron technology (see below).

How the Castings Were Made

The objects examined here all seem to have been cast in piece molds. This follows the well-known traditional Chinese bronze-casting technology of previous ages (Gettens 1969), wherein the molds were made of fine windblown loess sediments that did not distort during firing (Freestone, Wood, and Rawson 1989). However, during the first millennium C.E., the piece mold seems to have been largely supplanted in China by the lost-wax method for the casting of bronze (Kerr 1990, 22; see also Cowell and La Niece, this volume), yet, rather inexplicably, piece molding was adopted for the iron castings of the art metalwork, statuary, and vessels that we have investigated here as well as for the major statues still standing in the Chinese countryside. Good evidence for piece molding are the often very prominent flash lines on the surface of the casting where the molten iron penetrated the gaps between the reassembled mold pieces (figs. 1 and 2).

Some attempt could be made to disguise the flash lines by making them coincide with elements in the design, such as bands of decoration and changes in profile. Necessity dictated that the larger statues were cast from molds made up of dozens of pieces. These can give the finished casting a decidedly unfinished, "patchwork quilt" appearance. These flash lines, which are so prominent on so many Chinese bronze and iron castings, have caused much comment over the years. Wagner (2000, 204–5) discussed the flash lines on the large iron castings and concluded that the reason for their continued presence is that they must have been impossible to remove; in other words, the castings were of white cast iron that would have resisted both chisel and file (see above), and this was a good indicator of the structure of the iron without having to resort to metallography. Unfortunately, this eminently reasonable hypothesis is not always supported by the metallurgical investigations of the pieces concerned. In this project five gray cast-iron pieces were identified, of which three had no trace of flash lines, one had the remains of flash lines imperfectly removed, and on the final piece no attempt at all had been made to remove the flashes. Similarly, the Cangzhou lion is of gray cast iron but has very prominent flash lines that could have been removed without too much difficulty. Thus, Wagner's suggestion that one can make an educated guess as to whether a casting is of white or gray cast iron by the flash lines is unfortunately not possible.

More generally, the flash lines were often not removed from the earlier bronze castings where it would have been relatively easy to do so. It has been suggested by way of explanation that the statues may have been coated with gesso or some other surface preparation, but unless this was up to a centimeter thick in places, thereby obscuring all of the detail of the casting, the very prominent flash lines would still have shown through. It has also been suggested that there was an element of ritual, that the representation had to be demonstrably in its pristine, "as cast" condition.

Some of the iron castings are hollow cast, and in these cases it seems that the procedure would have been to carve the original model in clay and take piece molds from this. The model would then have been used to form the core. This was done by paring it down by an amount equivalent to the thickness of metal required in the final casting, and then it was fired. The fired piece molds could then be reassembled around the core and the metal cast.

The very large castings, such as the Cangzhou lion, stand many meters tall and contain many tons of iron. To have cast such a statue in one pour would have presented almost insuperable problems. Examination of the statue (Till and Swart 1993; Wu Kunyi, Li Jinghua, and Wang Minzhi 1984) has shown that it was cast in stages (fig. 8). The sections of the external mold were built around the core to a manageable height, and the iron was poured in from the adjacent melting furnaces. The walls of the mold were then raised, and more metal was poured in. Evidence for this are the very prominent seams between the different pourings of metal that resulted from the metal cooling in the mold in the time that elapsed while the mold walls were raised, which must have taken many minutes to perform. Thus, the metal of the first pouring would have partially frozen before the next pouring flowed over it, leading to very incomplete fusion and the seams, known in English as shuts. This method of casting large statues was not confined to iron; the most famous example of this technology is the giant bronze Buddha at Nara in Japan, which is probably still the largest casting ever made (Ishino 1983). There the shuts are very prominent, and the casting as a whole must be very weak, as is probably true of the Cangzhou lion. Some attempt was made to compensate for this weakening caused by the cold shuts in the lion by the insertion of wrought-iron pins between adjacent layers.

This is not the case with the technically much more accomplished cast-iron oxen that acted as anchors for the

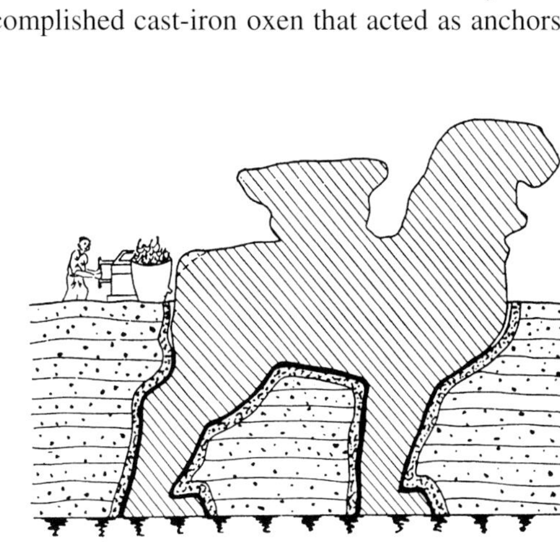

Figure 8. *Drawing showing how the Cangzhou Lion may have been cast in stages (Wu Kunyi, Li Jinghua, and Wang Minzhi 1984, 83)*

Pujin Bridge across the Yellow River, cast in 720 (Paludan 1994). They had to support the stress of the bridge cables and thus really had to be a coherent single mass, or they would have been pulled to pieces when a stress was applied. They are solid, cast in a single operation, and at seventy tons are among the heaviest iron castings ever made. They obviously had to be cast in situ and, sure enough, the remains of the over ninety melting furnaces, with channels running from them to the mold, have been excavated. The coordination of the melting and flow of such prodigious quantities of metal must command our admiration. No other castings of commensurate size are presently known from China or Korea, and certainly at the time of their production there can have been nothing remotely like them elsewhere.

By contrast, the coins were cast in strips in simple two-piece molds that were probably themselves made of cast iron, although ceramic molds have been found apparently for casting iron *wuzhu* coins, in the southeastern parts of Nanjing, and are now in the Shanghai Museum (Fan Weihong 2000/2). The swift dissipation of heat from a metal mold would bring about rapid cooling of the relatively small quantity of iron poured into the molds and would promote the formation of the white cast iron that we have observed in the coins themselves (fig. 5).

The Metallographic Examination

A selection of nineteen cast-iron objects was made from the collections of the Department of Oriental Antiquities (Wayman et al. forthcoming) (table 1). Fourteen were of the Ming or Qing dynasties (fourteenth to nineteenth century), and the other five were earlier, from the Tang and Song dynasties (eighth to thirteenth century). A selection of thirty-seven coins was made from the collections of the Department of Coins and Medals (Wayman and Wang forthcoming), all of the Song dynasty, and dated to between the eleventh and thirteenth centuries (table 2). The coins had the big advantage for our studies of being datable, often to within ten years, and in many cases the mint could be identified, enabling the provenance of the iron to be determined with some precision.

Samples were cut from the iron objects utilizing damaged areas where these existed, but the majority of the coins were intact, and so cut samples were not permissible. Instead, taper sections were polished on the edges of the coins. Several of the coins were fragmentary, and on these a section was polished both on a broken edge near the center of the coin and on the edge, and by comparing the interior and edge sections it was established that the structure at the polished edge was representative of the structure of the coin as a whole.

The cut sections and taper sections of the objects and coins were all examined in both the unetched and etched condition, using a Zeiss reflecting light microscope and a JEOL scanning electron microscope with an Oxford Isis EDX microanalytical system to provide analytical data mainly for the elements silicon, phosphorus, sulfur, and manganese. The carbon content of the iron was estimated, where possible, from the metallographic structure. More detail of the sample procedures and the examination procedures will be given in the forthcoming papers.

Five of the object castings were of gray cast iron, thirteen were of white cast iron, and one was a mottled cast iron. In every case the structure seems to have been largely determined by the carbon and silicon content, and to lesser degree by the sulfur and manganese contents. The six castings with more than 1% of silicon are the five gray cast irons and the one mottled iron. Conversely, these pieces all had low sulfur contents, and such sulfur as they do contain is likely to be largely in the form of manganese sulfides, a consequence of the manganese contents.

The phosphorus in the metal is likely to have come with the iron ores. Phosphorus does not have a great influence on whether gray or white cast iron forms but does have effects on other aspects of the structure and, more significantly for the casting of iron, both increases the fluidity and reduces the melting point of the iron. Read (1913, and reported in Rostoker and Bronson 1990, 11, as well as Needham 1958, 14 and 1980, 520) reported very high phosphorus contents, averaging about 2% in twelve iron castings he analyzed from Shaanxi, and correlated this with addition of small quantities of phosphatic material, which he apparently observed being made to the charge of the Shaanxi crucible process (see above). He went on to suggest, followed by Needham, that small additions of phosphatic material to lower the melting point may have been a contributory factor in the success of ancient Chinese iron-casting technology. Subsequent analysis of early Chinese castings has not revealed such unusually high phosphorus contents, and the whole idea seems inherently unlikely.

There has been some suggestion that some of the high sulfur contents encountered by Pinel, Read, and Wright (1938) could be an indication that the element was being deliberately introduced by the use of sulfur-rich iron ores or by adding minerals such as iron pyrites to the molten iron. The presence of sulfur, although detrimental to the working properties of the iron, does increase the fluidity of the molten metal, making it easier to cast (Rostoker and Bronson 1990, 22). Rostoker, Bronson, and Dvorak also suggested that the sulfur content in Chinese cast-iron bells that they examined (1984) could be a deliberate addition in order to improve the vibrational characteristics of the iron. However, the sulfur contents they found—0.29, 0.4, and 0.57%—seem similar to the levels encountered in the more general range of iron objects analyzed here (table 1) and are interpreted by us as an unconscious effect of smelting with fossil fuel. Thus, overall there is no convincing evidence that the Chinese sought to increase either the phosphorus or the sulfur content.

The coins are all of white cast iron, again consistent with their compositions. The silicon content is moderate to low, but many have quite high sulfur contents, and given that the coins are likely to have cooled quickly in their molds, white cast iron was to be expected. The most interesting feature in the composition of the coins is that they divide into two compositional groups, determined largely by their sulfur content, and that this in turn is related to where the coins were minted. The significance of this is discussed below.

Discussion

This survey has already yielded many important results and posed many interesting questions, not least the evidence for the use of fossil fuel. The primary evidence for this is the sulfur content of the iron. Iron ores containing sulfur were generally avoided in antiquity, and the sulfur is believed to come almost exclusively from the fuel. Charcoal contains only minute quantities, but coal and its derivative coke can contain several percent of sulfur. The charcoal-smelted Chinese iron objects of the first millennium B.C.E. have very low sulfur contents, but the majority of the iron analyzed for this project have sulfur levels that are similar to that found in the coke-smelted iron produced in Britain in the eighteenth century (Tylecote 1976, 1991). Note that some Chinese coals have remarkably low sulfur contents. For example, zinc is still smelted by the traditional process with coal in direct contact with the ore in the retorts (Craddock and Zhou 2003).

As noted above, there is documentary evidence for the use of coal and coke for the smelting of iron from the Tang dynasty, and certainly by the Song dynasty (eleventh and twelfth centuries), coal and coke were being produced on a very considerable scale for the smelting of iron (Hartwell 1966, 1967; Wagner forthcoming). The evidence from the contemporary castings and coins examined here both confirms and amplifies our understanding of the introduction of fossil fuels to the smelting of iron.

The castings fall into two apparent groups: those with high silicon (over 1%) that in general have rather low sulfur contents, and those with moderate silicon and sulfur contents. As the sulfur is associated with fossil fuel, this could be interpreted as the high-silicon, low-sulfur group being smelted with charcoal, while the remainder were smelted with fossil fuel. However, smelting at high temperature tends to lower the sulfur content of the iron, so in fact the high-silicon and -manganese and low-sulfur contents combine to indicate a high smelting temperature. Also, even though the sulfur is depleted in the high-silicon irons, it is still detectable in six of the eight pieces examined and higher than in the earlier charcoal-smelted iron, strongly suggesting that they had been smelted with fossil fuel.

The evidence of the small group of early castings (eighth to thirteenth century) is rather equivocal. The plaque carrying an inscription dating it to 718 has a high silicon and manganese content, but sulfur was not detected, suggesting that it was probably smelted at high temperature. The sulfur content is so low that it could be due to the high temperature or to the use of charcoal. The plaque is very thick and therefore will have cooled quite slowly, which would also have encouraged the formation of gray cast iron.

The standing ox of the ninth or tenth century (fig. 1) with low silicon but 0.2% of sulfur is more likely to have been smelted with fossil fuel. Two of the three other early pieces are likely to have been smelted with fossil fuel, the third probably with charcoal. Thus, it seems probable that both fossil fuel and charcoal were used as smelting fuels, a situation that has continued in China right through to the recent past.

The coins, with their closer dating and more precise provenance information, provide greater insight into the situation. The coins fall into two broad compositional groups: those with substantial traces of sulfur, and those where the sulfur was not detectable. These differences almost certainly reflect the different fuels used for smelting the iron. Furthermore, where the mints are known, those with high sulfur come from the northwest of the Song empire, from Shaanxi and the adjacent regions that were major coal producers. By contrast, the low-sulfur coins tend to come from over a thousand kilometers to the southeast, from regions that were the main charcoal producers. Thus, it seems that the choice of fuel to smelt the iron was at least to some degree dictated by local availability.

This leads to another question, namely the ability of the smelter to control the form of the iron produced. As already noted above, Chinese smiths casting woks in the nineteenth century went to great lengths to produce gray cast iron by slow cooling of the molds. It is probable that they knew, at least empirically, that iron smelted under different conditions was more or less likely to result in castings with properties more or less identifiable with white or gray cast iron, without, of course, knowing anything about composition. Iron smelted with charcoal (low silicon, manganese, and sulfur) was likely to give a metal that with slow, controlled cooling would be gray cast iron but at faster cooling rates would be white cast iron. Iron smelted with fossil fuel at relatively low temperatures (low silicon and manganese, high sulfur) could result in white cast iron no matter what was done, and conversely, iron smelted with fossil fuel at high temperatures (high silicon and manganese, low sulfur) could produce an iron that gave castings of gray cast iron relatively easily.

One of the differences between the iron-smelting technology of China and Europe is that smelting with charcoal continued in China until the present, long after fossil fuel was introduced, whereas in Europe fossil fuel rapidly replaced charcoal. A possible explanation lies in the ability to eliminate the sulfur from the iron. The first coke-smelted iron produced in the West, at Coalbrookdale by Abraham Darby in the early eighteenth century, for all its fame was only a qualified success. The high sulfur content of the iron limited its use to castings. This did not change until the mid-eighteenth century (Rehder 1987), when steam-powered piston bellows began to be used, and furnace sizes grew, ensuring that sufficiently high temperatures were regularly attained to remove the majority of the sulfur from the iron such that it could be routinely used to make wrought iron.

Some large furnaces, dated to the Song dynasty, have been found in north China (Han 1996), including a complete example at Handan in Hebei (Li Yuncai 1978; Wagner 1985, forthcoming) that stands about six meters tall, commensurate in size to the blast furnaces of eighteenth-century Britain. The large quantities of iron supplied on a regular basis by the mints documented at this time (Hartwell 1966, 39–40) may have encouraged the development of large furnaces such as these. Charcoal would not have been strong enough to support the weight of the ore charge in these tall furnaces, further encouraging the use of fossil fuel. The regular composition of the iron from the northern mints, with rather low sulfur compared to the other castings, may also suggest that they were smelted in large furnaces such as

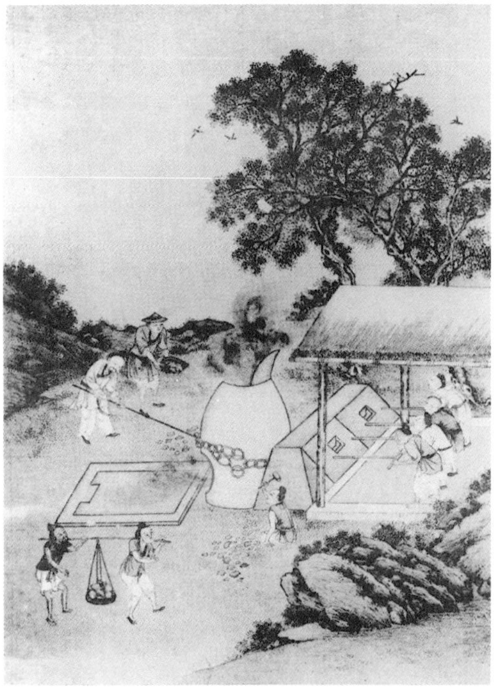

Figure 9. An illustration of a blast furnace in operation. Note the manually operated fan bellows. (Needham 1958, plate 15)

these, where a high temperature could be more easily maintained. Blowing air through such a large furnace must have been a problem and may explain why large furnaces do not seem to have become popular. In China, with the exceptions mentioned above, furnaces generally remained quite small (Needham 1958), and the bellows, although sometimes driven by water-wheels, were relatively low-powered (fig. 9). This meant that the sulfur content of the iron smelted with fossil fuel was probably usually too high for making good wrought iron, and thus the Chinese were obliged to continue with charcoal-smelted cast iron as the source material for their wrought iron and steel.

Conclusions

It seems that, as in Europe some centuries later, an ever-growing demand for iron developed in the Tang dynasty and stimulated the introduction of fossil fuel, which in turn allowed a further huge increase in production into the Song dynasty of the eleventh and twelfth centuries. In Europe fossil fuel replaced charcoal completely and quite quickly, certainly within 150 years of the first use of coke as a smelting fuel, whereas in China coal and coke fuel never replaced charcoal as the smelting fuel, and indeed all three continue in use to this day. Although the Mongol invasions of the thirteenth century seem to have cut short the large-scale production of iron, the smelting and casting technologies continued, with large bells (Rostoker, Bronson, and Dvorak 1984) and cannon (Wagner 2000) added to the repertoire by the Ming dynasty.

The nineteen castings and thirty-seven iron coins analyzed here, coupled with the very few metallographic examinations published already of other comparable pieces, have shown that the structures are perhaps more complex and less easy to predict from their outward appearance than was previously believed. They have enabled us to begin to appreciate more fully the true achievement of Chinese iron technology developing through three millennia.

Acknowledgment

We acknowledge our indebtedness to Jessica Rawson, who while keeper of the Department of Oriental Antiquities, greatly expanded the collection of Chinese iron castings, very much encouraged our scientific study, and has remained closely involved with the project.

References

Bronson, B. 1999. The transition to iron in ancient China. In *The archaeometallurgy of the Asian old world,* ed. V. C. Pigott, 177–93. Philadelphia: The University Museum, University of Pennsylvania.

Craddock, P. T. 1994. Iron and steel in ancient China: Origins and technical change. *Antiquity* 68(261):886–90.

——. 1995. *Early metal mining and production.* Edinburgh: Edinburgh University Press.

——. 2003. Cast iron, fined iron, crucible steel: Liquid iron in antiquity. In *Mining and metal production through the ages,* ed. P. T. Craddock and J. Lang, 231–56. London: BMP.

Craddock, P. T., and Zhou Weirong. 2003. Traditional zinc production in modern China. In *Mining and metal production through the ages,* ed. P. T. Craddock and J. Lang, 267–92. London: BMP.

Fan Weihong, 2000/2. The coin molds and iron coins of Su-Liang unearthed in Nanjing, Zhonggio Qianbi. *China numismatics* 69:4–7.

Freestone, I. C., N. Wood, and J. Rawson. 1989. Shang dynasty casting molds from north China. In *Cross-craft and cross-cultural interactions in ceramics,* ed. P. E. McGovern. Ceramics in Civilisation IV, 253–73. Westerville, Ohio: The American Ceramic Society.

Gassmann, G., U. Yalçin, and A. Hauptmann. 1995. Frühmittelalterliche Eisen produktion in Kippenheim, Südbaden: Ein "missing link" zwischen Rennverfahen und Roheisentechnologie. *Metalla* 2(2):43–52.

Gettens, R. J. G. 1969. *The Freer Chinese bronzes II: Technical studies.* Washington, D.C.: Smithsonian Institution.

Han, Rubin. 1996. The development of Chinese ancient iron blast furnace. In *The proceedings of the forum for the fourth international conference on the beginnings of the use of metals and alloys,* ed. K. Igaki, 151–74. Sendai: The Japan Institute of Metals.

Hartwell, R. 1966. Markets, technology, and the structure of enterprise in the development of the eleventh-century Chinese iron and steel industry. *Journal of economic history* 26:29–56.

——. 1967. A cycle of economic change in Imperial China: Coal and iron in northeast China, 750–1350. *Journal of the economic and social history of the Orient* 10:102–59.

Howard, E. D. 1958. *Modern foundry practice,* 3rd ed. London: Odhams Press.

Hua Jue-ming. 1983. The mass production of iron castings in ancient China. *Scientific American* 248(1):107–14.

Ishino, T. 1983. *How the great image of Buddha at Nara was constructed.* Tokyo.

Kerr, R. 1990. *Later Chinese bronzes*. London: Victoria and Albert Museum.

Li Jinghua. 1996. The excavation and study of iron smelting sites in Henan of Han dynasty, China. *Bulletin of the Metals Museum of the Japan Institute of Metals* 25:14–35.

Li Yuncai. 1978. Zhongguo gudai gaolu de qiyan he yanbian (The origin and development of the blast furnace in ancient China). *Wenwu* 2:18–27.

Mazundar, S. 1998. *Sugar and society in China*. Cambridge, Mass.: Harvard University, Asia Center.

Medley, M. 1989. *The Chinese potter*. London: Phaidon.

Needham, J. 1958. *The development of iron and steel technology in China*. London: Newcomen Society.

——. 1974. *Science and civilisation in China* 2 (May). Cambridge: Cambridge University Press.

——. 1980. The evolution of iron and steel technology in east and south east Asia. In *The coming of the age of iron*, ed. T. A. Wertime and J. D. Muhly, 507–41. New Haven: Yale University Press.

Paludan, A. 1994. The Tang dynasty iron oxen at Pujin Bridge. *Orientations* 25 (May):61–68.

Pinel, M. L., T. T. Read, and T. A. Wright. 1938. Composition and microstructure of ancient iron castings. *Transactions of the American Institute of Mining and Metallurgical Engineers* 131:174–94.

Read, T. T. 1913. The mineral production and resources of China. *Transactions of the American Institute of Mining, Metallurgy and Engineering* 43:3–53.

——. 1936. The largest and oldest iron castings. *The Iron Age* 136(6):18–20.

Rehder, J. E. 1987. The change from charcoal to coke iron in iron smelting, *Journal of the Historical Metallurgy Society* 21(2):37–43.

Rostoker, W., and B. Bronson. 1990. *Pre-industrial iron*. Philadelphia: Archaeomaterials Monograph 1.

Rostoker, W., B. Bronson, and J. Dvorak. 1984. The cast bells of China. *Technology and culture* 25(4):750–67.

Rostoker, W., B. Bronson, J. Dvorak, and G. Shen. 1985. Casting farm implements, comparable tools, and hardware in ancient China. *World archaeology* 15(2):196–210.

Till, B., and P. Swart. 1993. Cast iron statuary in China. *Orientations* 24 (August):40–45.

Tylecote, R. F. 1976. *A history of metallurgy*. London: The Metals Society.

——. 1991. Iron in the Industrial Revolution. In *The Industrial Revolution in metals,* ed. J. Day and R. F. Tylecote. London: The Institute of Metals.

Wagner, D. B. 1984. Some traditional Chinese iron production techniques in the twentieth century. *Journal of the Historical Metallurgy Society* 18(2):95–104.

——. 1985. *Dabieshan*. Scandinavian Institute of Asian Studies, Monograph Series 52. London: Curzon Press.

——. 1989. *Towards the reconstruction of ancient Chinese techniques for the production of malleable cast iron*. Paper 4. Copenhagen: University of Copenhagen, East Asian Institute.

——. 1993. *Iron and steel in ancient China*. Handbuch der Orientalistik, band 9. Leiden: E. J. Brill.

——. 1997. *The traditional Chinese iron industry and its modern fate*. Surrey: Curzon Press.

——. 1999. The earliest use of iron in China. In *Metals in antiquity,* ed. S. M. Young et al., 1–9. Oxford: Bar International Series 792.

——. 2000. Chinese monumental iron castings. *Journal of East Asian archaeology* 2(23–24):199–224.

——. Forthcoming. Blast furnaces in Son-Yuan China. Submitted to *East Asian science technology and medicine.*

Wayman, M. L., and H. Wang. Forthcoming. The metallurgy of ancient Chinese cast iron coins.

Wayman, M. L., J. Lang, C. Michaelson, and J. Rawson. Forthcoming. The metallurgy of Chinese cast iron statuary.

Wu Kunyi, Li Jinghua, and Wang Minzhi. 1984. The casting technique of the iron lion of Cangzhou. *Wenwu* 6:81–85.

Yalçin, Ü., and B. Lychatz. 1995. Frü- und hochmittelalterliche Eisengewinnung im Vorland der mittleren Schwäbischen Alb: Naturwissenschaftlich-technologische Aspekte zur Rekonstruktion der Eisenverhüttung. *Metalla* 2(2):53–61.

6 A Study of the Making of Bronze Weapons in the Wu and Yue States During the Eastern Zhou Period in China

Lian Haiping and Tan Derui

ABSTRACT This paper describes the study of bronze weapons made in the states of Wu and Yue during the Eastern Zhou period of China (770–221 B.C.E.). Three features of the weapons were the foci of investigation: bimetallic swords, the concentric-circle decorations found on some sword pommels, and the so-called lozenge pattern used to decorate weapon surfaces. Chemical analysis, metallographic analysis, and simulation experiments were used to determine the materials and methods used in the creation of weapons and their decoration.

Introduction

During the Eastern Zhou period of China, the casting techniques of bronze weapons in the Wu and Yue states in the southeast were far superior to that of the states in the Central Plains. Bimetallic swords, concentric-circle decorations on sword pommels, and the so-called lozenge pattern used to decorate weapon surfaces are representative of the superb techniques used in making bronze weapons at that time. The making of bimetallic swords demonstrates a sophisticated understanding of the relationship between alloy compositions and their properties. The concentric-circle decoration found on sword pommels of this period displays the skilled use of clay-mold casting techniques, while the lozenge-pattern decorations demonstrate the masterly use of surface treatment methods. The methods used to create the weapons and their decorations are discussed below.

Bimetallic Bronze Swords

The earliest bimetallic weapons from China date to the Shang dynasty (1600–1100 B.C.E.) and were composed of bronze and forged iron (Hebei Provincial Museum 1973). By the Eastern Zhou period, Chinese metalworkers knew well how to vary the contents of bronze alloys to achieve different properties. Formulas from the *Liu-ji* in the *K'ao Kung Chi, Zhou Li,* which dates to the Eastern Zhou period, specify the preferred copper and tin contents of bronzes according to the type of object to be made. The metalworkers fully understood the relationship between a bronze alloy and its properties, and the bimetallic bronze weapons produced during the Eastern Zhou perhaps best demonstrate that knowledge.

Bimetallic bronze swords were made by casting the spines and edges of the swords separately using different alloys of bronze; the use of different alloys of bronze results in a difference of color between the spine and the edge (plate 7). Because of this color difference, these swords are also called "two-tone swords." The Shanghai Museum has more than ten bimetallic swords in its collection. The authors were permitted to have fragments from four swords to sample for compositional analysis and metallographic study. Following the examination of the samples from the fragments, simulation experiments were performed in order to better understand the technology used in the making of the swords.

Plate 8 shows one of the fragments; this appearance is typical of all four fragments. On the fragment's surface, the spine is green and the edges are dark green or black with a smooth, shiny corrosion layer. In cross-section, it is apparent that the spine is more red in color than the edges, which are light yellow, or more white, in color. The difference in composition between the spine and edges is clearly indicated. It is also apparent that the spine is more porous while the edges show no porosity. Another feature seen in the cross-section is that the spine and edges have interlocking joins.

The chemical compositions of the spines and edges of the four fragments were determined using nondestructive x-ray fluorescence analysis; the results are shown in table 1. The spines and edges of the four fragments, while all made of bronze, differ significantly in their copper and tin contents. However, there is a fair degree of consistency between the results for the edges of the four fragments and the results for the spines of the fragments; exceptions to this are the lead contents in two of the spines. The copper-tin content distribution is perhaps best shown diagrammatically (fig. 1). The spines of the swords were made of alloys having relatively more copper and less tin, which would give a tougher, more flexible metal than the alloys used for the edges, whose high tin contents would result in hard, even brittle metals that would take and hold a good cutting edge.

***Table 1.** The composition of bimetallic swords (XRF)*

Analysis position		Composition content (%)						
		Cu	Sn	Pb	Fe	Ni	Zn	Sb
Fragment A	Spine	84.38	11.62	3.73	0.03	0.15		0.08
	Cutting edge	74.42	24.15	1.17	0.03	0.10	0.02	0.12
Fragment B	Spine	85.2	14.38	0.2	0.07	0.11	0.02	0.03
	Cutting edge	77.09	22.12	0.48	0.06	0.22	0.05	
Fragment C	Spine	78.46	8.13	13.14	0.05	0.14	0.004	0.086
	Cutting edge	74.52	23.72	1.42	0.07	0.27		
Fragment D	Spine	88.18	11.44	0.16				
	Cutting edge	74.36	23.0	0.59	1.91	0.15		

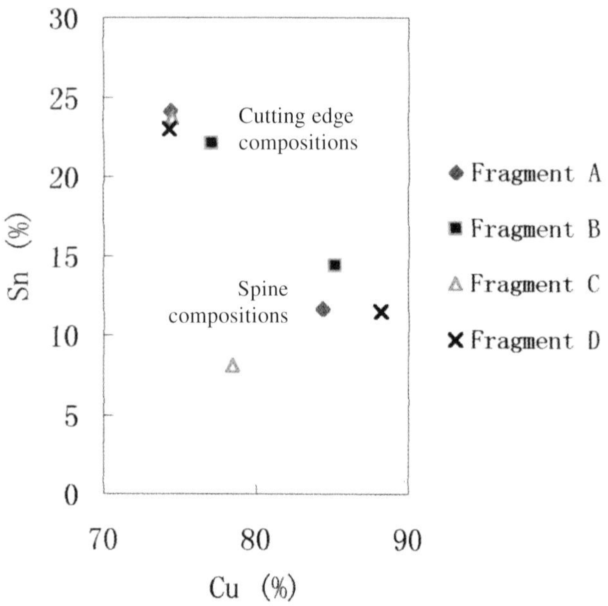

***Figure 1.** The copper and tin content distribution diagram of spines and cutting edges*

Pieces cut from the four fragments were mounted as metallographic samples, polished, and etched in an alcoholic ferric chloride solution. The microstructures of the four fragments were found to be similar to each other and the structure of only one fragment is shown here (fig. 2). The gray area at left is the structure of one of the edges, and the bright white area to the right is the spine. Clearly, the metallographic structures of the spine and edge are totally different. A typical cast-bronze dendritic structure is seen in the edge area. It is composed of alpha phase dendrites with alpha+delta eutectoid surrounding the dendritic arms. The dendrite crystallites are fine and compact. On the other hand, the structure of the spine is much coarser than that of the edge. The dendrites are barely visible, and there are many voids in the metal (fig. 3). Under high magnification, it can be seen that the spine is almost completely composed of alpha phase, and the edge of alpha phase is smooth. There is little alpha+delta eutectoid in the spine. The structures seen in the other three samples are the same; only the sizes of the dendrites seen in the edges and the alpha grains in the spine are a little different.

The properties of an alloy result from its structure and composition, and structure is due both to composition and how the alloy was made and worked. With a tin content lower than 27% the cast structure of a bronze is alpha and alpha+delta eutectoid. Within that alpha phase is a solid solution of tin in copper with less than 8% tin. Such an alloy possesses high plasticity and toughness. The delta phase in the eutectoid has a high tin content and is hard and brittle. The amount of delta phase in a bronze increases as the tin content increases. Bronze will become much harder and stronger and less tough when there is an increase in the delta phase. In general, the relation between tin content and bronze mechanical properties is that the higher the tin content, the more the hardness and strength will increase. Conversely, toughness and softness will decrease (Handbook 1978). Thus, the lower tin content bronze used for the

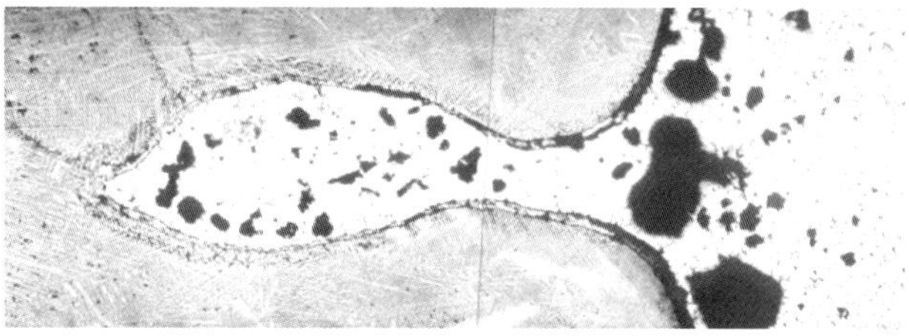

***Figure 2.** The microstructure of Fragment D*
Cutting edge structure is on left. Spine structure is on right and within the tenon. Etch: $FeCl_3$; ×30

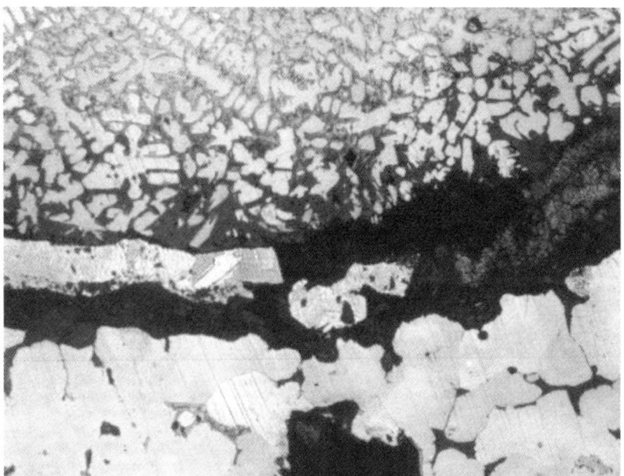

***Figure 3.** The microstructure of Fragment D seen in higher magnification at the upper tenon neck*
Cutting edge: upper side; spine: lower side. Etch: $FeCl_3$; ×200

spines of the swords give the swords good toughness, while the higher tin content bronze used for the edges gives high hardness and strength. Also, the microstructures of the edges indicate that they were not heat-treated after casting because the dendritic structures were unchanged. In contrast, the spines were apparently annealed at a high temperature for a long time after casting, because their dendritic structures are not clear but rather have become coarse. For bronze in an annealed condition, only the alpha phase appears when the tin content is about 14%. Thus, the spine structures of the four fragments are essentially composed of alpha phase with an equal distribution of tin. After annealing, the brittle delta phase in the spines was reduced, while the plasticity and toughness of spines was increased, making them still more resistant to shock.

Simulation Experiments

After some experimentation, we believe we successfully recreated the manufacturing technology used in the making of bimetallic swords. The steps were as follows. First, a two-part clay mold was made in order to cast the spine (fig. 4). For casting a reproduction sword, an alloy with a composition of 88% copper, 10% tin, and 2% lead was used. This alloy was chosen based on the results obtained from the analyses of the four fragments.

Second, the spine was annealed before the edge was cast on to it. Stress in the spine can make it twist or bend when it is heated. It was determined that the annealing temperature should be higher than 550°C. After annealing, it was necessary to straighten the spine while it was still hot if any bending or twisting had occurred. The surface was then cleaned to remove any oxides that had formed during heating.

Third, a two-piece clay mold for casting the edge was made, which incorporated the bronze spine (fig. 5). During the Eastern Zhou dynasty, the clay-casting mold was baked at more than 600°C for a long period before the metal was cast (Tan 1999). After baking the mold, the bronze was poured to form the edges. We used a bronze with the composition 78% copper, 20% tin, and 2% lead for the edges; again this alloy composition was based on the analyses of the four fragments. After solidification and shrinkage, the edges closed tightly around the flanges of the spine. Figure 6 shows the reproduction sword without a guard, grip, or pommel. On some ancient swords, the guard, grip, and pommel were made using separate castings. After the casting mold had cooled, the sword would have been removed and polished and the edge would be sharpened.

The sophisticated understanding of the mechanical properties of bronze that the ancient Chinese metalsmiths possessed is clearly displayed in their creation of bimetallic swords. The combination of bronze alloys made the swords both rigid and flexible, attributes that yield a superior weapon of this type.

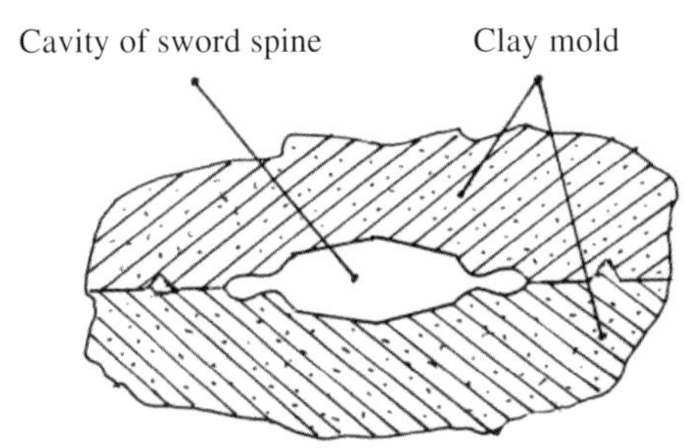

Figure 4. Diagram of the casting mold for the spine of a bimetallic sword

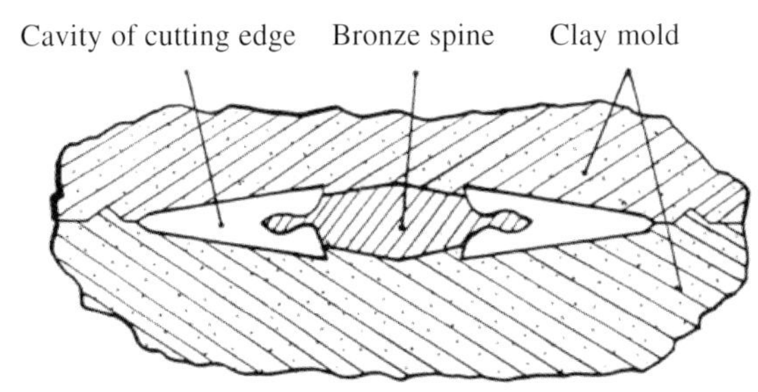

Figure 5. Diagram of the casting mold for the edges of a bimetallic sword

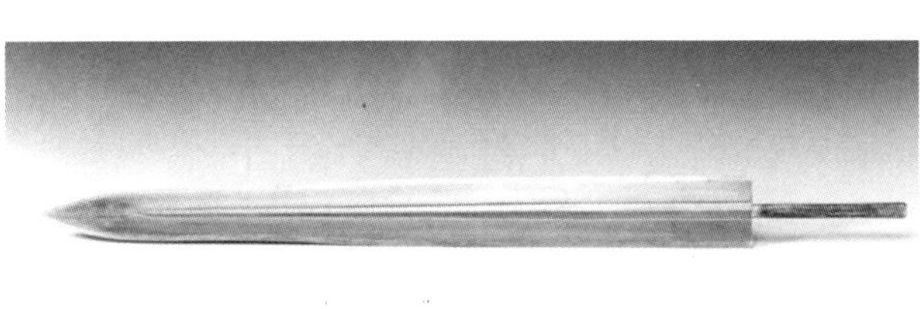

Figure 6. Reproduction bimetallic sword

Concentric-Circle Decorations on Sword Pommels

The next feature of Wu and Yue bronze weapons that will be described is the making of concentric-circle decorations on sword pommels (fig. 7). The concentric-circle decoration consists of many thin-walled, thread-relief annular ridges or rings with very regular distribution. The thickness

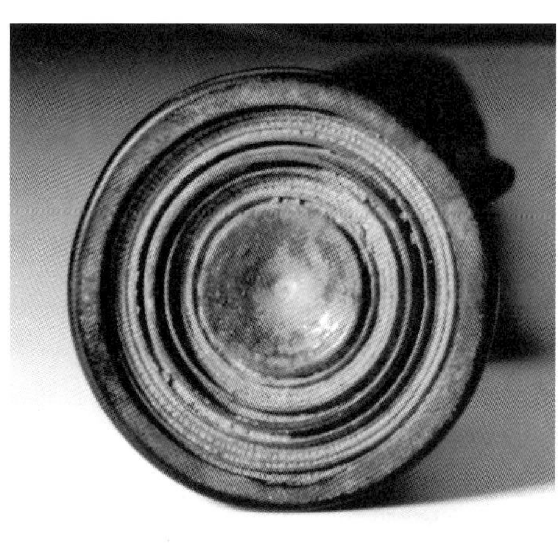

Figure 7. A concentric-circles pommel in ancient China

of the ridges is only about 0.2–0.8 mm; the height of the decoration is generally about 0.5–2.2 mm, and the gap between adjacent ridges is typically only 0.3–1.2 mm. The number of rings varies; at times, there are as many as eleven. Also, at times, the colors of pommel and grip are different, and junctions between the pommels and grips are found. In these cases, the pommels were made separately from the swords.

A group of pommels in the Shanghai Museum collection was studied. At the bottom of some of the grooves formed by the annular ridges, an extremely thin, rope-like pattern was observed. No tool marks could be seen in these areas, and we could only conclude that the decoration was cast. We believe that the lost-wax method would not result in a decoration with the sharpness of the thread relief. Also, it was very difficult to reproduce the concentric-circle patterns in a clay mold using a wax model. Thus, we believe that piece molds must have been used for making the pommels.

The thin-wall thread-relief rings on the pommels exhibit very exact concentricity. This feature suggests the use of some wheel-made method. We think that a pottery wheel may have been used to make the clay-casting molds used for the pommels. As early as the Longshan culture period of Neolithic age (2500–1900 B.C.E.), cups were made with very thin and even walls. The thickness of some of the walls is as little as 0.3 mm, and these cups were made using a wheel. Therefore, it was possible to use a wheel to make shapes in clay with the fineness one finds on the pommels, but it would be necessary to make separate casting molds for each pommel. One could not achieve the fineness of detail seen on the pommels if models of the pommels were used to make the casting molds (by taking impressions of the models).

To test this idea, several methods were tried for reproducing the pommel decoration. One method in particular yielded good results. First, the clay-mold materials were prepared with some baked clay and plant ash added to raw clay. The clay mixture was aged for a long time prior to use. Second, tools were made in wood that reproduced the type of decorations found on the pommels. Two of those tools are shown in figure 8.

A pottery wheel was used for making the clay mold for a pommel with thread-relief rings; this method is demonstrated in figure 9. The ropelike patterns were made by carving. After forming the decoration in negative on the upper half of the clay mold, it was joined to the lower half and baked. The bronze was then poured to produce the pommel. Figure 10 shows a reproduction of the bottom half of a casting mold for a pommel, and figure 11 shows a reproduction pommel. The pommels are just one more example of the superb skills in piece-mold casting that the ancient Chinese bronzecasters possessed.

Lozenge-Pattern Decoration

The last type of Eastern Zhou-period decoration found on weapons from the Wu and Yue states to be described is the so-called lozenge pattern found on some bronze weapons. Plate 9 shows the famous sword of the Yue king Gou Jian,

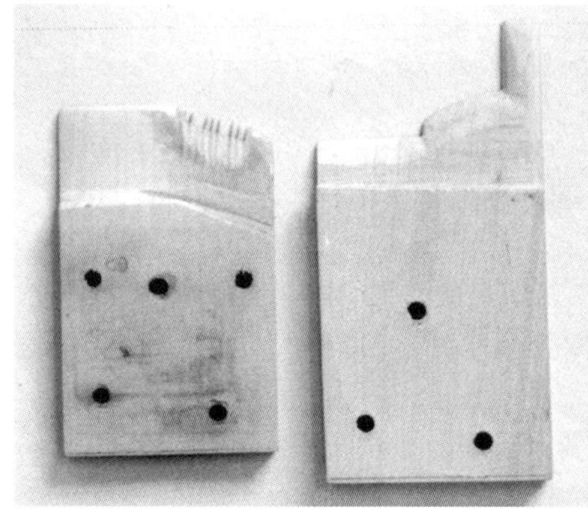

Figure 8. *Two mold plates, one with sawtooth shape, one with arc shape*

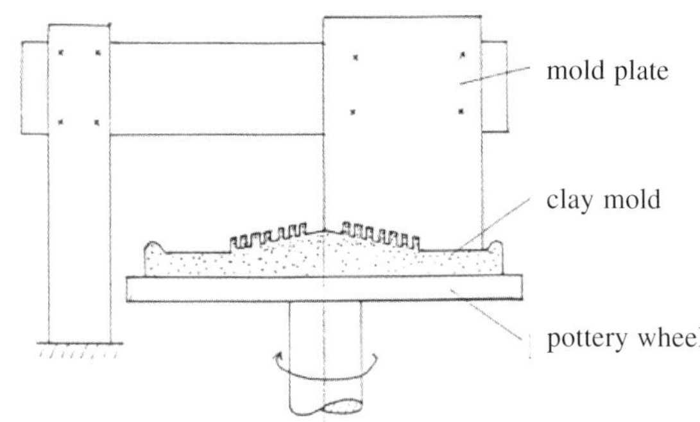

Figure 9. *Sketch of a pottery wheel for making a pommel*

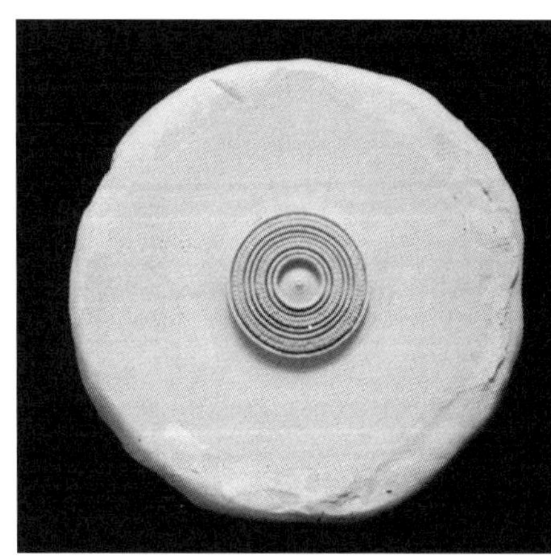

Figure 10. *A clay mold for a concentric-circles pommel*

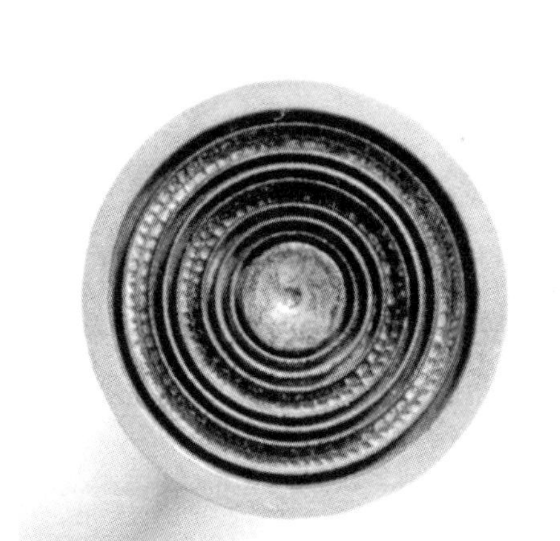

Figure 11. *A reproduction pommel*

Table 2. *The composition of the lozenge-pattern sword fragments*

	Composition (%)		
Fragments	Cu	Sn	Pb
Number 1	79.46	19.02	0.76
Number 2	80.34	19.66	–

which is one of the best examples of the use of the lozenge-pattern decoration. On this sword, the lozenge pattern consists of double lines that form diamond shapes, or lozenges. Plate 10 shows another example of a spear with this style of decoration. Bronze weapons of this type were significant symbols of power, and it is said in the historical literature that a sword of this type was worth several cities.

Two sword fragments with lozenge-pattern decorations were available to the authors for analysis and metallographic examination. While one of the samples had a smooth surface with a fine dark gray-black line design, the other was rough-surfaced and was extensively corroded in some areas.

The compositions of the bodies of the two sword fragments are both about 80% copper and 20% tin (see table 2). While similar to one another, the compositions are also typical for weapons of the Eastern Zhou period. Cross-sections from the two fragments have the same metallographic appearance (only one of which is illustrated here); the fragment shows a normal cast dendritic structure, with alpha phase dendrites and alpha+delta eutectoid (fig. 12).

At high magnification, we can compare the two different surface textures found on one of the sword fragments (fig. 13). There are some burnishing traces on the surface of the lines, but none is seen on the adjacent areas. The appearance of these areas is muddy or indistinct. After slight polishing, some clear dendritic structure and several cracks appear only on the area of the lines (fig. 14). The structure of the two areas must be different. Both areas were analyzed (table 3), and there are marked differences in the amounts of copper, tin, and oxygen found.

In order to get more information about the structure, samples were removed from the two fragments and prepared for metallographic study (fig. 15). The left side of the sample shown in figure 15 corresponds to the area of the lines, and the right of the image is the adjacent dark surface area. The structure on the left is the same as the body. However, on the dark surface areas there was a thin layer only a few microns thick. This layer also had a dendritic structure with eutectoid, as seen in figure 16, but its dendrites are much finer than those found in the body. Also, its secondary arms are often arranged only on one side of the primary arm, not like alpha dendrites (fig. 17). The composition of the fine dendritic area was analyzed using EDAX, and the results are shown in table 4. Also, the microhardness of this area was measured and is about three times as hard as the alpha dendrites in the body metal.

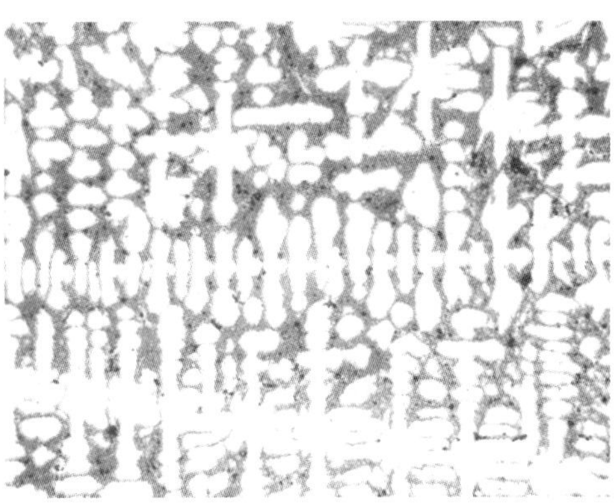

Figure 12. *The matrix structure of fragments with lozenge-pattern decoration (100×)*

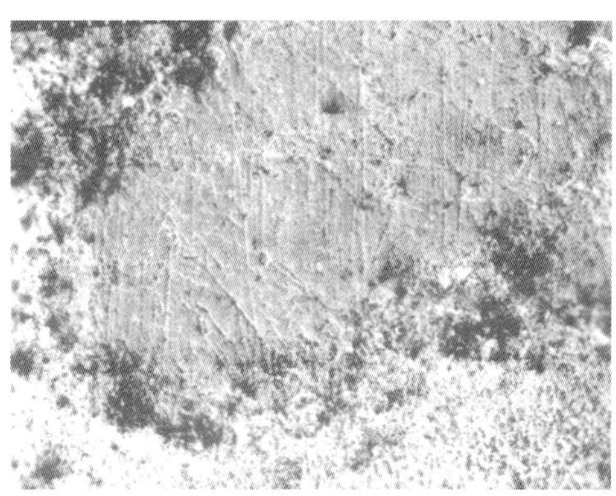

Figure 13. *Magnified details of the fragment's surface (untreated)*

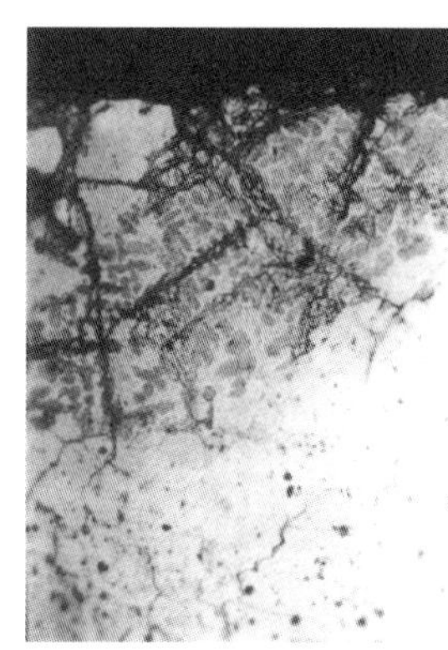

Figure 14. *Magnified details of the fragment's surface (after light polishing)*

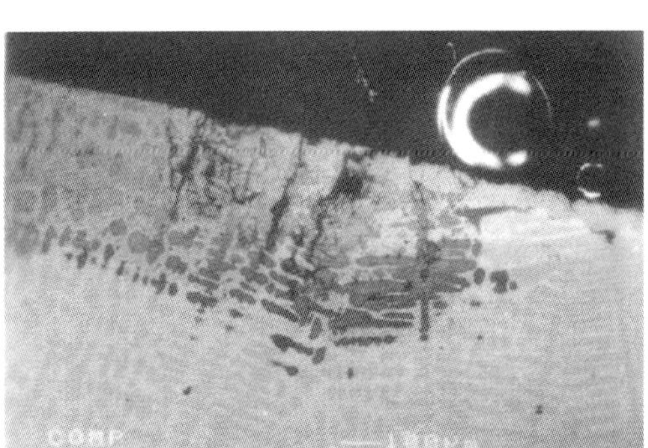

Figure 15. *Metallographic cross-section of the fragment with lozenge-pattern decoration (80×)*

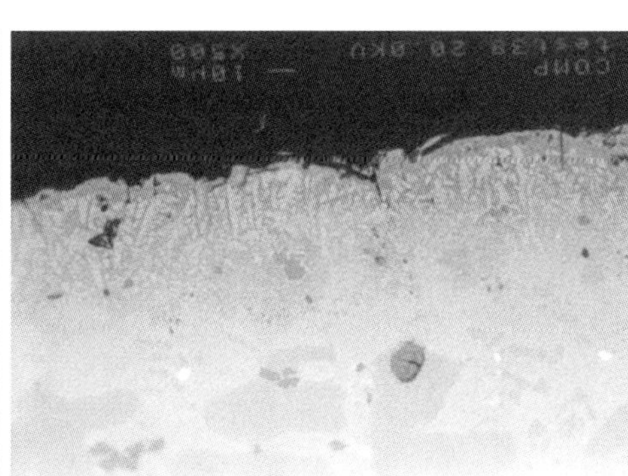

Figure 16. *Secondary electron image of fine dendrite layer (500×)*

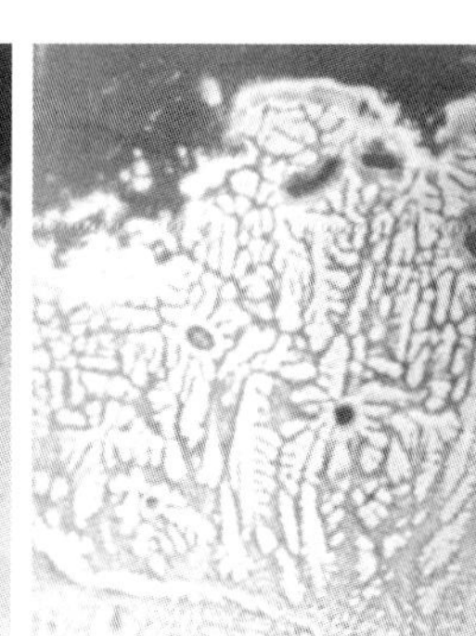

Figure 17. *Fine dendritic layer*

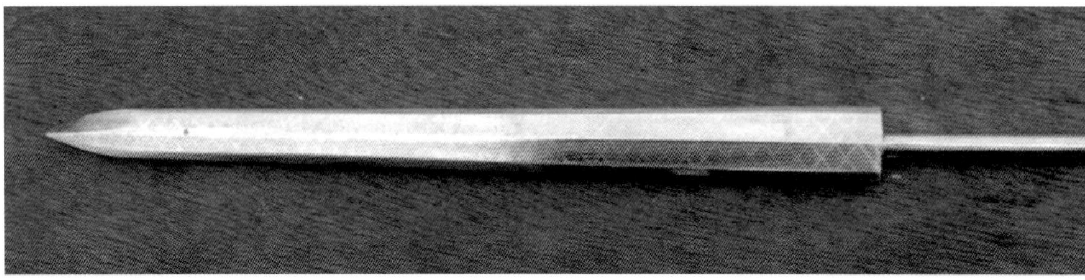

Figure 18. *A reproduction of lozenge-pattern sword*

Table 3. *Surface composition of Fragment 1 with lozenge-pattern decoration (EDAX)*

	Composition (%)				
Analyzed location	Cu	Sn	O	Si	Fe
The bright line area					
Point A	6.65	58.42	29.34	3.14	2.45
Point B	7.06	54.66	29.79	1.80	6.69
Point C	5.08	52.81	31.95	2.34	7.83
Average	6.26	55.30	30.36	2.43	5.66
The dark area					
Point A	3.70	40.73	48.88	3.50	3.18
Point B	4.07	42.02	48.27	2.77	2.87
Average	3.90	41.38	48.58	3.14	3.03

Table 4. *The composition of the fine dendrite layer on Fragment 1 with lozenge-pattern decoration (EDAX)*

The fine	Composition (%)			
dendrites	Cu	Sn	Si	O and others
Point A	66.46	32.95	0.16	0.33
Point B	67.23	32.46	0.13	0.16
Average	66.85	32.71	0.15	0.25

Discussion

The observations and tests discussed above indicate that the lozenge pattern on the sword surface is due to the selective formation of a layer with a fine dendritic structure in some areas of the surface. The composition and structure of this layer is quite different from the body of the sword. Also, it appears that this layer is attached to the body metal but is not mechanically inserted into it. The fine dendrites in the thin layer consist of 66.85% copper, 32.71% tin, 0.15% silicon, and 0.25% oxygen. The composition is very similar to the delta phase ($Cu_{31}Sn_8$), with 67.47% copper and 32.52% tin, and its microhardness is higher than the alpha phase. Hence, the fine dendritic layer consists of delta phase dendrites and alpha+delta eutectoid. The difference in the tin contents of the matrix and the fine dendritic layer, with the fine dendritic layer having the higher tin content, indicates that a selective tin-enrichment of surface took place during the decoration process.

One other feature should be noted. The areas of the lines and the adjacent dark areas have corroded very differently. On the fragment's cross-section, heavier corrosion is apparent where there is no thin, fine dendrite layer. This layer could block oxidation and the penetration of corrosive mediums into the body of the sword.

Simulation Experiments

Several methods to form the lozenge pattern were tried in experiments to duplicate the lozenge-pattern style of decoration. Eventually, a method was found for producing a tin-rich coating that gave a fine dendritic layer similar to that found on the fragments. X-ray diffraction analysis indicates that the fine dendritic layer consists of delta phase and alpha phase, with the major portion being delta phase. Figure 18 shows a reproduction sword with lozenge-pattern decoration. This sword could be induced to corrode in such a manner that its surface color would in the end be similar to that seen on ancient swords (Sun et al. 1992).

Acknowledgments

The authors wish to thank the staff of the Shanghai Museum—in particular, Xu Huikang for his great assistance in casting the reproductions of the swords and pommels and He Wenquan for performing the x-ray fluorescence analysis of the sword fragments.

References

Tan Derui. 1999. The Archaeology Institute of Chinese Academy of Social Sciences. *Acta Archaeologica Sinica* 2:211–55.

Hebei Province Museum, Relics Management Department. 1973. The Archaeology Institute of Chinese Academy of Social Sciences. *Archaeology* 5:266.

Handbook of casting nonferrous alloy. 1978. Beijing: The Publication of Machining Industry.

Sun Suyun, Ma Zhaozheng, and Jin Lianji, et al. 1992. Cultural relics publication. *Wenwu cultural relics* 12:79–89.

7 Metallurgical Analysis of Metal Cauldrons from Xinjiang, Northwest China

Mei Jianjun

ABSTRACT Recent archaeological discoveries in Xinjiang, eastern Central Asia, have attracted considerable interest throughout the world. One of the important signs of cultural changes in Xinjiang from the early first millennium B.C.E. is the appearance of metal cauldrons, which was clearly related to the rise of a nomadic way of life on the Eurasian steppe. Most of the recent studies of the cauldrons have been concerned mainly with origins, distribution, and types, with little attention being paid to their scientific aspects. In this paper, the results of the metallurgical analysis of five metal cauldrons from Xinjiang, northwest China, are presented. It has been revealed that four early cauldrons were cast with copper rather than tin bronze, while one cauldron definitely dating to a later period was made of a leaded tin-bronze alloy. It is interesting to note the existence of a small amount of antimony in two of the copper cauldrons. The cultural and technological backgrounds for the beginning and early use of copper and bronze cauldrons in Xinjiang as well as the dispersal of casting technology on the eastern Eurasian steppe are also discussed.

Introduction

The Xinjiang region, west of Gansu and Qinghai Provinces in northwest China, is well known as an important crossroads on the ancient Silk Road. It has been widely recognized as a pivotal region for understanding early cultural interaction between China and Central Asia (Shui 1993; Di Cosmo 1996). Over the last two decades, a number of significant archaeological discoveries have been made in Xinjiang, revealing considerable evidence for the existence of diverse Bronze Age and Iron Age cultures throughout the region. As far as the current archaeological data are concerned, the beginning of the Bronze Age in Xinjiang can be traced to the early second millennium B.C.E. The transition from the Bronze Age to the Iron Age took place around 1000 B.C.E., when iron appeared and came into common use gradually in Xinjiang (Wang 1985; Chen 1987, 1995; Debaine-Francfort 1988, 1989; Chen and Hiebert 1995; Mei and Shell 1998).

One of the important signs for cultural change in Xinjiang from the early first millennium B.C.E. is the appearance of metal cauldrons, which was clearly related to the rise of a nomadic way of life on the Eurasian steppe. As the first large-sized vessels of metal employed by the nomadic people, cauldrons played an important part in their daily life and religious activity. Most recent studies have been concerned mainly with origins, distribution, and types of cauldrons, with little attention being paid to their scientific aspects (Zhang and Zhao 1991; Wang and Qi 1995, 277–94; Érdy 1995; Guo 1999). This paper intends to present a metallurgical analysis of five metal cauldrons from Xinjiang. It is hoped that the scientific data will not only throw light on the manufacture of cauldrons but will also provide technical evidence for tracing cultural connections between Xinjiang and neighboring regions.

The Finds of Metal Cauldrons in Xinjiang

More than two dozen metal cauldrons have been found in Xinjiang, and most of them can be dated to the mid-first millennium B.C.E. (table 1). Except for two examples that were found in the mountain areas of southern Xinjiang, the majority of cauldrons were recovered from north of the Tian Shan Mountains in areas such as Yili, Balikun, and Altai (fig. 1; Wang and Qi 1995, 277–94; Gong 1997: 16; Yue et al. 1999, 252, 270). These areas have vast pastureland suitable for grazing sheep, cattle, horses, and other animals. From the concentrated distribution of cauldron finds in northern Xinjiang, it seems quite clear that the appearance and wide use of metal cauldrons was associated with the rise of a nomadic way of life in the region during the early first millennium B.C.E.

According to their handle and foot designs, these cauldrons can be roughly classified into four types. Type I is characterized by a pair of rounded handles standing upright from the rim of the mouth (fig. 2). Type II is distinguished by a pair of handles standing obliquely from the shoulder of the vessel (fig. 3: 1–5). Type III is a developed form of Type I, with two handles decorated with "mushroom" design (fig. 3: 6). This type is generally dated to the second to fourth century C.E., being much later than the other types (Érdy 1995, 45–46). In contrast to the above three types that have only a single stand, Type IV has three feet and is normally called a "tripod cauldron" (fig. 3: 7).

Several observations on the cauldron finds in Xinjiang may be mentioned here. First, while these cauldrons can be classified into four major types, their sizes, shapes, and decoration show quite a large range of variation; one cannot find two identical specimens. This suggests that the sources for these cauldrons were diverse and localized. Second, the

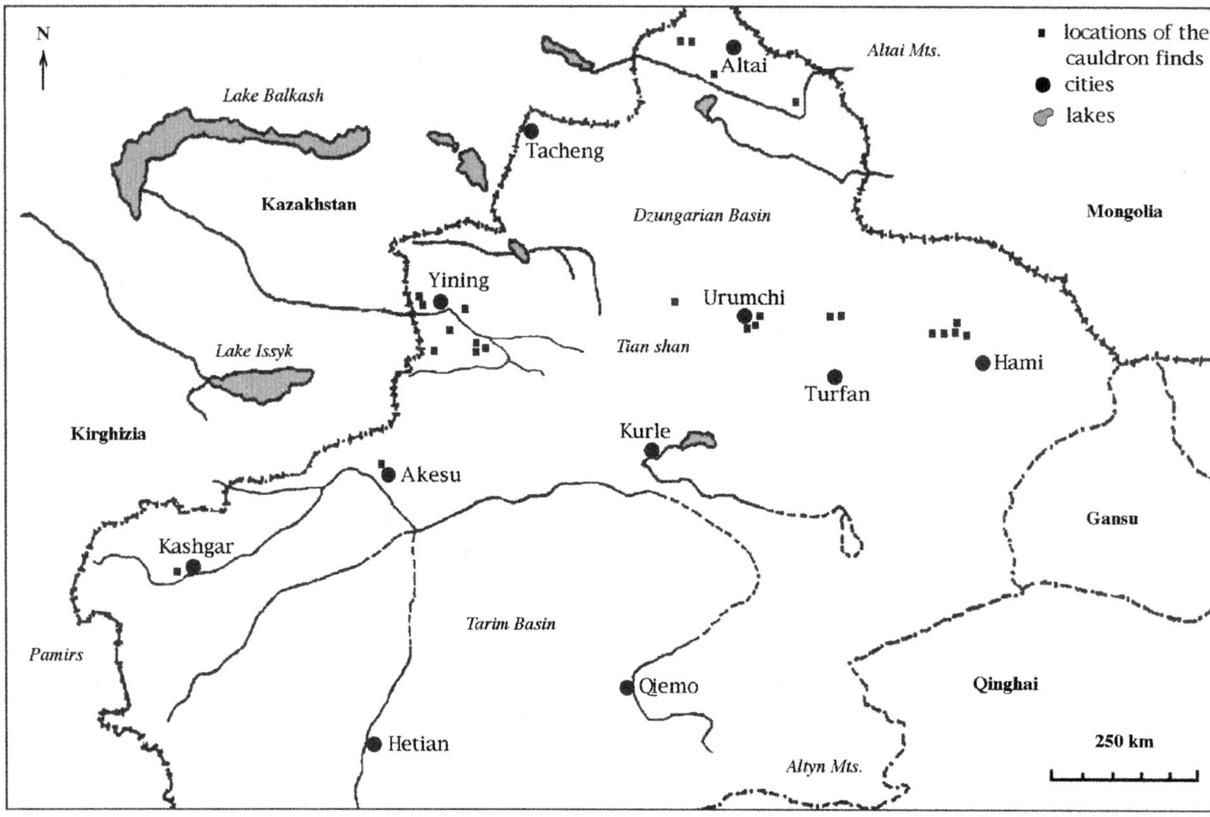

Figure 1. *The distribution of metal cauldron finds in Xinjiang, northwest China*

Table 1. *Early metal cauldrons recovered in Xinjiang, northwest China*

No.	Location of finds	Dimensions (cm)	Types	Figures
1	Tiereketi, Habahe, Altai	H 25.1; D (mouth) 20.7–21.8	I	2:1
2	Sha'erbulake, Fuyun Altai	H 37.2; D (mouth) 27–28	I (?)	2:2
3	Gongliu, Yili	H 61; D (mouth) 38	I	2:3
4	Nileke, Yili	H 34; D (mouth) 26.5	I	
5	Huocheng,Yili	H 56.4; D (mouth) 36.7–41.7	I	
6	Xiao'erbulake, Xinyuan Yili	H 76; D (mouth) 40.5	I	2:4
7	Wensu, Akesu	H 50; D (mouth) 33.2	I	2:5
8	Shufu, Kashgar	H 57; D (mouth) 46; D (foot) 17	I	2:6
9	Lanzhouwanzi, Balikun Hami	H 50.5; D (mouth) 33	I	2:7
10	Dahe, Balikun, Hami	H (preserved) 36	I	2:8
11	Nanwan, Balikun, Hami	H ca. 50	I	2:9
12	Hongshan, Balikun, Hami	H (preserved) 35.5; D (mouth) 22	I	2:10
13	Urumchi	H 51; D (mouth) 3	I	2:11
14	Ka'erzi, Qitai, Changji	H 56.5; D (mouth) 30	I	2:12
15	Taleqiate, Habahe, Altai	H 49; D (mouth) 40.4	II	3:1
16	Nanwan, Balikun, Hami	H (preserved) 26.7D (mouth): 29; D (belly): 32(b)	II	3:2
17	Xinyuan, Yili	H 57.5; D (mouth) 42	II	3:3
18	Zhaosu, Yili	H (preserved) 31; D (mouth) 47	II	
19	Shihezi	H ca. 30; D (mouth) ca. 30	II	3:4
20	Urumchi	H ca. 30	II	
21	Biliuhe, Qitai, Changji	H ca. 50	II (?)	3:5
22	Huocheng, Yili	H 48.5; D (mouth) 39.8–45.5	II	
23	Nanshan, Urumchi	H 57; D (mouth) 39	III	3:6
24	Bu'erjin, Altai	H ca. 30	?	
25	Xinyuan, Yili	H 44.5; D 42 (mouth)	IV	3:7

Key: H = height; D = diameter

Type I cauldrons from Xinjiang are comparable in form to those found in north China, Mongolia, and southern Siberia, which have been dated roughly to the eighth to fourth century B.C.E. (Liu 1987, 62; Chlenova 1994, 506; Takahama 1994, 1–6). The dates for the Xinjiang specimens are likely within a similar chronological range, but the cauldron with a pierced stand from Fuyun in Altai (fig. 2: 2) should be considered within a later date. Third, the cauldrons of Type II and Type IV typologically resemble those found in the Semirechiye region in Kazakhstan and

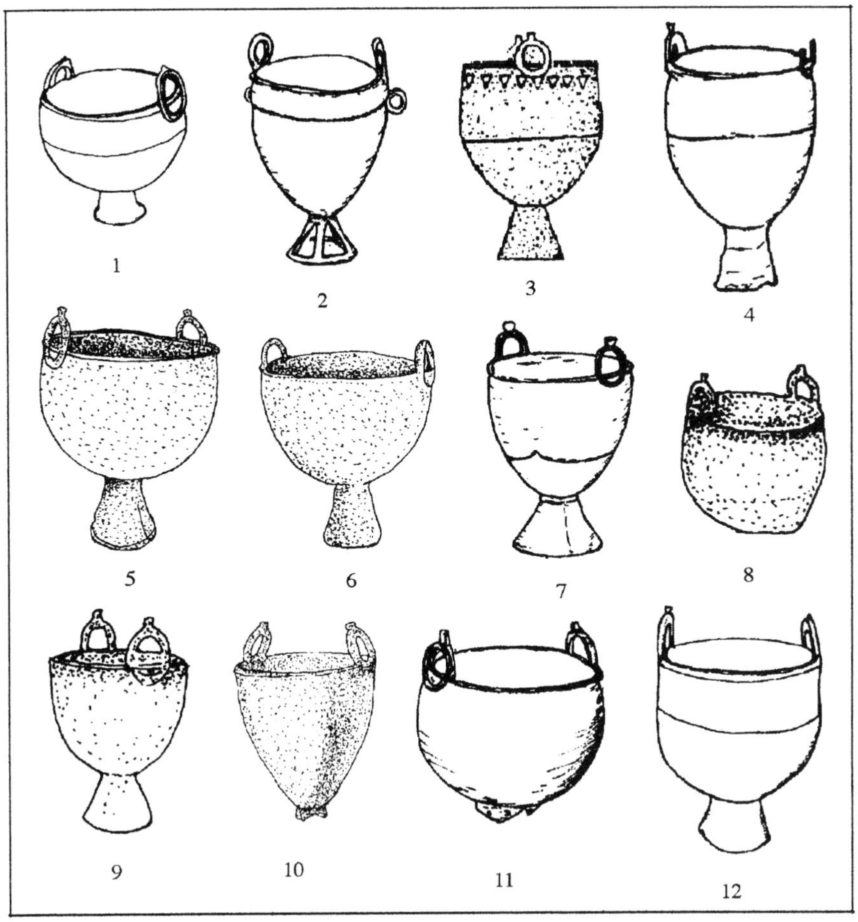

Figure 2. *The metal cauldrons of Type I recovered in Xinjiang*
1. from Tiereketi, Habahe, Altai; 2. from Sha'erbulake, Fuyun, Altai; 3. from Gongliu, Yili; 4. from Xiao'erbulake, Yili; 5. from Wensu, Akesu; 6. from Shufu, Kashgar; 7. from Lanzhouwanzi, Balikun, Hami; 8. from Dahe, Balikun, Hami; 9. from Nanwan, Balikun, Hami; 10. from Hongshan, Balikun, Hami; 11. from Urumchi; 12. from Ka'erzi, Qitai, Changji; (1, 2, 4, 7, 11, and 12 after Wang and Qi 1995, 290, fig. 11; 3 after Zhang and Zhao 1991, 42, fig. 1)

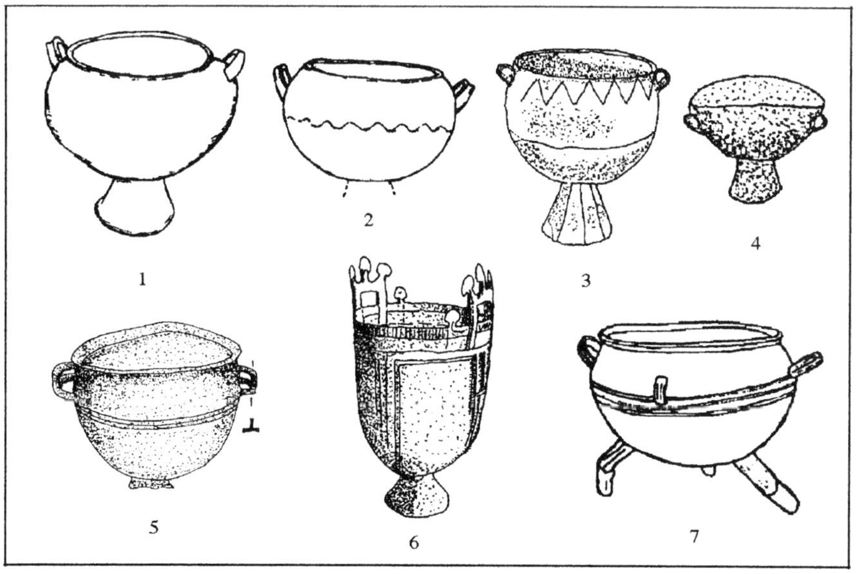

Figure 3. *The metal cauldrons of Types II, III, and IV recovered in Xinjiang*
1. Type II, from Taleqiate, Habahe, Altai; 2. Type II, from Nanwan, Balikun, Hami; 3. Type II, from Xinyuan, Yili; 4. Type II, from Shihezi; 5. Type II, from Biliuhe, Qitai, Changji; 6.Type III, from Nanshan, Urumchi; 7. Type IV, from Xinyuan, Yili; (1 and 2 after Wang and Qi 1995, 290, fig. 11; 7. after Li and Dang 1995, 44, fig. 3)

Kirghizia, which have been attributed by Russian scholars to the Saka culture of the seventh to fourth century B.C.E. (Akishev and Kushayev 1963, 111) or the fifth to third century B.C.E. (Bernshtam 1949, 353–54; 1952, 47). This resemblance suggests that the Xinjiang specimens could probably be considered within the Saka context and dated to the seventh to third century B.C.E. Among the Type II cauldrons, the one from Biliuhe in Qitai is worthy of special attention. It has a pair of "three-legged" handles standing horizontally on the shoulder (fig. 3: 5). Cauldrons with handles of this type were also recovered in southern Siberia, and it seems very probable that the Xinjiang specimen is related to those. Fourth, only one specimen of Type III cauldrons has been recovered so far in Xinjiang (fig. 3: 6). It has many close parallels in Eastern Europe, and it is tempting to think that there is possibly a connection. Érdy (1995, 46) suggests that the Xinjiang specimen "is an early local development in the vicinity of the Altai' in the second century C.E. rather than the one that was influenced by or connected to Europe. Finally, except for a few decorated with simple geometric patterns, such as lines and triangles, most cauldrons from Xinjiang are plain on the surface. This shows a clear contrast to those heavily decorated Scythian cauldrons of the seventh to fourth century B.C.E. from western Eurasia (Jacobson 1995, 189–92).

Metallurgical Analysis of the Cauldrons

With the kind support of Xinjiang archaeologists, five samples were obtained from five different cauldrons. The dates and origins of these cauldrons are given in table 2. The cauldron samples were first mounted in epoxy resin, then ground on silicon carbide papers, and finally polished on synthetic velvet. The polished sections were etched with an aqueous ferric chloride solution to reveal the microstructures and then studied and photographed using a metallographic microscope and a scanning electron microscope (SEM). The results of metallographic examinations are summarized in table 3. Elemental analysis was also conducted on the polished sections in the SEM using an energy-dispersive x-ray fluorescence analyzer with SEMQuant software for the quantitative analysis at the University of Cambridge. It should be noted that the energy-dispersive x-ray fluorescence analyses can only be taken as a general indication of the overall compositions of the samples. Table 4 presents the results of the elemental analyses.

From table 3 and figures 4–9 we can see that all five samples show a cast structure without any further working. The four earlier cauldrons of Types I, II, and IV (nos. 107, 108, 118, and 123) were made of copper with small amounts of minor elements, such as arsenic, antimony, and sulfur,

Table 2. *The five cauldron samples from Xinjiang*

Laboratory no.	Location	Dates	Types	Figures
107	Xinyuan, Yili	750–200 BC	IV	3: 7
108	Nanwan, Balikun, Hami	700–200 BC	II	3: 2
109	Urumchi	100–400 AD	III	3: 6
118	Tacheng	800–400 BC	I or II (?)	
123	Biliuhe, Qitai, Changji	800–400 BC	II	3: 5

Table 3. *Results of metallographic examinations of cauldron samples*

Laboratory no.	Microstructures and technologies	Figures
107	The microstructure of the sample consists of relatively large polygonal grains with extensive casting porosity, indicating that the tripod was as cast and not subsequently cold-worked. A number of inclusions are visible and have been identified by EDXA as oxides of Pb, Sb, and Bi (Pb 34.03%, Sb 31.31%, Bi 5.27%, O 12.66%, Cu 11.59%), SnO_2 (Sn 75.91%, O 19.92%, Cu 4.17%), and also copper sulfide (Cu 81.21%, S 18.79%). It is interesting to note that copper sulfide particles are sometimes associated with particles of cassiterite.	4, 5
108	Analysis of the metal shows a dendritic structure with a number of small white and gray-black inclusions as well as casting porosity. The coring is not extensive. The inclusions have been identified as lead globules and Cu-Fe sulfide particles.	6
109	The metallographic section reveals an undistorted dendritic structure with numerous small lead particles, indicating that this cauldron was as cast.	7
118	The metallographic section was cut from the stand. Its microstructure consists of undistorted dendrites, numerous gray-dark inclusions, and porosity, indicating that the cauldron was as cast. There is no indication of working. The EDXA indicates that the cauldron contains a relatively substantial amount of sulfur (1.28%) that is present as numerous copper sulfide inclusions (Cu 80.98%, S 19.02%).	8
123	A typical cast structure with interdendritic segregation is visible in the section. A number of gray-black and white particles have been identified by EDXA as copper sulfide and lead.	9

Table 4. *The results of elemental analyses of cauldron samples (wt %)*

Laboratory no.	Cu	Sn	Pb	As	Sb	S	O	Alloys
107	98.6	det.	det.	n.d.	det.	det.	1.8	Cu (O)
108	97.2	n.d.	det.	n.d.	1.2	0.5	Fe (det.)	Cu-Sb (S, Fe)
109	94.9	2.9	2.3	n.d	n.d.	Cl (det.)	det.	Cu-Sn-Pb
118	98.7	n.d.	n.d.	n.d.	n.d.	1.3	det.	Cu (S)
123	96.4	n.d.	n.d.	2.2	1.5	det.	n.d.	Cu-As-Sb

Key: n.d. = not detected; det. = detected.

present; only a Type III cauldron of a later date from Urumchi (no. 109) was made of a leaded tin-bronze alloy. Also, two of the earlier cauldrons (nos. 118 and 108) contain a substantial amount of copper sulfide inclusions, a possible sign of the use of a copper sulfide ore as the source of the copper. The reason for using copper rather than tin bronze to cast cauldrons is not clear, but it appears to be related to the availability of metal sources in the relevant localities. In this regard, it is worth noting that the two cauldrons containing 1–1.5% antimony (nos. 108 and 123) are from the areas close to each other along the northern foothills of the Tian Shan. This may hint at the availability of a local antimony-containing copper source in eastern Xinjiang. To clarify this issue, further investigation is required.

Discussions

The Background for the Appearance of Metal Cauldrons

Recent archaeological evidence has revealed that Xinjiang was in close contact with neighboring regions at least from the early second millennium B.C.E. (Shui 1993; Li 1999). In northwestern Xinjiang, the finds of a number of Andronovo-type objects attest to the significant existence of Andronovo cultural influence from the west (Li and Dang 1995; Mei and Shell 1999). A metallurgical analysis of some of these objects indicates that tin bronze containing 2–10% tin held a predominant place (Mei et al. 1998). These results are comparable to those obtained in the analysis of metal artifacts from Eurasia that showed that the majority of Andronovo metals are of tin bronze, with tin levels in the range of 3 to 10% (Chernykh 1992, 213). In eastern Xinjiang, the Bronze Age is characterized by diverse painted pottery cultures, such as Tianshanbeilu, Wupu, and Yanbulake, which show close connections with the cultural traditions found to the east in the Gansu and Qinghai regions. A large number of copper and bronze objects—mostly small ornaments, implements, and weapons—were unearthed in the region. Recent analyses reveal that tin bronze and arsenical copper were the major copper alloys used (Mei 2000, 37–49; Qian, Sun, and Han 2000). Artifacts made of copper with a significant amount of antimony have rarely been identified so far within the Bronze Age context in Xinjiang.

Interactions between Xinjiang and the neighboring regions were extensive during the first millennium B.C.E. The appearance and expansion of the Saka culture in the Yili River region was closely linked to the same process in neighboring Kazakhstan. The discovery of "animal-style" bronzes in Xinjiang, especially along the northern foothills of the Tian Shan, attests to the presence of cultural influences from the eastern Eurasian steppe. Painted pottery cultures distributed along the southern foothills of the Tian Shan also had a wide range of contact with both neighboring and distant cultures. It is upon this background that metal cauldrons came into use in Xinjiang, and their concentrated distribution in northern Xinjiang actually reflects the fact that the nomadic groups of people were the major players in the region during the first millennium B.C.E.

A preliminary analysis of more than twenty Iron Age metals from Xinjiang shows that while tin bronze continued to predominate among the metals in use during the first millennium B.C.E., the use of copper seems to have become more frequent than in the second millennium B.C.E. It has been suggested that this change may be linked to the exploitation of some new copper sources or the difficulty in securing the supply of tin (Mei 2000, 41–47). This technological background may help to understand why the four early cauldrons were cast with copper or copper containing small amounts of antimony, arsenic, and sulfur, rather than with tin bronze.

As mentioned previously, the cauldron from Biliuhe with a pair of "three-legged" handles is comparable to similar pieces recovered in southern Siberia. This typological correspondence seems to be strengthened by the elemental analysis, because the copper-arsenic-antimony alloy employed for manufacturing the Biliuhe cauldron was widely used in southern Siberia during the late first millennium B.C.E., especially for fabricating belt plaques in the "animal-style" (Devlet 1980, 32–33).

Intriguingly, the majority of cauldrons recovered in southern Siberia were also made of pure copper instead of tin bronze. According to Bogdanova-Berezovskaya (1963, 136, 153), among the twenty cauldrons analyzed, thirteen are pure copper, five are arsenical copper (As 1–1.5%), one is a tin bronze, and one a leaded tin-bronze alloy. However, in the cases of implements and weapons, such as knives, sickles, daggers, and mirrors, tin bronze and arsenical copper were preferred. It appears that copper rather than copper alloys was intentionally chosen for casting cauldrons in southern Siberia during the first millennium B.C.E. Whether this is also true for Xinjiang needs further research.

The Dispersal of Casting Technology on the Eastern Eurasian Steppe

On the surface of some cauldrons from Xinjiang, traces of the joint lines of casting molds can be seen, indicating that they were cast by using piece molds. It is well known that piece-mold casting is a typical Chinese technology widely employed to cast thousands of ritual bronzes during the Shang and Zhou periods (seventeenth through third century B.C.E.). So and Bunker (1995, 108) published a cauldron with characteristic Chinese-style decoration as well as two-piece mold marks on the surface. They dated it to the eighth century B.C.E. and considered it one of the earliest examples of its kind. Cauldrons with Chinese-style decoration but in smaller sizes were found also in Shaanxi and Gansu Provinces as well as in a museum collection in Japan (Wang Changqi 1991, 7–10; Takahama 1997, 169). These examples seem to provide a clue for understanding the early stages of adopting piece molds to the cauldron shape. So and Bunker (1995, 108) state that the Chinese decoration of the cauldron indicates a date of eighth century B.C.E. for the piece, which "suggests that the Chinese may have been among the first peoples to make this shape, which then traveled far into western Eurasia, becoming the signature article of the peoples in the Black Sea region."

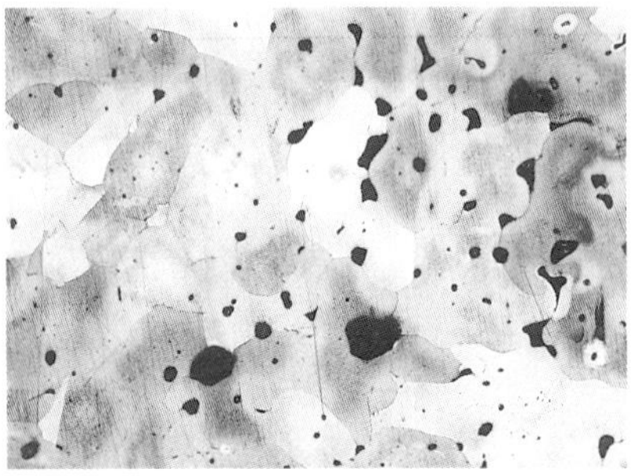

Figure 4. *Microstructure of the Xinyuan tripod cauldron (no. 107), showing relatively large polygonal grains and extensive casting porosity. ×225*

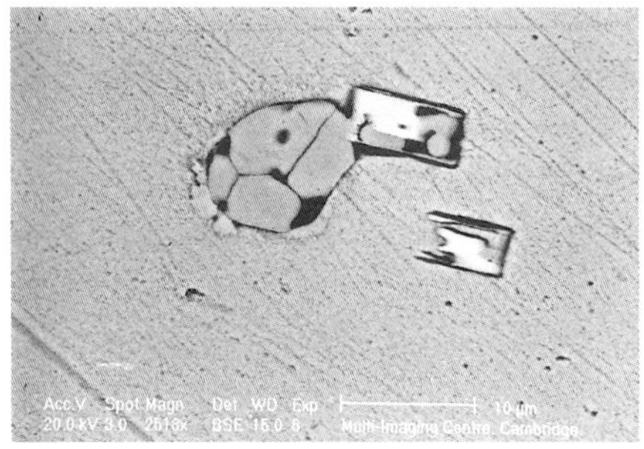

Figure 5. *BSE micrograph of the Xinyuan tripod cauldron (no. 107), showing the inclusions of copper sulfide (gray phase) and cassiterite (white phase) in the copper matrix*

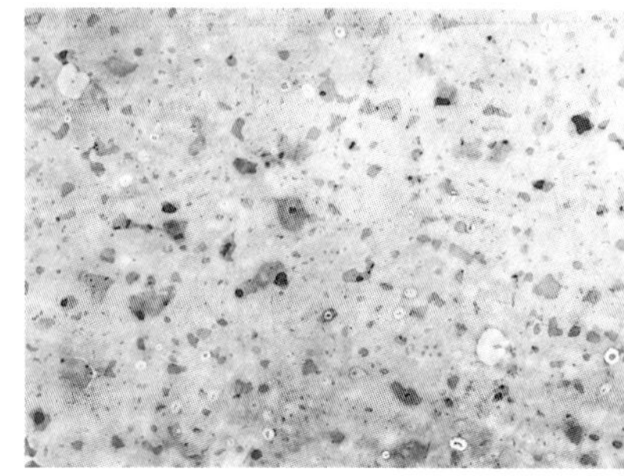

Figure 6. *Microstructure of the Nanwan cauldron (no. 108), showing a slightly cored cast structure with a large number of lead and Cu-Fe sulfide inclusions. ×225*

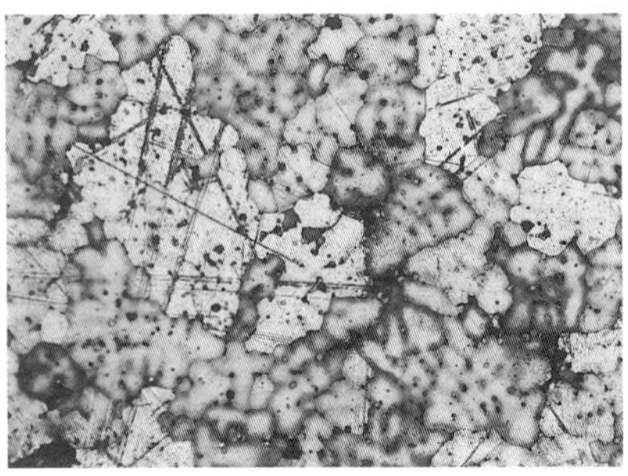

Figure 7. *Microstructure of the Urumchi cauldron (no. 109), showing an undistorted dendritic structure of cast. ×225*

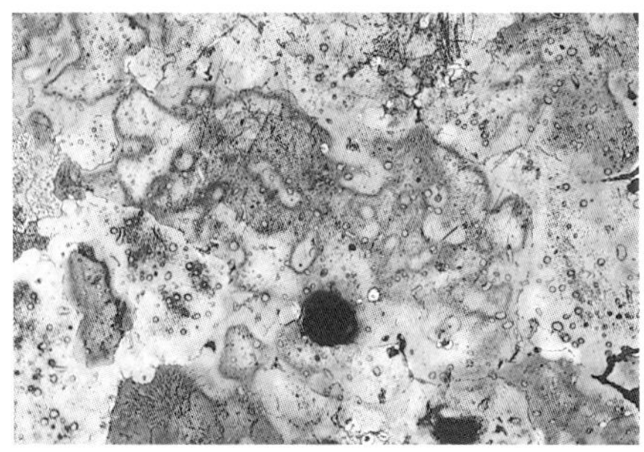

Figure 8. *Microstructure of the Tacheng cauldron (no. 118), showing a cast structure with dendritic segregation and numerous inclusions and porosity. ×225*

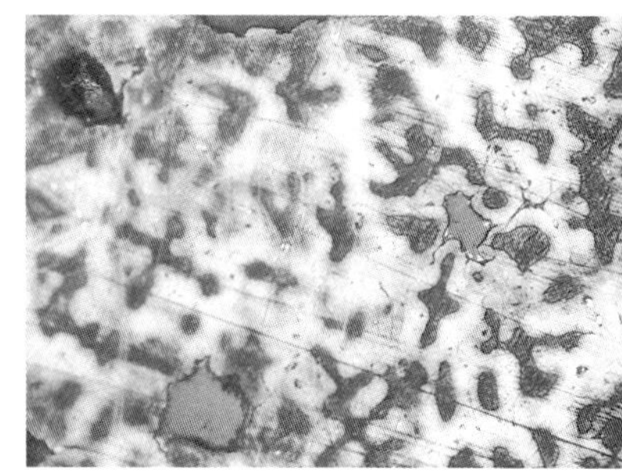

Figure 9. *Microstructure of the Biliuhe cauldron (no. 123), showing a typical cast structure with dendritic segregation and a number of copper sulfide inclusions. ×225*

While the casting technology may have originated in northern China, the shape of the cauldron may not necessarily have also originated there. Chlenova (1994, 506) suggests that the origin of the Scythian cauldrons goes back to those from Iran and partially from the Transcaucasian region and Asia Minor. Bunker (1997, 178) also considers that the cauldron shape and handles have no Chinese antecedents and traces the origins of the nomadic cauldron to Transcaucasia, where a cauldron made clearly in hammered metal was found. To get a better understanding of this issue, comparative research is needed to bring together all cauldron finds from both eastern and western Eurasian steppes.

With regard to the cauldron shapes, it should be noted that the tripod cauldrons used by the Saka people in the region of Semirechiye and Yili are quite similar in form to the three-legged *ding* vessels of the Shang and Zhou dynasties in the central plains of China (Bernshtam 1949, 351–52). This similarity may imply the presence of cultural influence from the east to Xinjiang and Semirechiye during the mid-first millennium B.C.E. (An 1996, 74).

Conclusions

The discovery of more than two dozen metal cauldrons in Xinjiang provides us with important evidence for understanding cultural changes in the region in the early first millennium B.C.E. The metallurgical analysis of five cauldron samples has revealed that four early cauldrons were made of copper or copper containing a small amount of arsenic, antimony, or sulfur, while the one cauldron of later date was composed of a leaded tin-bronze alloy. There is some evidence that indicates an increasing use of copper rather than copper alloys in Xinjiang from the beginning of the first millennium B.C.E. The influences of other cultures or changes in the availability of raw materials may help to explain the use of copper rather than tin bronze for casting cauldrons. It is interesting to note that two cauldrons contain 1–1.5% antimony and that one of them was made using a copper-arsenic-antimony alloy. The identification of a copper-arsenic-antimony alloy suggests a connection with southern Siberia. It has been suggested that most Xinjiang cauldrons were cast in piece molds and that the technology most likely was introduced from northern China.

Acknowledgments

I would like to thank Dr. Blythe McCarthy and Janet G. Douglas of the Freer Gallery of Art for kindly inviting me to take a part in the Fiftieth Anniversary Symposium on Scientific Research in the Field of Asian Art. I am also most grateful to Dr. Colin Shell, Professor Colin Renfrew, Professor Jessica Rawson, Professor Takahama Shu, and Dr. Hirao Yoshimitsu for their guidance, insight, and support of my work; to Professor Wang Bo and Dr. Li Xiao for kindly providing me with five cauldron samples; to Dr. S. V. Demidenko for generously sharing with me his wide knowledge of cauldron finds in Siberia; to Mr. Miklós Érdy for giving me his pioneering research paper on cauldrons; and to Dr. Hayakawa Yasuhiro for his help with preparing the figures for this paper. Finally, I wish to express my deep gratitude to the Department of Archaeology, the McDonald Institute for Archaeological Research, the University of Cambridge, and the Japan Society of the Promotion of Science for their considerable financial support of this research.

References

Akishev, K. A., and G. A. Kushayev. 1963. *Drevnyaya kul'tura Sakov i usuney d.r. ili.* (Ancient cultures of the Saka and Wusun in the Ili river basin). Alma-Ata: Nauka.

An, Zhimin. 1996. Talimu pendi ji zouyuan diqu qingtong wenhua yicun (The remains of Bronze Age cultures in the Tarim Basin and surrounding areas). *Kaogu* (Archaeology) 12:70–77.

Bernshtam, A. N. 1949. Osnovnye etapy istorii kul'tury Semirech'ya i Tyan'-Shanya (The main stages of the history and cultures of the Semirechiye and Tian Shan regions). *Sovetskaya arkheologiya* (Soviet archaeology) 11:337–78.

——. 1952. Istoriko-arkheologicheskie ocherki tsentral'nogo Tyan'-Shanya i Pamiro-Alaya (Historical-archaeological essay on the central Tian Shan and Pamiro-Alai mountain regions). *Materialy i issledovaniya po arkheologii SSSR* 26, Akademii Nauk SSSR, Moscow.

Bogdanova-Berezovskaya I. V. 1963. Himicheskiysostav metallicheskih predmetov iz Minusinskoy kotloviny (Chemical composition of tools from Minusinsk valley). In *Novie metodi v arheologicheskih issledovaniyah* (New methods in the archaeological research), 115–58. Moskow: Leningrad.

Bunker, Emma C. 1997. (ed.). *Ancient bronzes of the Eastern Eurasian steppes from the Arthur M. Sackler Collections.* New York: The Arthur M. Sackler Foundation.

Chen, Ge. 1987. Guanyu Xinjiang xinshiqi shidai wenhua de xinrenshi (New ideas on the Neolithic cultures in Xinjiang). *Kaogu* (Archaeology) 4:343–51.

——. 1995. Xinjiang yuangu wenhua chulun (A preliminary research on early cultures in Xinjiang). *Zhongya Xuekan* (Journal of Central Asian studies) 4:5–72.

Chen, K. T., and F. T. Hiebert. 1995. The late prehistory of Xinjiang in relation to its neighbors. *Journal of world prehistory* 9(2):243–300.

Chernykh, E. N. 1992. *Ancient metallurgy in the USSR: The early Metal Age.* Trans. Sarah Wright. Cambridge: Cambridge University Press.

Chlenova, N. L. 1994. On the degree of similarity between material culture components within the "Scythian World." In *The archaeology of the steppes: Methods and strategies,* ed. Bruno Genito, 499–540. Napoli: Istituto Universitario Orientale.

Debaine-Francfort, C. 1988. Archeologie du Xinjiang des origines aux Han: Premiere partie. *Paleorient* 14(1):5–29.

——. 1989. Archeologie du Xinjiang des origines aux Han: II eme partie. *Paleorient* 15(1):183–213.

Devlet, M. A. 1980. *Sibirskie poyasnye Plastiny, II v. do n.e.–I v. n.e.* (Siberian openwork belt plaques, second century B.C.–first century A.D.). Arkheologiya SSSR, D4-7. Moscow: Nauka.

Di Cosmo, N. 1996. Ancient Xinjiang between Central Asia and China. *Anthropology and archeology of Eurasia* 34(4):87–101.

Érdy, Miklós. 1995. Hun and Xiong-nu type cauldron finds throughout Eurasia. *Eurasian studies yearbook* 67:5–94.

Gong, Guoqiang. 1997. Xinjiang zhaoqi tongqing chulun (A brief discussion of the early copper artifacts of the Xinjiang region). *Kaogu* (Archaeology) 9:70–77.

Guo, Wu. 1999. Qingtong fu zai ouya dalu de chuchuan (The early spread of bronze cauldrons in Eurasia). *Ouya Xuekan* (Eurasian studies) 1:122–50.

Jacobson, Esther. 1995. *The art of the Scythians.* Leiden: E. J. Brill.

Li, Shuicheng 1999. Cong Kaogu (Archaeology) faxian kan gongyuanqian liangqianji dongxifang wenhua de pongzhuang yu jiaoliu (The cultural exchange between the East and the West during the second millennium B.C.: An archaeological perspective). *Xinjiang wenwu* (Cultural relics in Xinjiang) 1:53–65.

Li, Xiao, and Dang Tong. 1995. Zhunge'er pendi zhouyuan diqu chutu tongqi chutan (A preliminary study of copper and bronze artifacts unearthed from the surrounding regions of the Zhunge'er Basin). *Xinjiang wenwu* (Cultural relics in Xinjiang) 1:40–51.

Liu, Li. 1987. Tongfu kao (A study of copper cauldrons). *Kaogu yu wenwu* (Archaeology and cultural relics). 3:60–65.

Mei, Jianjun. 2000. *Copper and bronze metallurgy in late prehistoric Xinjiang: Its cultural context and relationship with neighboring regions.* BAR International Series 865. Oxford: Archaeopress.

Mei, Jianjun, and Colin Shell. 1998. Copper and bronze metallurgy in late prehistoric Xinjiang. In *The Bronze Age and early Iron Age peoples of eastern Central Asia,* ed. Victor H. Mair, vol. 2, 581–603. Washington, D.C.: Institute for the Study of Man.

——. 1999. The existence of Andronovo cultural influence in Xinjiang during the second millennium B.C. *Antiquity* 73.281:570–78.

Mei, Jianjun, Colin Shell, Li Xiao, and Wang Bo. 1998. A metallurgical study of early copper and bronze artefacts from Xinjiang, China. *Bulletin of the Metals Museum* 30 (1998-II):1–22.

Qian, Wei, Sun Shuyun, and Han Rubin. 2000. Gudai shentong yanju zongshu (A review on the studies of ancient arsenic copper). *Sciences of conservation and archaeology* 12.2:43–50.

Shui, Tao. 1993. Xinjiang qingtong shidai zhu wenhua de bijiao yanjiu (A comparative study of the Bronze Age cultures in Xinjiang). *Guoxue Yanjiu* (Studies in sinology) 1:447–90.

So, Jenny F., and Emma C. Bunker. 1995. *Traders and raiders on China's northern frontier.* Seattle and London: Smithsonian Institution, in association with the University of Washington Press.

Takahama, Shu. 1994. Chugoku no kama (The cauldrons of China). *Newsletter on steppe archaeology* 4:2–9.

——. 1997. *Mounted nomads of Asian steppe: Chinese northern bronzes.* Tokyo: Tokyo National Museum.

Wang, Binghua. 1985. Xinjiang diqu qintong shidai kaogu wenhua shitan (A preliminary analysis of the archaeological culture of the Bronze Age in the region of Xinjiang). *Xinjiang Shehui Kexue* (Social sciences of Xinjiang) 4:50–61.

Wang, Bo, and Qi Xiaoshan. 1995. *Sichouzhilu caoyuan shiren yanjiu* (A research on the grassland stone figures along the Silk Route). Urumchi: Xinjiang Renmin Chubanshe.

Wang Changqi. 1991. Xi'anshi wenguanhui cang E'erduoshi qingtongqi jiqi tezheng (The Ordos bronzes in the collection of the Committee of Cultural Relics of Xi'an City). *Kaogu yu wenwu* (Archaeology and cultural relics) 4:6–11.

Yue, Feng, et al., eds. 1999. *Xinjiang wenwu guji daguan* (A grand view of Xinjiang's cultural relics and historic sites). Urumchi: Xinjiang Meishu Sheyin Chubanshe.

Zhang, Yuzhong, and Zhao Derong. 1991. Yili hegu xinfaxian de daxin tongqi ji youguan wenti (The new discoveries of copper artifacts in the Yili valley and their relevant problems). *Xinjiang wenwu* (Cultural relics in Xinjiang) 2:42–48.

8 Tinned Belt Plaques of the Sixth to Fifth Century B.C.E. from Gansu Province, China: A Technical Study

Ma Qinglin and D. A. Scott

ABSTRACT Tinned bronze belt plaques, excavated from the southeast of Gansu Province, dating to the sixth to fifth century B.C.E. were analyzed using optical metallography, x-ray diffraction (XRD), and electron probe microanalysis (EPMA) in order to determine the nature of the tinned surfaces and how they were produced. The thickness of the tinned layers, which often include two or three separate phase regions, ranged from 2 μm to 35 μm. X-ray diffraction analysis indicated that there are three or four intermetallic phases present in these tinned layers: eta phase, η (Cu_6Sn_5, 61.0 wt. % tin, $Cu_{6.26}Sn_5$, 59.9 wt. % tin); epsilon phase, ε (Cu_3Sn, 38.2 wt. % tin); delta phase, δ ($Cu_{31}Sn_8$, 32.6 wt. % tin); and ($\alpha + \delta$) eutectoid in the tinned layer. The delta phase present was shown to be of variable composition, representing a mixture of compounds, including $Cu_{31}Sn_8$, $Cu_{81}Sn_{22}$, $Cu_{40.5}Sn_{11}$, $Cu_{41}Sn_{11}$, and $Cu_{327.92}Sn_{88.08}$. EPMA showed evidence for the following layered structure in the tinned surface, from ε, η, δ, and in some cases, the $\alpha + \delta$ eutectoid. Replication experiments revealed that the mercury-amalgam tinning process produces similar tinned layers to that obtained with tinning by wiping or dipping.

Introduction

The complex nature of the interaction of tin applied to the surface of copper or bronze artifacts has been the subject of a considerable amount of research in order to determine how objects were tinned in antiquity and the nature of the evidence that can be derived from their study (Meeks 1986, 1993a, 1993b; Tylecote 1985; Oddy and Bimson 1985; Dinsdale 1978; Britton 1975; ITRI 1983). This background of metallographic and experimental work is of great assistance in pursuing further studies of tinned bronzes from different regions of the world.

The key issues are particularly associated with the interpretation of the technological processes by which some of these complex-surface microstructures were formed. It is important to characterize the metallographic structures, clarify the technological choices, and account for the highly protective nature of these tinned layers.

From previously published work, the earliest uses of dipped or wiped tinning recognized date from about the fifth (450) century B.C.E.—for example, the tinned Celtic bronze spurs found in France (Hedges 1964) and the Hellenistic low-tin bronze mirrors of the fifth to third century B.C.E. (Craddock 1981). Chinese tinned bronzes before the fifth century B.C.E. are mentioned on the basis of documentary and archaeological site reports (Needham 1962; Zhou 1957; Gansu Provincial Museum 1972), and some research relevant to this topic has been published (Bunker 1990; Han and Bunker 1993; Kossolapov and Twilley 1994; Ma et al. 1999; Ma, Hu, and Li 1999; Tan et al. 2000; Anheuser 2000).

The Book of Songs (ShiJing) collates 305 songs dating from the sixth to fifth century B.C.E. and clearly depicts gilded bronze ornaments that must have been in use during the eighth to seventh century B.C.E. in the areas of today's Shaanxi and Gansu Provinces (Legge 1967). One Chinese chemical historian has maintained that gilded bronze harnesses illustrated in the Book of Songs were also of tinned bronze (Zhang 1958).

Recently, some tinned bronzes dating to the later part of the Spring and Autumn period (770–476 B.C.E.) and the early period of the Warring States (475–221 B.C.E.) were excavated from Li county, Gansu Province, China. Archaeologists believe the capital of the early Qin state was located in Li county, since several large tombs and many important artifacts dated to that period have been found from these areas (Dai 2000). The tinned objects include weapons, tools, and small ornaments, some of them being tinned on one side and others on both sides. These excavated tinned bronze artifacts provided a good opportunity for us to study the metallurgy of the tin coatings that had been employed. In our present paper, the analyzed samples are tinned belt plaques dating from the sixth to fifth century B.C.E. and are shown in figure 1. The ornamentation, style, and utilitarian features of these objects have already been discussed (Ningxia Archaeology Institute 1995; So and Bunker 1995).

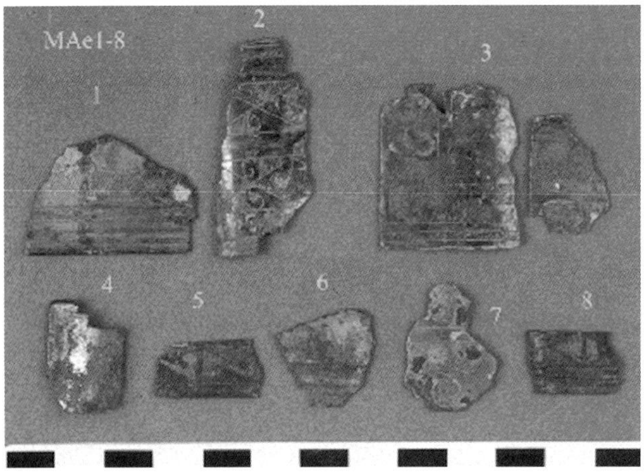

Figure 1. *Tinned belt plaques, MAe1–MAe8*

Analytical Work

Samples, approximately 3 mm × 3 mm, were removed from fragmentary material representative of the different types of tinned bronze belt plaques. The samples were cleaned thoroughly in acetone or ethanol and dried in air. The phase composition of the tinned surfaces was analyzed using XRD. To identify the elemental composition and microstructure of the tinned layers and bronze substrates, cross-section samples were prepared according to standard metallurgical procedure (Scott 1991).

Wipe-tinned and mercury-amalgam tinned samples were prepared for replication experiments. For the amalgam tinning work, 40% tin and 60% mercury, by weight, were mixed together and heated below 232°C for five minutes to form an amalgam. The amalgam-tinned samples were prepared below 200°C, and the wipe-tinned below 350°C, employing a worked and annealed 6% tin 94% copper standard alloy substrate, 3 mm thick. The samples were then heated to different temperatures for fixed periods of time, and the phases formed on the surface, determined with in situ XRD measurements, phase composition by EPMA, and microstructure by optical metallography. The in situ XRD measurements were made possible by the use of a Siemens (Bruker) D5005 x-ray diffractometer in detector scan mode with attached Gobles mirror assemblage employing copper K alpha radiation at 40 kV and 40 mA. A Cameca SX100 electron microprobe was used to obtain wavelength dispersive (WDS) elemental maps with analysis conditions of 20 kV and 10 mA. Dwell time was 50 msec/pixel.

The x-ray fluorescence spectrometer employed was a Kevex 0750A freestanding instrument, with the operational components contained in a lead-lined room and employing an x-ray flux from a barium/strontium secondary target (STXRF) at 50 kV 3.3 mA for two hundred seconds acquisition. Analytical data for the XRF results are shown in table 1.

Results and Discussion

Under bench binocular microscopy at ×40 magnification, the surfaces of the tinned belt plaques were carefully examined. The examination revealed that surface decoration on the plaques was made by inscribing using a chisel that probably had a U-shaped cutting profile. Some of the tinned layers on the artifacts were partially detached because of corrosion of the underlying bronze substrate. Some examples are shown in figures 2 and 3. Cross-sectional studies were facilitated by optical metallography, with figure 4 showing a typical photomicrograph of a cross-section through the tinned layer and bronze substrate of sample MAe1 at ×100, which reveals that the inscribed decoration has been made before tinning. The tinned layer can be seen to be quite uniform in thickness. In the bronze substrate, the worked and annealed grain structure can be seen, with equiaxed grains and straight twin lines from which we can infer that annealing was the final working operation. Extensive corrosion has occurred in the bronze substrate beneath the tinned layer, while the tinned layer itself was only slightly corroded.

The photomicrograph of figure 5 illustrates the typical substrata that developed within the tinned surface layers of sample MAe5 and reveals that there are three or four intermetallic phases in the tinned layer. In some regions, three phases are manifest, and in others four are manifest. From the top sublayer to the bottom sublayer, the tin-copper

Table 1. *XRD results on the tinned surface; micro-XRF compositional analyses of the tinned belt plaques. (The composition of the samples includes the composition of the bronze alloy and tinned layer or layers.)*

Sample no.	Analysis area	Cu wt. %	Sn wt. %	Pb wt. %	Ag wt. %	Thickness of tinned layer and XRD
MAe1	Cross-section	83.98	12.68	3.03	0.31	
	Tinned surface	74.35	23.46	1.91	0.28	20–35 microns ε, δ
MAe2	Cross-section	86.32	12.82	0.83	0.13	
	Tinned surface	62.68	35.55	1.57	0.2	16–32 microns δ
MAe3	Cross-section	84.93	11.93	2.92	0.19	
	Tinned surface	78.32	20.23	1.28	0.17	20–30 microns ε, δ
MAe4	Cross-section	84	13	2.73	0.27	
	Tinned surface	75.25	21.86	2.66	0.23	3–9 microns η, ε, δ
MAe5	Cross-section	80.44	16.77	2.34	0.45	
	Tinned surface	78.92	19.63	1.24	0.21	20–25 microns η, ε, δ
MAe6	Cross-section	83.94	13.59	2.2	0.27	
	Tinned surface	68.06	26.41	5.26	0.27	2–7 microns η, ε, δ
MAe7	Cross-section	82.34	14.83	2.64	0.19	
	Tinned surface	59.6	38.27	1.82	0.31	20–32 microns η, ε, δ
MAe8	Cross-section	81.49	15.36	2.84	0.31	
	Tinned surface	57.97	34.21	7.29	0.35	8 microns δ

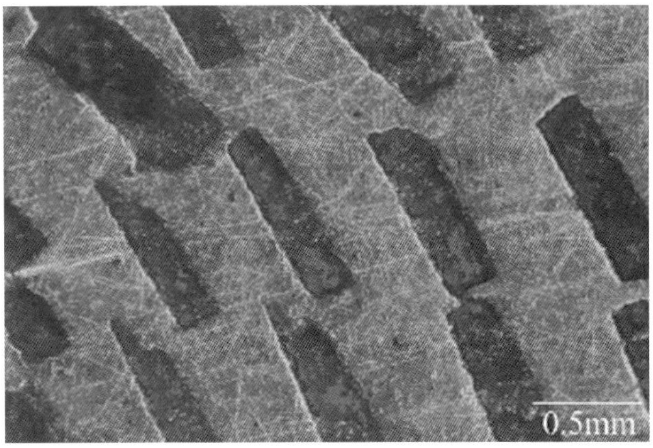

Figure 2. *Tool marks on the belt plaque MAe1. The surface was inscribed with a rectangle or sharp chisel that probably had a V- or U-shaped cutting profile. This was inscribed before tinning. The length and width of the chisel is about 1 mm × 0.2 mm*

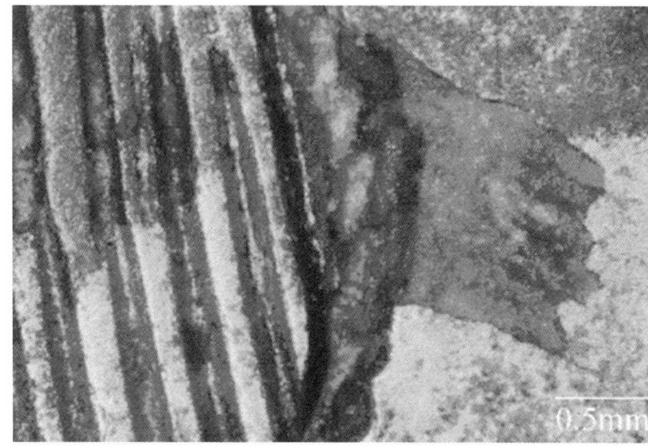

Figure 3. *Tinned belt plaque MAe7. The tinned layer covers the entire surface, including the engraved area. In some areas, the tinned layer was detached, and the bronze corrosion can be seen*

compounds are η-phase (Cu_6Sn_5, 61.0 wt. % tin, JCPDS 2-0713, $Cu_{6.26}Sn_5$, 59.9 wt. % tin, JCPDS 47-1575), ε-phase (Cu_3Sn, 38.2 wt. % tin, JCPDS 1-1240), and δ-phase ($Cu_{31}Sn_8$, 32.6 wt. % tin, $Cu_{41}Sn_{11}$, 33.38 wt. % tin) together with (α + δ) eutectoid (Owen and Williams 1936). The η compound is light gray in color, occurring on the outermost surface of the tinned layer, and is present only in some areas. The ε compound is darkest gray/blue, δ compound is mid-gray, and under the ε phase, the (α + δ) eutectoid is contiguous with the δ phase but contains a large amount of the α compound, globular lead inclusions, and porosity. EPMA results of the areas marked in figure 5a for object MAe5 revealed details of the different phases: area 1 is the η phase, area 2 is the ε phase, and area 3 is the δ phase.

In other tinned areas of this sample, no η and ε phases occur above the δ phase, but only the δ and (α + δ) eutectoid. This variation is partially due to differences in the tinning process and to some corrosion during burial. The persistence of the δ phase is not uncommon, due in part to its high corrosion resistance (Tan and Wu 1997).

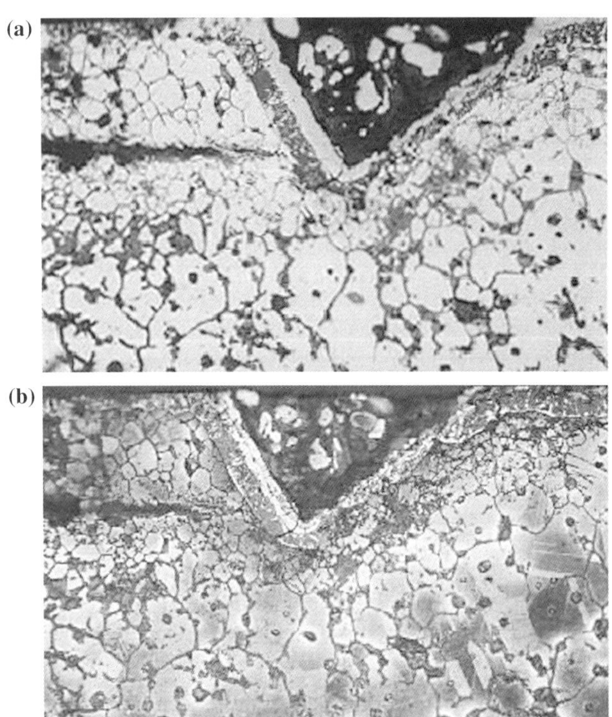

Figure 4. *(a) Polished cross-section of MAe1, inscribed before tinning. On the engraved area there are two or three sublayers. The polished surface contains only one or two sublayers. ×100 (b) Etched in alcoholic ferric chloride, the worked and annealed grain structure with some twinning is visible. It is very clear there are three sublayers in the engraved area. These are the ε, δ, and (α + δ) eutectoid. ×100*

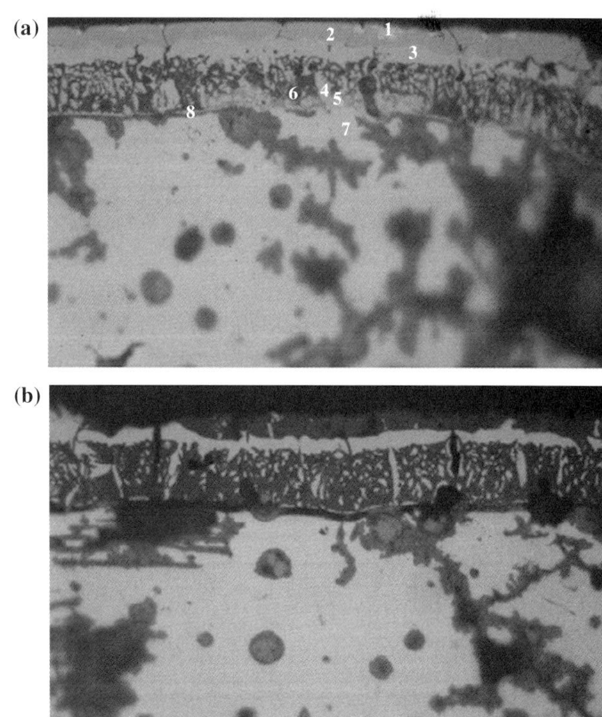

Figure 5. *(a) Polished and etched section of MAe5. There are three or four sublayers in the tinned layer: η, ε, δ, and (α + δ) eutectoid. The η phase is light gray in color on the outermost surface of the tinned layer. The ε phase is darkest gray/blue, the δ phase is mid-gray, and under the ε phase the (α + δ) eutectoid is contiguous with the δ phase but contains a large amount of α compound, globular lead, and some porosity. (b) The main phase in the tinned layer is the (α + δ) eutectoid. A small area of ε phase can be seen on the upper-left corner of the tinned layer*

Similar morphological characteristics were observed in the cross-sections of other tinned plaque samples. There is a considerable variation in thickness of the tinned layer, however, from 2 to 35 μm. The thickness of the tinned layers probably has a direct relationship with the extent of polishing after tinning, which is why the principal tin-copper compounds are the δ and (α + δ) eutectoid in the tinned areas, which have been extensively smoothed. In modern industrial practice, the average thickness of tinned layers varies from 1 to 12 μm for hot-wiped coatings and from 1.5 to 25 μm for hot-dipped coatings (Britton 1975, 199). As mentioned above, the thickness of the tinned layer for ancient European examples varied from 2 to 10 μm (Meeks 1986). These results show that the tinned layers on the surface of ancient Chinese tinned bronzes are thicker than that of their European counterparts and closer to that of modern thick-tinned food containers.

Elemental Composition

The tin contents in the bronze alloys themselves are in the range from about 11.9 to 16.7 wt. %, and the copper contents range from 80.4 to 86.3 wt. %. The range of tin content in the tinned surface layers, as measured with XRF, varied from 19.6 to 38.2 wt. %, with copper contents in the range from 57.9 to 78.9 wt. %. The tin content on the tinned layers is higher than that of the bronze substrates, as shown in table 1. The copper and tin contents of the tinned surfaces determined by XRF is affected by the depth of penetration of the x-ray beam, and therefore, although the results do provide evidence of a tinning process, they do not provide an exact measure of the composition of the surface phases.

The analytical work revealed a curious feature for sample MAe1, which is shown in the WDS elemental distribution maps of figure 6. The apparent conjunction of lead and tin at the surface, especially in the triangular depression seen in figure 6, might suggest that a lead-tin alloy was used as the surface coating, which has also been postulated by Han and Bunker (1993). However, in this case the elemental distribution maps indicate that the lead, phosphorus, oxygen, and chlorine appear to be coincident in the area of the depression on the surface, suggesting that the occurrence of substantial amounts of lead in this area is a consequence of the corrosion of the lead globules from the alloy itself rather than a surface deposition associated with the tinned layer. Lead is well known to have the ability to precipitate

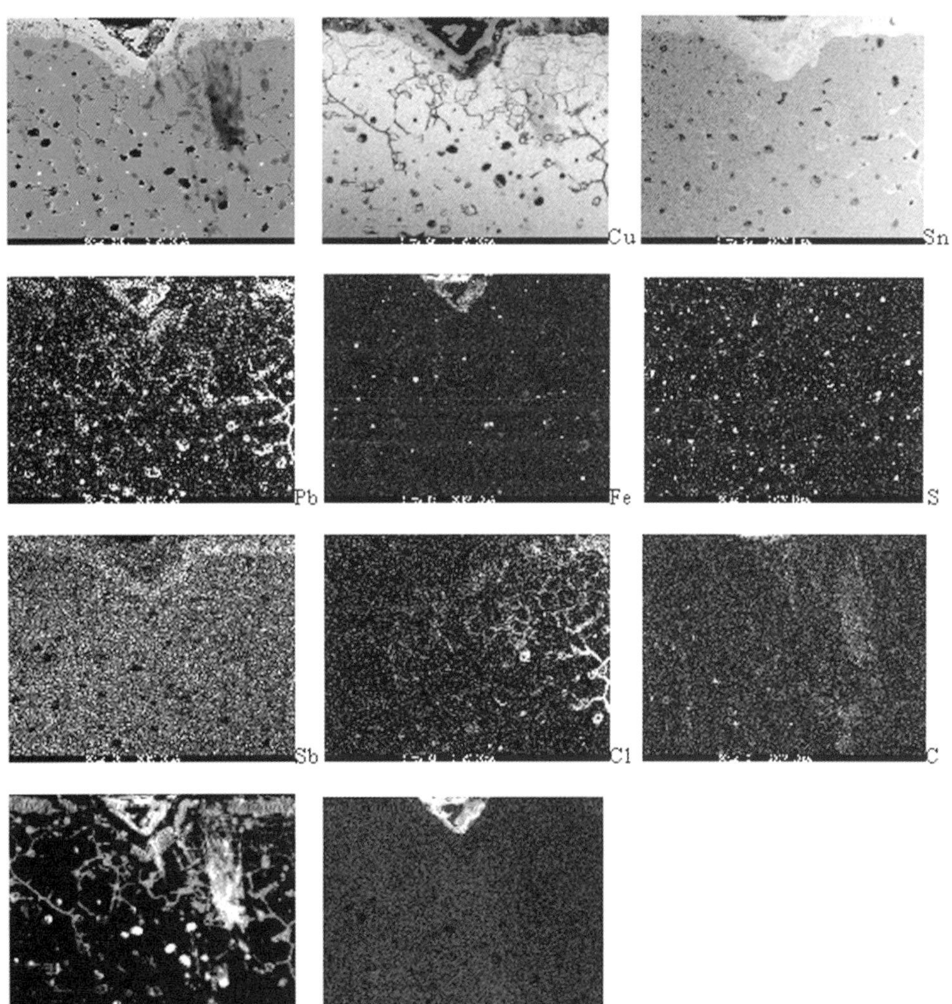

Figure 6. *WDS elemental maps of a cross-section of MAe1. A backscattered electron image of the section is included at the upper left. The resolution possible with WDS mapping allows phosphorous to be distinguished from lead. It also allows phosphorus to be distinguished from mercury, which was not seen in this sample*

insoluble salts at the surface of bronze objects, and that is what appears to have occurred here, rather than the lead content being due to corroded lead derived from the surface tinning layer.

X-ray Diffraction Analysis

As a consequence of the hot tinning of bronze substrates, as found by Daniels (1936), Hedges (1960), and Meeks (1986), there are four different surface zones (surface tin, η, ε, and δ) that can be found in the tinned layer determined from experimental samples. δ phase is not formed on normal tinned surfaces, but it can occur on heat-treated tinned surfaces, and in this case it forms at the boundary between ε and the substrate bronze in a new tinned layer. There was no surface tin on the outermost tinned layer of ancient tinned bronzes, because the surface tin, if it existed, would have been converted to a corrosion product, such as cassiterite or romarchite (SnO_2 or SnO, respectively) in the tinning process or during burial. Part of the corrosion products had been absorbed by the soil of the buried surroundings; others were still present on the tinned surface.

The analytical work reported here confirms that the ε and η phases in the tinned layer can be found in ancient examples of tinned bronzes, where the presence of either the ε or η phase is an indicator of a hot-tinned surface as the phases form due to diffusion between the copper and tin at elevated temperatures but do not form from corrosion processes. Further surface heating of the hot-tinned applied layer at temperatures around 350°C would lead to the formation of the ε phase alone in the tinned layer (Kay and Mackay 1976; Meeks 1986).

In the tinned layers of these ancient Chinese belt plaques, some examples examined metallographically show the presence of the four phases (η, ε, δ, and [α + δ] eutectoid), while others reveal three phases (ε, δ, and [α + δ] eutectoid) or even just two (the δ and [α + δ] eutectoid). Some of these data are tabulated in table 1, where XRD confirmed the presence of the η, ε, and δ phases in some of the tinned layers.

In previously reported work, Meeks (1986, 1993a) and Oddy and Bimson (1985) showed that usually only the ε and η phases or ε phase can be detected within the tinned surfaces of ancient objects by using XRD. We found that the η phase is sometimes too thin to be detectable by XRD, either because the object was heat-treated after tinning, leaving the ε and δ phases as the principal compounds in the tinned layer, or because most of it was lost after grinding and polishing to create a smooth appearance. Sometimes the η or ε phases can be found in the inscribed areas of the tinned belt plaque surface, since on these areas the thickness of the η and ε was not affected by polishing. Table 2 shows the XRD data for the tinned belt plaque MAe1, where the existence of the ε phase in the engraved area is confirmed compared with the bulk area of the surface of the tinned artifact.

Table 3 and figure 7 show the phases determined by XRD for the surface of the tinned bronze belt plaque MAe7, which confirmed the conclusions reached by optical examination that three intermetallic phases (η, ε, and δ) are present in the surface layer.

In these tinned bronze or tin-enriched bronze surfaces, the delta phase, δ ($Cu_{31}Sn_8$), is sometimes difficult to assign to one XRD dataset as a result of experimental or fundamental variations in the x-ray data determined from individual samples. Thus, the ICDD files have four entries for the delta phase, which are given as $Cu_{81}Sn_{22}$ (JCPDS 31-486), $Cu_{40.5}Sn_{11}$ (JCPDS 31-485), $Cu_{41}Sn_{11}$ (JCPDS 30-510), or $Cu_{327.92}Sn_{88.08}$ (JCPDS 30-511). The usual way to refer to the delta phase in terms of composition for this intermetallic constituent is as $Cu_{31}Sn_8$ (Owen and Williams 1936; Meeks 1986).

It is possible to infer that a tin-enriched bronze was produced with hot tinning if the η or ε phase can be detected by XRD. However, it is difficult to determine the mode of formation of the tin-enriched surface if the delta phase is the only copper-tin compound making up the surface layer. In such cases it is essential to observe the metallographic structure of the polished cross-section in order to decide if the tin-enriched surface could be a consequence of tin sweat from the alloy interior or whether the delta phase is derived from a deliberate tinning process.

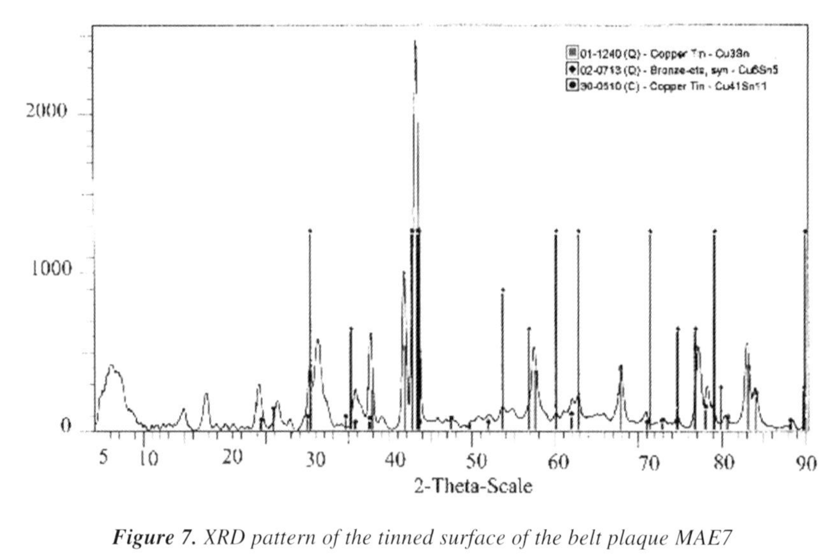

Figure 7. *XRD pattern of the tinned surface of the belt plaque MAE7*

Table 2. *XRD results of an engraved area and a polished area of the tinned bronze belt plaque MAE1*

MAe1 Engraved area		Cu_3Sn 01-1240		$Cu_{41}Sn_{11}$ 30-510		MAe1 Polished area	
D-spacing	I/I_0	D-spacing	I/I_0	D-spacing	I/I_0	D-spacing	I/I_0
3.48	4			3.46	9	3.48	6
3.4	12			2.99	5	3.01	5
2.77	4						
2.76	4						
2.59	3			2.59	5		
2.53	1						
2.52	1					2.46	30
2.38	23	2.38	20	2.4	4		
2.16	35	2.16	40			2.19	18
2.11	100			2.11	100	2.12	100
2.09	97	2.08	100				
1.91	4			1.91	4	1.92	3
1.85	3						
1.83	4			1.83	3	1.83	4
1.76	3			1.76	2		
1.6	17	1.6	27		2	1.61	7
1.49	9			1.49	6	1.5	11
1.37	12	1.38	30			1.38	6

Table 3. *XRD data of the tinned surface of belt plaque MAE7 and the JCPDS standard of η, ε, and δ phases*

MAe7		Cu_6Sn_5(η)	2-0713	Cu_3Sn (ε)	1-1240	$Cu_{41}Sn_{11}$(δ)	30-510
D-spacing	I/I_0	D-spacing	I/I_0	D-spacing	I/I_0	D-spacing	I/I_0
3.69	1					3.69	3
3.38	7					3.46	9
3.21	2						
2.95	16	2.96	100			2.99	5
2.86	25						
2.77	7					2.59	5
2.5	25	2.55	50			2.51	2
2.38	25			2.38	20	2.4	4
2.3	20						
2.16	38	2.09	100	2.16	40	2.11	100
2.09	100	2.08	100	2.08	100	2.07	2
1.97	4					1.91	4
1.82	3					1.83	3
1.79	4					1.76	2
1.75	4						
1.7	6	1.71	70				
1.67	6						
1.6	21	1.62	50	1.6	27		
1.53	4	1.54	100				
1.51	5						
1.49	8					1.49	6
1.48	9	1.48	100				
1.37	15			1.38	30		
1.32	4	1.32	100			1.32	2
1.29	3					1.29	3
1.26	3	1.27	50			1.27	2
1.23	21	1.24	50	1.24	40	1.23	10
1.21	11	1.21	100				
1.2	5	1.2	20				
1.19	4			1.19	5		
1.16	22			1.16	30		
1.15	11			1.15	17		
1.1	2					1.11	3
1.09	1			1.08	7		

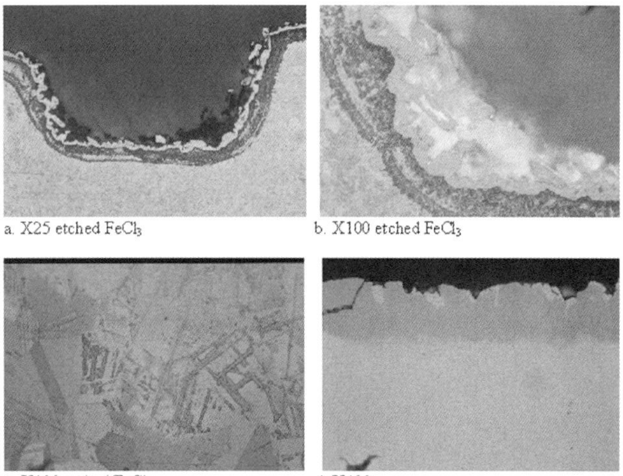

a. X25 etched $FeCl_3$ b. X100 etched $FeCl_3$ c. X100 etched $FeCl_3$ d. X400

Figure 8. *Replication experiment of mercury-amalgam tinning*
(a) A tin-mercury amalgam (40% Sn + 60% Hg) was prepared below 232°C. The powdered amalgam was then placed on the surface of an engraved bronze plaque and heated between 170 and 200°C. The sample was next heated at 320°C for twenty minutes to allow the amalgam to cover the bronze. Then the sample was heated at 360°C for twenty minutes and 420°C for sixty minutes. Paper was used as a brush to form a smooth surface. XRD revealed $Hg_{0.1}Sn_{0.9}$, *η,* $Cu_{6.26}Sn_5$. *Etched in* $FeCl_3$. *×25*
(b) This higher-magnification view of the sample in 8(a) shows that only a thin tinned layer is formed. The bronze alloy under the tinned layer is heavily damaged because of the great penetration of the mercury-tin amalgam at high temperatures. Etched in $FeCl_3$. *×100*
(c) A small fragment removed from the sample shown in 8(a) was then further heated at 450°C for another thirty minutes. The mercury-tin amalgam can be seen to have penetrated deeply into the bronze alloy. In this situation, no ε phase formed in the tinned layer. Etched in $FeCl_3$. *×100*
(d) A second small fragment removed from the sample shown in 8(a) was further heated at 420°C for sixty minutes and then at 450°C for thirty minutes. This creates three phases in the tinned layer. The light gray layer is η, and dark gray is ε. The δ is very thin between the ε phase and the bronze alloy. The thickness of the tinned layer in this sample is 26–32 μm. ×400

Replication of Wipe Tinning and Mercury-Amalgam Tinning

When the microstructures of these belt plaques are compared with the reference data from the work of Meeks, we may reasonably infer that these tinned belt plaques could be fabricated using hot tinning. As illustrated in figures 2, 3, and 4, these belt plaques were tinned after inscribing the fine line decoration. The engraved area is clearly visible, as it is not covered fully by the tin or copper-tin compounds. It is very difficult to produce finely engraved surfaces that are evenly covered with a hot tinned surface because the engraved lines have a very narrow cross-section that does not allow the tin to penetrate. One method of ensuring complete coverage of the engraved surfaces is to use amalgam tinning, and in order to investigate this technique in greater detail, the laboratory samples referred to earlier were prepared for amalgam tinning as well as hot tinning to compare the microstructural detail of the two methods. Figures 8 and 9 and tables 4, 5, and 6 show the metallographic microstructures and XRD from the experimental work.

In summary, either method could be used to produce similar tinned microstructures on the bronze alloy, the principal differences being that the amalgam-tinning process requires higher temperatures and longer times than hot tinning in order to get the ε and δ phases to develop. Additionally, mercury-tin amalgam has a greater depth of penetration into the alloy substrate in the temperature range from 350 to 450°C, and substantial internal diffusion may take place. The microstructure of the substrate bronze contiguous with the tinned surface may be heavily affected during the amalgam tinning. The mobility of this amalgam-tinned surface means that it can easily diffuse to cover areas of engraved design where no amalgam was applied to the surface. One advantage of the amalgam-tinning process is that it is easy to achieve an overall silvery appearance on the bronze object. Even with the use of a pine-resin flux, the same is not true for hot tinning, where it is very difficult to achieve complete coverage of all small depressions or inscribed areas of design, as shown in figure 9.

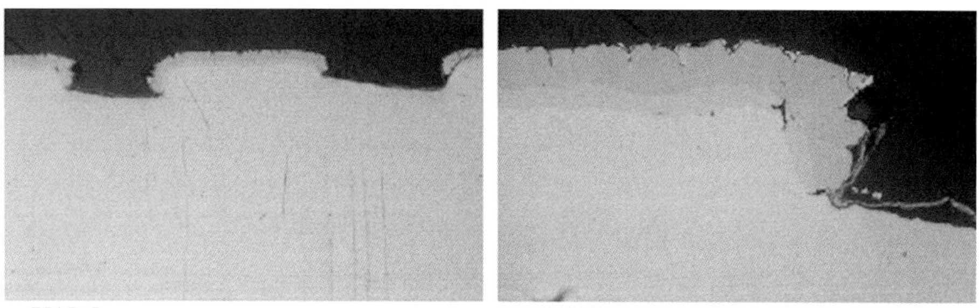

a. X12.5 b. X100

Figure 9. *Wipe-tinning experiment*
An engraved bronze was placed into melted tin at 320°C (using pine resin as a flux). An iron bar was used to polish the melted tin to cover the surface of the bronze. Next it was heated at 450°C for twenty minutes and then at 510°C for sixty minutes
(a) It is difficult to obtain a tinned layer in the engraved area. ×12.5
(b) There are three phases in the tinned layer. The main tin-copper compounds are ε and δ. Only a small amount of η phase is found covering the ε phase. The thickness of the tinned layer in this sample is 60 μm (ε 40 μm, δ 20 μm). Etched in $FeCl_3$. *The ε and δ phases were found with XRD, but the η phase can be seen only with microscopy. ×100*

Table 4. *Analysis of wipe- and amalgam-tinning samples from replication experiments using XRD and micro-XRF*

Temperature	Wipe-tinning sampling	XRD results	Amalgam-tinning sampling	XRD results
250°C			250°C, 10 mins	η, $Hg_{0.1}Sn_{0.9}$
320°C			320°C, 20 mins	η, $Hg_{0.1}Sn_{0.9}$
350°C	350°C, 10 mins	η, ε, Sn		
360°C			320°C, 20 mins 360°C, 20 mins	η, $Hg_{0.1}Sn_{0.9}$ Hg, 12.06%
420°C	350°C, 10 mins 420°C, 5 mins	η, ε, Sn	320°C, 20 mins 360°C, 20 mins 420°C, 60 mins	η, $Hg_{0.1}Sn_{0.9}$
			320°C, 20 mins 360°C, 20 mins 420°C, 100 mins	η, $Hg_{0.1}Sn_{0.9}$
450°C	350°C, 5 mins 410°C, 40 mins 450°C, 30 mins	η, ε, Sn	350°C, 10 mins 410°C, 40 mins 450°C, 30 mins	η, ε $Hg_{0.1}Sn_{0.9}$ Hg, 1.86%
520°C	350°C, 5 mins 450°C, 30 mins 520°C, 5 mins	η, ε, trace δ, Sn	250°C, 10 mins 520°C, 5 mins	η, ε $Hg_{0.1}Sn_{0.9}$
525°C	350°C, 5 mins 450°C, 30 mins 520°C, 5 mins 525°C, 10 mins	η, ε, trace δ, Sn	250°C, 10 mins 520°C, 5 mins 525°C, 10 mins	η, ε $Hg_{0.1}Sn_{0.9}$
	350°C, 5 mins 450°C, 30 mins 520°C, 5 mins 525°C, 40 mins	ε, δ	250°C, 10 mins 520°C, 5 mins 525°C, 40 mins	η, ε $Hg_{0.1}Sn_{0.9}$ Hg, 4.46%
550°C	350°C, 5 mins 450°C, 30 mins 550°C, 5 mins	ε, δ, and trace η	250°C, 10 mins 550°C, 5 mins	η, ε $Hg_{0.1}Sn_{0.9}$
	350°C, 5 mins 450°C, 30 mins 550°C, 15 mins	ε, δ, and trace η	250°C, 10 mins 550°C, 10 mins	η, ε trace $Hg_{0.1}Sn_{0.9}$
	350°C, 5 mins 450°C, 30 mins 550°C, 30 mins	ε, δ	250°C, 10 mins 550°C, 30 mins	η, ε, trace δ Hg, 0.233%
	350°C, 5 mins 450°C, 30 mins 550°C, 40 mins	ε, δ	250°C, 10 mins 550°C, 40 mins	η, ε, trace δ Hg, 0.174%
			250°C, 10 mins 550°C, 50 mins	ε, δ Hg, 0.157%

The thickness of the tinned layer formed by these different methods is dependent on the amount of molten tin or amalgam that is placed on the bronze alloy, the duration of the heating, and the temperature employed.

Mercury and Corrosion Characteristics of Tinned Belt Plaques

It is hard to distinguish between the microstructures and XRD data obtained from either hot-tinned surface application or from amalgam tinning. In recent years, more research has been conducted in order to investigate this problem. Kossolapov and Twilley (1994) investigated a bronze dagger dating to the Warring States period that was covered with a pattern of more or less regular dark spots. They concluded that the trace amounts of mercury detected in the Cu-Sn hypereutectiod areas of the dark spots on the dagger surface were a consequence of the use of mercury-amalgam tinning. Using x-ray fluorescence analysis, Kossolapov and Twilley determined that about 1% of mercury existed in these areas. These conclusions were questioned by Anheuser (2000), who refined to about 0.2–0.34% by EPMA and maintained that the trace amounts of mercury in the dark spots originated from contamination derived from the burial environment. He is of the opinion that no amalgam tinning existed before the thirteenth century.

In China, scholars in the Shanghai Museum have also undertaken extensive investigations into the subject of amalgam tinning (Tan and Wu 1997). These authors compared the microstructure and composition of a lozenge pattern on a sword dated to the Eastern Zhou period (770–256 B.C.E.) and carried out replication experiments using a tin-rich paste coating that contains a mercury-tin amalgam with the addition of some organic materials as flux. Using this technique

Table 5. *XRD results of a sample made using a tin-mercury amalgam tinning method (360°C, twenty minutes; 410°C, forty minutes; 450°C, thirty minutes)*

MAamal450 30 mins		$Hg_{0.1}Sn_{0.9}$ 48-1546		$Cu_{6.26}Sn_5$ 47-1575		Cu_6Sn_5 02-713		Cu_3Sn 01-1240		SnO_2 41-1445	
D-spacing	I/I_0	D-spacing	I/I_0	D-spacing	I/I_0	D-spacing	I/I_0	D-spacing	I/I_0	D-spacing	I/I_0
3.37	6									3.347	100
2.96	9	2.99	50	2.96	100	2.96	100				
2.75	8	2.78	100								
2.64	5									2.643	75
2.52	9			2.54	5	2.55	50				
2.38	14							2.38	20	2.369	21
2.31	3									2.309	4
2.16	7			2.1	82	2.09	100	2.16	40		
2.09	100			2.08	64	2.08	100	2.08	100	2.1189	1
2.04	2	2.03	80								
1.76	3			1.71	18	1.71	70			1.7647	57
1.6	14	1.6	15	1.62	8	1.62	40	1.6	27	1.675	14
1.53	2	1.49	15	1.54	10	1.54	100	1.38	30	1.5934	6
1.37	6	1.39	10							1.4984	11
										1.4392	12
										1.4155	14

Table 6. *XRD results of a sample made using a wipe-tinning method (320°C, ten minutes; 450°C, thirty minutes)*

MAhottin45021		$Cu_{6.26}Sn_5$ 47-1575		Cu_6Sn_5 02-713		Cu_3Sn 01-1240		$Cu_{41}Sn_{11}$ 30-0510		Sn 40-673	
D-spacing	I/I_0	D-spacing	I/I_0	D-spacing	I/I_0	D-spacing	I/I_0	D-spacing	I/I_0	D-spacing	I/I_0
3.4	7							3.46	9		
2.97	46	2.96	100	2.96	100			2.997	5	2.915	100
2.79	18									2.793	90
2.55	11	2.54	5	2.55	50			2.595	5		
2.39	9					2.38	20	2.403	4		
2.17	48	2.1	82			2.16	40				
2.09	100	2.08	64	2.09	100	2.08	100	2.119	100	2.062	34
2.01	9			2.08	100			1.916	4	2.017	74
1.72	4	1.71	18	1.71	70						
1.65	3									1.659	17
1.62	3	1.62	8	1.62	40						
1.6	9					1.6	27				
1.54	8	1.54	10	1.54	100						
1.48	9	1.48	20	1.48	100			1.498	6	1.484	23
1.44	7									1.458	13
1.38	3					1.38	30			1.442	20

they were able to successfully reproduce the same kind of decorative effect as found on the Eastern Zhou sword.

From the XRF results of the amalgam-tinning experiments, shown in table 4, the mercury content was shown to be about 0.157%, even if the sample was heated for fifty minutes at 550°C. As mentioned above, a longer heating time is needed if the ε and δ phases are to form during the amalgam-tinning process, as compared to hot tinning. If the amalgam-tinned surfaces are heated for longer periods, the mercury content will be reduced to very low levels within the coating, substantially less than 0.1%.

Further work will be undertaken to characterize the technological choices involved in tinning of bronze surfaces. Particularly important is to address the issue of whether amalgam tinning was employed in ancient China at this period. There are two methods to resolve this; one is to find evidence for the presence of mercury in the tinned layer and in the bronze alloy because of the tendency of the amalgam process to produce deep penetration (Chase pers. comm.), and the other is to investigate in detail the characteristics of the corrosion processes that have occurred within the bronze alloy under the tinned layer, which may preserve evidence for the mode of application of the tinned surface. These are the avenues that we are currently pursuing in the next phase of this research.

Conclusions

From the analytical results, we can see that there are three or four intermetallic phases in the tinned layer of these Chinese bronze belt plaques from the sixth to fifth century B.C.E. excavated from Gansu Province. These layers are the η, ε, δ and (α + δ) eutectoid. The thickness of the tinned layer was found to vary from 2 μm to 35 μm. The replication experiments revealed that it was possible to achieve very similar microstructures using either hot tinning or amalgam tinning, although higher temperatures and longer heating times were necessary for the amalgam coatings compared with hot tinning when the ε and δ phases emerged. It is possible for the tin amalgam to penetrate deeply into the substrate bronze alloy. One example of a lead-tin alloy being used as a coating rather than tin alone was found during the study of these belt plaques. Further research is required to establish the precise nature of the tinning process that has been used in ancient China, and this is what we intend to address in the next phase of this work.

Acknowledgments

We would like to express our gratitude to Professor Hu Zhide, Lanzhou University, Chemistry Department, and Zuixiong Li, Dunhuang Academy, for past cooperative work; to Dr. Neville Agnew, the Getty Conservation Institute, for supporting one author as a visiting scientist in Getty Conservation Institute; and to Professor Zhang Penchuan, the former deputy director of Gansu Provincial Museum, for helping to find samples. Thanks are also due to Eric Doehne and David Carson, the Getty Conservation Institute, for operating the electron microprobe and the environmental scanning electron microscope. This work was carried out under the auspices of a visiting scholarship during 2001, based at the GCI Museum Research Laboratory.

References

Anheuser, K. 2000. Amalgam tinning of Chinese bronze antiques. *Archaeometry* 42(1):189–200.

Britton, S. C. 1975. *Tin verses corrosion.* Report no. 510. Middlesex: International Tin Research Institute. 19–25.

Bunker, C. E. 1990. Ancient ordos bronzes with tin-enriched surfaces.*Orientations* 1:78–80.

Craddock, P. T. 1981. Report on the composition of nineteen Etruscan bronze mirrors from Danish museums. In *Corpus speculorum Etruscorum, Denmark 1,* ed. H. S. Roberts, 131–33. The Danish National Museum. Odense: Odense University Press.

Dai, Chunyang. 2000. Several problems concerning the royal graveyard of the Qin state at Dabuzi, Li-Xian, Gansu. *Wenwu* 5:74–80.

Daniels, E. J. 1936. The hot-tinning of copper. *Journal of the Institute of Metals* LVIII(1):199–210.

Dinsdale, P. M. 1978. Hot tinning for the food industry. In *Tin and its uses,* no. 116, 9–11. Middlesex: International Tin Research Institute.

Gansu Provincial Museum. 1972. The Western Chou tombs at Pai-Ts'Ao-P'o in Ling-T'ai country, Kansu Province. *Kaoguxuebao* 3:99–130.

Han, Rubin, and E. C. Bunker. 1993. Research for the ancient ordos bronzes with tin-enriched surfaces. *Wenwu* 9:80–96.

Hedges, E. S. 1960. *Tin and its alloys.* Chap. 5. London: Edward Arnold.

——. 1964. *Tin in social and economic history.* London: Edward Arnold.

ITRI. 1983. *Tin and tin alloy coatings.* No. 625. Middlesex: International Tin Research Institute. 15–19.

Kay, P. J., and C. A. Mackay. 1976. The growth of intermetallic compounds on common basis materials coated with tin and tin-lead alloys. In *Transactions of the Institute of Metal Finishing* 54:68–74.

Kossolapov, A., and J. Twilley. 1994. A decorated Chinese dagger: Evidence for ancient amalgam tinning. *Studies in conservation* 39:257–64.

Legge, J. 1967. *The book of poetry (Shih Ching).* New York: Paragon. 141–42.

Ma, Qinglin, Hu Zhide, and Li Zuixiong. 1999. Tinned bronzes in ancient China. *The National Palace Museum monthly of Chinese art (Taiwan)* 198:102–14.

Ma, Qinglin, Su Boming, Hu Zhide, Li Zuixion, and Chen Gengling. 1999. Evaluation of microstructure of tinned coatings on ancient Chinese bronzes. *Journal of Lanzhou University* 35(4):67–71.

Meeks, N. 1986. Tin-rich surfaces on bronze: Some experimental and archaeological considerations. *Archaeometry* 28(2):133–62.

——. 1993a. Surface characterization of tinned bronze, high-tin bronze, tinned iron, and arsenical bronze. In *Metal plating and patination: Cultural, technical, and historical developments,* ed. S. La Niece and P. Craddock, 247–75. Oxford: Butterworth-Heinemann.

——. 1993b. Patination phenomena on Roman and Chinese high-tin bronze mirrors and other artifacts. In *Metal plating and patination: Cultural, technical, and historical developments,* ed. S. La Niece and P. Craddock, 63-84. Oxford: Butterworth-Heinemann.

Needham, J. 1962. *Science and civilization in China.* Vol. 4, part 1. Cambridge: Cambridge University Press.

Ningxia Archaeology Institute. 1995. The tombs of Yujia Zhuang, Penbao, Ningxia Hui autonomous region. *Kaoguxuebao* 79–106.

Oddy, W. A., and M. Bimson. 1985. Tinned bronze in antiquity. In *Lead and tin: Studies in conservation and technology,* Occasional Paper no. 3, 33–39. United Kingdom Institute of Conservation.

Owen, E. A., and E. C. Williams. 1936. Further study of copper-tin alloys by x-ray analysis. *Journal of the Institute of Metals* LVIII (1):283–97.

Scott, D. A. 1991. *Metallography and microstructure of ancient history metals.* Malibu, Calif.: The Getty Conservation Institute/The J. Paul Getty Museum. 63–66.

So, Jenny F., and Emma C. Bunker. 1995. *Traders and raiders on China's northern frontier.* Seattle and London: Smithsonian Institution, in association with the University of Washington Press. 177–78.

Tan, Derui, and Wu Laiming. 1997. The forming mechanism of *shui-yin-qin* on the surface of mirrors in East Han dynasty. *Science of conservation and archaeology* 8(1):1–8.

Tan, Derui, Wu Zejia, Lian Haiping, et al. 2000. A study into the making technique of the lozenge pattern on bronze weapons in the Eastern Zhou period. *Kaoguxuebao* 111–46.

Tylecote, R. F. 1985. The apparent tinning of bronze axes and other artifacts. *Journal of the Historical Metallurgy Society* 19(2):169–75.

Zhang, Zhigao. 1958. Getting the original meaning of the word *wo* from the ancient Chinese tinned bronze. *Kaoguxuebao* 3:73–74.

Zhou, Wei. 1957. *The weaponry history of ancient China.* Peking: Book Store of Sanluan. 169–70.

9 The Manufacturing Technique of Ferrous Artifacts Excavated from Xianbei Gravesite at Lamadong Village of Liaoning Province[1]

Chen Jianli, Han Rubin, and Wan Xin

ABSTRACT Tombs of the Xianbei people dated to the third to fourth century C.E. in Lamadong, Beipiao City, Liaoning, were excavated in the 1990s. Because of the importance of the Xianbei in the history of China and of the artifacts unearthed from these tombs, the archaeological project was distinguished as one of the top ten archaeological finds in 1996. In order to study the manufacturing techniques of these artifacts and the relationship between iron technology in northeast China and that in central China together with the spread of advanced technology and its economic and social effects, thirty-two iron and steel samples, including weapons, tools, and farming implements, were selected and examined by metallographic analysis. The results show that these artifacts were made with a relatively high level of skill and that the beginning of iron use and manufacture by Xianbei people was later than in central China, which strongly influenced their techniques and styles. Xianbei tribes played important roles in the spread of advanced iron technology from central China to neighboring regions.

Introduction

Xianbei is the name given to one of the large tribes of ancient northern China. According to Chinese historical records, the Xianbei were a stock-raising group of the Donghu of Eastern Hu, a tribal alliance whose homeland was in the Xianbei Mountains. During the Western Han period (206 B.C.E. to 24 C.E.), the Eastern Hu were defeated by the Xiongnu and then became differentiated into two groups: the Wuhuan to the south and the Xianbei further north (Chen Shou 1959; 30: 832). During the late first century C.E., as the Xiongnu were defeated by the Han Chinese and escaped to the west, the Xianbei tribes migrated into areas formally occupied by the Xiongnu (Fan Ye 1965; 90: 2991). Later, the dispersed Xianbei tribes were united by Tanshihui (Fan Ye 1965; 90: 2989) and gradually became a dominant power. Around 207 C.E., the Wuhuan tribes were compelled by Caocao's army to move to the central plain, vacating areas that were then occupied by the Xianbei (Fan Ye 1965; 90: 2984). The Xianbei then became the main minority tribe in northern China and a power strong enough to threaten the central plains of China. They eventually divided into two branches: the Eastern Xianbei and the Tuoba Xianbei. During the late third century to the end of the sixth century C.E. (i.e., Sixteen States and Northern-Southern dynasties period), the Xianbei moved into northern territories and were the first among northern ethnic tribes to establish states, including the four "Yan" States—Southern Liang, Wei, Qi, and Zhou—of the Northern dynasties period. They were in contact not only with nomadic tribes from the grasslands but also with the Han Chinese in the central plains. Finally, most of the Xianbei merged with the Han people.

The Xianbei had no writing of their own, and records of their early activity in Chinese historical documents are few. By the time they sought to record their own history in Chinese literature, it was relatively late. Archaeological materials of the Xianbei and related tribes are essential to learning about their activities and the role they played in Chinese history.

Archaeological Remains of the Xianbei

The discovery of archaeological remains of the Xianbei and related tribes is one of the most significant achievements since the establishment of the People's Republic of China. In 1980, Chinese archaeologists discovered inscriptions carved on a rock wall in the Gaxiandong cave in Daxinganling, Inner Mongolia (Mi Wenping 1981). According to these inscriptions, the cave was a "stone house" in which the ancestors of the Xianbei lived, and the area included the Xianbei Mountains of historical record.

To date, over forty cemeteries of the Xianbei, located in different areas, have been excavated in Inner Mongolia, Jilin, Liaoning, and Shaanxi Provinces (fig. 1). There are differing opinions about some cemeteries, but no habitation sites are known. The Xianbei remains are distributed over three areas: the Hulunbeier region, western Liaoning, and the Yinshan region. In general, the Xianbei graves of Western Liaoning region belonged to Eastern Xianbei tribes and have yielded ferrous artifacts, including weapons, tools, farming implements, and household utensils. Ferrous artifacts began to be used in the third century B.C.E.; their

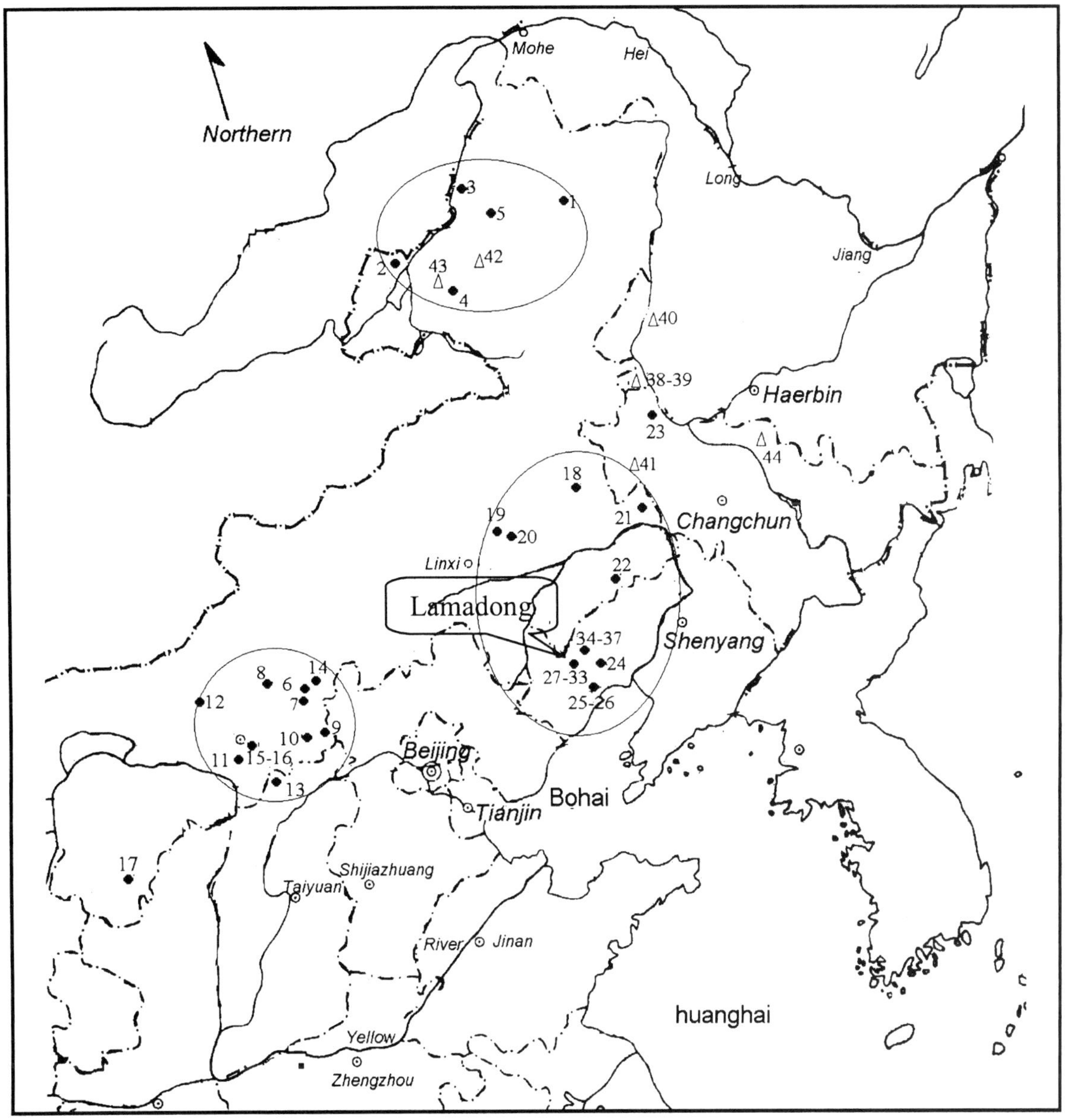

Figure 1. *Archaeological sites of Xianbei tribes excavated*

1 Gaxiandong
2 Zhalainuoer
3 Qika
4 Yihewula
5 Labudalin
6 Zhaojiafang
7 Sandaowan
8 Erlanhugou
9 Bagou village
10 Xiaheigou
11 Pitiaogou
12 Bailing Temple
13 Shanjiabao
14 Shangdu County
15 Meidai Village
16 Tianmiliang
17 Wushen County
18 Beimanitu
19 Nanyangjiayingzi
20 Susitai
21 Liujiazi
22 Xinshengtun
23 Tombs in Fishery of Daan county
24 Baoan Temple
25 Tomb of Ligui
26 Qianshan
27 Wangzifen Hill
28 Tiancaogou
29 Tombs of Fengche Duwei
30 Tombs in Brickyard of Shiertai
31 Sanhecheng
32 Tomb of Cuiyu
33 Yuantaizi
34 Fangshen village
35 Lamadong
36 Tombs of the coupe of Fengsufu
37 Cangliangjiao
38 Tombs in Brickyard of Tailai county
39 Tombs in Zhandou of Tailai county
40 Sanjianzi in Qiqihaer
41 Xinglongshan in Tongyu
42 Yimin station and Menggenchulu
43 Wangong
44 Laoheshen

use clearly increased over time and spread during the second century C.E. after moving westward. The Xianbei finds have provided much evidence about the society, economy, culture, and customs of the Xianbei tribes. This article discusses ferrous artifacts excavated from the tombs at Lamadong, Beipiao, in Liaoning Province.

Lamadong Graves

Tombs of the Xianbei tribes in Lamadong, Beipiao, Liaoning Province were excavated five times from 1992 to 1998 (Institute of Cultural Relics and Archaeology, Liaoning Province 2001). Because of the importance of Xianbei tribes in the history of China and of the artifacts unearthed from these tombs, the archaeological project was rewarded as one of the top ten archaeological finds of 1996.

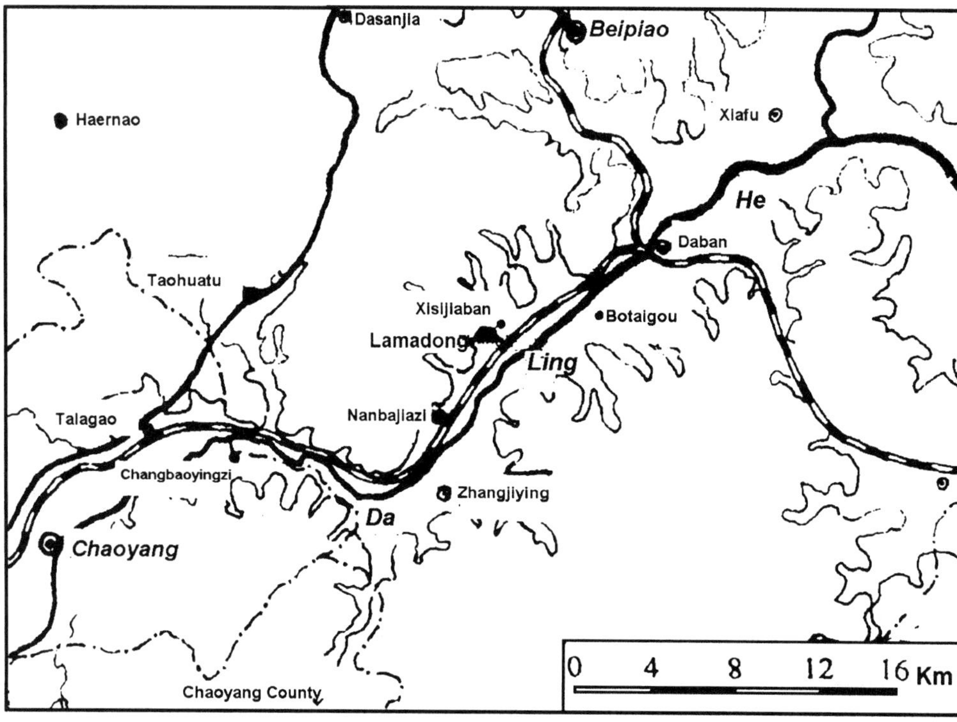

Figure 2. Map of the Lamadong graves and vicinity in Beipiao County

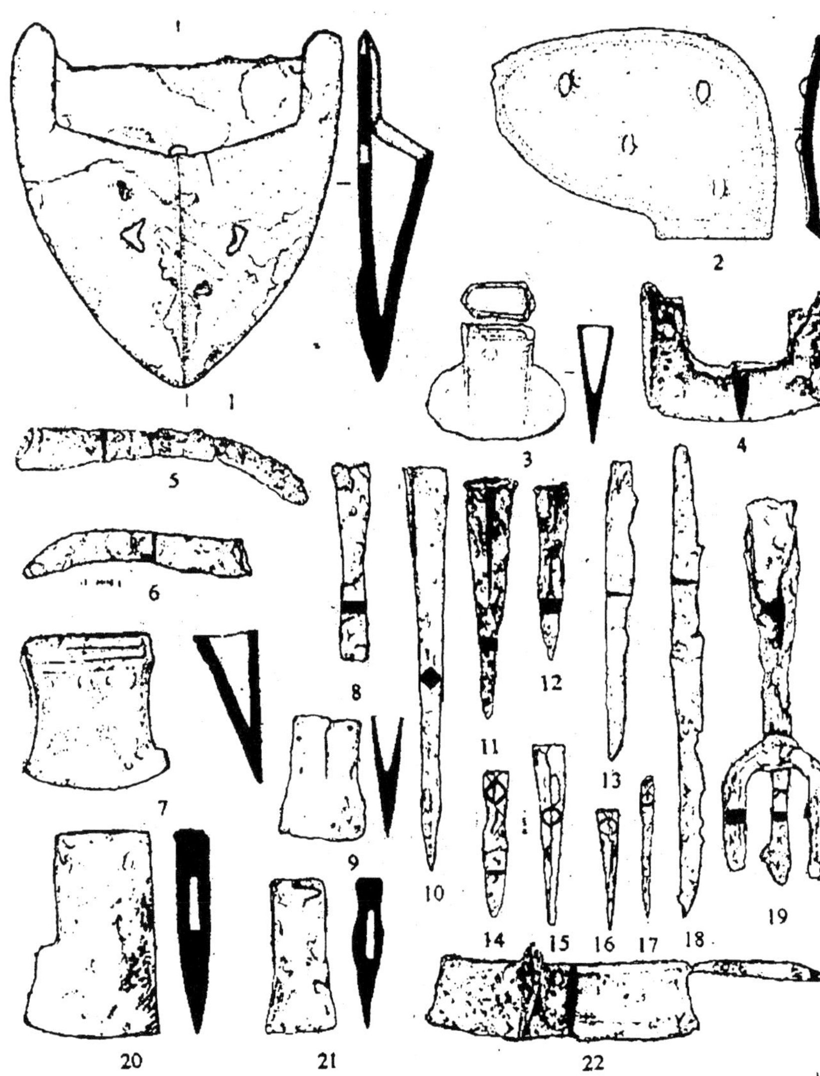

Figure 3. Ferrous tools and farming implements excavated from the Lamadong graves

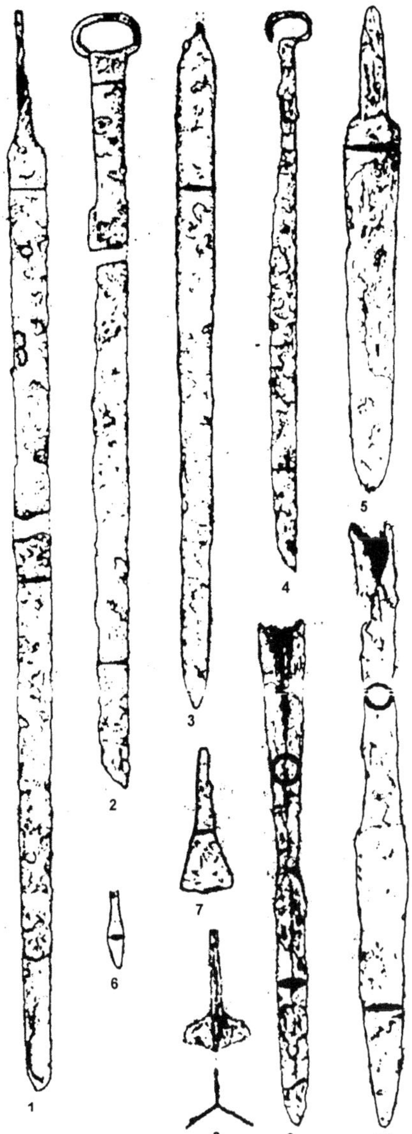

Figure 4. *Ferrous weapons excavated from the Lamadong graves*

The Lamadong graves are situated about 1.5 km north of Beipiao city and 1.2 km southeast along the Daling River (fig. 2), with a total area of 6,600 m^2. Three hundred and fifty-five tombs of the Xianbei tribes have been found and excavated. They were arranged in twelve dense lines, each line having thirty to over forty tombs, and the stratigraphic relationship was essentially undisturbed. The archaeological team of Liaoning Province determined from the burial customs, artifacts, and cultural context of the tombs that they belonged to the Xianbei people and were dated to the late third century to the mid-fourth century C.E.

There are some differences between Lamadong and other Xianbei tombs. Most of the excavated Lamadong tombs were uniform with rectangular pit graves, wooden coffins, and single supine burials oriented to the northeast. Most of the pottery was buried in the earth filling the pit, and a few animal bones were found, metatarsal bones of oxen being placed in the coffin emblematically. The important distinguishing feature is the discovery of buried iron weapons, including long or short swords, spears, arrowheads, ring-handle knives, and so forth (fig. 4), or farming implements, including ploughshares, moldboards, sickles, U-shaped implement caps, shovels, and so forth along with tools (fig. 3). There are over 2,400 ferrous artifacts from the Lamadong graves. Most of the artifacts in each tomb were used habitually during the person's lifetime and should be paid great attention. They show that the grave's occupant was both a soldier and a peasant and combined farming with warfare.

In order to study the manufacturing techniques of ferrous artifacts and the relationship between those in Xianbei and in the central plain, and the spread of advanced technology with its economic and social effects, thirty-two samples from twenty-four tombs and six kinds of typical artifacts were selected and examined by metallographic analysis. They are listed in table 1.

Most of the samples were selected from broken parts of artifacts; we used a 0.1 mm Mo wire-cutter machine to cut off a small piece, then fixed the cut piece back into the original object after the analysis. The samples were mounted, polished, and etched with 4% nital (4% concentrated nitric acid in ethyl alcohol). A summary of the main results follows.

Results and Preliminary Discussion

The objects are described in table 1, which includes the analysis number (four digits) and the tomb designation (prefix "M").

White Cast Iron

These artifacts include two ploughshares, one U-shaped implement cap, and one cauldron made by casting pig iron. The microstructure of ploughshare 7136 (tomb M5) is hypoeutectic with less than 4.3% carbon (fig. 5); that of ploughshare 7134 (M20:4) is hypereutectic with more than 4.3% carbon (fig. 6). A kind of cauldron (7114; M11:2) is seriously corroded but still retains the pig-iron structure (fig. 7).

The earliest white-iron fragments examined from Tianma-Qucun, Shaanxi Province, are dated to the eighth to seventh century B.C.E. (Han 2000). This innovation, the use of liquid iron saturated with carbon and melted around 1200°C, took place in China within only one to two centuries of the introduction of iron making. This may be compared with millennia taken elsewhere and was one of the great inventions of China.

We have recently examined the earliest known cast-iron ploughshare excavated from SiQi county, Houma, Shaanxi Province. It is made of white cast iron and is dated to the fourth to third century B.C.E. (Duan Hongmei 2001). Metallographic examination has also been performed on the earliest known U-shaped implement cap from a Warring States period, which was excavated in Changsha, Hunan Province, in 1957 (Hua Jueming, Yang Gen, and Liu Enzhu 1960).

Before the first century C.E., cast iron had begun to be used for tools and farm implements in the Xianbei tribal regions. Their widespread application introduced a green revolution in agriculture.

Table 1. *Samples and test results*

Style	Artifact type	Tomb	Sample number	Location of sample	Conclusion
Tool	Axe	M4:46	7112	Cutting edge	Decarburized steel from pig iron
Tool	Axe	M205:11	7131	Cutting edge	Decarburized steel from pig iron
Tool	Chisel	M186:16	7141	Tip of cutting edge	Steel, decarburized and forged
Tool	Chisel	M218:49	7140	Tip of cutting edge	Puddled steel, cold worked
Tool	Chisel	M368:8	7139	Tip of cutting edge	High- and low-carbon puddled steels forged together to give a high-carbon edge and low-carbon matrix
Tool	Drill	M370:10	7143	Tip of cutting edge	Two pieces of puddled steel forged together
Tool	Sickle	M4:7	7116	Cutting edge	Puddled steel, forged and quenched
Tool	Sickle	M118:5	7138	Cutting edge	Decarburized steel
Tool	Sickle	M129:5	7137	Cutting edge	Decarburized steel
Tool	Adze	M70:6	7130	Socket	Malleable cast iron
Farming implement	Shovel	M88:13	7132	Cutting edge	Two pieces of puddled steel forged together
Farming implement	Shovel	M155:2	7133	Cutting edge	Decarburized cast iron
Farming implement	U-shaped implement cap	M20:4	7134	Cutting edge	White cast iron
Farming implement	Ploughshare	M5	7136	Broken part	White cast iron
Farming implement	Ploughshare	M42:9	7135	Broken part	White cast iron
Weapon	Spear	M4:44	7107	Cutting edge	Decarburized steel from pig iron
Weapon	Spear	M216:4	7120	Cutting edge	Puddled steel, forged
Weapon	Spear	M218:48	7122	Cutting edge	Decarburized steel from pig iron
Weapon	Spear	M322:16	7121	Cutting edge	Steel made by infusion carburization
Weapon	Sword	M218:10	7118	Cutting edge	Puddled steel, forged
Weapon	Sword	M309:13	7119	Cutting edge	Puddled steel, forged and cold worked
Weapon	Sword	M374:15	7117	Cutting edge	Puddled steel, forged and surface hardened by carburizing
Weapon	Knife	M308:20	7124	Cutting edge	Puddled steel, forged
Weapon	Knife	M374:21	7123	Cutting edge	Puddled steel, forged and surface hardened by carburizing
Weapon	Arrow head	M33:11	7125	Tip of arrow head	Puddled steel, forged
Weapon	Arrow head	M379:13	7126	Tip of arrow head	Malleable cast iron
Other	Plate	M4:49	7101	Broken part	Puddled steel, forged
Other	Ring of prop stand	M6:16	7115	Broken part	Decarburized steel from pig iron
Other	Coffin nail	M9:28	7142	Tip of nail	Puddled steel, forged
Other	Corner cover for wooden coffin	M9:29	7103	Broken part	Puddled steel, forged
Other	Cauldron	M11:2	7114	Broken part	White cast iron
Other	Ring-handled object	M48:1	7144	Handle	Steel, decarburized from pig iron and forged

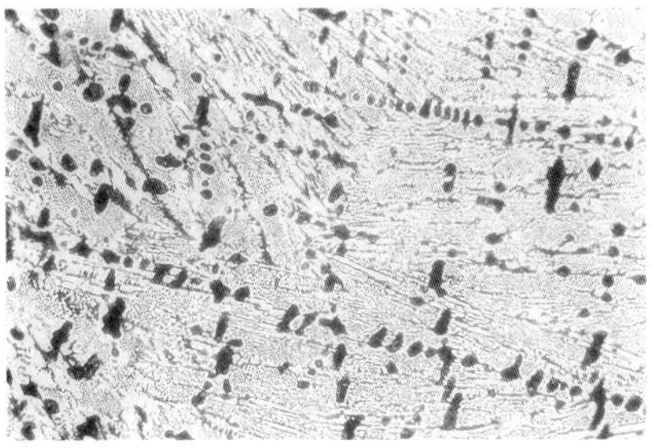

Figure 5. *7136 ploughshare (M5), hypoeutectic white iron. ×100*

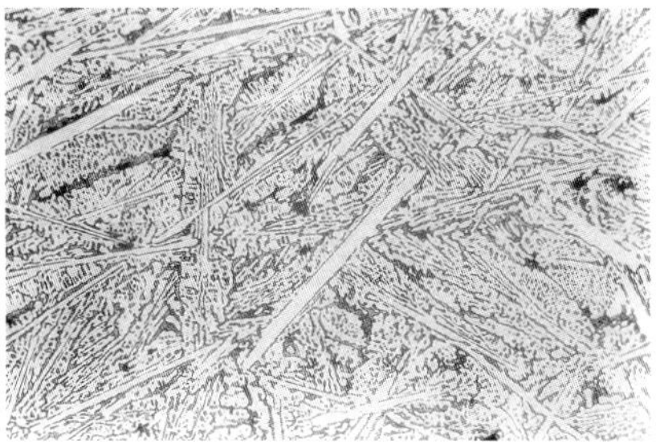

Figure 6. *7134 ploughshare (M20:4), hypereutectic white iron. ×100*

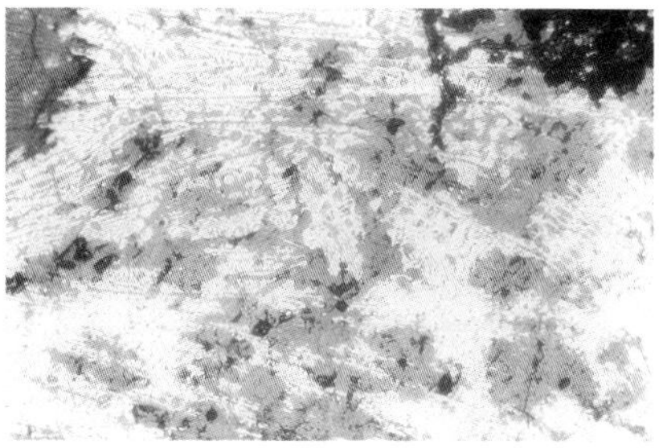

Figure 7. *7114 cauldron (M11:2), corroded pig iron structure. ×100*

Annealing Techniques

Disadvantages of white cast iron are its brittleness and hardness. These were overcome in the early fifth century B.C.E. in China by the invention of a malleabilizing process that made use of white cast iron (Li Zhong 1975). The atmosphere in an annealing furnace could be either oxidizing or reducing, this being the decisive factor for the product: decarburized iron, malleable iron, or steel, all made from cast iron by decarburization in the solid state.

Sample 7133 from a shovel (M155:2) was made of white cast iron with a ledeburite core and decarburized surface; here, prolonged annealing caused the decomposition of cementite into graphite nodules and created malleable iron with good toughness. The arrowhead (7126; M379:13) and the adze (7130; M70:6) were found to be of black malleable iron with some snowflake graphite embedded in a ferrite matrix (figs. 8 and 9). The microstructures are direct evidence that these artifacts have been made by an annealing process.

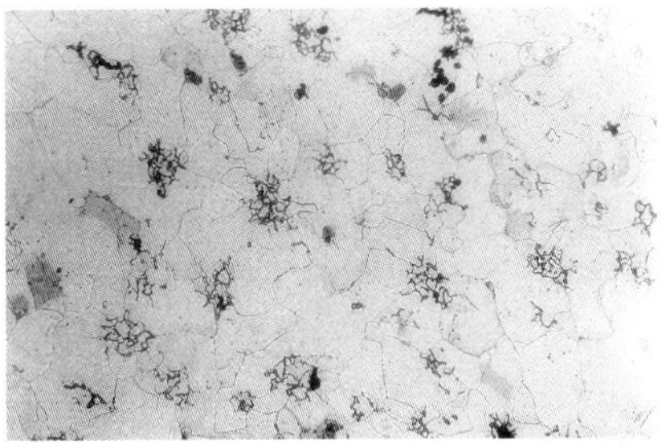

Figure 8. *7126 arrowhead (M379:13), black heart malleable iron, some snowflake graphite embedded in ferrite matrix. ×100*

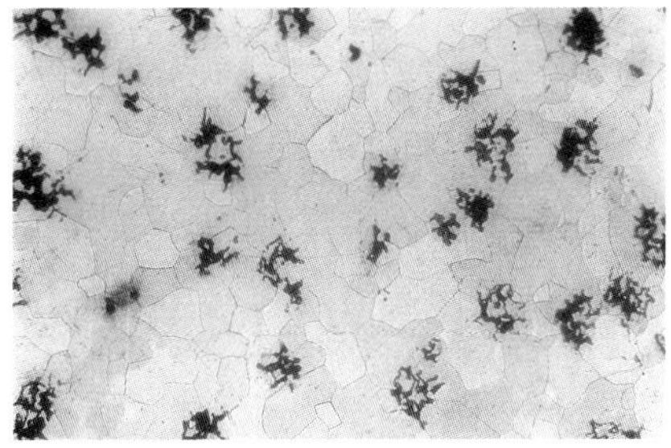

Figure 9. *7130 adze (M70:6), black heart malleable iron. ×100*

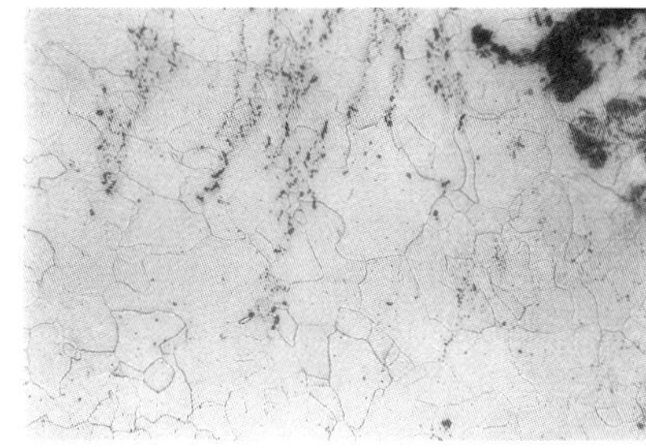

Figure 10. *7137 sickle (M129:5), decarburized steel from pig iron with ferrite grains. ×100*

Steel Making from Cast Iron by Decarburization in the Solid State

Examination of ferrous artifacts shows that a simple economic process of steel making was in practice by the fifth century B.C.E. Thin or narrow plates or tools were made by casting, followed by an oxidation anneal into steel containing spheroidal carbide without any formation of graphite. Many artifacts of this nature, such as a pickaxe in Dengfeng Henan, a ring-handle knife in Changzhi, Shanxi, and an arrowhead in Mancheng, have been excavated and examined. Though they were from different localities in the central plain, they possessed similar structures free from inclusions and with mold joins evident. Cast-iron plates (with mold materials) subjected to processing in the solid state and objects forged from the steel plates so obtained have been discovered in ancient iron-working sites such as Dengfeng, Guxing, Tieshengou, and Nanyang (Han 1998).

Nine ferrous objects unearthed from the Lamadong graves have also been made by this method. They consist of two axes, two sickles, two spears, a chisel, a ring-handled object, and a prop stand, all apparently made by casting. The metallographic structures of the cutting edges of the tools and weapons show different carbon contents. The sickle 7137 (M129:5) and the spear 7177 (M4:44) are both ferrite with different grain sizes (figs. 10 and 11); in the ferrite grains of spear 7107 (M4:44), the sample shows a ghosting structure that was identified as a phosphorus-rich region (fig. 12).[2] The axe 7131 (M205:11) is ferrite with a small amount of pearlite remaining at grain boundaries and a carbon content of 0.2% (fig. 13). The axe 7112 (M4:46) has fine pearlite and ferrite uniformly distributed and a carbon content of 0.4% (fig. 14). The spear 7122 (M218:48) is pearlite with net ferrite and a carbon content of 0.7% (fig. 15). These structures are uniform and the materials clean, with extremely few inclusions; some porosity from casting was seen.

The making of steel from cast iron by decarburization was a simple and economic process, making the product readily available with a high enough quality for its purposes. The plates could be transported and traded easily, and it was possible for the Xianbei people to use the materials to manufacture further products. The chisel 7141 (M186:16), ring of the prop stand 7115 (M6:16), and ring-handled object 7114 could be this kind of product. This has shown that an advanced steel-making technique was well developed and applied to production at the Lamadong site.

Puddled Steel

Puddling was a more advanced steel-making process that first appeared in the second century B.C.E. and was very popular by the second century C.E. In the puddling process, pig iron was first heated in air with stirring. When preferential oxidation of the carbon raised the temperature from the heat of combustion, the iron changed to a pasty state. As the temperature increased, the carbon content of austenite gradually fell, and oxidation of the silicon and manganese in the pig iron formed silicon inclusions along with iron oxide. If the stirring was continued in the semisolid state, the result was low-carbon wrought iron, but it was also possible to stop before decarburization was complete, obtaining medium- or high-carbon steel. The steels from puddled pig iron were stacked forged, probably with repeating hammering (Li Zhong 1975).

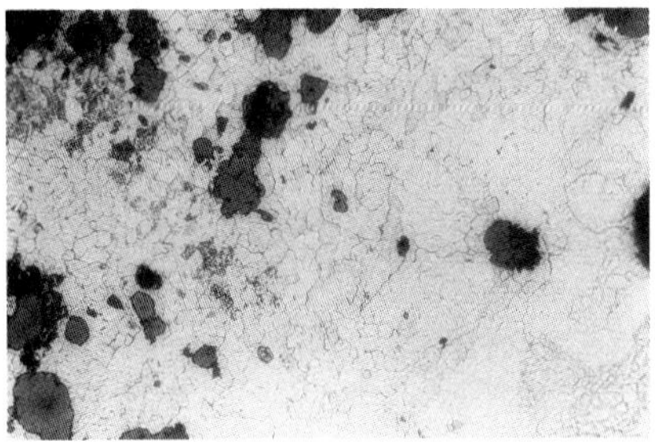

Figure 11. *7107 spear (M4:44), decarburized steel from pig iron with ferrite and different grain size. ×100*

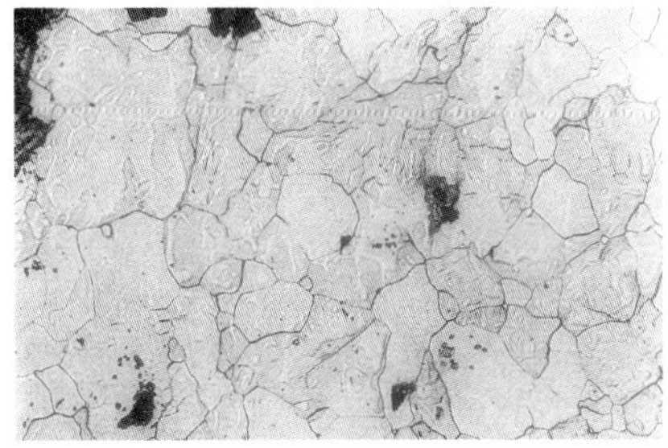

Figure 12. *7107 spear (M4:44), in the ferrite, grains exist ghosting structure and identified as phosphorus segregation. ×200*

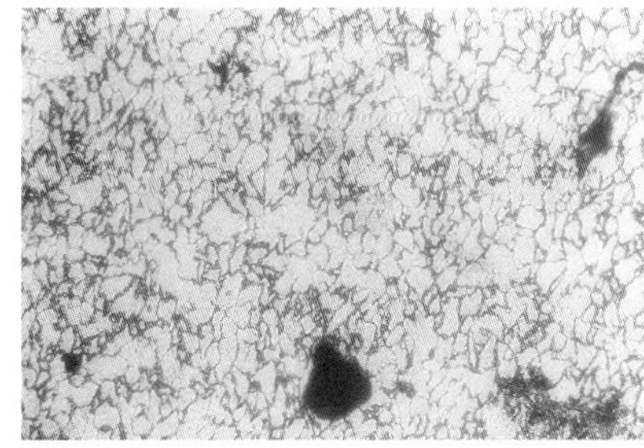

Figure 13. *7131 axe (M205:11), decarburized steel from pig iron with ferrite and pearlite remaining at grain boundaries, carbon content 0.2%. ×125*

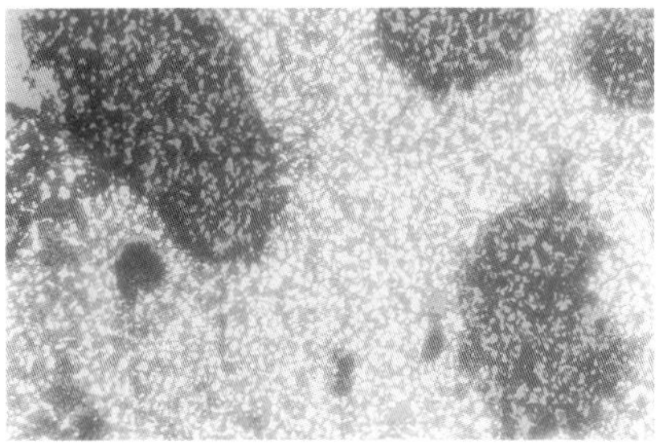

Figure 14. *7112 axe (M4:46), decarburized steel from pig iron with fine pearlite and ferrite uniformly distributed, and carbon content 0.4%. ×200*

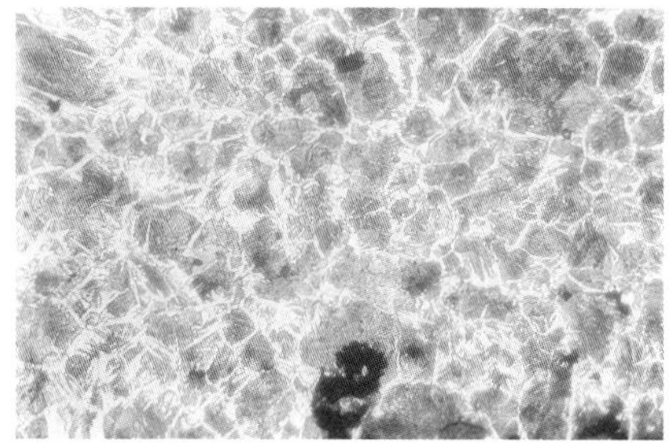

Figure 15. *7122 spear (M218:48), decarburized steel from pig iron with pearlite and net ferrite, carbon content 0.6%. ×125*

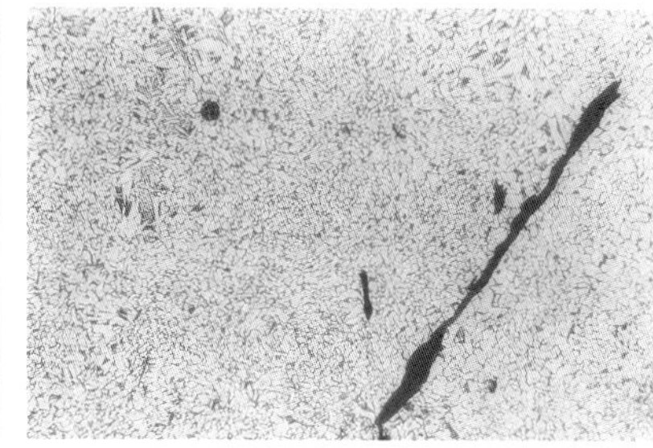

Figure 16. *7118 sword (M218:10), puddled steel forged, fine ferrite and pearlite, carbon content 0.2–0.3%, elongated inclusions. ×100*

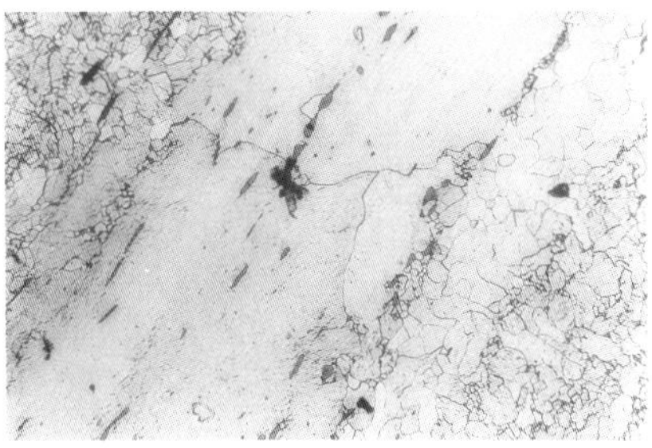

Figure 17. *7124 knife (M308:20), puddled steel forged, ferrite with different grain size and grains of pearlite with ferrite boundaries, carbon content 0.1–0.2%, banding structure and elongated inclusion. ×100*

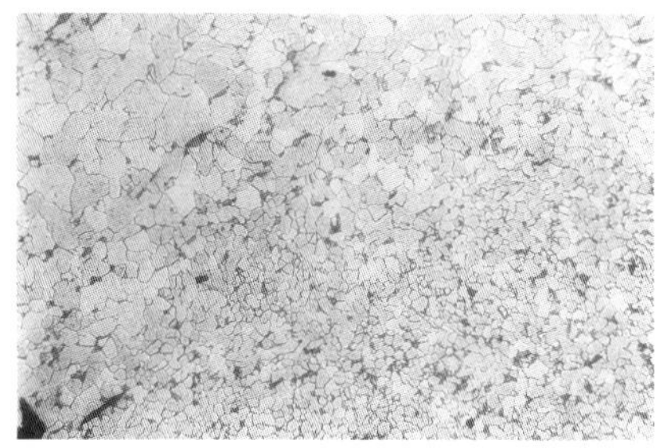

Figure 18. *7123 knife (M373:21), puddled steel forged, ferrite and pearlite, carbon content 0.1–0.2%, and elongated inclusion. ×100*

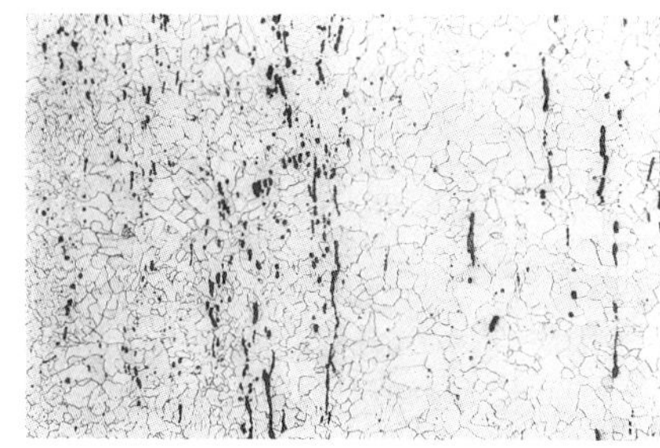

Figure 19. *7120 spear (M216:4), puddled steel forged, ferrite with different grain size and many elongated inclusions. ×100*

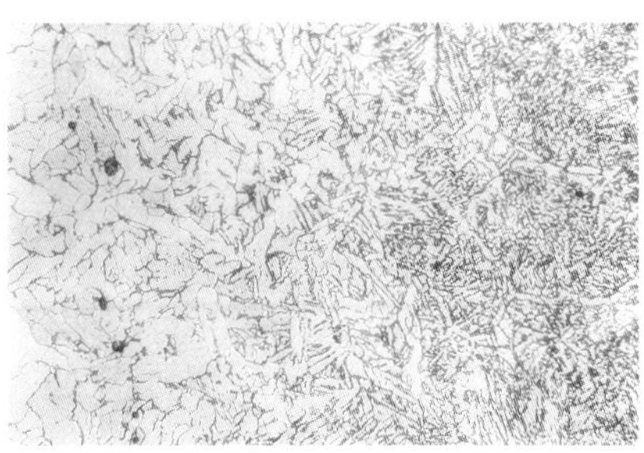

Figure 20. *7120 spear (M216:4), puddled steel forged with surface hardened by carbonizing and Widmannstatten structure, carbon content 0.3–0.4%. ×100*

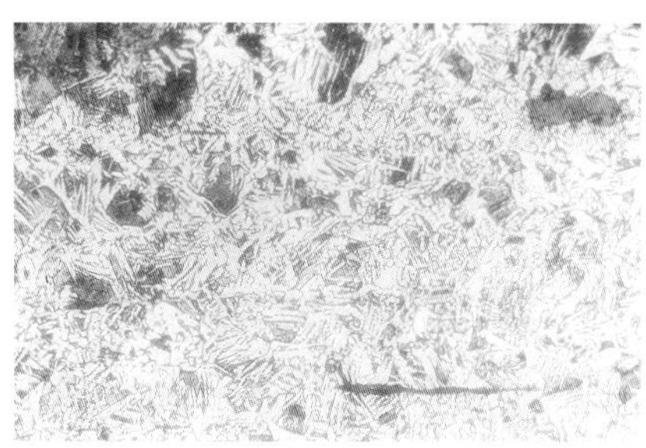

Figure 21. *7117 sword (M374:15), puddled steel forged and Widmannstatten structure, carbon content 0.3–0.6%. ×100*

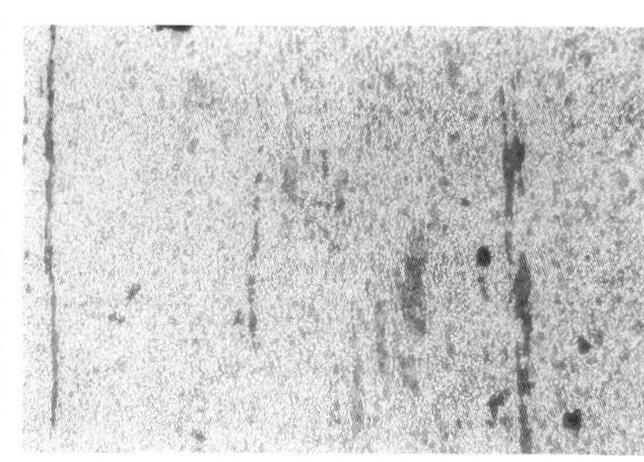

Figure 22. *7143 drill (M370:10), two pieces of puddled steel forged together with fine ferrite and pearlite, different carbon content 0.2–0.6%, many elongated inclusions. ×100*

Examination of fourteen artifacts from the Lamadong graves showed that they were made from steels of different carbon contents produced by puddling, followed by forging and folding until the desired product was achieved. In the steel matrix are a large number of inclusions elongated in the direction of working. The artifacts include seven weapons (one spear, three swords, two knives, and one arrowhead), four tools (a sickle, shovel, chisel, and drill), one plate, and one corner cover for a wooden coffin.

Metallographic samples from the cutting edges of sword 7118 (M218:10) (fig. 16), knife 7124 (M308:20) (fig. 17), and knife 7123 (M374:21) (fig. 18) show that they have a non-uniform structure composed of fine-grained ferrite and pearlite with different carbon contents and grain sizes.

As can be seen from figures 19 and 20, the samples from spear 7120 (M216:4), with a surface hardened by carbonizing, contain different carbon contents consisting of pearlite and ferrite with a tendency to a Widmannstatten structure. Sword 7117 (M374:15) is puddled steel whose lower carbon contents give a Widmannstatten structure (fig. 21). Numerous very thin silicate stringer inclusions are found. Since they were made from pig iron, which had been molten, the chemical composition of these inclusions, including aluminum, silicon, phosphorus, potassium, and manganese, was shown to be uniform distributed by scanning electron microscopy with energy dispersive x-ray analysis (SEM-EDXA).

Chisel 7140 (M218:49) and drill 7143 (M370:10) were found to be made by forge welding two pieces of steel or wrought iron with different carbon contents (fig. 22). Results from chisel 7139 (M368:8) showed a sandwich of one piece of steel in the center between two pieces of wrought iron on the outside; inclusions in the lower-carbon zones and the high-carbon zone have different elemental compositions (fig. 23).

Arrowhead 7125 (M33:11) shows a large inclusion believed to be a forge-welding inclusion (fig. 24). These can

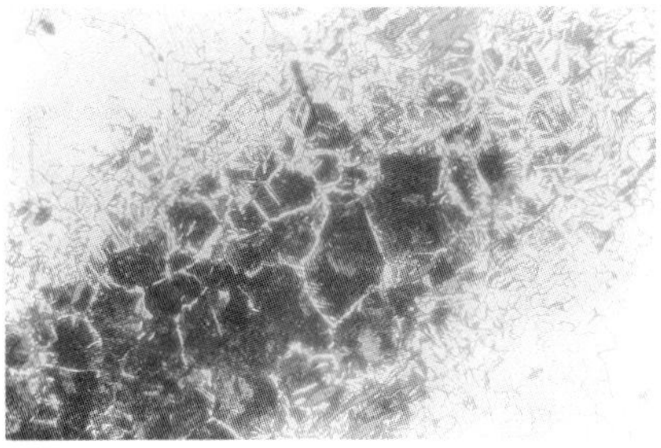

Figure 23. *7139 chisel (M368:8), two pieces of puddled steel sandwiched together, ferrite in the edge and pearlite in the center. ×100*

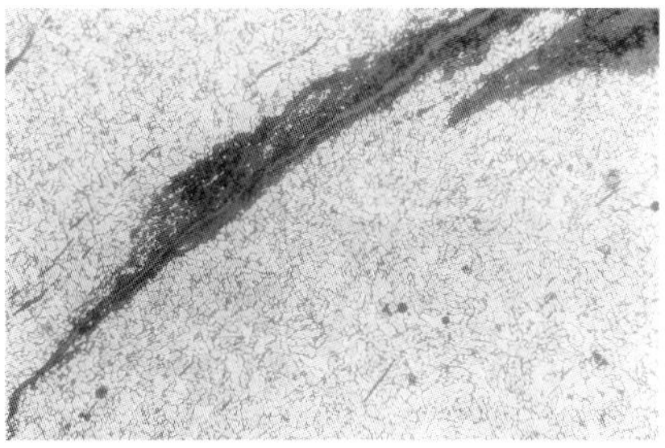

Figure 24. *7125 arrowhead (M33:11), two pieces of puddled steel forged together. The large inclusion is a forge-welding one, carbon content 0.2%. ×100*

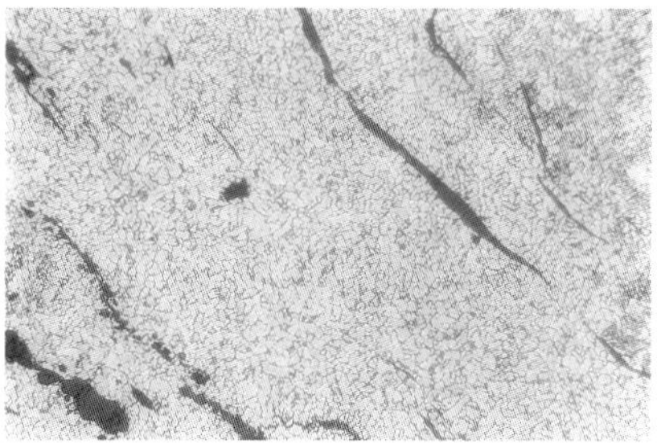

Figure 25. *7125 arrowhead (M33:11), fine ferrite and pearlite, carbon content 0.2% folding produced bands of inclusions. ×100*

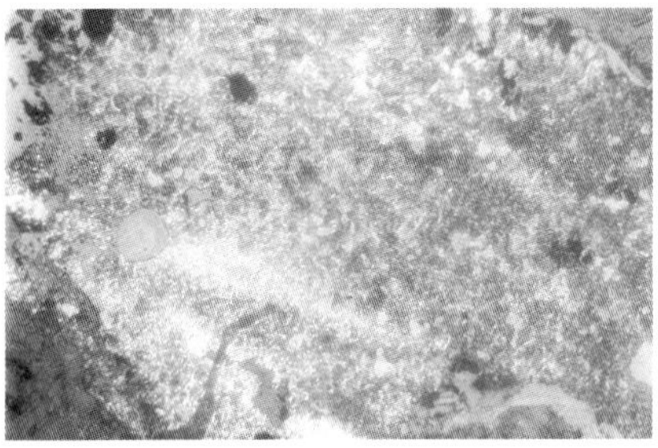

Figure 26. *7116 sickle (M4:7), quenched steel, martenstic structure. ×320*

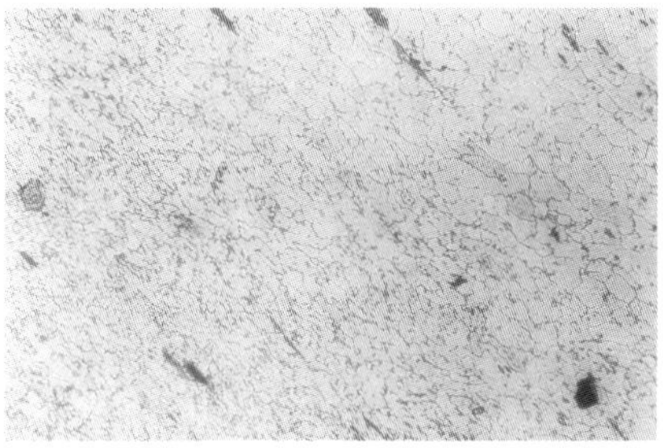

Figure 27. *7140 chisel (218:49), ferrite grains of an obvious elongation, carbon content 0.1–0.2%, and elongated silicate inclusions. ×100*

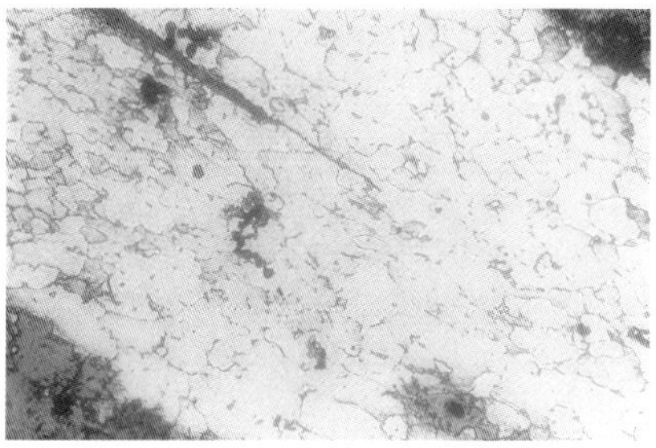

Figure 28. *7103 covered corner of wooden coffin (M9:29), elongation ferrite grains and pearlite with their boundaries, elongated inclusions. ×100*

usually be distinguished by their location, their much higher FeO content, and their production by stacking steels of different carbon contents and folding to produce the bent inclusions (fig. 25).

The structure of sickle 7116 (M4:7) shows that steel of lower carbon was used; it is partially martensitic and was quenched (fig. 26). In figures 27 and 28, the ferrite grains have an obvious elongation. This indicates that the last steps in the fabrication of chisel 7140 (M218:49) and corner cover 7103 for a wooden coffin (M9:29) were hardened by cold hammering.

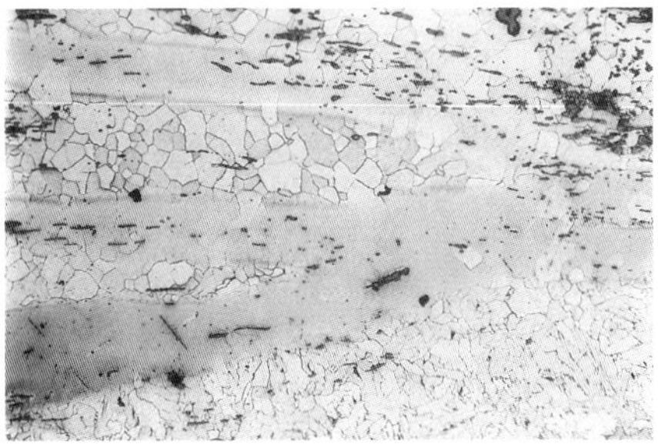

Figure 29. *7132 shovel (M88:13), ferrite uneven grain size and folded bend effect. The etched lighter area is rich in arsenic. ×50*

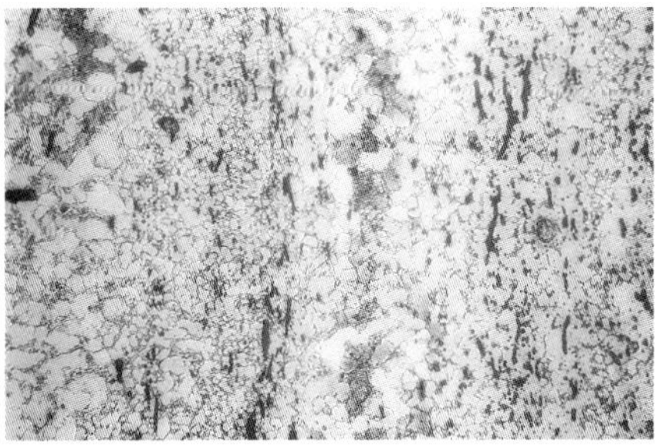

Figure 30. *7101 plate (M4:49), puddled steel forged, uneven ferrite grain size, and pearlite with their boundaries, elongated inclusions, carbon content 0.2%. ×100*

It can be seen from figure 29 that shovel 7132 (M88:13) was made of wrought iron, which was first surface carburized and folded, then piled and forge welded. This is clearly illustrated by the uneven grain sizes, distribution of carbon content, and bending effects seen throughout the cross-section. There are distinct bright bands, and judging from the alternation of light and dark, this region appears to be composed of fourteen layers. The region that etched lighter was identified as an arsenic-rich layer with a microhardness higher than the dark region. These results show that the structure was a result of arsenic segregation similar to phosphorus.

Steel plate 7101 (M4:49) was made by puddling pig iron containing large quantities of inclusions having non-uniform structures with differing carbon contents and grain sizes (fig. 30). The possibility exists that the raw material for puddling came from the central plains, as this one piece of puddled-steel plate was found without a corresponding smelting site or iron-working foundry.

Conclusions

Pig iron, decarburized cast iron, malleable iron, steel made from cast iron by decarburization in the solid state, and puddled steel were found among these thirty-two ferrous artifacts that included farm implements, weapons, tools, and plates. The results indicated that most of these artifacts buried in the owner's tomb were made with a relatively high level of skill. An important conclusion is that the malleabilizing technique adapted to the making of farm implements and weapons created a very significant condition for the rapid development and expansion of economic and military superiority of the Xianbei people.

The beginning of iron manufacture and use by the Xianbei tribes was later than in central China, and their techniques and styles were very influential. It is clear that exchanges of iron and steel techniques between the Chinese and others were happening as communication developed, at latest after the Warring States period and certainly during the Han dynasty. Xianbei people, who had already learned the metallurgical techniques from the Hans, were defeated and moved westward in the second century C.E. Xianbei tribes played important roles in the spreading of the advanced iron technology of central China to neighboring people through wars, migration, gifts, robbery, and so forth, by way of Gaogoli or through the Korean Peninsula and to Japan. A multidisciplinary study of the Lamadong graves should furnish more new evidence and information.

Cooperation in archaeometallurgical research is essential if mutual influences and their impact on the historical progression of ferrous technology are to be exhaustively understood. We hope that this paper will show the little we could do to help this project.

Acknowledgments

We wish to thank the Institute of Cultural Relics and Archaeology, Liaoning Province, for inviting us to study these artifacts excavated from the Lamadong graves, for allowing us to cooperate with their archaeologists Xin Zhanshan, Zhang Keju, Tian Likun, Wan Xin, et al., and for supporting several periods of travel and research.

We are grateful to our colleagues at the Institute of History of Metallurgy and Materials, University of Science and Technology Beijing: Sun Shuyun, Li Xuihui, and Li Yanxiang, who gave us support and help, and most especially to Professor T. Ko, who instructed us during the project, and to Liu Jianhua, Yao Jianfang, and Qiu Baoqin, who helped with evaluation of the metallographic and composition analyses of the ferrous artifacts.

Notes

1. The content of this paper forms part of the Ph.D. dissertation of Dr. Chen Jianli. The project was in cooperation with the Institute of Cultural Relics and Archaeology, Liaoning Province.
2. Over forty ferrous artifacts unearthed with ghosting structures and banding structure caused by phosphorus segregation have been studied systematically using optical microscopy, scanning electron microscopy (SEM), TEM, microhardness, and other methods. The results will be reported elsewhere.

References

Chen Shou. Ca. 233–97 C.E. (reprinted 1959). *Sanguo zhi* (Record of the Three Kingdoms period). Reprinted, Beijing: Zhonghua Shuju. 832.

Duan Hongmei. 2001. Survey and study of ferrous artifacts of the Warring States period unearthed from Shanxi. Ph.D. diss., Institute of the History of Metallurgy and Materials, University of Science and Technology, Beijing.

Fan Ye. Ca. 398–445 C.E. (reprinted 1965). *Hou Han shu* (History of the Eastern Han dynasty). Reprinted, Beijing: Zhonghua shuju. 2984, 2989, 2991.

Han Rubin. 1998. Iron and steel making and its features in ancient China. *Bulletin of the Metals Museum* 30:23–37.

——. 2000. *The examination of ferrous artifacts excavated from Tianma Qucun, 1980–1989.* Vol. 3. Beijing: Science Press. 1178–80, plates 7 and 8.

Hua Jueming, Yang Gen, and Liu Enzhu. 1960. Preliminary report on the metallographic examination of some iron artifacts of the Warring States and Han periods. *Kaogu Xuebao* 1:73–88.

Institute of Cultural Relics and Archaeology, Liaoning Province. 2003. Annual excavation report of the Lamadong graves, Beipiao City, Liaoning, for 1998. *Kaogu Xuebao* 4: in press.

Li Zhong. 1975. The development of iron and steel technology in ancient China. *Kaogu Xuebao* 2:1–22.

Mi Wenping. 1981. A preliminary study of the discovery of the stone caves of Xianbei nationality. *Wenwu* 2:1–7.

Metalworking Traditions

10 A Study of Later Chinese Metalwork

Michael Cowell, Susan La Niece, and Jessica Rawson

ABSTRACT The construction, composition, and decoration of some seventy items of later Chinese religious figurines and vessels in the British Museum collections were investigated. All were made by lost-wax casting. Brass, or zinc-containing alloys were more common from the fifteenth to sixteenth century. Trace elements in the brass (e.g., high cadmium content) may identify artifacts of late date. Mercury, as opposed to leaf, gilding was more frequently used on artifacts with low lead contents. The range of decorative pigments was similar to that on related Himalayan religious figurines. A cobalt blue pigment on a fourteenth-century figurine is similar to material imported into China prior to the mid-fifteenth century and widely used to decorate porcelain.

Introduction

This project was inspired by an earlier technical study of the Chinese copper alloy *cash* coinage (Bowman, Cowell, and Cribb 1989). The coins revealed interesting developments in the application of brass in China during the sixteenth century and later, which were more clearly defined by analyzing such closely datable artifacts. Other metalwork of this period, mainly religious items such as figurines and vessels, were examined for comparison with these data but also to add further knowledge on the construction and decoration of these items.

There have been a number of technical studies that include at least some Chinese religious metalwork (Beguin and Liszak-Hours 1982; Craddock 1981; Jett and Douglas 1992; Kerr 1990; Nieuwenhuysen and Adams 1984; Sayre, Jett, and Joel 1992; Werner 1972). Although not specifically covering Chinese figurines, the intensive study by Reedy (1997) provides valuable comparative material. Some of these surveys have examined relatively large numbers of items, but the later periods (i.e., Ming dynasty and later) have not been well covered, and it is this period that is strongly represented in the British Museum's collections.

Historical Background

China has had two very different forms of bronze casting. The first is the well-known ancient tradition, originating in the Shang period (sixteenth to eleventh century B.C.E.) and continuing to the Han dynasty (206 B.C.E.–220 C.E.), and used for casting ritual vessels. These containers were made for offering food and wine ceremonially to the ancestors. They were cast by what is known as the piece-mold system and carried, in some instances, extraordinarily fine decoration. During the Han dynasty, bronze ritual vessels seemed to have fallen into decline and seem to have eventually gone out of use. Bronzes of various categories continue to be made—particularly incense burners, mirrors, and belt hooks—but the vessels were ever less important. Offerings of food and drink to ancestors seem to have been made in everyday utensils, mainly of lacquer or ceramic. This change may have made a major contribution to the decline of the early bronze-casting tradition. Economic difficulties may also have assisted this decline.

A completely new bronze-casting tradition was initiated with the introduction of Buddhism, from no later than the fourth century C.E. With this religion, brought into China from Central Asia and places further west, came an alien tradition of bronze casting, which we call lost-wax casting. The Chinese had already used lost-wax casting, probably introduced from Central Asia in the sixth century B.C.E., but only on a limited scale. Now, lost-wax casting became the norm, and all of the later bronzes discussed here were made by this method, although piece molding continued for major iron castings.

The first types of bronzes to be made in this way were Buddhist figures for worship. In addition, figures for other religions, particularly Daoism, came to be made in this way. A limited number of bronze vessels continued to be made, and some of these may have been made by lost-wax casting, although these have not yet been examined.

A new production of bronze vessels came about during the tenth to twelfth century as the Chinese looked back to their ancient tradition. They started, for political and cultural reasons, to collect ancient bronzes, and these were imitated by contemporary bronze casters for use at court and by members of the elite. Although the bronzes copied had originally been vessels for wine and food, they were now deployed for different purposes: for altar vessels to hold incense or flowers. This major change came about because, as indicated above, the traditional uses of the ancient forms had ceased during the Han period, when vessels in other materials took their place. Thus, there was very little call for food containers in the shape of ancient *ding* or *gui;* these were now used for incense. Similarly, the wine vessels, *hu* and *gu,* were now used for flowers.

These revived categories of bronzes were made also by lost-wax casting. They reproduced ancient vessel categories but with some distortions of shape and decoration. These changes came about, in large measure, because craftsmen worked from wood-block illustrations in catalogues and manuals. Large numbers were cast and used in many parts of the country from the eleventh century, at least, to the twentieth. Copies of these revived bronze forms were also made in a wide variety of ceramics, most especially porcelain decorated in underglaze blue and green-glazed celadon wares. This tradition has been only summarily studied. A few articles have described individual pieces or groups of pieces, and some catalogues of exhibitions have been published.

This paper (see also Rawson 2001) provides a background to this industry by examining some technical aspects of the bronzes collected over the last three decades by the British Museum. The seventy bronzes examined include both altar vessels and figures. Because the evidence of this bronze production has not been systematically recorded in China itself, other metal objects have been included in the study, especially coinage.

Previous Work on Chinese Coinage

The cash coinage is an important reference material for comparison with other cast Chinese metalwork over the period from the second century B.C.E. to the nineteenth century C.E. Analysis of examples from the British Museum collections are reported by Bowman, Cowell, and Cribb (1989) and Cowell et al. (1993). They are typical examples of cast metalwork for the periods investigated, and the results showed that brass or other zinc-containing alloys were not used for coinage in China until the sixteenth century; prior to that, leaded bronze was employed. A complete change of alloy occurred during the first quarter of the sixteenth century, in the Hongzhi period (1503–5), from leaded bronze to, initially, leaded brass, and then to a purer brass. Judging by the amounts of zinc in the brass coins, but also considering the amounts of other alloying elements, it was suggested that the coinage alloy was produced by speltering (i.e., adding zinc metal to copper) rather than by the older cementation or calamine process; from the very high zinc contents (over 30%), this seems to have occurred at the latest by the end of the sixteenth century. Zhou and Fan (1993), Zhou et al. (1996), and Zhou (2001), however, have maintained that while the manufacture of metallic zinc in China may have commenced in the mid-sixteenth century, it was not used for brass manufacture until the early seventeenth century, as recorded in the *T'ien-Kung K'ai-Wu* (Sung 1966). Other work on the coinage dealt with determining the chronology of some seventeenth-century coins by comparing alloy changes with documentary evidence, and the trace metal composition, which showed changes in the metal supply (Cowell and Wang 1998).

Clearly, the coin data cannot provide a specific framework for the composition of other metalwork, since the former were produced to officially prescribed standards of composition, unlike the latter. Nevertheless, there are general points, such as alloy trends, the use of brass, and, possibly, distinctive trace element patterns, that could be usefully compared with other classes of metalwork. The coinage provides a chronological indication of the available metal stock.

Analytical Techniques

All of the artifacts were analyzed quantitatively, and, where appropriate, multiple analyses were made of individual components. The techniques used for metal analysis were x-ray fluorescence (XRF), atomic absorption spectrophotometry (AAS), and inductively coupled plasma emission spectrometry (ICPAES). In addition, x-ray diffraction (XRD) was used to identify decorative minerals and pigments. For fully quantitative analysis, either AAS or ICPAES was used with samples of typically 10–20 mg. Details of the sample preparation and the operation of AAS for copper-based alloys are given by Hughes, Cowell, and Craddock (1976), and the ICPAES is similarly described by Hook (1998). The complete results are shown in table 1. The precision and accuracy of both AAS and ICPAES are typically ±1–2% for major elements (e.g., copper, tin, lead, and zinc) and about ±5–20% for minor and trace elements depending on concentration. XRF (see Cowell 1998) was only used to provide approximate analyses for artifacts and components.

Construction

Most of the figurines were radiographed to investigate construction details. This, coupled with visual examinations, confirms that all seem to have been manufactured by lost-wax casting with clay cores. In a few cases, part of the core remains, but usually this only applies to the head or appendages. The quality of the castings range from very good, with almost no flaws, to very poor, where there are numerous voids due to incomplete filling of the mold, probably through premature solidification. Where flaws occurred they have usually been repaired by working in a strip of metal or casting-in. Some of the repairs are very similar to the body and are probably contemporary with manufacture of the artifact. Dissimilar repairs may be of later date, but a different alloy may been chosen for easier casting or working.

Armatures are often used to support the clay cores in lost-wax casting but, in the examples examined here, only one figurine (1973,7-26,81) showed evidence of this, between the body and head. However, this is more likely a repair, since the head had become detached from the body, and thus we must conclude that the use of armatures was not common in Chinese casting practice. There are several examples of the use of chaplets as a support to maintain the separation of the clay core and outer mold walls during casting. These become incorporated into the metal wall of the figurine. It is important that chaplets do not melt during casting, to maintain the relationship of the core and mold. Most are of iron that will not

Table 1. Quantitative analyses by atomic absorption (AAS) and inductively coupled plasma atomic emission spectrometry (ICPAES)

BMRL	Reg. No.	Description	Part	Period	Date	Cu	Sn	Pb	Zn	Fe	Ni	As	Sb	Bi	Co	Mn	Cd	Ag	Au	Total	Gilding
29678W	1953,7-13,4	Figure, Bodhisattva manjusri	Body	?	?	85.0	0.60	1.80	11.9	0.49	0.030	< 0.03	0.05	0.06	< 0.002	0.0005	< 0.0008	0.080	< 0.003	100.0	Hg
51705Q	1992,7-11,1	Figure, seated Daoist		Liao	1032, inscribed	76.1	9.97	15.9	0.46	0.29	0.070	0.28	0.16	0.01	0.029	< 0.0005	< 0.002	0.132	0.015	103.4	
31511Y	1959,7-13,1	Figure, Guanyin	Body	Liao	11–12thC	92.3	4.56	1.81	0.022	0.22	0.141	0.04	< 0.02	0.02	0.024	0.0004	< 0.0004	0.189	< 0.006	99.3	Hg
31510P	1947,7-12,392	Figure, Guanyin	Body	Jin	12–13thC	90.2	< 0.23	0.65	8.1	0.20	0.018	0.05	0.02	< 0.01	0.002	0.0043	0.0007	0.020	< 0.007	99.3	Hg
5752-3-W	1994,1-29,1	Incense burner, tripod	Body	Jin	12–13thC	68.8	7.20	23.6	0.018	0.13	0.037	0.27	0.13	0.09	0.008	0.0006	< 0.001	0.067	< 0.001	100.3	
33552Q	1986,7-13,1	Vessel, altar-vase Hu shape	Body	Song	12–13thC	69.4	6.43	22.9	0.009	0.05	0.035	0.20	0.14	0.10	0.008	0.0004	< 0.0004	0.065	< 0.006	99.3	
29677Y	1986,10-10,1	Figure, Bodhisattva	Body	Yuan	13–14thC	69.8	7.40	21.5	0.014	0.10	0.042	0.21	0.10	0.08	0.042	< 0.001	< 0.0009	0.076	< 0.003	99.4	
31506S	1972,3-1,1	Figure, Dharmapala standing	Base (peg)	Yuan	13–14thC	96.1	< 0.13	0.07	0.003	0.90	0.098	0.67	0.01	0.01	0.015	0.0008	< 0.0003	0.038	0.009	97.9	Hg
31506S	1972,3-1,1	Figure, Dharmapala standing	Middle skull	Yuan	13–14thC	96.4	< 0.28	0.08	0.004	0.97	0.106	0.75	0.03	0.02	0.014	0.0022	< 0.0006	0.042	0.008	98.4	Hg
31506S	1972,3-1,1	Figure, Dharmapala standing	Body (back peg)	Yuan	13–14thC	95.4	< 0.08	0.02	0.002	0.72	0.083	2.23	0.02	0.00	0.019	0.0003	< 0.0003	0.039	0.002	98.5	Hg
42539R	1991,2-27,1	Vessel, altar vase Hu		Yuan	13–14thC	64.0	7.04	26.4	0.027	0.16	0.035	0.24	0.16	0.09	0.009	< 0.0008	0.0003	0.060	< 0.004	98.1	
40743u	1991,7-19,1	Figure, seated Bodhisattva	Body	Yuan	1339, inscribed	71.1	13.40	13.2	< 0.005	0.10	0.009	0.23	0.14	0.06	< 0.003	< 0.002	0.0000	0.056	< 0.002	98.3	
31501R	1973,7-26,81	Figure, Buddha standing	Body	Ming	1396, inscribed	85.1	4.77	6.06	0.61	0.20	0.043	0.25	0.13	0.06	0.003	0.0005	< 0.0003	0.085	0.007	97.3	Hg
51704S	1993,4-5,1	Vessel, altar vase	Base ring	Song	13thC	67.0	7.89	23.7	0.060	0.18	0.025	0.23	0.14	0.07	0.010	0.0009	< 0.0016	0.058	< 0.004	99.4	
29679U	1908,4-20,4	Figure, Buddha	Body	Ming	1403-24, inscribed	85.0	0.23	1.40	12.7	0.28	0.032	0.11	0.08	0.05	< 0.002	0.0006	< 0.0009	0.100	0.006	100.0	Hg
51706Z	1994,1-29,2	Dish		Yaun-Ming	14–15thC	84.3	7.41	8.02	1.2	0.19	0.230	0.41	0.30	0.03	0.012	< 0.0005	< 0.0013	0.078	0.004	102.1	
5752-2-R	1993,10-5,1	Dish	Body	Yuan-Ming	14–15thC	75.9	6.70	16.5	0.033	0.54	0.071	0.44	0.59	0.06	0.006	< 0.0007	< 0.002	0.066	< 0.002	100.9	
51703U	1981,6-27,1	Incense burner		Yuan-Ming	14–15thC	73.5	4.85	12.9	6.8	0.27	0.115	0.38	0.39	0.02	0.008	< 0.0004	< 0.0011	0.059	< 0.003	99.3	
18586T	1883,10-20,8	Perfume still, archaistic	Foot	Yuan-Ming	14–15thC	81.0	5.50	10.6	< 0.011	0.03	0.128	0.11	0.20	0.02	0.024	< 0.001	< 0.001	0.112	0.007	97.7	
18586T	1883,10-20,8	Perfume still, archaistic	Bell end of tube	Yuan-Ming	14–15thC	82.8	7.06	6.62	< 0.017	0.83	0.214	0.15	0.18	0.03	0.340	< 0.002	< 0.0016	0.109	< 0.008	98.3	
18586T	1883,10-20,8	Perfume still, archaistic	Body (from base)	Yuan-Ming	14–15thC	75.7	6.73	12.1	< 0.016	0.03	0.132	0.07	0.24	0.03	0.036	< 0.002	< 0.0014	0.409	0.010	95.5	
18586T	1883,10-20,8	Perfume still, archaistic	Straight tube	Yuan-Ming	14–15thC	82.1	7.20	8.20	0.000	0.84	0.210	0.24	0.16	0.00	0.330	0.0000	0.0000	0.108	0.006	99.4	
33551S	1988,4-16,1	Vessel, altar-vase	Body	Yuan?-Ming	14–15thC	76.1	6.54	13.3	0.32	0.06	0.076	0.80	0.42	0.08	0.004	0.0008	< 0.0003	0.090	< 0.004	97.7	
33551S	1988,4-16,1	Vessel, altar-vase	Ring on handle	Yuan?-Ming	14–15thC	79.5	6.65	12.9	0.35	0.20	0.082	0.78	0.42	0.09	0.006	0.0012	< 0.0004	0.092	< 0.006	101.1	
51609W	1993,10-28,1	Vessel, arrow vase		Ming?	14–15thC	70.0	3.00	26.0	0.85	0.43	0.340	0.88	0.35	0.03	0.002	< 0.0003	< 0.0011	0.120	0.006	102.0	
38321R	1936,11-18,59	Vessel, vase Hu, archaistic		Yuan-Ming	14–16thC	73.2	6.40	16.0	4.10	0.27	0.084	0.28	0.27	0.08	0.038	< 0.001	0.0002	0.107	0.040	100.8	
33556T	1895,12-20,13	Censer	Body	Ming	14–17thC	74.0	1.50	23.2	0.28	0.26	0.029	0.33	0.42	0.03	0.002	< 0.0004	< 0.0004	0.031	< 0.005	100.1	
31507Q	1893,3-20,164	Figure	Body	Ming	14–17thC	78.4	4.65	8.39	3.9	0.58	0.101	0.35	0.44	0.05	0.007	0.0004	< 0.0004	0.069	0.005	97.0	Hg
31509X	1952,12-19,17	Figure, female	Body	Ming	14–17thC	62.1	6.97	19.4	8.7	0.16	0.071	0.53	0.27	0.06	0.010	< 0.0003	0.0016	0.082	0.005	98.4	
31509X	1952,12-19,17	Figure, female	Base stand	Ming	14–17thC	67.9	7.05	13.6	7.8	0.15	0.071	0.47	0.29	0.07	0.009	< 0.0005	0.0002	0.113	0.012	97.6	
33561Z	1883,10-20,13	Incense burner	Body	Ming	14–17thC	67.9	0.10	2.83	25.9	0.29	0.121	0.05	< 0.01	0.01	0.009	0.0002	0.0037	0.022	< 0.0037	97.2	Hg
38319Q	1973,7-26,83	Vessel, vase Hu, archaistic	Body	Ming	14–17thC	93.7	0.61	1.86	0.11	2.52	0.043	0.74	0.10	0.02	0.007	< 0.001	0.0001	0.108	0.028	99.8	
33558P	1894,1-8,21	Incense burner	Body (foot)	Ming-Qing	14–20thC	80.5	< 0.16	2.19	17.2	0.26	0.127	0.21	0.04	0.02	0.006	< 0.0004	0.0043	0.008	< 0.006	100.5	

BMRL	Reg. No.	Description	Part	Period	Date	Cu	Sn	Pb	Zn	Fe	Ni	As	Sb	Bi	Co	Mn	Cd	Ag	Au	Total	Gilding
31512W	1986,8-5,1	Figure, bearded man	Body	Ming	1436, inscribed	72.9	0.22	2.78	22.0	0.10	0.084	0.53	0.14	0.01	0.016	0.0012	0.0009	0.016	< 0.0061	98.8	
42395S	1992,2-1,2	Figure, seated Buddha		Ming	1459, inscribed	62.4	6.55	28.3	< 0.01	0.09	0.038	0.22	0.17	0.10	0.009	< 0.0007	0.0002	0.063	< 0.004	97.9	
31500T	1942,4-17,1	Figure, Buddha standing	Body	Ming	1467, inscribed	71.1	0.73	6.56	18.0	0.26	0.019	0.12	0.07	0.02	0.002	0.0003	0.0002	0.040	< 0.005	96.9	Hg
31499X	1988,8-8,1	Figure, lohan tiger tamer	Body	Ming	1481, inscribed	79.1	3.50	14.5	0.25	0.67	0.256	0.67	0.27	0.02	0.004	0.0013	< 0.0004	0.087	0.006	99.3	Leaf
38314P	1989,5-25,1	Bell	Body	Ming	1493, inscribed	73.7	7.82	17.9	0.37	0.06	0.064	0.32	0.21	0.06	0.008	< 0.001	0.0001	0.160	0.032	100.6	
33555V	1985,10-11,1	Vessel, altar vase	Body	Yuan	14thC	79.6	1.91	14.3	0.41	0.52	0.030	0.18	0.25	0.05	0.002	0.0004	< 0.0004	0.038	< 0.006	97.3	
33555V	1985,10-11,1	Vessel, altar vase	Escut-cheon	Yuan	14thC	67.9	3.89	15.7	11.5	0.42	0.174	0.36	0.20	0.04	0.009	< 0.0003	0.0004	0.030	0.011	100.2	
33555V	1985,10-11,1	Vessel, altar vase	Ring on handle	Yuan	14thC	77.0	6.60	14.7	0.20	0.54	0.161	0.32	0.18	0.04	< 0.003	< 0.0005	< 0.0005	0.063	< 0.008	99.8	
38316W	1893,11-1,15	Incense burner	Body	Ming	15–16thC	67.3	9.74	27.5	0.014	0.07	0.134	0.22	0.29	0.05	< 0.006	< 0.002	0.0002	0.153	0.012	105.5	
31508Z	1988,10-8,1	Figure, seated	Body	Ming	1563, inscribed	69.5	6.03	17.4	3.1	0.50	0.070	0.40	0.19	0.06	0.010	0.0123	< 0.0007	0.063	< 0.01	97.3	Leaf
42393W	1991,10-28,24	Figure, standing Buddha		Ming	1568, inscribed	68.3	4.36	17.7	5.6	0.25	0.289	0.30	0.15	0.07	0.009	< 0.0007	0.0002	0.134	< 0.003	97.1	
5752-4-Y	1993,4-16,1	Figure, standing Buddha	Body	Ming	1578, inscribed	69.1	2.60	26.9	0.11	0.02	0.632	1.04	0.49	0.01	0.002	0.0005	< 0.001	0.159	0.006	101.1	Leaf
42538T	1991,10-28, 13B	Altar vessel, li ding shape		Ming	15thC	73.8	5.61	13.1	5.5	0.19	0.063	0.30	0.17	0.06	< 0.006	< 0.001	0.0003	0.082	< 0.005	98.8	
29602s	1987,11-10,1	Figure, Avalokitesvara, Tibetan style		Ming	15thC	84.4	0.11	2.28	12.7	0.27	0.030	0.08	0.06	0.06	< 0.003	0.0010	0.0012	0.092	< 0.002	100.1	Leaf
31502P	1985,10-14,1	Figure, Buddha seated	Body	Ming	15thC	73.0	5.72	15.2	1.9	0.59	0.082	0.45	0.17	0.05	0.009	< 0.0003	< 0.0003	0.131	0.007	97.3	Leaf
31502P	1985,10-14,1	Figure, Buddha seated	Cast in repair	Ming	15thC	63.4	6.10	23.9	0.707	0.60	0.109	0.36	0.18	0.05	0.006	0.0012	< 0.0006	0.130	< 0.009	95.6	Leaf
31502P	1985,10-14,1	Figure, Buddha seated	Strip on base	Ming	15thC	73.7	5.65	15.2	1.9	0.57	0.082	0.41	0.17	0.05	0.009	0.0003	< 0.0003	0.128	0.007	97.9	Leaf
42394U	1991,10-28,21	Figure, seated Daoist		Ming?	15thC?	72.5	7.89	19.7	1.3	0.14	0.075	0.40	0.20	0.08	0.013	< 0.0003	< 0.0011	0.083	< 0.003	102.4	Leaf
51610Z	1993,10-28,2	Vessel, arrow vase		Ming?	15–16thC?	83.9	1.05	8.32	8.3	0.21	0.065	0.18	0.06	0.01	0.005	< 0.0003	< 0.001	0.039	< 0.003	102.2	
42535Z	1938,5-24,731	Figure, Bodhisattva Avalokitesvara		Ming-Qing?	15–16thC?	62.0	0.55	2.18	31.2	0.50	0.032	0.35	0.11	< 0.03	< 0.008	< 0.001	0.0012	0.028	< 0.007	97.0	
33813V	1986,7-30,2	Candlestick	Body	Ming	16–17thC	64.7	4.81	25.7	1.4	0.23	0.063	0.36	0.20	0.06	0.004	0.0005	< 0.0002	0.159	0.012	97.7	
38317U	1986,10-9,1	Ewer	Body	Ming	16–17thC	74.8	3.11	13.4	5.6	0.10	0.076	0.93	0.52	0.11	0.006	< 0.002	0.0009	0.110	< 0.005	98.7	
31505U	1987,10-7,5	Figure, Manjusri on lion	Body	Ming	16–17thC	72.8	0.71	20.6	0.001	0.04	0.087	0.66	2.43	0.03	0.002	0.0011	< 0.0004	0.103	< 0.005	97.5	
31505U	1987,10-7,5	Figure, Manjusri on lion	Chaplet	Ming	16–17thC	98.9	< 0.25	0.05	0.004	0.11	0.008	< 0.05	< 0.03	< 0.01	0.003	0.0005	< 0.0005	0.020	< 0.008	99.1	
31504W	1987,10-7,4	Figure, Samantabhadra	Body	Ming	16–17thC	78.7	0.85	14.0	0.002	0.02	0.098	0.74	2.60	0.02	0.002	0.0009	< 0.0004	0.106	< 0.006	97.1	
31504W	1987,10-7,4	Figure, Samantabhadra	Chaplet	Ming	16–17thC	98.6	< 0.14	0.30	0.001	0.12	0.007	< 0.03	0.03	0.01	0.001	0.0003	< 0.0003	0.025	< 0.004	99.1	
31498Z	1988,11-8,3	Figure, Zhenwu seated	Body	Ming	16–17thC	67.2	3.64	13.2	12.1	0.82	0.101	0.72	0.19	0.06	0.010	0.0010	0.0006	0.060	< 0.007	98.1	
33811Z	1988,10-7,1	Vase, archaistic, inlaid	Body	Ming	16–17thC	74.6	11.55	13.3	0.005	0.24	0.032	0.17	0.05	0.06	0.020	0.0002	< 0.0002	0.097	0.006	100.1	
33811Z	1988,10-7,1	Vessel, vase, archaistic		Ming	16–17thC	74.4	11.23	13.2	0.009	0.22	0.026	< 0.09	< 0.06	0.06	0.023	< 0.003	0.0003	0.116	0.055	99.3	
29441V	1983,11-17,1	Figure, seated Buddha	Body	Ming	1623, inscribed	55.5	3.30	36.5	0.70	0.67	0.439	0.22	0.11	0.06	0.006	< 0.001	< 0.0009	0.090	0.000	97.6	
29441V	1983,11-17,1	Figure, seated Buddha	Head	Ming	1623, inscribed	56.8	3.80	35.0	0.70	0.64	0.495	0.29	0.12	0.06	0.007	< 0.001	< 0.001	0.095	0.000	98.0	
40742W	1990,5-30,1	Plaque	Body	Qing	1648, inscribed	57.7	0.88	4.60	34.8	1.00	0.086	0.58	0.83	0.03	0.036	< 0.003	0.0080	0.033	< 0.003	100.6	

(*Table 1 continued*)

BMRL	Reg. No.	Description	Part	Period	Date	Cu	Sn	Pb	Zn	Fe	Ni	As	Sb	Bi	Co	Mn	Cd	Ag	Au	Total	Gilding
38313R	1989,6-8,1	Bell	Body	Ming	16thC	68.0	0.54	11.3	15.3	0.62	0.122	< 0.07	0.81	< 0.02	< 0.007	< 0.002	0.0035	0.056	0.049	96.8	
33554X	1986,4-12,1	Brush Rest	Body	Ming	16thC	78.0	5.53	13.4	0.045	0.06	0.140	0.45	0.28	0.04	0.028	< 0.0003	< 0.0003	0.078	< 0.005	98.1	
51611X	1993,11-10,1	Incense burner		Ming	16thC	71.7	1.68	10.2	18.0	0.81	0.137	0.43	0.07	0.02	0.008	< 0.0003	< 0.0011	0.027	< 0.003	103.1	
33810Q	1987,11-9,1	Censer, West Zhliding shape	Body (foot)	Qing	17–18thC	69.9	6.43	19.7	2.2	0.45	0.117	0.76	0.25	0.08	0.008	0.0003	< 0.0003	0.101	0.009	100.0	
33559Y	1985,5-10,1	Dish, chrysanthemum shape	Body	Qing	17–18thC	69.0	0.73	3.46	26.0	0.52	0.066	0.13	0.13	0.02	0.006	< 0.0004	0.0007	0.024	< 0.005	100.2	
31517X	1952,12-19,23	Figure, man kneeling	Body	Ming-Qing	17–18thC	82.9	3.85	7.76	3.2	0.69	0.040	0.20	0.15	0.07	0.004	0.0004	0.0004	0.093	< 0.006	98.9	
31520W	1952,12-19,25	Figure, seated	Body	Ming-Qing	17–18thC	79.0	3.51	7.74	7.4	0.04	0.034	0.75	0.07	0.03	0.002	0.0009	0.0006	0.050	< 0.007	98.6	
33555V	1985,10-10,1	Vessel, altar, Ding shape	Body (leg)	Qing	17–18thC	73.3	4.33	16.1	5.7	0.35	0.100	0.33	0.25	0.04	0.006	< 0.0004	< 0.0004	0.091	0.014	100.7	
38320T	1883,10-20,9	Vessel, Guang, archaistic		Ming-Qing	17–19thC	76.8	3.01	14.4	2.5	0.30	0.118	0.91	0.20	0.05	0.012	< 0.002	0.0003	0.076	0.005	98.4	
31519T	1883,10-20,3	Figure, group of three	Body	Qing	17–20thC	67.8	7.21	13.6	8.9	0.43	0.157	0.60	0.21	0.05	0.010	0.0004	0.0050	0.076	0.006	99.0	
31516Z	1885,12-27,5	Figure, Guanyin	Body	Qing	17–20thC	68.0	5.53	18.3	4.5	0.79	0.108	0.28	0.22	0.05	0.007	0.0003	< 0.0003	0.072	0.007	97.9	Leaf
31514S	1883,10-20,1	Figure, male	Body	Qing	17–20thC	71.6	2.84	6.40	18.0	0.69	0.115	0.69	0.11	0.03	0.019	0.0015	< 0.0002	0.054	< 0.01	100.6	
31514S	1883,10-20,1	Figure, male	Base plinth	Qing	17–20thC	70.0	5.33	17.6	3.9	0.73	0.098	0.83	0.20	0.06	0.008	0.0007	< 0.0003	0.092	0.008	98.9	
31513U	1938,5-24,726	Figure, Shide	Body	Qing	17–20thC	67.2	7.13	18.0	2.8	1.18	0.132	0.37	0.19	0.06	0.022	0.0005	< 0.0002	0.081	< 0.004	97.2	Leaf
31515Q	1893,11-1,2	Figure, TaiQing	Body	Qing	17–20thC	64.8	6.59	20.4	3.7	1.08	0.520	0.80	0.20	0.05	0.032	< 0.0003	< 0.0003	0.142	0.012	98.3	Leaf
33560Q	1985,10-21,1	Vase, tree trunk shape	Body (foot)	Qing	17–20thC	75.2	0.74	18.9	2.0	0.09	0.106	0.39	0.81	0.14	0.023	< 0.0003	< 0.0003	0.041	< 0.004	98.4	
33814T	1986,8-8,1	Vessel, altar vase, Fang zun	Body	Qing	17–20thC	61.7	< 0.09	3.31	31.9	0.46	0.042	< 0.02	0.02	0.01	0.006	< 0.0002	0.0070	0.012	< 0.004	97.5	
38318S	1936,11-18,39	Vessel, bull shaped	Body	Qing	17–20thC	89.3	2.19	6.95	0.005	0.02	0.169	0.10	0.04	0.03	0.040	< 0.002	0.0001	0.158	0.052	99.1	
33557R	1894,1-8,24	Vessel, Ding	Body (foot)	Qing	17–20thC	66.1	2.49	11.7	15.3	0.78	0.281	0.71	1.42	0.08	0.047	0.0003	0.0010	0.056	0.010	99.0	
42544X	1992,6-12,1	Figure, Liu-Hai on toad	Toad	Qing	1723, inscribed	56.3	0.68	5.02	35.9	0.83	0.058	0.30	0.36	0.05	0.036	< 0.0008	0.0072	0.027	0.030	99.6	
42544X	1992,6-12,1	Figure, Liu-Hai on toad	Figure	Qing	1723, inscribed	58.2	0.67	3.56	35.2	0.64	0.055	0.25	0.33	0.04	0.026	0.0705	0.0062	0.027	< 0.004	99.0	
38315Y	1989,3-9,2	Bell	Body	Qing	1732, inscribed	56.8	0.23	4.81	36.3	0.51	0.179	< 0.04	0.09	0.02	0.007	< 0.002	0.0023	0.051	< 0.005	99.0	
31518V	SL 1185	Figure, man seated	Body	Ming-Qing	17thC	68.1	2.48	7.51	18.7	0.65	0.133	0.27	0.36	0.03	0.010	0.0004	0.0003	0.040	< 0.0062	98.3	
42396Q	1992,2-1,1	Figure, seated Zhenwu		Ming	17thC	65.9	8.86	22.3	0.043	0.34	0.056	0.44	0.20	0.09	0.013	< 0.0006	0.0002	0.064	< 0.003	98.3	
42536X	1939,1-18,1	Figure, mandala		Qing	18thC	93.4	< 0.1	0.80	5.3	< 0.1	< 0.1	0.20	0.20	0.00	0.000	0.0000	0.0000	< 0.05	0.000	99.9	Hg
5752-1-P	1983,4-20,1	Vessel, You, archaistic		Qing	18thC	86.3	1.83	7.76	3.2	0.09	0.060	1.07	0.12	0.01	< 0.005	< 0.0007	< 0.0019	0.074	< 0.005	100.6	

melt at the pouring temperature of bronze or brass. However, some are copper-based, and two were analyzed (e.g., on 1978,10-7,4 and 5) (plate 11). In both cases they are of unalloyed (and well-refined) copper, which has a higher melting point (1050°C) than that of the body metal of the artifact (about 850°C). As with the figurines, the vessels are also lost-wax castings and show a similar range of features, such as remaining core, poor castings, and chaplets.

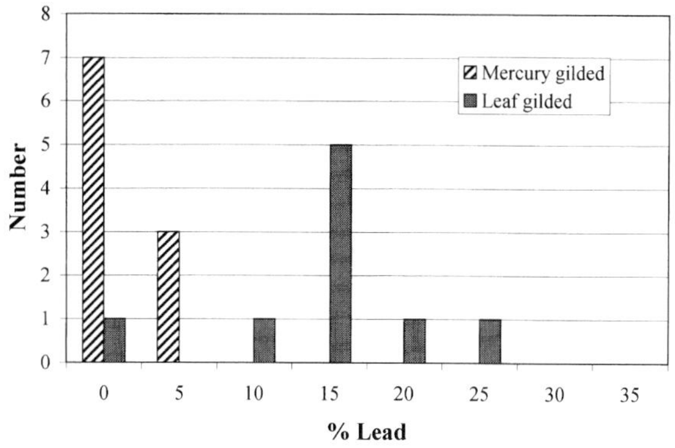

Figure 1. Lead content of gilded artifacts

Decoration

Many of the artifacts have applied decoration in the form of patination, gilding, pigments, and inlays. Patination is primarily confined to the vessels, and the materials identified are summarized in table 2. A characteristic feature of these artifacts is the absence of natural corrosion products. Instead, many have a coating of a dark brown lacquer (not yet analyzed), and it has not been possible to establish if this was contemporary with manufacture. Some of the finer pieces have a thin red patina (identified as cuprite) that seems to have been chemically induced rather than painted on.

About a quarter of the artifacts have identifiable gilding. A combination of microscopy and analysis (by XRF or emission spectroscopy for mercury) was used to establish whether leaf or mercury (fire) gilding was involved. Analysis alone is insufficient, as some artifacts have applications of cinnabar (mercury sulphide) in association with the gilding. The alloy compositions of the objects differ, depending on which gilding method was used. As has been noted before, for artifacts from the Roman to the medieval period (Oddy, Bimson, and La Niece 1981), high lead contents were avoided for fire gilding, as this can spoil the finish. Figure 1 summarizes the lead contents of the gilded artifacts and confirms that artifacts with fire gilding generally have lower lead contents than those with leaf gilding. Nevertheless, as found by Jett (1993), objects with relatively high lead contents, over 8%, are often still fire gilded.

A number of the figurines have applications of pigment, and table 2 summarizes those found, but they may not be contemporary with the manufacture of the objects. The range is similar to that found on Himalayan metalwork (Oddy, Bimson, and La Niece 1981; Reedy 1997) and other similar Chinese artifacts (Beguin and Liszak-Hours 1982). However, a pigment encountered here, not reported on Himalayan figurines, is cobalt blue. This was also found on a nineteenth-century Chinese figurine by Beguin and Liszak-Hours (1982), mixed with lazurite (lapis). It was identified on the head of a figurine dated by inscription to 1396 (1973,7-26,81) (plate 12). Cobalt was widely used in China, and elsewhere, as a colorant for ceramic glazes and glass. Analyses of these show that prior to the mid-fifteenth century, the Chinese were probably importing their cobalt from Persia, where it occurs principally as cobalt arsenides and arseno-sulphides, but from the mid-fifteenth century onward, local sources, based on manganese minerals, were frequently used (Cowell and Fukang 2001). Scanning electron microscopy (SEM) analysis indicates that the pigment on the figurine is in the form of a glass frit, and XRF analysis detected nickel, arsenic, and traces of copper in addition to cobalt, but no manganese. This is consistent with the imported form of cobalt used up to the fifteenth century, suggesting that it is contemporary with the manufacture of the figurine. The use of cobalt blue on the metalwork shows an interesting connection with ceramic production and decoration.

Table 2. Analysis of applied pigments

Reg. No.	Description	Period	Date	Part	Color	Pigments
1973,7-26,81	Figure, Buddha standing	Ming	1396, inscribed	headdress	gray-blue	cobalt glass frit and hydro-cerussite
1942,4-17,1	Figure, Buddha standing	Ming	1467, inscribed	headdress eyes and lips	gray-blue red	azurite and calcite cinnabar?
1985,10-14,1	Figure, Buddha seated	Ming	15thC	headdress top of head	dark green red	calcite and ?? (XRD unidentified) cinnabar
1983,11-17,1	Figure, seated Buddha	Ming	1623, inscribed	headdress headdress forehead	bright blue dull blue red	lazurite azurite and hydrocerussite cinnabar? (XRF)
1893,11-1,2	Figure, TaiQing	Qing	17–20thC	headdress inside	blue red	azurite cinnabar? (XRF)
1893,3-20,164	Figure	Ming	14–17thC	headdress	black	gypsum base and carbon black?
1972,3-1,1	Figure, Dharmapala standing	Yuan	13–14thC	eye of skulls, etc., red	red	cinnabar and beeswax
1988,10-8,1	Figure, seated	Ming	1563, inscribed	drapery, etc., red	red	cinnabar
1885,12-27,5	Figure, Guanyin	Qing	17–20thC	inside and on gilding	red	cinnabar? (XRF)

Composition

Alloy Types

All of the artifacts are copper-based with varying combinations of tin, lead, and zinc alloying components. The lead contents are often high, frequently exceeding 10%; the tin contents are lower, and only a few exceed 8%. The zinc contents range up to 33%. Some artifacts are almost certainly spelter brasses, made by direct alloying of zinc metal with copper or bronze, but relatively few contain the higher amounts of zinc encountered in the coins. All these alloys are suitable for artifact production by lost-wax casting; the lead contents will increase the fluidity of the molten alloy, reaching a maximum with about 2% lead (Gregory 1932, 332) and will also slightly reduce its freezing temperature. In many cases the lead content is well in excess of this amount and may have been added primarily to economize on copper and tin.

A multivariate statistical analysis of the major elements (copper, tin, lead, and zinc) using principal component analysis did not reveal separate alloy groups, unlike for the earlier Chinese bronzes examined by Jett and Douglas (1992). However, they can be broadly categorized as leaded bronzes, brasses, and "mixed alloy." The leaded bronzes are alloys of copper and tin with variable amounts of lead (up to 35%) and, at most, traces of zinc (less than 1%). The brasses are defined here as essentially copper-zinc alloys (arbitrarily defined here as having more than 10% zinc), also often containing small amounts of lead and tin, but with zinc being the dominant alloying metal. The mixed alloys lie between these two and are essentially quaternary alloys of copper with lead, tin, and zinc. This distribution is consistent with the previous analyses of later Chinese religious metalwork (e.g., Beguin and Liszak-Hours 1982) but different in certain respects from the contemporary coinage (Cowell et al. 1993).

The data are shown in the three-dimensional scatter plot in figure 2, which also includes representative examples of post-fourteenth-century coinage (Cowell et al. 1993). The coins fall predominantly into two groups: leaded bronzes and brass. There are fewer examples of mixed alloy, but note the similarity between some brass coins and religious artifacts. The sharper division of the coinage alloys is almost certainly due to the official control over their composition. The alloys used for the religious metalwork would, on the other hand, depend on more practical constraints, such as the availability and cost of materials, the required properties of the alloy, function, and workshop traditions.

Comparison with analyses of related Himalayan (Tibetan and Nepalese) metalwork (figurines representing various deities) of similar date range (e.g., Craddock 1981) shows that there are significant differences, particularly in the zinc and lead contents. Thus, a much higher proportion of the Himalayan metalwork has zinc contents above 30%—that is, most are definite spelter brasses. Conversely, the Chinese metalwork usually has much higher amounts of lead. Since the two groups of metalwork were produced using the same techniques, the differences in composition must depend on traditions or the availability of raw materials rather than technology or function.

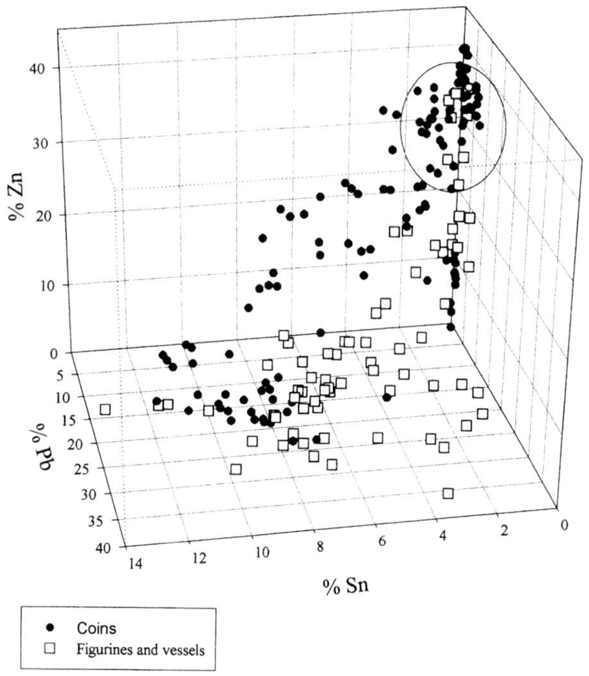

Figure 2. *Alloy composition: Comparison with coins*

Alloy Chronology

As noted above, the coinage alloy changed during the mid-sixteenth century from leaded bronze to brass. Not surprisingly, because there was no central control over their production, the religious metalwork shows no such dramatic changes. However, there is a rapid change from bronze to brass, with a preference for alloys with a higher zinc content dating from the fifteenth to sixteenth century onward. This is clearer when earlier artifacts are considered; thus, Jett and Douglas's analyses (1992) included 107 artifacts from the fifth to the nineteenth century C.E., with eleven dating to the Ming or Qing dynasties. All those prior to the Ming dynasty (i.e., pre-fifteenth century) were made of bronze or leaded bronze, none contained more than traces of zinc, and of the later pieces, only two contain significant amounts of zinc.

To examine the chronological trends more objectively, the data have been summarized as histograms by alloy and date (see fig. 3). This shows, respectively, those with major amounts of zinc (above 10%, defined as brasses), those with intermediate amounts (1–10%, usually mixed alloy), and those with at most traces (less than 1%, mainly bronzes and leaded bronzes). Unfortunately, the metalwork cannot generally be dated precisely and can often only be assigned a date range on grounds of style. Those artifacts with inscribed and reliable dates or with date ranges of a single century contribute one unit to the appropriate date column. Those with wider ranges, but not more than two centuries, have been split between columns by contributing half a unit to each appropriate date column. The histograms are therefore approximate and intended only as a guide to the preva-

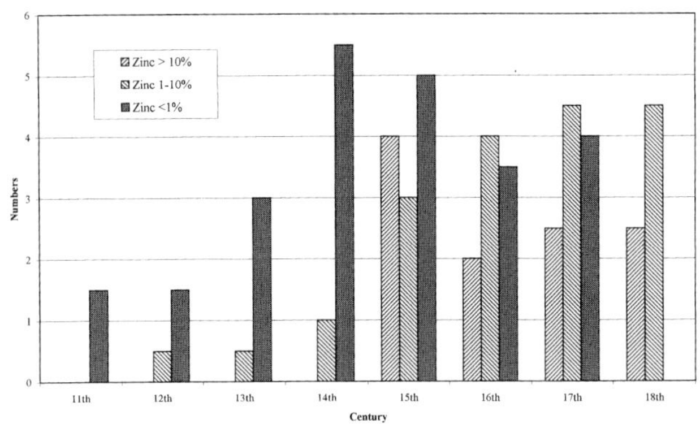

Figure 3. *Alloy type of artifacts by date*

lence and trend of zinc-containing alloys. It can be seen that brasses are not represented in these data before the fifteenth century, and even alloys with minor amounts of zinc are infrequent before this. Zinc-containing alloys begin to dominate in the fifteenth century. Leaded bronze is the preferred alloy before the fifteenth century but continues to be regularly used alongside brass and mixed alloy until the seventeenth century.

The period of initial use of brass for these artifacts thus appears to be slightly earlier than that found for the coinage and for some other collections; Jett and Douglas (1992) do not record any Ming-dynasty items with more than 10% zinc, although Beguin and Liszak-Hours (1982) show two brass artifacts dated to the fifteenth century. Here, four brass artifacts were assigned to the fifteenth century: 1908,4-20,4; 1986,8-5-1; 1942,4-17,1; and 1987,11-10,1. Although three have inscribed dates or reign marks, apparently added during their manufacture, it is possible that an earlier mark was deliberately applied when the artifact was manufactured; some support for this is provided by the relatively high cadmium content of 1986,8-5,1, which implies a seventeenth-century origin (see below). Another factor to consider is that at least one artifact (1987,11-10,1) (plate 13) shows strong Tibetan influence in its style, and an alloy more appropriate to that form (i.e., brass) may have been used. The composition of this fifteenth-century group may be a true reflection of the practice at the time, but the other factors cast some doubt on this; it is a question that can only be clarified by further data on objects from well-dated contexts.

Trace Metal Components

The data on the coinage show that there were some short-term trends in the trace metal composition that sometimes correlate with contemporary accounts of changes in the raw metal sources. For example, during the seventeenth to eighteenth century, there were systematic changes in the concentrations of antimony, nickel, and silver, which correlate with the reintroduction of imports of copper from Japan (Cowell and Wang 1998).

Japanese copper seems to have been a regular import as indicated in the *T'ien-Kung K'ai-Wu* (Sung 1966, 247) and is characterized by relatively low levels of trace elements compared with Chinese copper (Craddock and Hook 1997). In the late seventeenth century, Japanese copper became more readily available, and this is reflected in the composition of the coinage, which shows lower trace element levels. The religious metalwork of this period shows only a few cases where antimony, nickel, and silver are uniformly low and similar to that in Japanese copper. This may be due to more prevalent recycling, a multiplicity of sources, and mixing of different metal stocks for this class of metalwork compared with the coinage

Another trace element that is seen to vary systematically in the coinage is cadmium. This metal is commonly associated with zinc ores and has similar chemical and physical properties to zinc. Cowell et al. (1993) first observed that the cadmium content of brass coins produced after 1621 was significantly higher than before that date, and Zhou and Fan (1994) have confirmed this. Typically, cadmium levels are less than 10 ppm before 1621 and usually exceed 20 ppm or reach several hundred ppm after 1621. Zhou and Fan (1994) attribute this to the introduction of spelter brass, but other evidence—the amounts of zinc in earlier coins—suggests that this occurred much earlier than the 1620s; an ingot of Chinese zinc, recovered from a wreck dating to 1613, also has a high cadmium content of 0.04% (Craddock and Hook 1997). The high cadmium content of the brass may be related either to some subtle change in zinc production methods or, possibly, a different source of zinc. Whatever the reason, high traces of cadmium seem to be diagnostic of early seventeenth-century and later brass and may therefore be useful for suggesting the earliest date of manufacture for other Chinese brass artifacts, particularly in conjunction with the trace contents of antimony, nickel, and silver.

Table 1 shows several artifacts with the high cadmium contents typical of the post-1621 coinage. The more precisely dated artifacts have cadmium levels completely consistent with the coins; thus, the vessel 1986,8-8,1, which is thought to be mid-seventeenth century or later, has an expected high cadmium content coupled with low concentrations of antimony, nickel, and silver, suggesting the use of Japanese copper.

Even artifacts with broader date ranges are generally consistent, and if the coin data can be applied to this class of metalwork, then more precise date ranges may be suggested for them. Thus, the high cadmium contents of figurine 1952,12-19,17 and incense burner 1894,1-8,21 suggests that both are seventeenth century or later. Similarly, the low cadmium contents of two figurines (1883,10-20,1; SL1185) suggests that they date before the early seventeenth century, that is, the early part of their attributions. It should be noted, however, that this interpretation does not take into account the likely recycling of early low-cadmium scrap metal or the losses of cadmium during remelting. Thus, a high cadmium content should be a more reliable indicator of a late date than a low cadmium content is of an early date.

Stylistic Groups

A group of high-quality artifacts, executed in a "grand style," was identified. These are generally manufactured from a distinctive alloy containing a high zinc content (over

25%) with moderate amounts of lead (1–5%) and a consistently low tin content (below 1%). Most are mid-seventeenth century or later in date and contain the high cadmium typical of the later coinage brasses (plate 14). The British Museum examples are listed in table 3.

Beguin and Liszak-Hours (1982) report the analyses of several figurines with similar compositions, and Jett and Douglas (1992) likewise include a similar Ch'ing-dynasty figure (post-1644) (table 4).

The high zinc content of all artifacts in this category implies that most, if not all, were manufactured using spelter brass. Their composition is very similar to the coinage for the period from the mid-seventeenth to the mid-eighteenth century; it is possible that melted-down coins were used as direct raw material for their manufacture, which would have added to their prestige. It is also possible that the same individuals producing coins were also involved in figurine manufacture and had easy access to the appropriate raw materials, although of course slightly different processes were involved.

Table 3. *Objects from the British Museum with high zinc and cadmium contents*

Reg. No.	Description	% Zn	% Pb	% Sn	Cd (ppm)
1883,10-20,13	Incense burner	25.9	2.8	0.1	37
1938,5-24,731	Seated figure	31.2	2.2	0.5	12
1985,5-10,1	Dish	26.0	3.5	0.7	7
1986,8-8,1	Vase	31.9	3.3	< 0.1	70
1989,3-9,2	Bell	36.3	4.8	0.2	23
1950,5-34,1	Plaque	34.8	4.6	0.9	80
1992,6-12,1	Liu-Hai on toad	35.9	5.0	0.7	72

Table 4. *Objects from other collections with high zinc contents*

No.	Reg. No.	Description	% Zn	% Pb	% Sn
BL47	MG18262	Standing figure	35.8	1.3	0.4
BL49	MG17465	Standing figure	37.1	2.9	0.4
BL50	MG18617	Figure of lion	35.7	2.0	0.2
BL51	EG2147	Standing figure	31.8	1.4	0.3
BL53	MG18542	Seated figure	33.5	3.3	0.9
JD	F1917.262	Animal figure	26.7	2.6	0.3

Conclusions

These analyses show that the Chinese were comparatively late in adopting brass, or zinc-containing alloys, for this class of metalwork, even though they were familiar with its manufacture. In common with the coinage, brass was not extensively used until the fifteenth to sixteenth century, which contrasts with Tibetan and Nepalese practice, wherein brass was employed from the twelfth century and even earlier in north India.

The alloys used for the religious metalwork are generally more variable than those used for the contemporary coinage. The zinc content is lower, and mixed alloys are more frequent. In common with the coinage, it tends to have higher zinc contents, but this is based on a small number of analyses. While the use of brass is more likely from the fifteenth century onward, the amount of zinc or other alloying components does not assist with defining date of manufacture. The trace-element composition of the brasses, particularly the cadmium content, may be more useful in this respect—certainly in identifying artifacts that seem to have been manufactured in the seventeenth century or later.

Acknowledgements

We would like to thank Duncan Hook and Paul Craddock for contributions to this project.

References

Beguin, G. and J. Liszak-Hours. 1982. Objects Himâlayens en métal du Musée Guimet. *Annales du laboratoire de recherche des musées de France* 28–82.

Bowman, S. G. E., M. R. Cowell, and J. Cribb. 1989. Two thousand years of Chinese cash: An analytical survey. *Historical metallurgy* 23(1):25–30.

Cowell, M. R. 1998. Coin analysis by energy dispersive x-ray fluorescence spectrometry. In *Metallurgy in numismatics 4,* ed. W. A. Oddy and M. R. Cowell, 448–60. London: Royal Numismatic Society.

Cowell, M. R., and Zhang Fukang. 2001. In *Ming ceramics: A catalogue of late Yuan and Ming ceramics in the British Museum,* ed. J. Harrison-Hall. London: British Museum Press.

Cowell, M. R., and H. Wang. 1998. Metal supply for the metropolitan coinage of the Kangxi period. *Numismatic chronicle* 158:185–96.

Cowell, M. R., J. Cribb, S. G. E. Bowman, et al. 1993. The Chinese cash: Composition and production. In *Metallurgy in numismatics 3,* ed. M. M. Archibald and M. R. Cowell, 185–98. London: Royal Numismatic Society.

Craddock, P. T. 1981. The copper alloys of Tibet and their background. In *Aspects of Tibetan metallurgy,* ed. W. A. Oddy and W. Zwalf, 1–32. London: British Museum Press.

Craddock, P. T., and D. R. Hook. 1997. The British Museum collection of metal ingots from dated wrecks. In *Artifacts from wrecks,* Oxbow Monograph 84, ed. M. Redknap, 143–54. Oxford: Oxbow.

Gregory, E. 1932. Metallurgy. London: Blackie and Son.

Hook, D. R. 1998. Inductively coupled plasma atomic emission spectrometry and its role in numismatic studies. In *Metallurgy in numismatics 4,* ed. W. A. Oddy and M. R. Cowell, 237–52. London: Royal Numismatic Society.

Hughes, M. J., M. R. Cowell, and P. T. Craddock. 1976. Atomic absorption techniques in archaeology. *Archaeometry* 18:19–37.

Jett, P. 1993. A study of gilding of Chinese Buddhist bronzes. In *Metal plating and patination,* ed. S. La Niece and P. T. Craddock, 193–200. Guildford: Butterworth-Heinemann.

Jett, P., and J. G. Douglas. 1992. Chinese Buddhist bronzes in the Freer Gallery of Art: Physical features and elemental composition. *Proceedings of the Materials Research Society symposium* 267, 205–24.

Kerr, R. 1990. *Later Chinese metalwork.* London: Victoria and Albert Museum.

Nieuwenhuysen, P., and F. Adams. 1984. Elemental analysis of Himalayan metal statues. *Historical metallurgy* 18: 105–8.

Oddy, W. A., M. Bimson, and S. La Niece. 1981. Gilding Himalayan images: History, tradition, and modern techniques. In *Aspects of Tibetan metallurgy,* ed. W. A. Oddy and W. Zwalf, 87–101. London: British Museum Press.

Rawson, J. 2001. The many meanings of the past in China. In *Die Gegenwart des Altertums, Formen und Funktionen des Altertumsbezugs in den Hochkulturen der Alten Welt*, ed. D. Kuhn and H. Stahl, 397–421. Heidelberg: Edition Forum.

Reedy, C. 1997. *Himalayan bronzes: Technology, style, and choices*. Newark: University of Delaware Press.

Sayre, E. V., P. Jett, and E. C. Joel. 1992. A technical examination of the Buddhist bronzes in the Freer Gallery of Art, part B: Stable lead isotope analysis. *Proceedings of the Materials Research Society symposium*, 267, 225–39.

Sung, Ying-Hsing. 1966. T'ien-Kung K'ai-wu. *Chinese technology in the seventeenth century*. University Park: Pennsylvania State University Press.

Werner, O. 1972. *Spektralanalytische und Metallurgische Untersuchungen an Indischen Bronzen*. Leiden: Brill.

Zhou, Weirong. 2001. The emergence and development of brass smelting techniques in China. *Bulletin of the Metals Museum* 34:87–98.

Zhou, Weirong, and Xiangxi Fan. 1993. A study of the development of brass for coinage in China. *Bulletin of the Metals Museum* 20:35–45.

——. 1994. Application of zinc and cadmium for the dating and authenticating of metal relics in ancient China. *Bulletin of the Metals Museum* 22:16–21.

——. 1996. Chinese traditional zinc smelting technology and the history of zinc production in China. *Bulletin of the Metals Museum* 25:36–47.

11 Medieval Islamic Metal Technology

Susan La Niece

ABSTRACT Islamic rule extended over a large part of Asia and North Africa and into Spain in the medieval period. There is evidence for some continuity from earlier metalworking traditions but also for innovations, notably the introduction of sand casting and the popularity of brass vessels, richly inlaid with silver, gold, or copper and a black organic material. The great majority of the vessels are made of brass, with a small group made of high-tin bronze. The evidence for lost-wax casting and piece-mold casting as well as that for spinning is examined. This paper summarizes the findings of an analytical and technological study of the British Museum's collections of Islamic copper alloy vessels.

Introduction

The term *Islamic* is here used in the widest possible sense to include objects made in the extensive area under Islamic rule from the first year of the Islamic era, 622. At its peak Islamic influence extended from Spain in the west to Afghanistan in the east, centered on the regions of the Middle East. The focus of this analytical and technological study is copper-alloy vessels, most of which were intended for secular use. The finest of these are richly inlaid with silver, gold, or copper and a black organic material (plate 15).

Islamic metal technology has been the subject of research for some years. The translation and interpretation of contemporary texts has opened up important source material (Allan 1979; al-Jazari 1974). The Islamic metalwork in the Freer Gallery of Art has been extensively studied and published (Atil, Chase, and Jett 1985), including a valuable technical examination of each object and analysis by x-ray fluorescence spectrometry (XRF). Now it is not uncommon for catalogues to include some chemical analysis as well as comment on method of manufacture—for example, Allan (1999). Archaeological excavations are also producing important new evidence, as discussed below. A technical study of late medieval Veneto-Saracenic bowls and incense burners (Ward et al. 1995) and analyses of over two hundred Islamic copper-alloy objects in the British Museum by atomic absorption spectrophotometry (AAS) and inductively coupled plasma atomic emission spectrometry (ICPAES) have also been published (Craddock, La Niece, and Hook 1998). The published database of fine metalwork is now substantial, and new discoveries of evidence for manufacturing methods from study of the collections in the British Museum are refining our understanding of Islamic metal technology. It is the aim of this paper to present an overview of that technology and to correct some of the current misconceptions.

Islamic Brass Technology

The composition of Islamic metalwork has been discussed in detail, and analyses have been published elsewhere (Craddock 1979; Craddock, La Niece, and Hook 1998; Ward et al. 1995). The technology was inherited from earlier cultures, especially the Roman, Byzantine, and Sasanian empires. The favored copper alloy of the Islamic world was brass, an alloy of copper with zinc. Brass had been introduced by the Romans, who used it alongside bronze (an alloy of copper with tin). Analyses confirm that bronze is rare after the early Islamic period (Craddock, La Niece, and Hook 1998). Only specialized groups of objects, such as mirrors and a small class of vessels made with a very high tin content (20–25%), continued to be made of bronze (Melikian-Chirvani 1974; Lakpour 1997; Ackermann and La Niece forthcoming). Unalloyed copper was used, as it always had been, for simple hammered wares, but most of the fine vessels that were shaped by hammering were relatively pure brasses. The levels of lead seen in the worked items of Islamic metalwork are almost always below 2% and probably occurred as an accidental impurity. The presence of lead can cause splitting when brass is worked, so it was avoided for vessels made by hammering. For casting, lead was added to the brass alloy, both for economy and to improve casting properties.

The transition from bronze to brass in the Islamic world has been explained by the difficulty in obtaining adequate supplies of tin (Craddock, La Niece, and Hook 1998). In early medieval times, most tin seems to have come to the Islamic world from southeast Asia and later also from Europe (Allan 1979, 27–28). Unlike tin, sources of zinc were readily available within the Islamic world, in Iran, Afghanistan, and Anatolia (Allan 1979; Craddock, La Niece, and Hook 1998). The golden color of brass may have been a factor in the preference for brass over bronze. Furthermore, zinc is a more effective deoxidant of copper than tin, and brass has better properties for working than bronze in that annealed brasses will withstand a remarkable degree of deformation by cold work.

A trend toward consistently low levels of iron in Islamic brass and more erratic levels in contemporary European brass has been attributed to the different treatment of the zinc ores in Europe from those in the Middle East (Craddock, La Niece, and Hook 1998). The trace elements in Islamic metalwork show only relatively minor changes with time, with the notable exception of nickel (Craddock, La Niece, and Hook 1998). The nickel content of brass was found to increase to a median value of 0.16% in vessels dated to the fifteenth and sixteenth centuries, compared to a median value of 0.06% in those of the preceding centuries. Nickel is an impurity of copper ore, and this change in concentration suggests that a different source of copper, or of processing technology, was introduced in the fifteenth century. A trend toward increasing nickel content from the eleventh to the sixteenth century has also been observed in European copper alloys (Werner 1977, 1981, 1982; Pollard 1996). This suggests that by the late medieval period, the Islamic and European worlds were sharing some of the more important sources of copper. Indeed, there is documentary evidence from both European and Muslim sources that Europe was the main supplier of copper to the eastern Mediterranean at least in the fifteenth and sixteenth centuries (Braunstein 1977).

Casting

Casting was extensively used in vessel manufacture, as it always had been. Various methods could have been available to medieval Islamic metalworkers, but contemporary descriptions of casting are sparse, and archaeological evidence for molds is rare (Allan 1979). Consequently, we have little real knowledge of Islamic casting technology. A simple two-part mold may have been used for casting components such as handles, but more complex molding arrangements were required for casting hollow items such as vessels. It is unusual to find definitive evidence of the manufacturing method on well-finished articles, where decoration and inlay work have obliterated casting features from the surface, but careful examination of unfinished interiors can yield clues.

Lost-wax casting is an important historical method of casting (Maryon 1971; Reeves 1962). It would be surprising if the tradition were unknown in the Islamic world, but it has not been established how widespread it was. A group of Umayyad objects (eighth century) excavated in Jordan (Garenne-Marot et al. 1998) and Fatimid objects (late tenth/early eleventh century) from Caesarea (Shalev in press) are thought to be lost-wax castings. Mention in a poem of wax and clay in connection with the manufacture of metal objects has been used to suggest that lost-wax casting was employed in Islamic Iran (Giuzalian 1968). Perhaps more convincingly, lost-wax casting of a brass door-knocker was mentioned by the court inventor al-Jazari, writing in the city of Amid, near Mosul from 1198 to 1200, with the comment that the casting was "as done by the masters of this craft" (al-Jazari 1974). A similar candlestick in the collection of the Freer Gallery (Atil, Chase, and Jett 1985, no. 15) also shows marks on the interior consistent with work on a wax model (Paul Jett, pers. comm.). Clearly al-Jazari was not the only one practicing lost-wax casting.

Examination of the objects themselves confirms that lost-wax casting was indeed used to make some vessels—for example, a late-thirteenth-century candlestick base, probably from the Jaziran area (east Anatolia, north Iraq, west Iran) is cast in one piece and has features which are best explained as originating in a multicomponent wax model (fig. 1). Lost-wax casting is a method that is particularly suited to relief decoration. However, this is not a feature of cast Islamic vessels, with the exception of mortars and some early ewers. For example, an eighth/ninth-century ewer from Iran was cast in one piece, with cast low relief decoration and additional chased details (fig. 2). Chaplets, which are metal pins pushed through the wax model into the core to hold it in place while the molten metal is poured in, are characteristic of these early ewers. The chaplets can be seen in the walls of vessels, often projecting into the interior cavity where the core had been but filed flush on the outside.

Another type of mold for casting vessels is the piece mold. The technique of making a mold that could be dismantled in several sections to extract the casting was developed in China in the Bronze Age (Barnard 1961), although by 1000, lost-

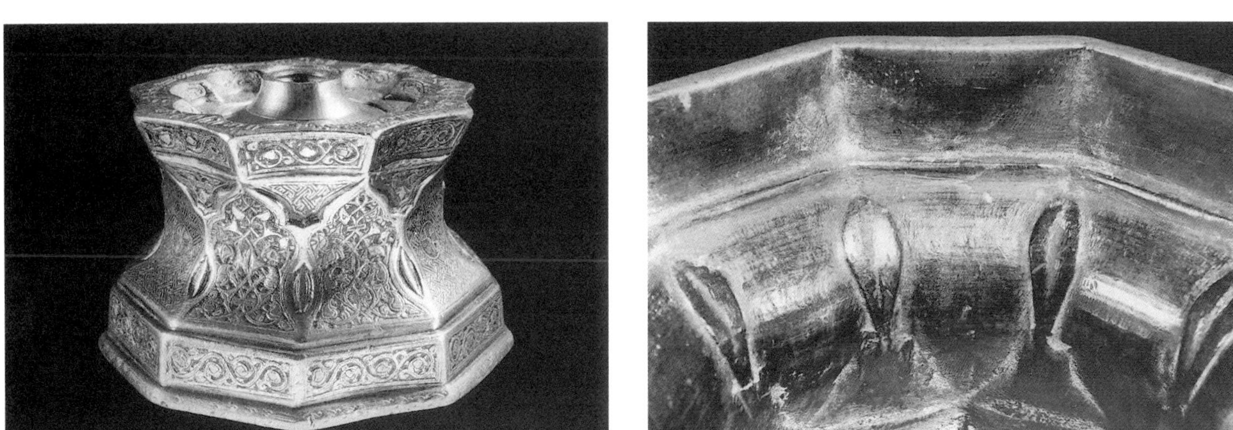

Figure 1. *A lost-wax cast candlestick base of inlaid brass and view inside upended candlestick. The complex form was built up from several pieces of wax (e.g., the recessed almond-shaped features were created by cutting the wax model and pressing ovals of wax into the holes from the inside). The scrape marks on the inner surface are features from the wax model that are preserved in the metal of the casting. W. Iran, late 13th century* c.e. *(BM 1881.1-27.1), height 7.3 cm. Copyright British Museum*

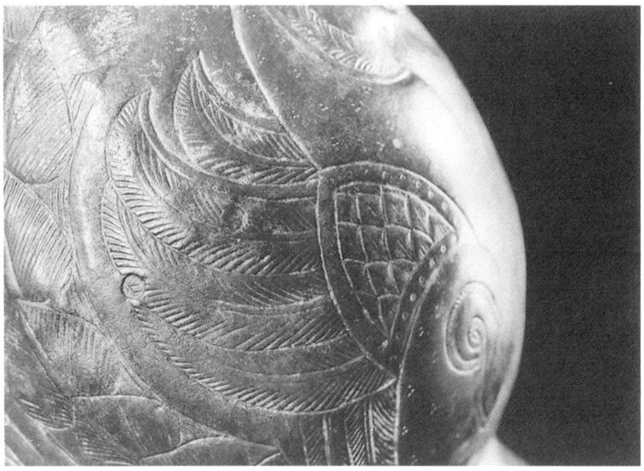

Figure 2. Two round casting chaplets, ca. 4 mm diameter, embedded in the wall of a cast brass ewer with low relief decoration. On the outside surface, the chaplets are partly concealed by chasing, and they project into the interior cavity. Iraq/Iran, 9th century C.E. (BM1959.10-23,1). Copyright British Museum

wax casting and sand casting seemed to have superseded piece molds (Kerr 1990). Many of the forms of cast Islamic items such as candlesticks are very standardized, and piece-mold casting would have been a practical means of manufacturing these. It is assumed that this method would have been known in the Islamic world, but evidence is rather scarce as yet (Allan 1979). The position of joins between abutting mold segments can often be seen on Chinese vessels, as little further work was done on the metal after casting, but such joins are not found on Islamic vessels. The Chinese used metal spacers to separate the core from the outer casing of the piece mold (Gettens 1969). This is a similar idea to the chaplet pins, which were pushed through the wax model and into the core in lost-wax casting, but the chaplets used in Chinese piece-mold castings are flat pieces of bronze and do not project into the space left by the core. The projecting chaplets do not seem to occur in later Islamic castings, but neither have the fully embedded spacers been reported, so the case for complex piece molds is inconclusive.

Sand casting is an important and well-described method of brass casting today—for example, in India (Untracht 1968) and in Cairo (Shalev in press). Sand mixed with a binder, such as clay or oil, and packed into a two-part frame known as a casting flask forms the mold. A removable model, often of wood, is embedded in the sand between the two halves of the casting flask to create the void for casting, and channels are made for the molten metal to be poured in. Sand casting is well suited to the manufacture of flat items, such as plaques and mirrors, but the method is also used to make hollow items by inserting a core into the cavity. The core is made in a two-part mold called a core box (Hasluck 1904). It can be secured by chaplets, by a rod fixed across the casting flask, or by a core print, which is where the core is joined to the outer part of the mold through a natural gap in the object (e.g., the openings at the foot and the socket of a candlestick). Advantages of the method include reusable patterns and low porosity in the casting, because the sand mixture allows gases to escape.

Sand casting is not thought to have been practiced in Europe until the fifteenth century, and even then it was uncommon (Biringuccio 1959). Sand molds were used in China from at least the seventeenth century for casting *cash* (Burger 1976) and mirrors (Kerr 1990). Bidri, the inlaid and black patinated zinc alloy wares of India, were also cast in sand molds. These include hollow vessels that are datable to at least the mid-eighteenth century, and the technique may be as early as the fifteenth century (Stronge 1985). A recent archaeological find makes it clear that the method was being used by Islamic metalworkers well before it was known in Europe. A small metal frame of the type used for sand casting and dated to the tenth century by coin evidence was excavated, along with casting debris, at Tiberias by Professor Yizhar Hirschfeld, Hebrew University of Jerusalem (Dr. M. Ponting, Nottingham University, pers. comm.). Later evidence is also known. One half of a small stirrup-shaped casting flask made of copper alloy was found in the thirteenth to early fourteenth century at Uvek, a town on the Volga, at the northern extremities of Islamic influence, north of the Caspian Sea (Natasha Eniosova, Archaeological Department, Moscow State University, pers. comm.). To reinforce this archaeological evidence, al-Jazari described the process of sand casting brass latticework plaques for a door to a palace (al-Jazari 1974).

A much larger, circular frame (31 cm diameter × 24 cm deep) in the collections of the British Museum appears to be a sand-casting flask. It was itself made by casting, in leaded bronze, complete with pins to locate the two halves precisely, handles to lift them, relief patterns on the inner surface to provide keying to hold the sand in place, and an opening for the molten metal to be poured in (fig. 3). Its date and provenance are unknown except that it was donated to the museum with a group of Persian objects, and it has cast in relief on the interior "work of . . . [unclear] Mohamed" in Arabic script. Intriguingly, the plastic appearance of the relief keying indicates that the flask was probably made by lost-wax casting.

Identifying objects that may have been made by sand casting is difficult, although excavations of Fatimid levels (tenth to twelfth century) at Caesarea have produced small brass items that are thought to have been cast by this method

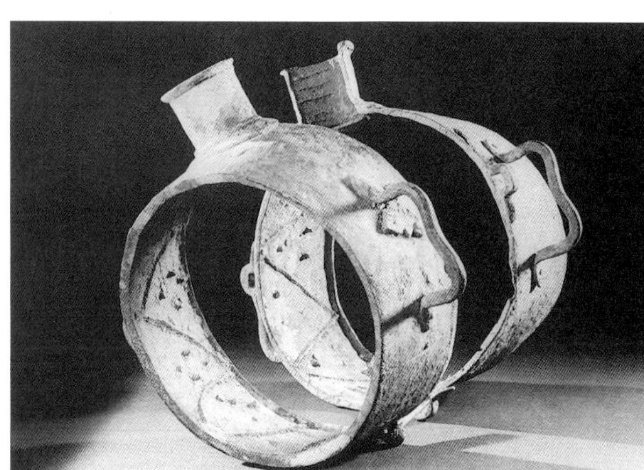

Figure 3. Leaded bronze, two-part sand-casting flask of unknown date, complete with lifting handles and locating pins. An opening in the sand allowed molten metal to be poured in via the funnel at the top. The relief features on the inside were designed to prevent the sand from shifting. (BM1969.9-24.4) 24 cm deep, 31 cm diameter. Copyright British Museum

(Shalev in press). Apart from the example described above, most of the known Islamic sand flasks are rather smaller than casting flasks used in modern foundries and too small to have been used for casting vessels, although they may have been suitable for casting al-Jazari's door plaques.

There is a group of candlesticks, dated to the thirteenth and fourteenth centuries by the inscriptions on some, that may have been made by sand casting (fig. 4). They have been attributed to Anatolia (Melikian-Chirvani 1982), and more than seventy are known from collections today; four are in the British Museum. All of those examined have a number of features in common. They have similar dimensions (approximate height: 20 cm; diameter: 19.5 cm; weight: ca. 1.8 kg; metal thickness: ca. 3 mm). They were cast in one piece and have turning marks both inside and out, where the as-cast surface was finished. The two main distinguishing features are first, a thin metal wall separating the hollow candle socket from the hollow base, which is pierced by a small hole about 8 mm in diameter, slightly off-center, and second, that inside both of these cavities is a linear discontinuity in the casting (fig. 5). There is no such line visible on the outside, which is highly finished and decorated with inlays and chasing. The interior features indicate that the cores were mass-produced in two-part molds (core boxes) that were not always well fitted, leaving a seam down the side of the cores that was then imprinted on the cast metal in negative. As the hole in the thin metal between the two candlestick cavities is too small for a core print, perhaps there was a rod running through the center, piercing the cores like a kebab stick, to hold them in place. The design of these candlesticks is impractical to produce by lost-wax casting, because the thin metal wall between the two cores would be difficult, if not impossible, to model in wax. They may have been cast in piece molds, but the features left by the core, the low casting porosity, and the unusually large number of similar items made all point to sand casting.

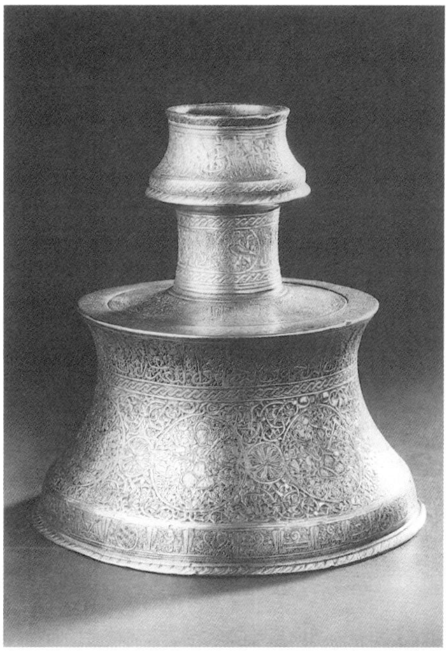

Figure 4. *Brass candlestick cast in one piece, with inlaid silver astrological symbols. Anatolia, second half of 13th century* C.E. *(BM1885.10-10.8), height 20 cm. Copyright British Museum*

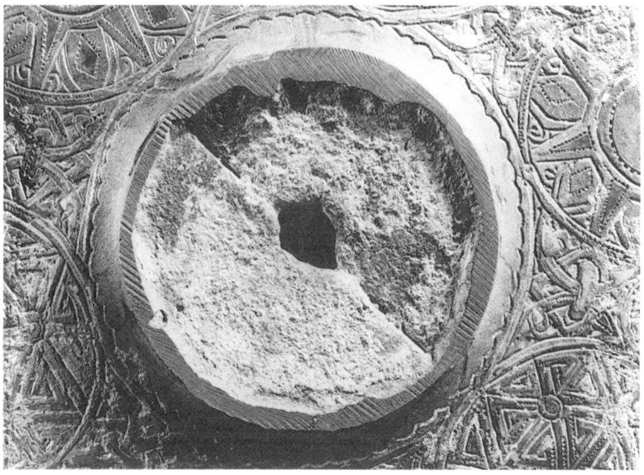

Figure 5. *Detail of the top of a damaged candlestick of the same type as that shown in figure 4. The damage allows a view of the bottom of the candle socket. There is a small hole (ca. 3 mm diameter) where a rod linked and supported the casting cores (now removed) for the hollow candle socket and the base. The linear feature across the center indicates the use of a two-part core that had a thin, projecting seam where the two halves were joined. Copyright British Museum*

Raising and Spinning

A substantial proportion of Islamic vessels were hammered to shape. Indeed, craftsmen beating brass into a variety of forms can still be seen today in the markets of the Middle East and North Africa, and the hammering techniques of raising, sinking, and so forth are well known (Maryon 1971). It is rarely possible to section fine vessels to ascertain from the metallographic structure whether they were worked or cast. Nevertheless, there is often sufficient evidence from surface examination and radiography to detect hammer impressions and the characteristic irregular thickness of hammered metal. The metal composition is also diagnostic in that the lead content of hammered brass is usually below 2%. It is simplistic to expect a vessel to be made exclusively by one method; combinations of casting a basic shape, followed by finishing to a greater or lesser extent with hand tools, is a commonsense approach adopted throughout history. However, the evidence in the medieval period for the more mechanized method of shaping metal, known as spinning, is less well established.

Spinning is a method of shaping a flat disc of metal into a hollow vessel using a spinning lathe. A metal disc is secured to the rotating spindle of the lathe and forced with a rounded, bright steel, hardwood, or bloodstone tool against a wooden or metal form called a chuck until the metal takes the shape of the chuck (Crawshaw 1909; Tuells and Painter 1910). Simple shapes, such as plates, cones, and domes, are suitable for manufacturing by this method, and even complex shapes can be made using a multisectioned chuck that can be dismantled to allow for its removal from inside the finished article. Metal spinning was an important manufacturing method in nineteenth-century Europe but was overtaken by pressing and stamping for industrial applications. Today it is largely a technique for specialist craftsmen (Reagan and Smith 1936).

The date at which metal spinning was first introduced has never been satisfactorily established. Several claims to have identified the technique in premedieval metalwork have been made—for example, a group of Iron Age bronze bowls from Britain (Coghlan, Watson, and Voce 1951). In the Islamic world, some vessels of the thirteenth century or later have been described as spun (for example, Melikian-Chirvani 1982; Atil, Chase, and Jett 1985). Several of the shapes encountered in the repertoire of Islamic metalwork would be ideal for spinning—for example, spherical incense burners and conical candlestick bases—and it has been argued that one such Mamluk inlaid candlestick base shows evidence of spinning in the broad concentric ripples in its metal wall (Craddock and Lang 1983).

There are no known contemporary Islamic or European references to spinning, but there is evidence for the use of lathes dating back to about 700 B.C.E. These were driven by a cord wound around the workpiece or the lathe spindle and pulled back and forth by an assistant, or perhaps by the craftsman himself, with the aid of a bow (Woodbury 1972). The earliest lathes were for wood turning, but Theophilus, writing in early twelfth-century Europe, described the use of a lathe to turn down clay-casting cores for censers and bells, for rotating the cast metal bell while it was smoothed with sandstone, and for turning cast-tin cruets (Theophilus 1979). However, he makes no mention of using a lathe for spinning metal.

A metal-spinning lathe is considerably more sophisticated than a wood-turning lathe or even a lathe for turning metal. Metal turning is a finishing process whereby a cutting tool or abrasive is held against a metal article while it is rotated on the lathe. This can be done to smooth the surface of a casting, to incise a line, or to true the mouth of a circular vessel. None of these processes requires the amount of force necessary to shape a hollow vessel from a flat disc of metal. A metal-spinning lathe must be capable of continuous rotary motion and variable speeds from 300 r.p.m. up to at least 1000 r.p.m. (Untracht 1968). It must be firmly mounted to take the considerable pressure exerted on the metal by the craftsman and to have sufficient power to keep up the rotation against this pressure. A continuous-drive metal-turning lathe, powered by means of a hand-cranked wheel and cord, is documented by a European engraving of 1568, and by this date both waterpower and draught animals were also being used to power lathes (Woodbury 1972). The documentation from the Islamic world is not as good. Al-Jazari, writing between 1198 and 1200, describes a crank—admittedly only in connection with a pump—but the principle of continuous rotary motion was known (Coomaraswamy 1924). Thus, at least as early as the sixteenth century, the technology required for metal spinning was available, but is there any evidence that it was utilized in the Islamic world?

Brass, which was widely used in the Islamic world, is suitable for spinning, with frequent annealing to relieve work hardening. The thickness of brass recommended by modern handbooks for spinning is in the range of 0.4 to 0.8 mm (Crawshaw 1909; Johnson 1960). Many of the Islamic vessel forms are suited to spinning; however, the

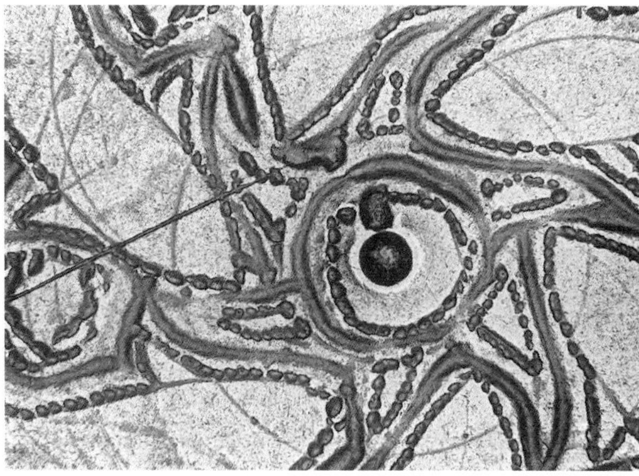

Figure 6. *Magnified detail inside a cast brass Mamluk bowl. Egypt or Syria, 14th century C.E. Six fish radiating around a central sun disc were inlaid with silver (now missing), although the punched pits for the keying remains. In the middle are two centering pips (dark spots). One was used in the laying out of the design (faint compass drawn arcs across the fish). The other may have been to clamp the bowl to a lathe for finishing (BM 1878.12-30.690). Copyright British Museum*

thickness of the metal may well preclude some. Thicker metal can be spun but requires greater power of the lathe and strength of the craftsman. Some of the spherical incense burners are 1 mm or less in thickness, but the metal of most of the trays and conical candlestick bases is at least 2 mm thick. There is some doubt as to whether it would have been feasible to spin this thickness of metal without a modern, motor-powered lathe.

Examination of the vessels themselves may provide clues to manufacture. The metallographic structure of a spun brass vessel would typically show a worked and annealed microstructure, but this would be the same as for any hammered brass. There may be more directionality in the orientation of inclusions, but this could only be determined by destructive sectioning of the object, which is obviously out of the question. Lathe-centering pips and concentric scratches on the surface have often been interpreted as evidence for spinning, but there are other possible explanations for these marks. A lathe may be used to rotate a vessel for several finishing processes: for turning—that is, paring down the surface with a sharp tool; for polishing with abrasive; or for planishing the surface with a rounded tool or hammer. Centering marks may also be used in laying out a decorative design (fig. 6).

The candlestick examined by Craddock and Lang (1983) has broad concentric ripples in the metal that are visible on the interior under raking light. They are also visible in a radiograph, although the inlaid exterior surface is quite even (figs. 7a and 7b). However, the metal is over 2 mm thick. While perhaps it is not impossible that it was shaped by spinning, it is well beyond the recommended thickness for spinning on a modern, motor-powered lathe. Furthermore, the spinning tool would have been in contact with the outside surface, while the inner, uneven face would have been against the smooth former if it had been spun. The concentric markings do indeed suggest the use of a lathe in its manufacture but probably in finishing rather than shaping the candlestick.

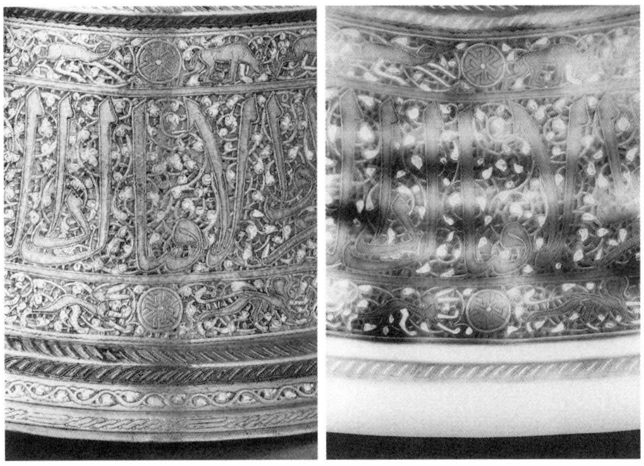

Figure 7. Detail of the base of an inlaid brass candlestick and radiograph of the same area showing broad, irregular ripples in the metal on the inner surface, which indicates that it was planished on a rotating lathe. The silver inlays appear white. Second half of 13th century, Damascus (BM 1866.12-29.62). Copyright British Museum

Broad ripples are not a feature diagnostic of spinning. Concentric lines are created by spinning, but these are very fine because only a small area of the spinning tool can be in contact with the metal if creasing is to be avoided. Creasing of the metal from the base to rim of the vessel, not rippling in the direction of rotation, is the fault most usually reported by spinners. The creasing is caused by uneven pressure from the spinning tool and poor contact with the former behind.

When spinning was introduced into the manufacturing industry of Birmingham in the early nineteenth century, it resulted in a considerable saving in labor costs, as shapes could be mass-produced with greater speed than traditional hand raising (Timmins 1866). The demand for labor-saving devices and industrial scale output is hard to identify with the workshops of the medieval Islamic world, which were making brass vessels to commissions from wealthy patrons. There is less information on utilitarian metalwork for the masses, but these items appear to have been either beaten or cast to shape. The spherical incense burners, with their thin walls, are the most likely candidates for spinning, but hammer impressions can often be seen in radiographs of them (fig. 8). The technology for spinning was available, but neither the demand for it nor any conclusive evidence for the products has yet to be convincingly demonstrated in the medieval period. In any event, it seems unlikely that objects with walls thicker than 2 mm were spun before the advent of industrial motor-driven lathes in the nineteenth century, and any concentric marks on the heavier Islamic pieces are more likely to be explained by finishing work with a cutting or planishing tool, albeit while being rotated on a lathe or similar device.

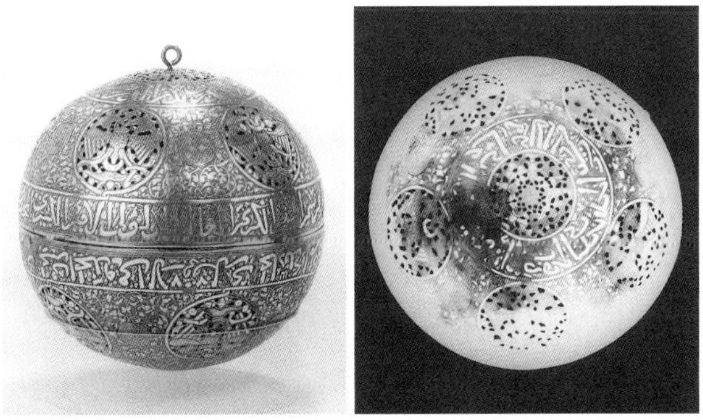

Figure 8. Spherical brass incense burner, pierced and inlaid with silver. The inscriptions include the name and titles of the Mamluk amir, Badr al-Din Baysari. Damascus, 1264–79 C.E. (BM 1878.12-30.682), diameter 18.4 cm. The radiograph of the upper hemisphere has a mottled appearance, indicating the variation in metal thickness typical of hammered work. Copyright British Museum

Conclusions

Brass was the predominant alloy of the Islamic world, with bronze confined to specialist classes of metalwork. The evidence for the casting technology is far from complete, but it is possible to draw some conclusions from examination of the objects as well as from the literary and archaeological evidence discussed above. Lost-wax casting was known both in Iran and Jordan in the early Islamic period, and there is a firsthand account of the method in the Jazira around 1200, so it is appears to have been widespread. Evidence for piece-mold casting is equivocal, but the method cannot be ignored. Sand casting, on the other hand, was certainly being practiced as early as the tenth century, well before it is thought to have been used in Europe.

Brass is particularly suited to being hammered out into thin sheets, and this was utilized in the manufacture of fine inlaid objects, but there is no substantive evidence for the shaping of brass vessels by spinning, although both cast and raised objects were often finished on a rotating lathe. The technical study of objects is still yielding information about Islamic metal technology; the picture is becoming clearer and some of the common misconceptions can be corrected.

Acknowledgments

I would like to thank my colleagues Rachel Ward, Paul Craddock, Duncan Hook, Venetia Porter, Michael Wayman, Antony Simpson, and Tony Milton, who have all contributed to this research. I would also like to thank Natasha Eniosova, Matthew Ponting, Irit Narkiss, and Sariel Shalev. I would particularly like to thank Paul Jett for his observations about pieces in the collection of the Freer Gallery.

References

Ackermann, S., and S. La Niece. Forthcoming. Metals and mansions: A thirteenth-century astrolabe analysed. *Annals of science*.

Allan, J. 1979. *Persian metal technology, 700–1300 C.E.* Oxford: Ithaca Press.

——. 1999. *Islamic metalwork: The Nuhad Es-Said Collection.* London: Philip Wilson Publishers, Ltd.

al-Jazari, I. al-R. 1974. *The book of knowledge of ingenious mechanical devices.* Trans. D. R. Hill. Boston: D. Reidel. 191–95.

Atil, E., W. T. Chase, and P. Jett. 1985. *Islamic metalwork in the Freer Gallery of Art*. Washington, D.C.: Smithsonian Institution.

Barnard, N. 1961. *Bronze casting and bronze alloys in ancient China*. Japan: Monumenta Serica Monograph XIV.

Biringuccio, V. 1959. *The pirotechnia*. Book 8. Trans. C. Stanley Smith and M. Teach Gnudi. New York: Basic Books.

Braunstein, P. 1977. Le marché du cuivre à Venise à la fin du moyen-age. In *Schwerpunkte der Kupferproduktion und des Kupferhandels in Europa, 1500–1650,* ed. H. Kellenbenz, 78–94. Cologne: Böhlau.

Burger, W. 1976. *Chi'ing cash until 1735*. Taiwan: Mei Ya Publications.

Coghlan, H. H., W. Watson, and E. Voce. 1951. Ancient Mining and Metallurgy Committee: Examination of a fragment of the Welwyn bowl in the British Museum. *Man* 64–65:37–39.

Coomaraswamy, A. K. 1924. *The treatise of al-Jazari on automata*. Boston. 17 Pl. VII.

Craddock, P. T. 1979. The copper alloys of the medieval Islamic world: Inheritors of the classical tradition. *World archaeology* 11(1):68–79.

Craddock, P. T., and J. Lang. 1983. Spinning, turning, and polishing. *Journal of the Historical Metallurgy Society* 17(2):79–81.

Craddock, P. T., S. C. La Niece, and D. R. Hook. 1998. Brass in the medieval Islamic world. In *2000 years of zinc and brass*, Occasional Paper 50, ed. P. T. Craddock, 73–114. London: British Museum.

Crawshaw, F. 1909. *Metal spinning*. Chicago: Popular Mechanics. Reprinted, 1991, Bradley, Ill.: Lindsay Publications.

Garenne-Marot, L., R. Bertholon, B. Bell, and N. Lacoudre. 1998. Métal et techniques de fabracation d'un lot d'objects en alliage base-cuivre de l'époque omeyyade (VIIIe siècle). In *Les métaux antiques: Travail et restauration,* ed. G. Nicolini and N. Dieudonné-Glad, 29–38. Montagnac: Monique Mergoil.

Gettens, R. J. 1969. *The Freer Chinese bronzes*. Vol. II: Technical studies. Washington, D.C.: Smithsonian Institution. 99–107.

Giuzalian L. T. 1968. The bronze qalamdan (pen-case) from the Hermitage collection. *Ars Orientalis* 7:95–119.

Hasluck, P. N. 1904. *Metalworking: A book of tools, materials, and processes for the handyman*. London: Cassell.

Johnson, H. V. 1960. *Metal spinning, techniques and projects*. Milwaukee: Bruce Publishing.

Kerr, R. 1990. *Later Chinese bronzes*. Far Eastern Series. London: Victoria and Albert Museum. 90–104.

Lakpour, S. 1997. *Sefid Ruy*. Teheran: National Cultural Heritage Organization.

Maryon, H. 1971. *Metalwork and enamelling*. 5th revised ed. New York: Dover.

Melikian-Chirvani, A. S. 1974. The white bronzes of early Islamic Iran. *The Metropolitan Museum journal* 9:123–51.

——. 1982. *Islamic metalwork from the Iranian world*, London: HMSO.

Pollard, A. M. 1996. *Archaeological chemistry*. London: Royal Society of Chemistry. 218.

Reagan, J. E., and E. E. Smith. 1936. *Metal spinning*. Milwaukee: Bruce Publishing. Reprinted, 1991, Bradley, Ill.: Lindsay Publications.

Reeves, R. 1962. *Cire perdue casting in India*. New Delhi: Crafts Museum.

Shalev, S. In press. The Fatimid metal hoard from Caesarea: A case study in archaeology, metallurgy, and anthropology. *Caesarea maritima excavation reports*.

Stronge, S. 1985. *Bidri ware, inlaid metalwork from India*. London: Victoria and Albert Museum.

Theophilus. 1979. *On divers arts*. Trans. J. G. Hawthorne and C. Stanley Smith. New York: Dover.

Timmins, S. 1866. *The resources, products, and industrial history of Birmingham and the Midland hardware district*. London: Robert Hardwicke.

Tuells, C., and W. A. Painter. 1910. *Metal spinning*. Machinery's Reference Series no. 57. Reprinted, 1994, Bradley, Ill.: Lindsay Publications.

Untracht, O. 1968. *Metal techniques for craftsmen*. London: Doubleday.

Ward, R., S. La Niece, D. R. Hook, and R. White. 1995. Veneto-Saracenic metalwork: An analysis of the bowls and incense burners in the British Museum. In *Trade and discover: The scientific study of artefacts from post-medieval Europe and beyond,* Occasional Paper no. 109, ed. D. Hook and D. Gaimster, 235–58. London: British Museum.

Werner, O. 1977. Analysen mittelalterlicher bronzen und messinge I. *Archäologie und naturwissenschaft* 1:44–120.

——. 1981. Analysen mittelalterlicher bronzen und messinge II & III. *Archäologie und naturwissenschaft* 2:106–120.

——. 1982. Analysen mittelalterlicher bronzen und messinge IV. *Berliner beitrage zur archäometrie* 7:35–174.

Woodbury, R. S. 1972. *History of the lathe to 1850*. Cambridge, Mass.: MIT Press.

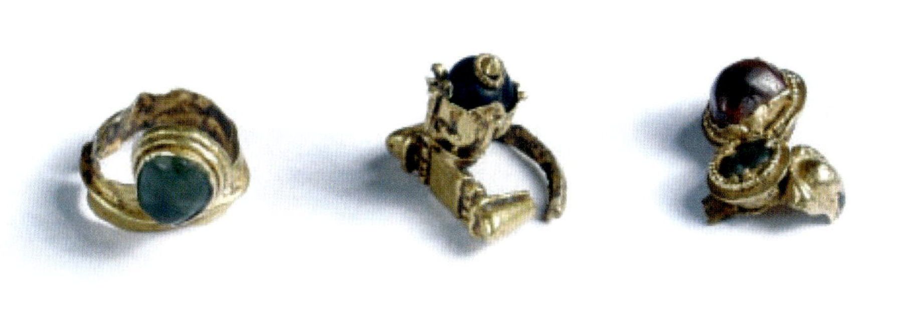

***Plate 1.** Three small gold earrings, or "hawk tags," Java, Early Classic period, private collection*

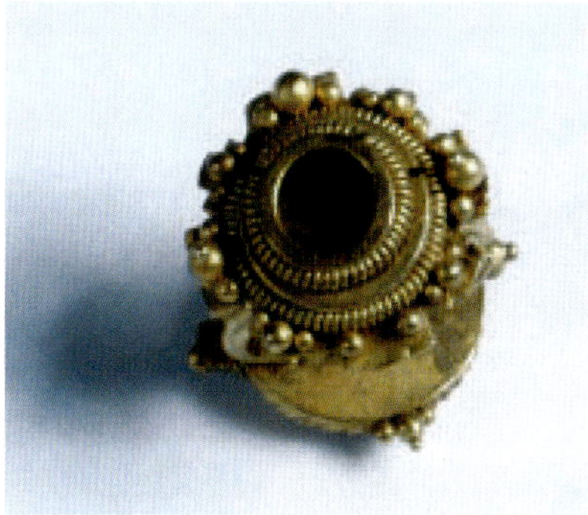

***Plate 2.** Gold earring, or "hawk tag," Java, Late Classic period, private collection*

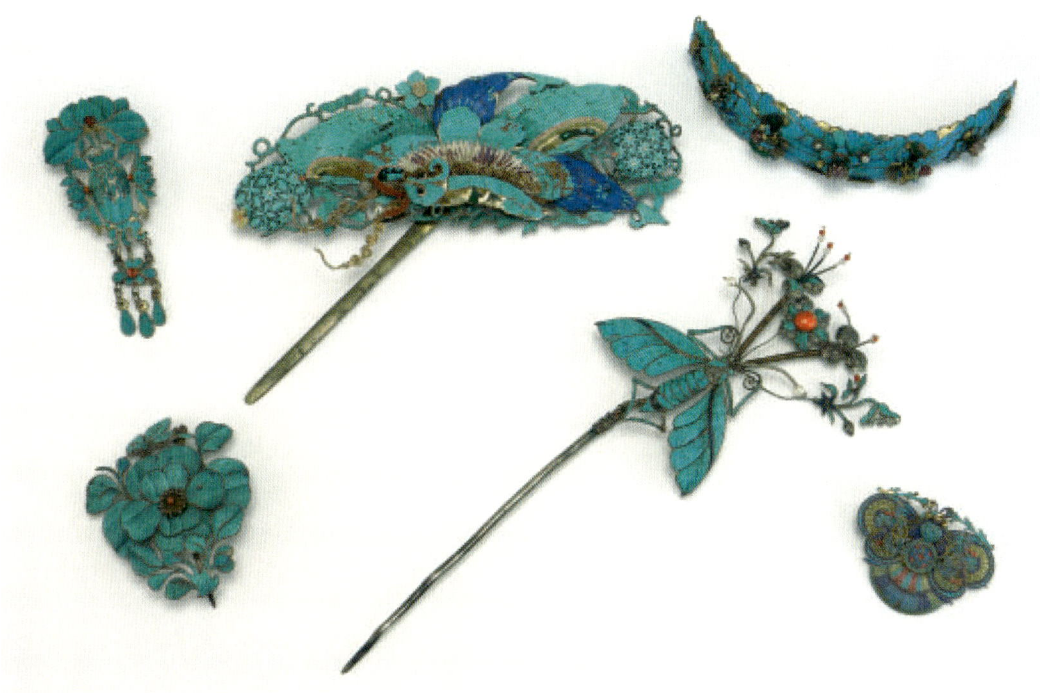

***Plate 3.** Jewelry from the Arthur M. Sackler Gallery with applied feather decoration (*clockwise from top left: *S2001.3, RLS1997.48.4644, S2001.4, RLS1997.48.4625, RLS1997.48.4645, RLS1997.48.4635)*

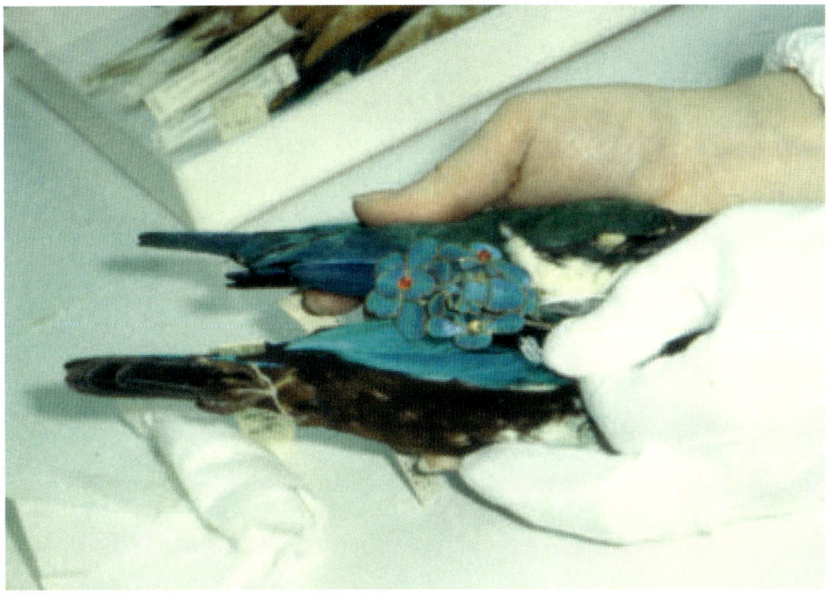

Plate 4. Halcyon chloris *(top) and* Halcyon smyrnensis *(bottom) specimens with feathered pin (RLS1997.48.4638)*

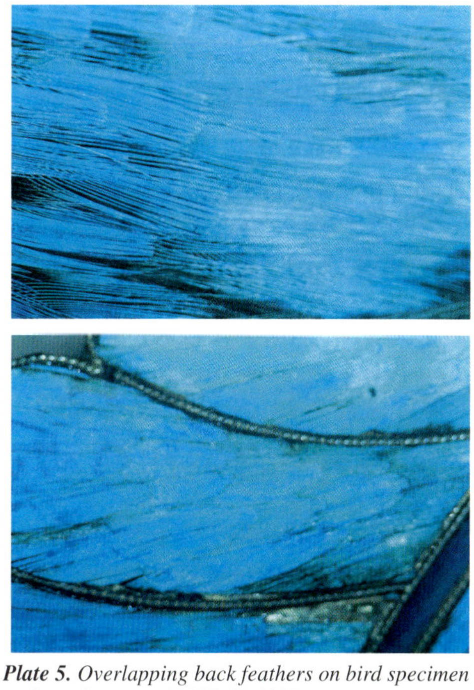

***Plate 5.** Overlapping back feathers on bird specimen and on object (RLS1997.48.4645)*

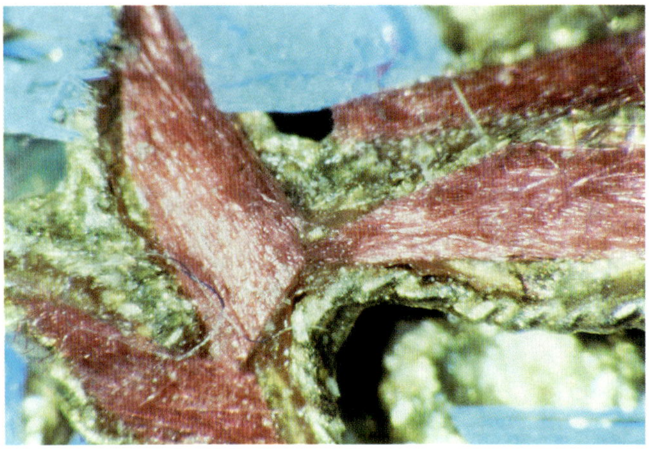

Plate 6. *Detail of red silk floss inlay used in addition to purple Indian Roller and blue kingfisher feathers on a butterfly-shaped pin (RLS1997.48.4625)*

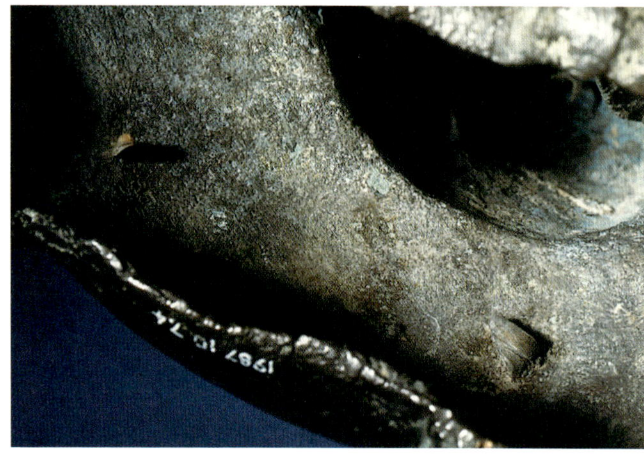

Plate 11. *Copper chaplets visible inside the base of a figurine tag (reg. 1987-10-07,04)*

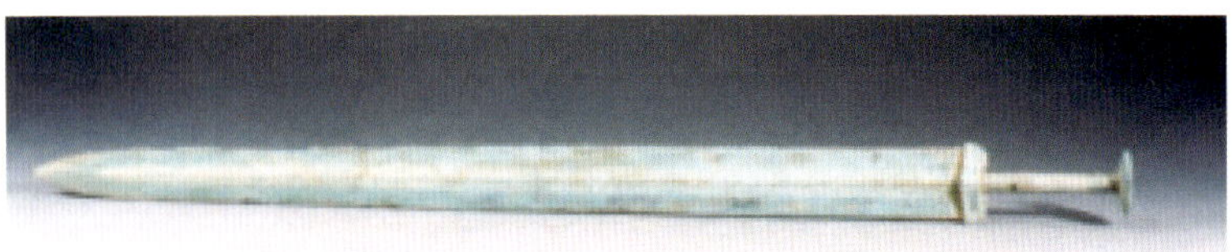

Plate 7. *A bimetallic sword from ancient China*

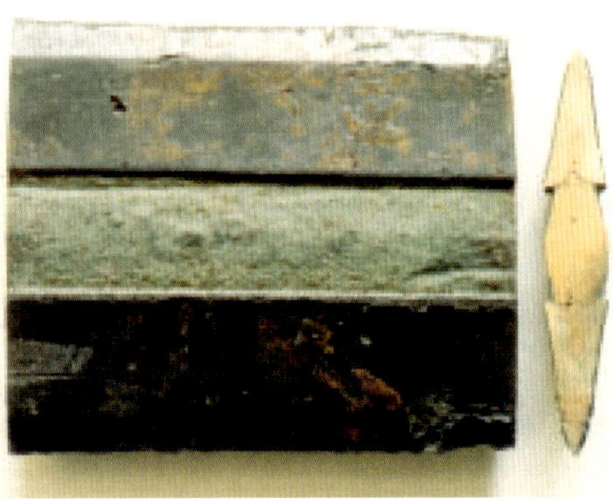

Plate 8. *A fragment of bimetallic sword and its cross-section*

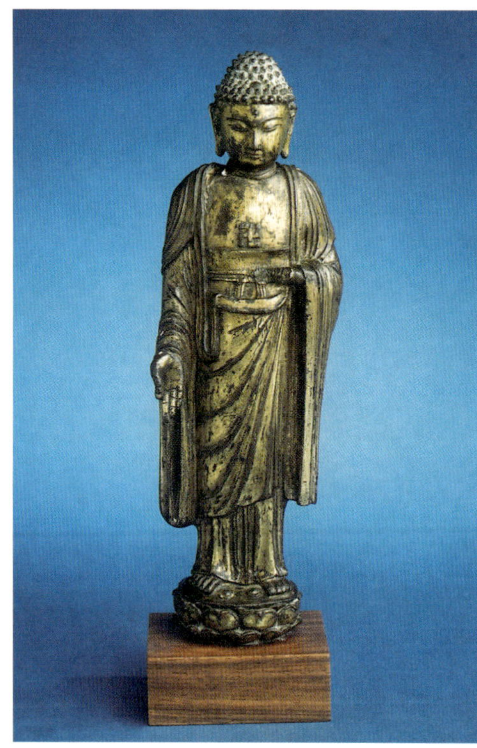

Plate 12. *Figurine with cobalt blue pigment on the head (reg. 1973-07-26,81)*

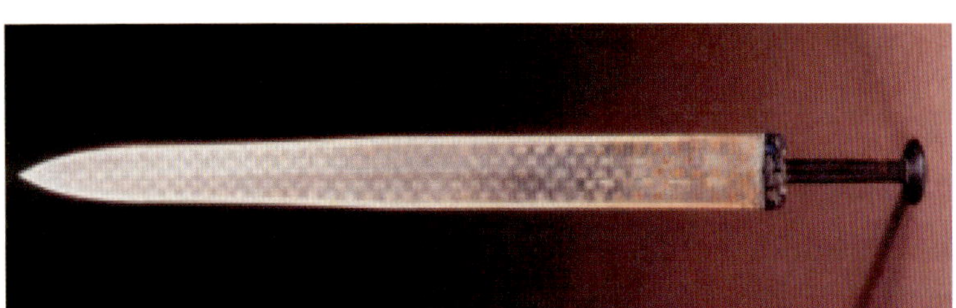

Plate 9. *The sword of Yue king Gou Jian*

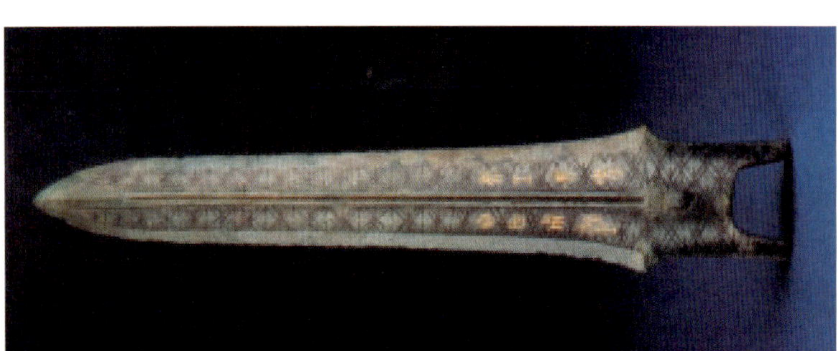

Plate 10. *A second example of a lozenge-patterned weapon*

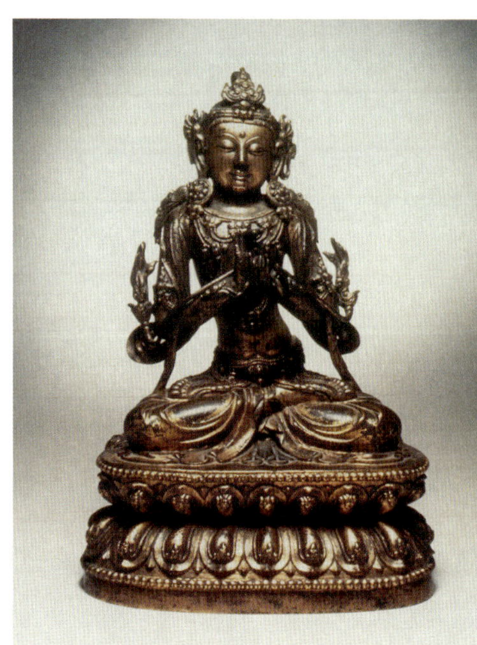

Plate 13. *Figurine of Sino-Tibetan style (reg. 1987-11-10,01)*

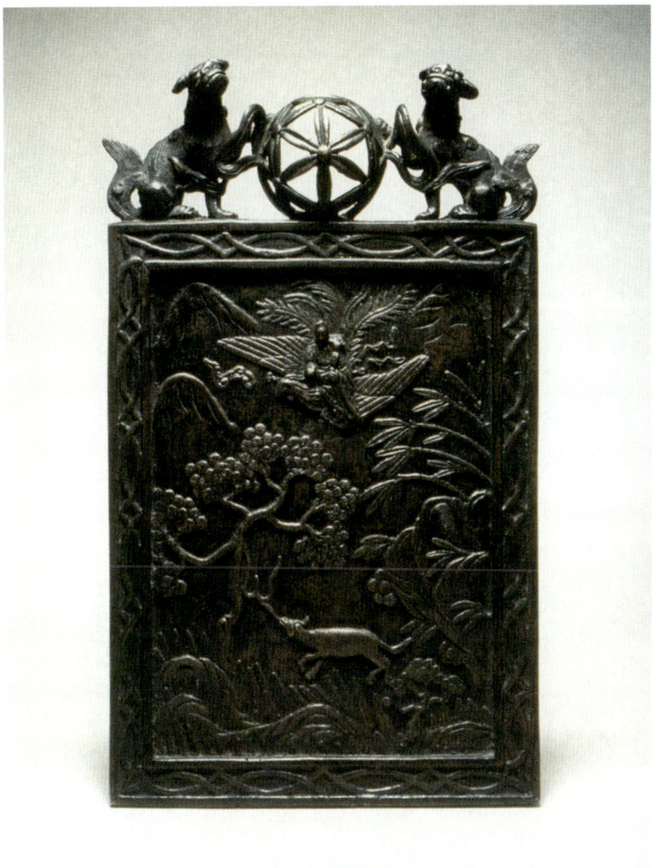

Plate 14. *"Grand-style" artifact, plaque dated 1648 C.E. (reg. 1990-05-30,01)*

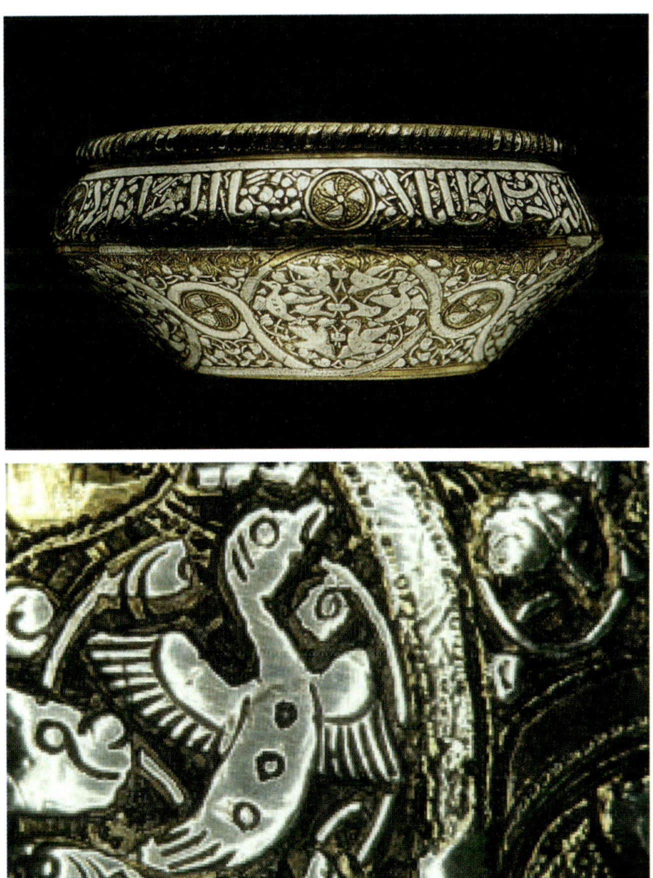

Plate 15. *A Mamluk brass bowl and detail of the inlay. Egypt or Syria, 13th century C.E. (BM 1878.12-30.691) rim diameter 11.2 cm. Thin silver sheet was keyed into tiny pits punched into the brass around the edge of the inlay area. Details were gently chased onto the surface of the silver. Copyright British Museum*

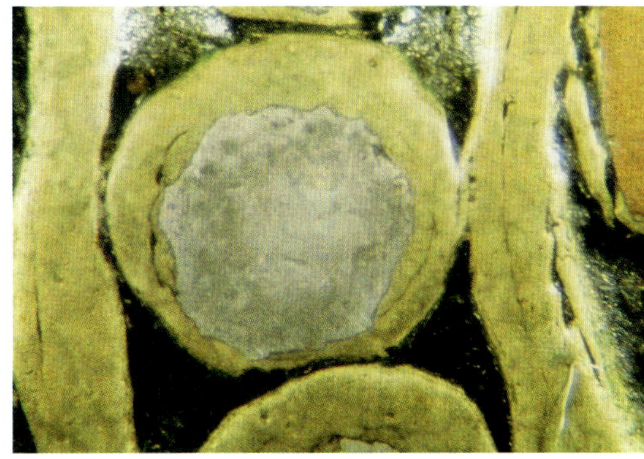

Plate 16. *Inlaid individual silver dot in the front of the throne cushion. Faintly inscribed marks encircling the dot and following the ridges on either side are planning lines*

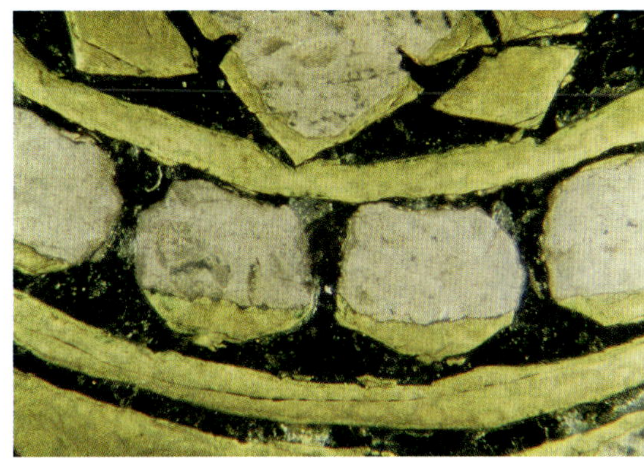

Plate 17. *Inlaid silver circle that has subsequently been divided into dots for the side of the throne cushion. Planning lines are faintly visible on the brass*

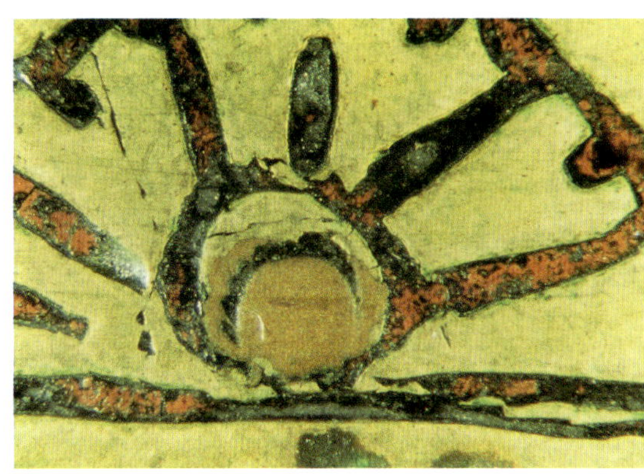

Plate 18. *Inlaid copper from the throne surface, showing the use of deeply incised lines to accentuate the design in a step subsequent to the copper inlay. Red-pigmented resin colors the incised decoration in the brass*

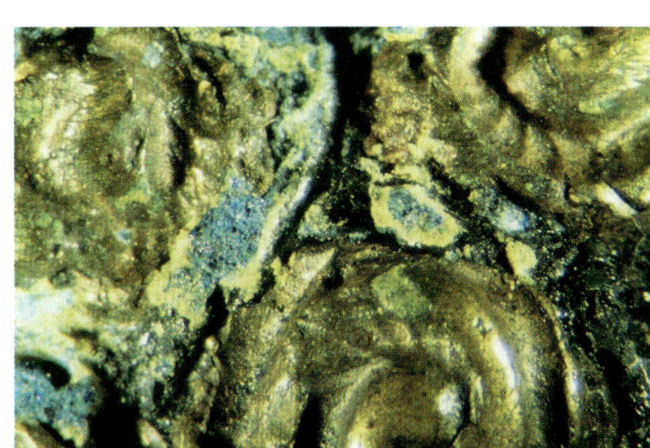

Plate 19. *Traces of original polychrome decoration in the Buddha's hair with a clear sequence of black resin overlaid by tan- and azurite-colored tempera paints*

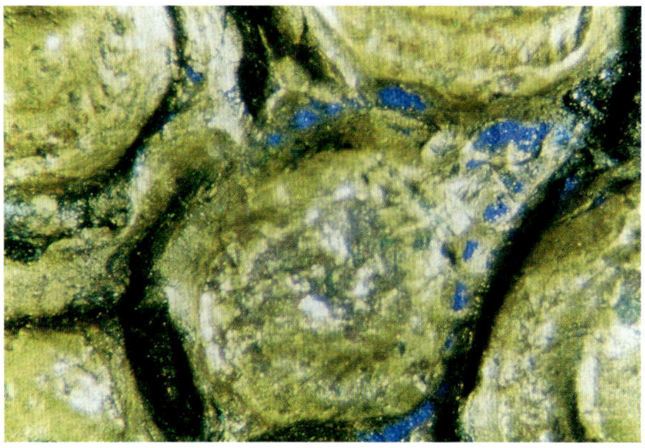

Plate 20. *Restoration in the form of synthetic ultramarine applied directly atop brass in the hair of the Buddha*

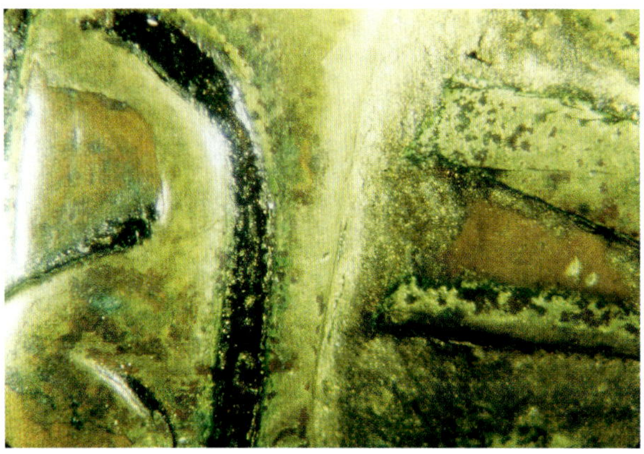

Plate 21. *Simulated copper inlay using brown-pigmented resin nearby copper inlay. Copper is on the right. Remnants of later gold powder paint obscure some edges*

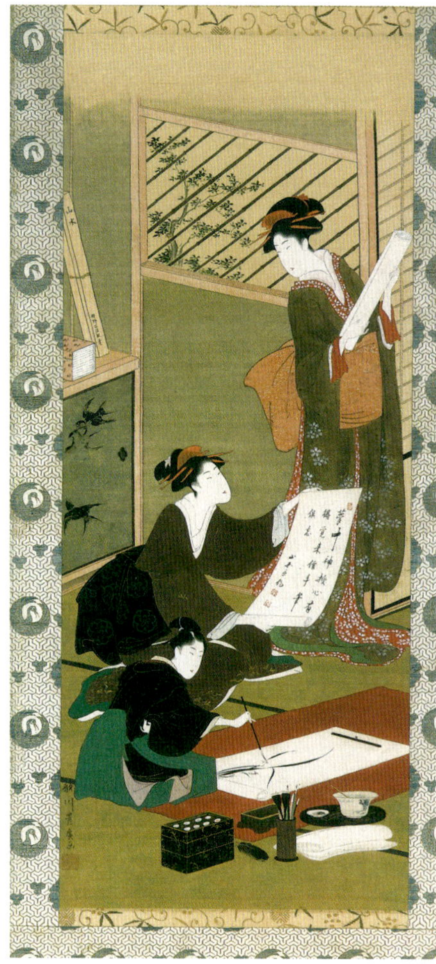

Plate 22. *Utagawa Toyohiro (1773–1828),* Two of the Four Accomplishments: Painting and Calligraphy. *Japan, Ukiyo-e school, painting on silk. Freer Gallery of Art, Smithsonian Institution, Washington, D.C. Gift of Charles Freer, F1903.57*

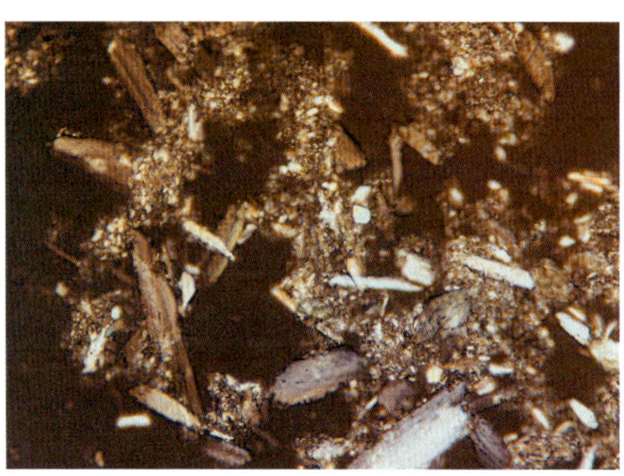

Plate 23. *Shell white, photomicrograph. From a painting in the style of Katsushika Hokusai (1760–1849). Mounted in Canada balsam, n = 1.53. ×300, between partly crossed polars. Freer Gallery of Art, Smithsonian Institution, Washington, D.C. Gift of Charles Lang Freer, F1898.95*

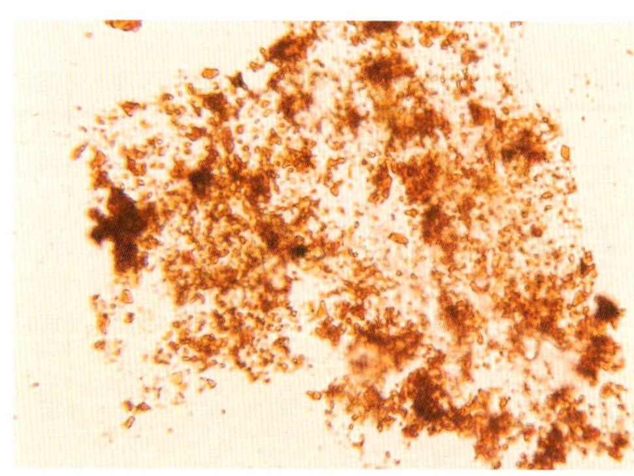

Plate 25. *Red iron oxide, photomicrograph. From a sketch by Katsushika Hokusai (1760–1849). Mounted in Canada balsam, n = 1.53. ×800, in transmitted light. Freer Gallery of Art, Smithsonian Institution, Washington, D.C. Gift of Charles Lang Freer, F1904.22*

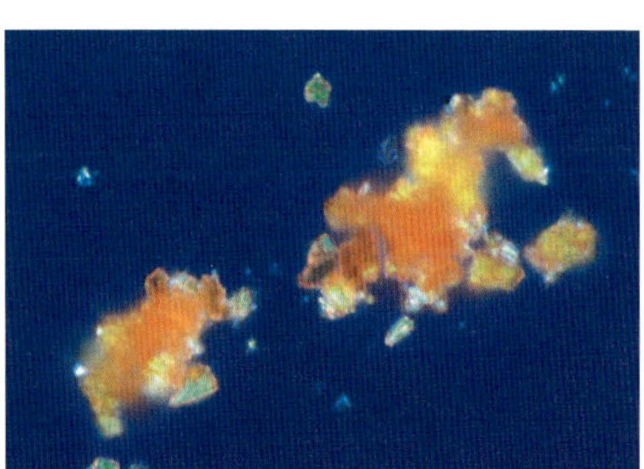

Plate 24. *Red lead, photomicrograph. From a painting in the style of Katsushika Hokusai (1760–1849). Mounted in Canada balsam, n = 1.53. ×800, between crossed polars. Freer Gallery of Art, Smithsonian Institution, Washington, D.C. Gift of Charles Lang Freer, F1904.134*

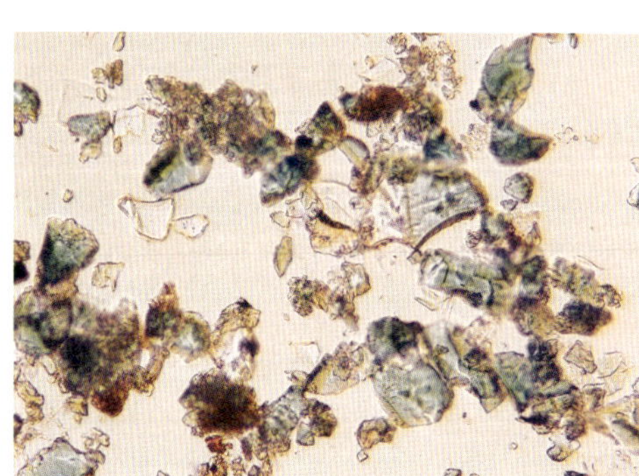

Plate 26. *Azurite, photomicrograph. From a six-fold screen in seventeenth-century style. Mounted in Canada balsam, n = 1.53. ×200, in transmitted light. Freer Gallery of Art. Smithsonian Institution, Washington, D.C. Gift of Charles Lang Freer, F1903.208*

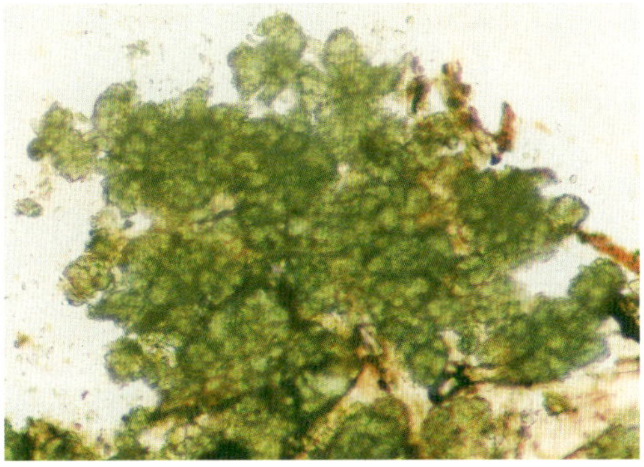

Plate 27. *Copper green (atacamite), photomicrograph. From a painting by Torii Kiyonaga (1752–1815). Mounted in Canada balsam, n = 1.53. ×265, in transmitted light. Freer Gallery of Art, Smithsonian Institution, Washington, D.C. Gift of Charles Lang Freer, F1902.101*

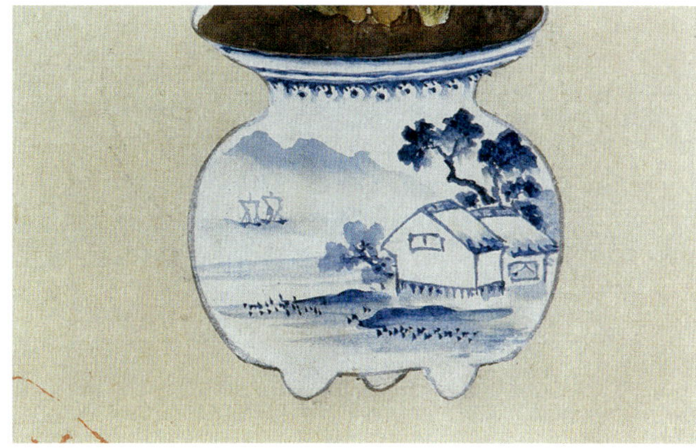

Plate 30. *Detail of plate 29*

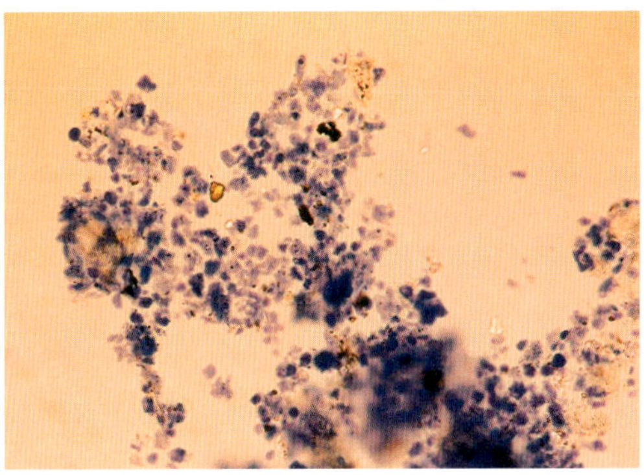

Plate 28. *Ultramarine, artificial, photomicrograph. From a nineteenth-century painting by Hiroshige II, nineteenth century. Mounted in Canada balsam, n = 1.53. ×400, between partly crossed polars. Freer Gallery of Art. Smithsonian Institution, Washington, D.C. Gift of Charles Lang Freer, F1903.106*

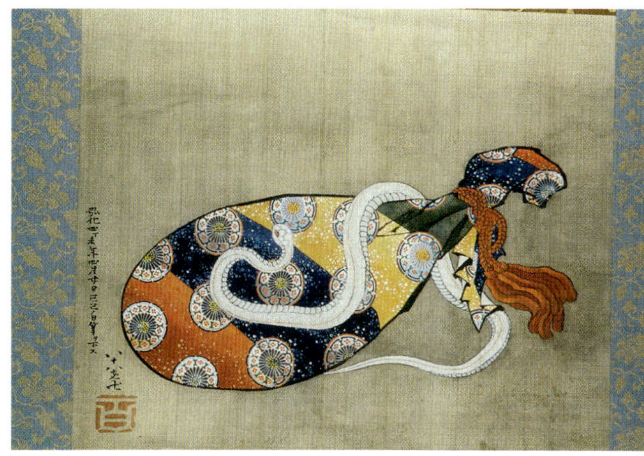

Plate 31. *Hokusai* Lute and White Snake of Benten (Sarasvati) *hanging scroll*

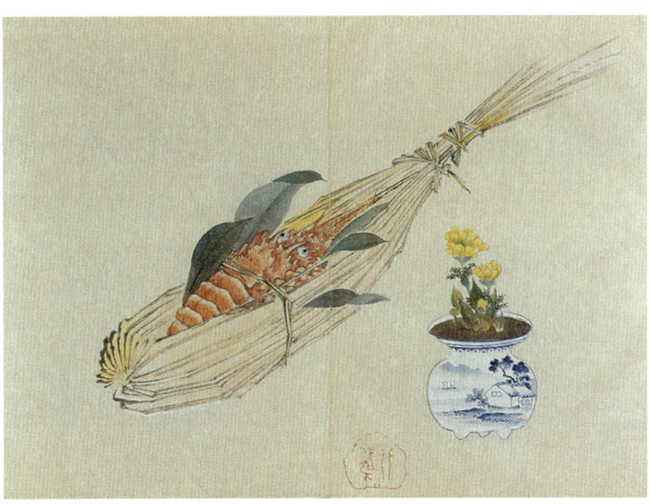

Plate 29. *Painting from album by Hokuga (Japan, 1st half 19th century)*

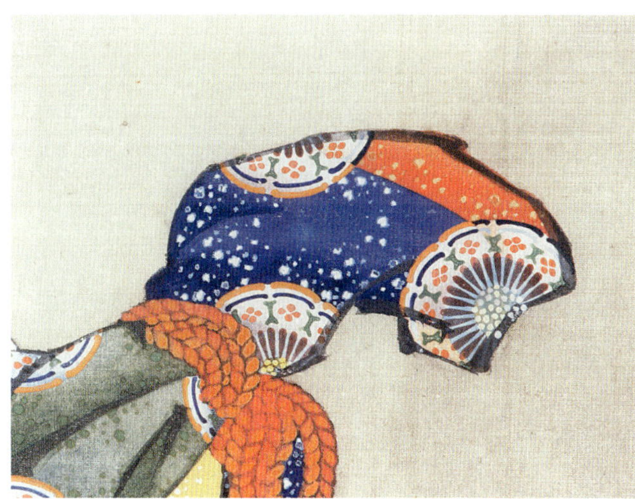

Plate 32. *Detail of plate 31*

Fluorescence Maxima

	Excitation λ	Emission λ
Carminic Acid	380 365 300	590
Lac Dye	400 320	600

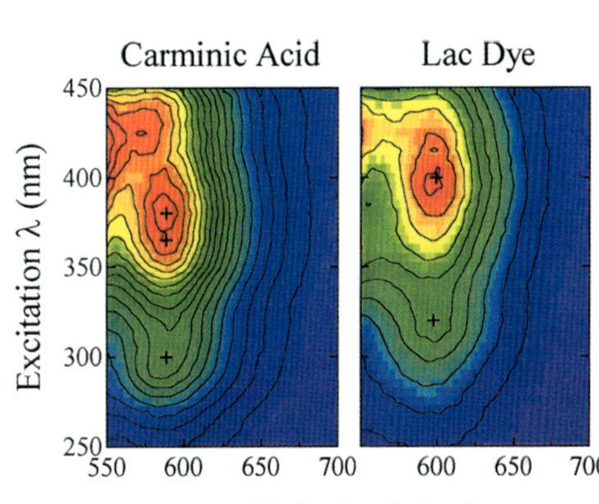

Plate 33. *Fluorescent spectra, lac dye and carminic acid*

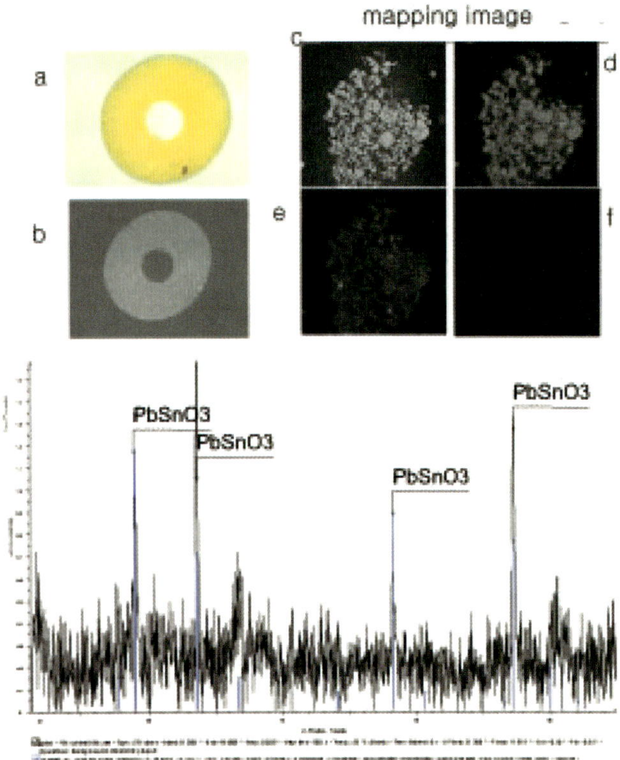

Plate 34. *Particle embedded in yellow glass ($PbSnO_3$). (a) PM image (b) SEM(BSE) image (c) image (d) Pb (e) Sn (f) O*

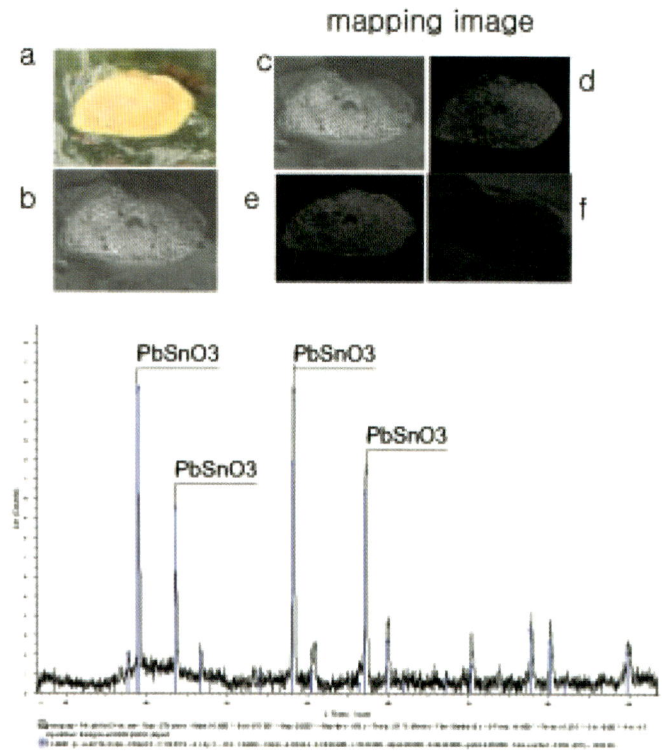

Plate 35. *Particle embedded in green glass ($PbSnO_3$). (a) PM image (b) SEM(BSE) image (c) image (d) Pb (e) Sn (f) O*

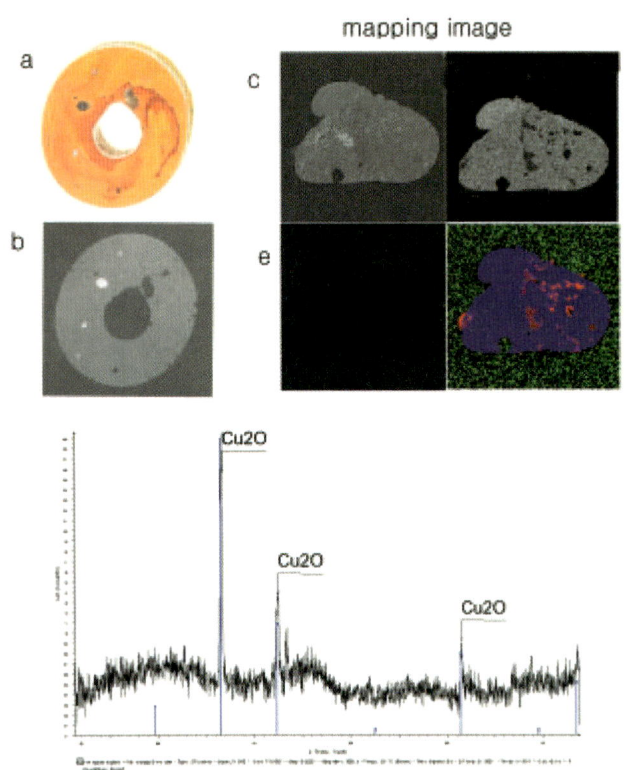

Plate 36. *Particle embedded in orange glass (Cu_2O). (a) PM image (b) SEM(BSE) image (c) image (d) Cu (e) O (f) map*

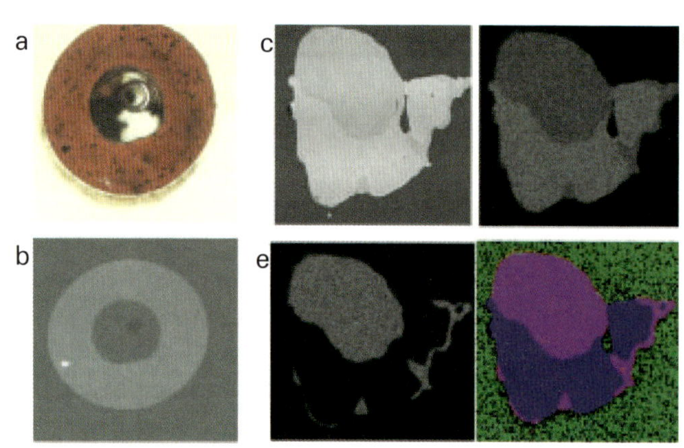

Plate 37. *Particle embedded in red glass (Cu_2S?). (a) PM image (b) SEM(BSE) image (c) image (d) Cu (e) S (f) map[B(Cu), R(S), G(O)]*

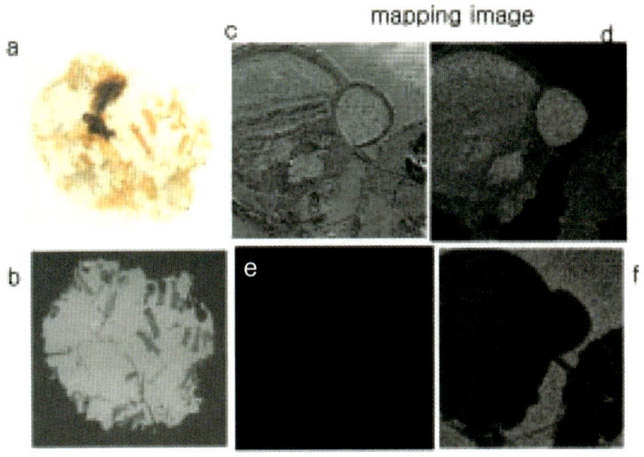

Plate 38. *Particle embedded in dark-green glass (Fe, O). (a) PM image (b) SEM(BSE) image (c) image (d) Fe (e) O (f) Ca*

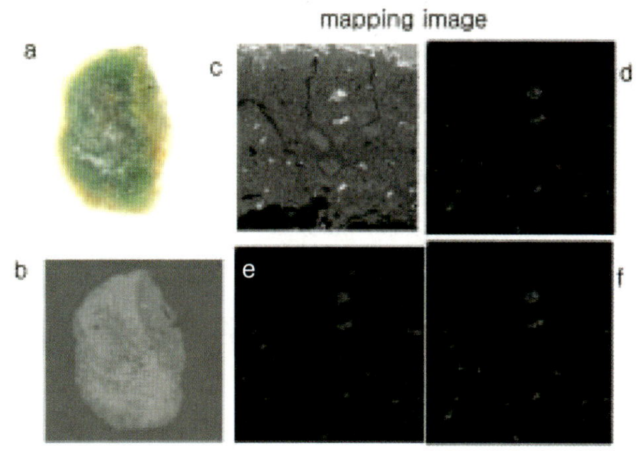

Plate 39. *Particle embedded in dark-green glass ($Pb_2Sb_2O_7$?). (a) PM image (b) SEM(BSE) image (c) image (d) Pb (e) Sb (f) O*

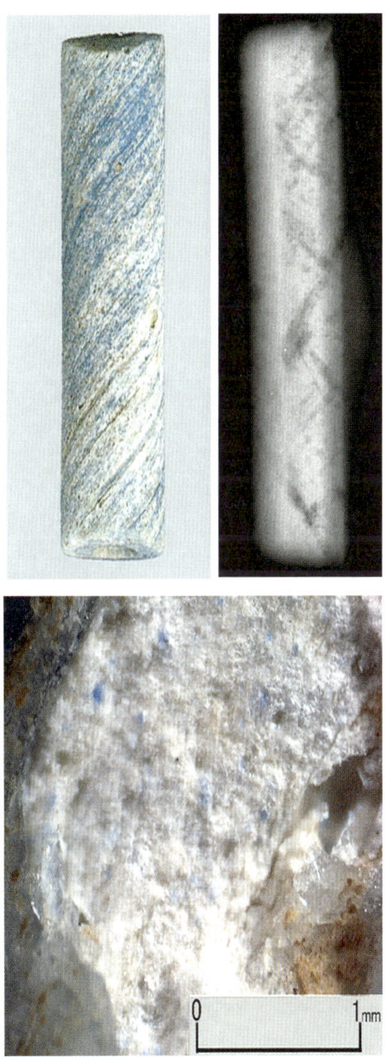

Plate 40. Above left: *Light blue tubular bead, moderately weathered, 20.3–22.8 mm in length, Arimoto site in Okayama Prefecture (1st–2nd cent. C.E.).* Above right: *CR image. The density of less weathered portion is 3.76–4.19 g/cm³ by X-CT.* Below: *Micro-photograph*

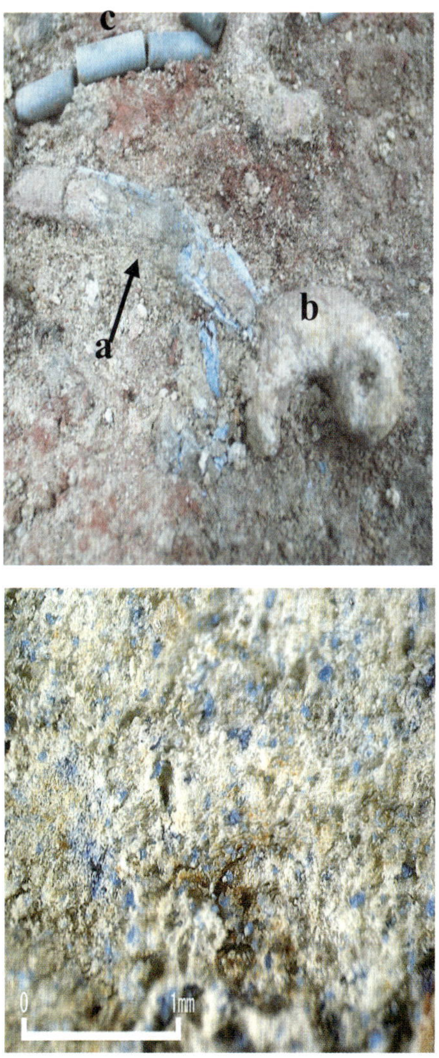

Plate 41. Above: *Akasaka-Imai site in Kyoto Prefecture. (a) Light blue lead-barium tubular beads, heavily weathered, about 20 mm in length. (b) Lead-barium green comma-shaped beads. (c) Green stone tubular bead.* Below: *Micro-photograph. Han blue was discovered from light blue tubular bead*

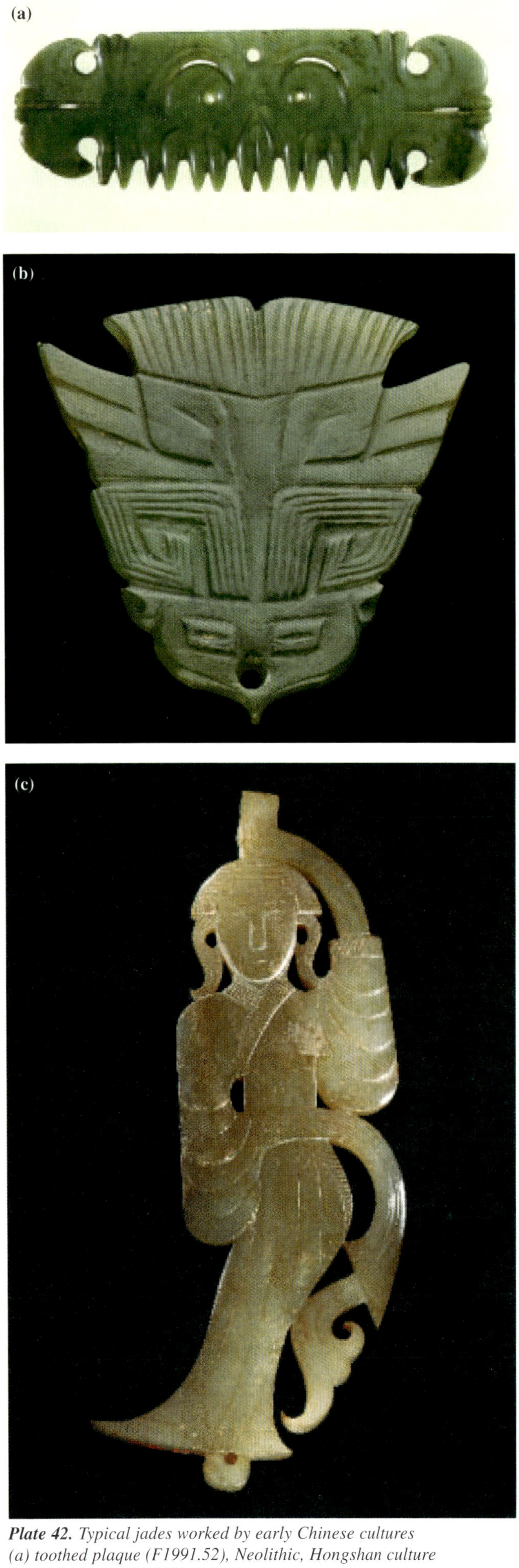

Plate 42. *Typical jades worked by early Chinese cultures*
(a) toothed plaque (F1991.52), Neolithic, Hongshan culture
(b) bird pendant (S1987.567), Shang dynasty
(c) dancing figure (F1938.16), Eastern Zhou dynasty

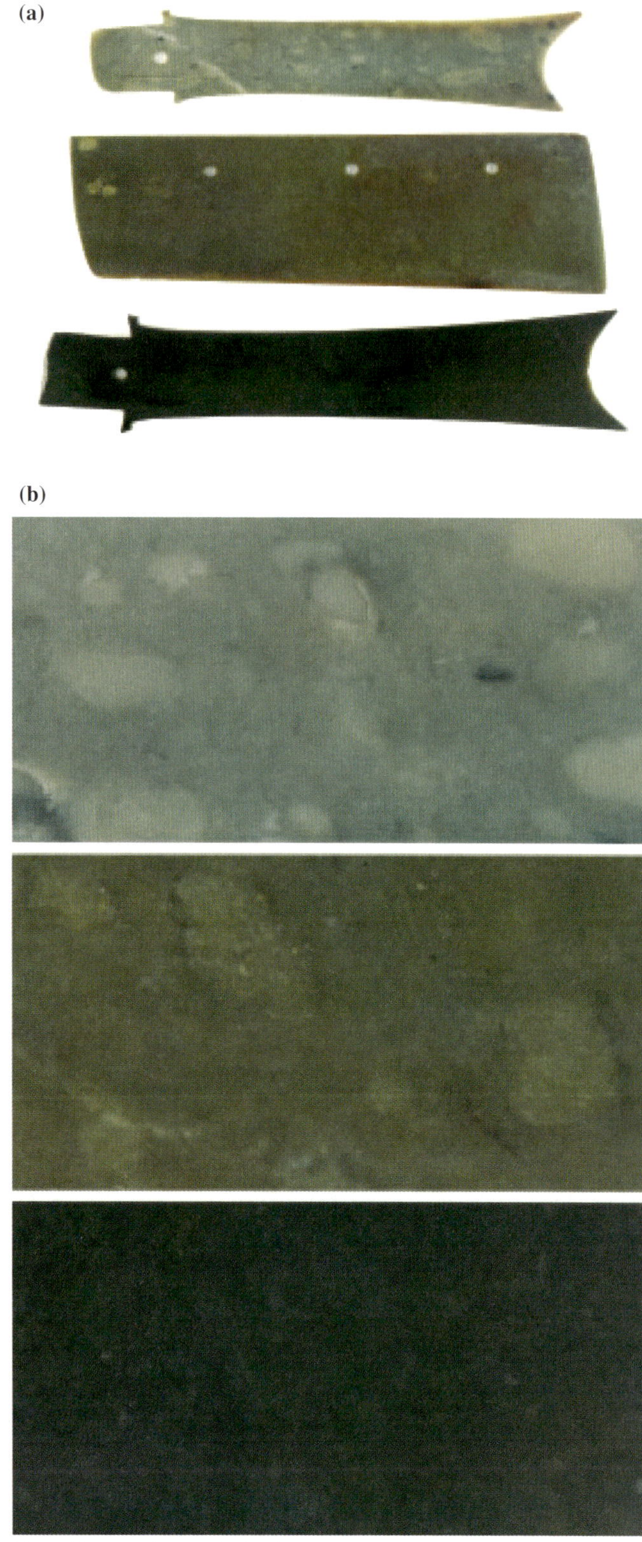

Plate 43. *(a) Three jade blades with a subnodular structure, indicating a conglomerate parent rock for the nephrite (F1916.162, F1917.24, F1916.491)*
(b) Close-up of these jades (image width = 5 cm). These blades tend to be gray to almost black in color, becoming generally darker with increasing iron and manganese content. The first contains FeO = 1.42%, MnO = 0.23%; the second contains FeO = 2.32%, MnO = 0.24%; and the third contains FeO = 10.57%, MnO = 0.76% (by wt.)

12 A Scientific Study of Identical Bronze Mirrors from Japan

Ryu Murakami, Masaaki Sawada, W. T. Chase, and P. Jett

ABSTRACT Two identical bronze mirrors from Japan dating roughly to the third/fourth century C.E. were studied and compared in order to determine their method of manufacture. Questions concerning this general type of mirror and its history are many. The study hoped to address and answer at least the question of what type of replicative procedure was used to create the mirrors' identical decoration. The main techniques employed in the study were radiographic and microscopic examination, but other forms of physical examination, including surface analysis by x-ray fluorescence spectroscopy, were also used. The possible methods of casting that might have been used to make the mirrors are also reviewed.

Introduction

Two ancient bronze mirrors found in Japan are linked by their identical design but separated by their different histories. One of the mirrors was acquired in Japan in 1909 by Charles Lang Freer and was part of the original collection of the Freer Gallery of Art. The other mirror was found in 1989 during the excavation of the Yukinoyama tumulus in Shiga Prefecture, Japan.

The mirrors are of a type known as "triangular-rim" mirrors, and their decoration consists of deity and animal figures, bosses, and annular bands of calligraphy and geometric forms. In Japan there is an ongoing controversy regarding the origins of this type of mirror and its role in Japanese history (Kondo 1988; Okamoto 1999). The controversy extends to a lack of agreement about how such mirrors were made.

A fuller understanding of the mirrors requires study of both their technical and historical nature. Questions concerning technique may be addressed by the comparison of the differences and similarities between mirrors that appear identical; in this way, much can be revealed about the methods of manufacture. To that end, these two mirrors were subjected to close examination, with the observations of that examination related here along with the conclusions reached about the mirrors' method of manufacture.

Histories of the Two Mirrors

Freer Mirror

Charles Lang Freer traveled to Japan in the fall of 1909 to attend a memorial service for Ernest Fenollosa, held at Miidera Temple in Otsu, Shiga Prefecture (fig. 1) (Lawton and Merrill 1993, 149–51). It was during this trip that Freer acquired the mirror shown in figure 2(b) from Mitsuaki Tanaka through Samurai Shokai in Yokohama. Twenty-two years after Freer acquired the mirror, it was studied and published by Professor Suegi Umehara, the preeminent scholar of East Asian bronzes at that time (Umehara 1931).

Although the origins of the Freer mirror are unknown, it must have been removed from a tumulus. Extensive

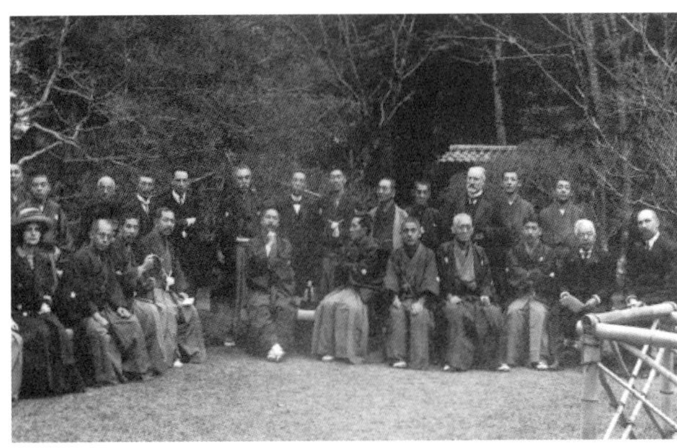

Figure 1. *The people attending the memorial service for Ernest F. Fenollosa, held in Miidera temple in 1909. The person sitting at the right is Mr. Freer*

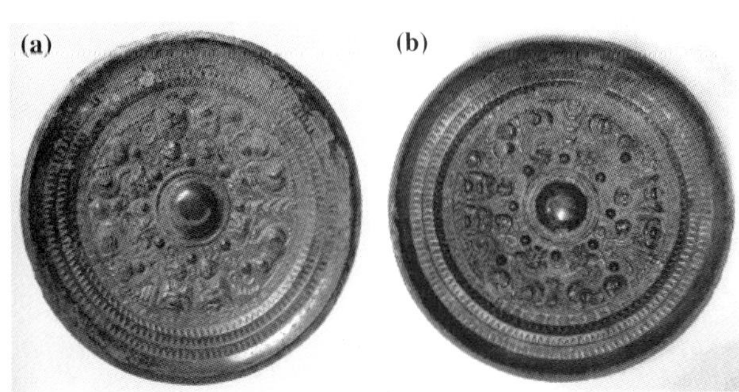

Figure 2. *The identical mirrors. (a) Yukinoyama mirror. (b) Freer mirror (F1909.340)*

investigation in the Freer archives by one of the authors, Ryu Murakami, did not produce any further information about the history of the mirror.

Yukinoyama Mirror

The Yukinoyama mirror (fig. 2a) is so called because it was found in a tumulus located at the top of Yukinoyama Hill in southern Shiga Prefecture. The tumulus is a "keyhole-shaped" burial mound, or *zempōkōenfun,* typical of the Kofun period (250–600 C.E.) in Japan (Pearson 1992, 197). The site was excavated during four seasons over the course of three years from 1989 to 1992 (Tsude, Fukunaga, and Sugii 1996). Evidence from the site confirmed a Kofun-period date of about the early fourth century.

Among the discoveries at the site was an undisturbed pit-style burial chamber. Within the chamber was a stone coffin that contained five bronze mirrors and over sixty bronze arrowheads (fig. 3). It was soon realized that one of the mirrors was identical to the Freer mirror published by Umehara almost sixty years earlier (Umehara 1931).

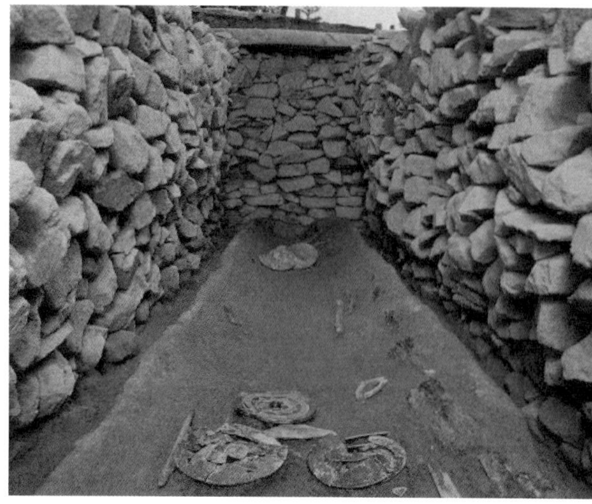

Figure 3. *The inside of the stone chamber of the Yukinoyama tumulus*

The Comparison of Both Mirrors

Figure 4 shows a detailed drawing with dimensions of the Yukinoyama mirror (Tsude, Fukunaga, and Sugii 1996). The Freer mirror has precisely the same dimensions. The weights of both mirrors are shown in table 1. Their weights are also very similar.

The analytical results of both mirrors are tabulated in table 2 (Murakami and Sawada 1996; Murakami, Koezuka, and Sawada 1996). In this research, we were impeded by the fact that we could not examine the mirrors together because both mirrors are restricted to their present locations. They were analyzed with nondestructive x-ray fluorescence (XRF) analyzers at both the laboratories in Freer Gallery and the Nara National Institute of Cultural Properties (Murakami and Sawada 1996). The analytical results show that both mirrors are typical high-tin bronzes containing some lead. Although surface analysis using XRF can only yield approximate results for cast bronzes, the mirrors appear to be very similar in composition.

Radiographs of both mirrors were taken (figs. 5[a] and [b]) and studied by overlapping the two in order to more easily discern any differences between the radiographic images (fig. 6). After aligning the designs on both mirrors,

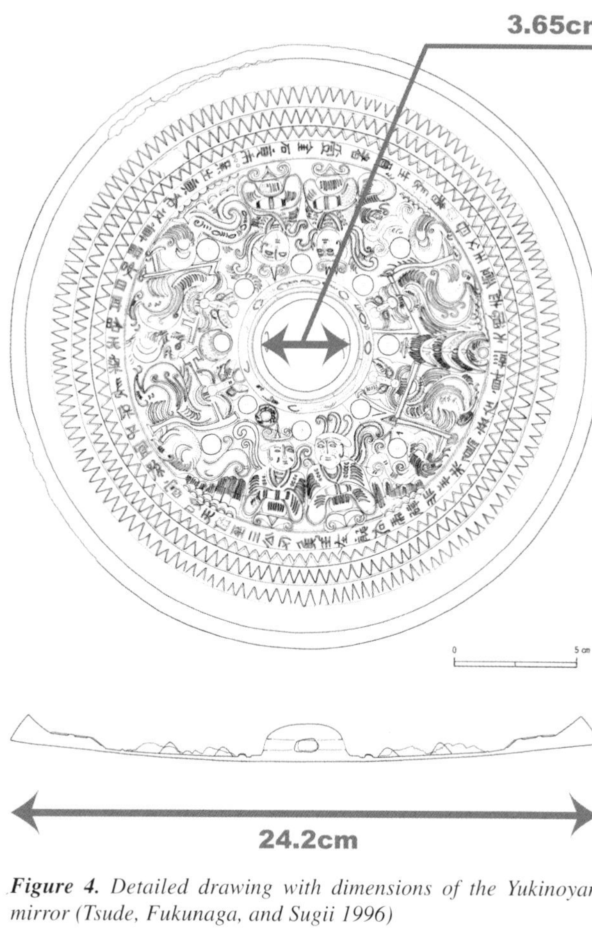

Figure 4. *Detailed drawing with dimensions of the Yukinoyama mirror (Tsude, Fukunaga, and Sugii 1996)*

Table 1. *The weights of both the Yukinoyama mirror and the Freer mirror*

	Weight (g)
Freer	1587.4
Yukinoyama	1617.0

Table 2. *The analytical results by nondestructive x-ray fluorescence analyses of both the Yukinoyama mirror and the Freer mirror*

		Cu	Sn	Pb	As	Ag	Hg	Bi	Fe
Freer	Mirror Surface	35.0	56.0	6.1					3.0
F1909.340	Mirror Surface	34.9	55.6	5.7					3.8
	Mirror Back	41.4	46.8	7.9					4.0
Yukinoyama	Mirror Surface	48.0	41.0	6.2	0.6	0.2	0.1	0.1	4.2
	Mirror Surface	36.0	55.0	7.8	0.9	0.3		0.1	0.7
	Mirror Back	52.0	40.0	4.5		0.2			3.3
	Mirror Back	47.0	43.0	5.9	0.6	0.3	0.2		2.8

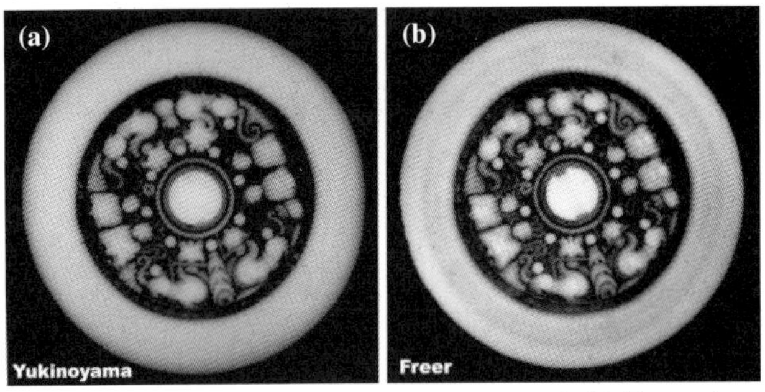

Figure 5. X-ray radiograph. (a) Yukinoyama mirror. (b) Freer mirror

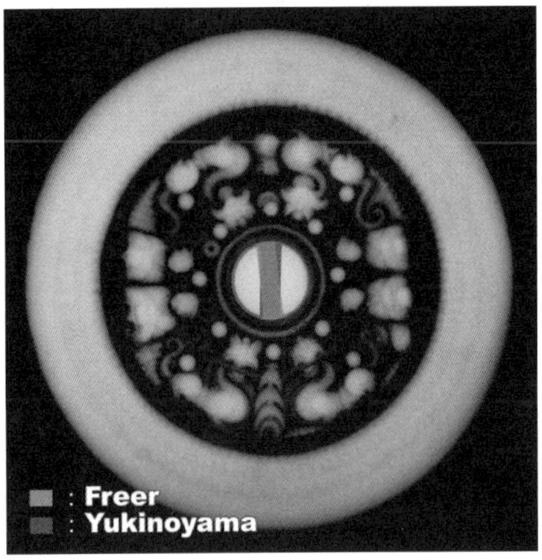

Figure 6. The overlapped images of the radiographs

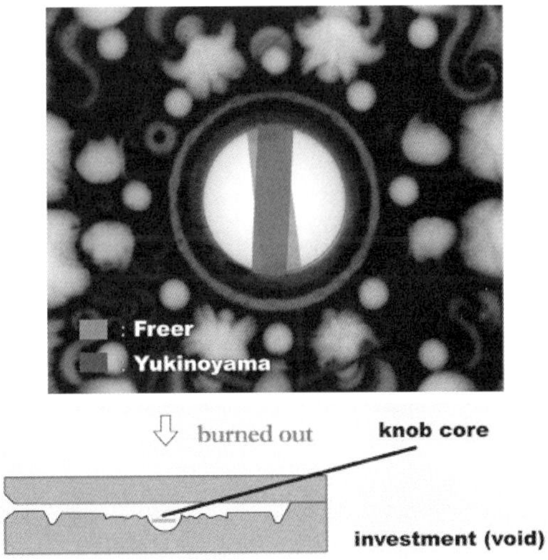

Figure 7. Magnified image of the center bosses

it can be clearly seen that both mirrors are identical with the exception of the center boss (fig. 7). The position of the hole in the boss is slightly different on both mirrors. This is one of the more important clues we have for determining the possible manufacturing process.

The Manufacturing Methods

Now possible manufacturing methods for these mirrors will be considered. It is clear that the mirrors were cast; surface features, metal thickness, and alloy type all indicate casting. Furthermore, the metallographic examination of other mirrors has always revealed cast structures. (The typical bronze mirror alloy—high in tin, and containing some lead—is difficult to fabricate by hammering but easy to cast.) The question of how they were made thus involves questions of casting and mold making.

The mirrors have a great deal of fine detail on their surfaces. Molds can be made either by carving or by taking impressions from a model; lost-wax molds are made by investing a wax positive of the object in a refractory material and then burning out the wax. In all cases, at the end of the mold-making process, one has a heat-resistant void into which to pour the metal. The discussion below is concerned less with materials than with process, as some of the steps can be done with a number of different materials. For example, the prototype for the section-molding process could be made of metal or wood rather than clay. Here, the term *clay* simply refers to a ceramic modeling material with some plasticity and good refractory qualities. Actual compositions of the ceramic material used are still a matter of conjecture.

For this discussion, ancient casting techniques are divided into three classes: direct molding, section molding, and the lost-wax method.

Direct Molding

Molds can be made by directly carving stone or a block of clay (see fig. 8a). A clay mold would have been fired hard and a core pulled from the fired mold. The core would have then been shaved down to establish the casting space. Finally, the mold would be reassembled and the bronze poured. The initial establishment of the shape could also be made with an undecorated model that gave the basic shape of the mirror to be cast. The model would have been pressed into wet clay, as in figure 8b. Fine decoration would then be added by carving it into the mold, and the core would be made and the piece cast as outlined above.

To use this method to produce duplicate pieces would require a very stable mold material with good mechanical properties. Nonetheless, molds would probably not last for very many castings. On the mirrors in question, there is no evidence to suggest the progressive deterioration of the mold surface that would be seen in reused molds. While reproduced cracks are seen on the mirrors, they are slight. Last, if one were using the direct-molding process to make a mirror, the hole within the boss might be made by wedging a piece of mold material of the correct length into the deeply concave space for the boss in the mold.

Section Mold

Section molding is another possible technique whose use in Asia is probably best understood because of the intensive

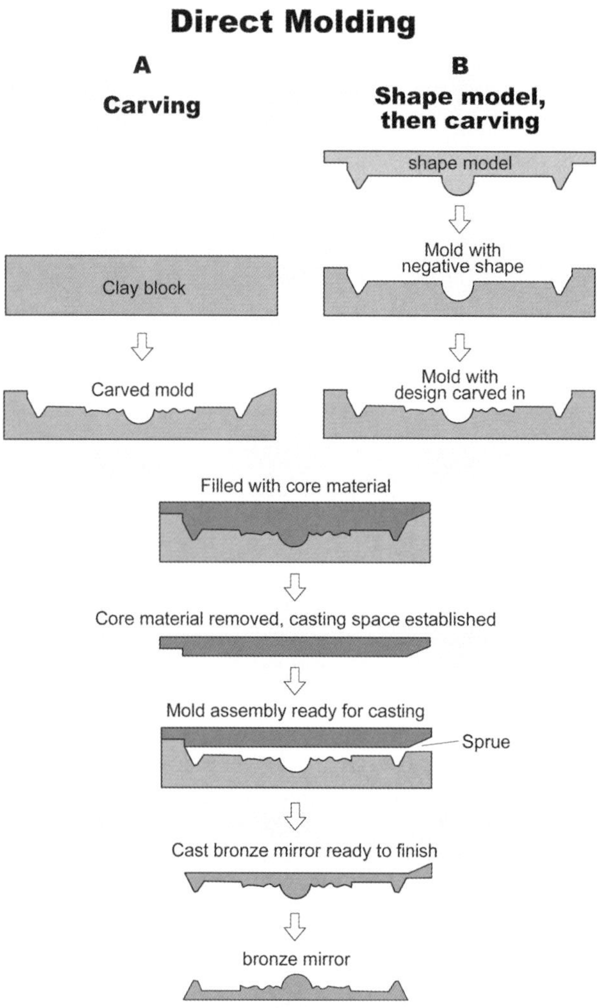

Figure 8. Process of direct molding

investigations of ancient Chinese bronzes. (See, for example, the discussion of the fragments of molds and models found at the site of the Houma Foundry in Shaanxi, China, in the Institute of Archaeology of Shanxi Province [1996].) The use of the section-molding technique for casting a mirror is illustrated in figure 9. The shape model is positive, and the mold pulled from it is negative. The core is made (the same technique as that applied in the section on direct molding above can be used), and the bronze is then cast into the mold. In the section-molding technique, multiple objects are produced by multiple numbers of individual molds pulled from the same model. Often, the individual molds have their own peculiarities. Since the original model is a positive, cracks and layout lines in the original model will show up as positive grooves, not ridges, on the final bronze. The hole in the boss could be made in the same way as in the direct-molding method described previously.

Lost-Wax Method

The lost-wax method begins with the production of a wax-positive prototype of the object. The wax positive can be unique, with the artist directly shaping the wax to be cast. An alternative method—that used to produce multiple identical castings—is to use a negative matrix, with all of the detail in it, into which molten wax is poured. After the wax hardens and is removed from the matrix, it is then covered with mold material. The mold assembly is heated, the wax burned out, and the mold material fired. Finally, the molten bronze is poured in. The process is illustrated in figure 10.

Various materials are used for investment, including clay, plaster of paris (a modern material), and even cattle dung mixed with water (Chase 1975). The quality of the first layer of the investment is critical to the finished casting. Typically, this layer is painted on as a slurry in order to closely reproduce the details of the wax in the investment. In modern practice, when a poured-in plaster-of-paris slurry is used, it is easy to entrap air bubbles. The entrapped air bubbles are then reproduced in the casting as tiny spheres on the surface of the bronze.

When an original negative matrix is used to create multiple wax positives for casting, details in the matrix, including features such as cracks, grooves, or layout lines, also will be reproduced in the casting. However, these features will be reproduced in the positive—that is, a crack in the matrix will appear as a fine ridge in the casting.

In the case of mirrors and the possible use of lost-wax casting as the method of manufacture, one detail in particular is of interest—the hole within the boss and how it would have been produced. One method that might have been used is the placement of a bar of mold material into the concave space for the boss in the matrix that, when the wax was poured in, would hold the bar in place. When the investment material surrounded the wax, the bar would be in contact with the investment and be held in place when the wax was burned out. This procedure might leave the bar placed at any angle relative to the boss and the surrounding design elements.

Evidences of the Lost-Wax Method: Details on the Surfaces of Two Mirrors

Irregularities in Ring-Shaped Design Elements

In the Yukinoyama mirror, the outside band of one of the ring-shaped elements has an irregularity in it (fig. 11a). Also, the edge of the ring looks "softer" (i.e., less sharp) than it does on the Freer mirror (fig. 12a). The circular scratches on the raised boss beside the ring-shaped element demonstrate clearly that the boss was finished after casting. The scratches do not seem to extend down into the cutoff area on the ring-shaped element. It was initially thought that this deformation was due to tooling on the wax prior to investment. Since it is so close to an area that was finished after casting, the case is less clear, but there is still the possibility that this ring-shaped element could be evidence of the lost-wax method.

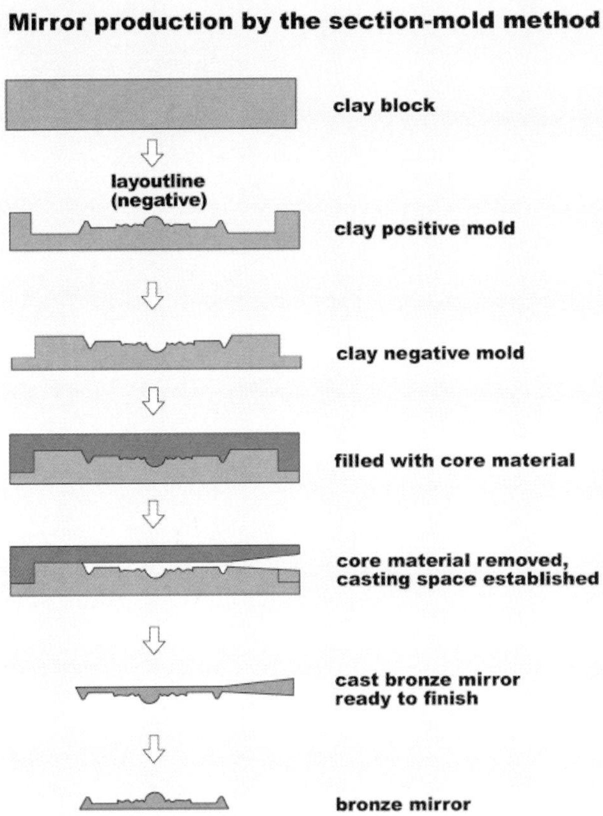

Figure 9. *Process of section molding*

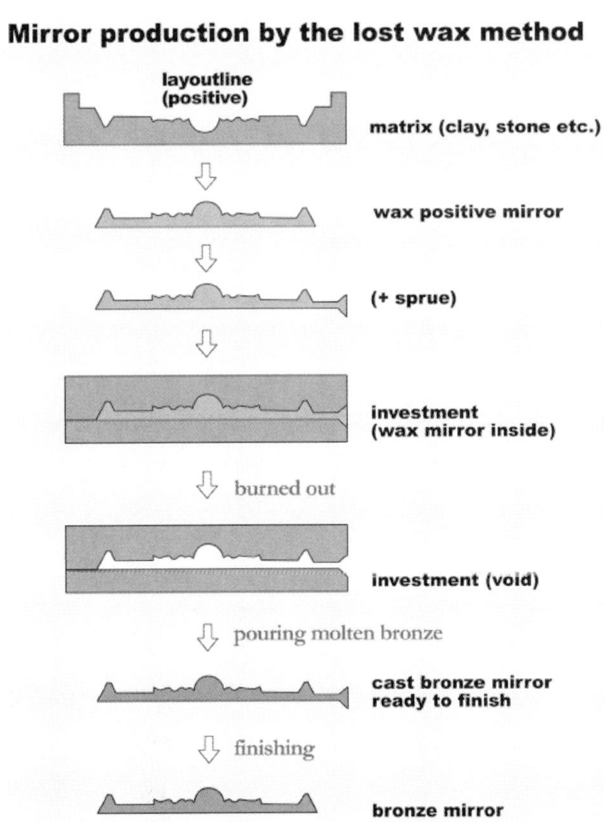

Figure 10. Process of the lost-wax method

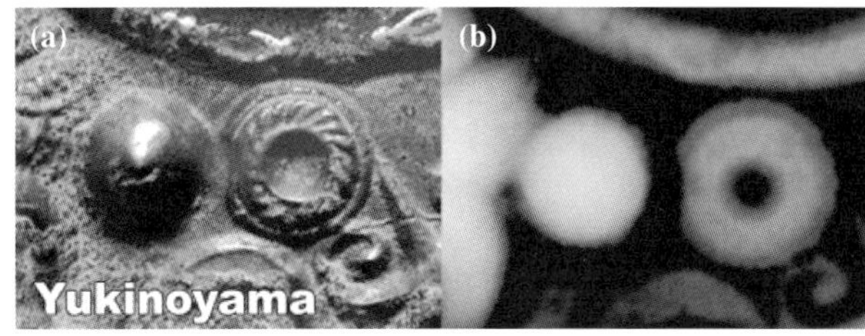

Figure 11. *Ring-shaped design element on the Yukinoyama mirror. (a) Microscopic image. (b) X-ray radiograph*

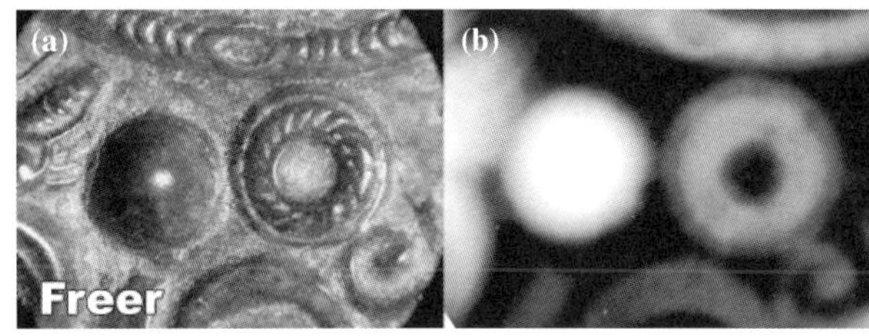

Figure 12. *Ring-shaped design element on the Freer mirror. (a) Microscopic image. (b) X-ray radiograph*

Cracks on the Sawtooth Design Element

At a number of places on the mirrors, elements that look like cracks can be seen. One particularly clear instance runs across one of the three annular sawtooth bands that run between the rim and the annular band of calligraphy (fig. 13). Under magnification, it is clear that these are ridges rather than cracks. They often bifurcate, as seen on the right side of the image. This particular feature was found reproduced quite exactly on both mirrors. Another instance is seen in the area between the two figures, one of which wears

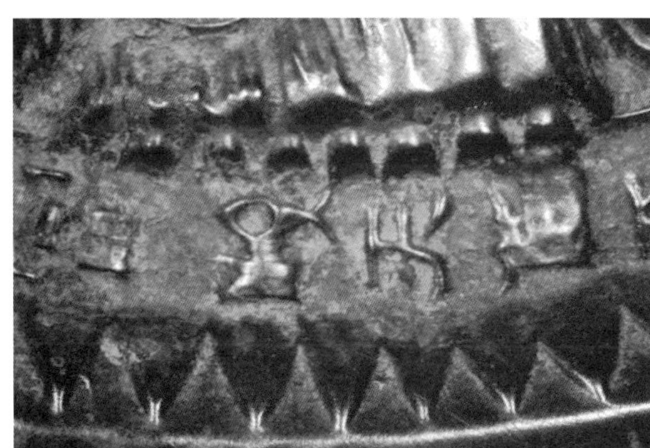

Figure 13. *Sawtooth design element on the Freer mirror*

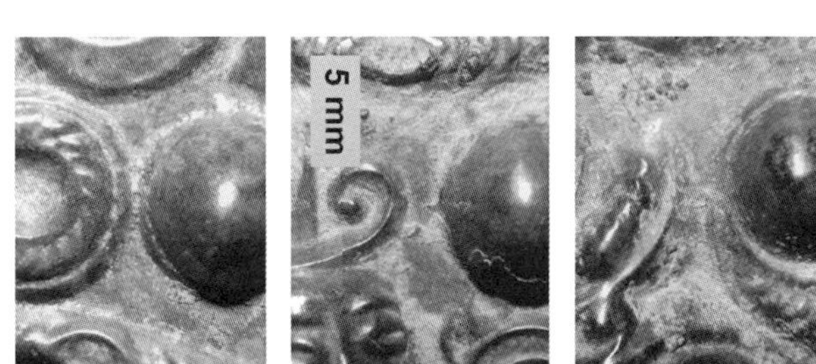

Figure 14. *Layout line used to align the positions of the bosses on the Freer mirror*

a hat (fig. 11). The cracks give the definite feeling that they were made in a brittle substance, like ceramic.

Layout Lines

On both mirrors, a circle of bosses surrounds the central boss. On the Freer mirror, in the spaces between design elements, a small raised line can be seen (fig. 14). This is on the lowest part of the surface, so it would have been on the highest part of the mold. We suggest that these are remnants of a layout line used to align the positions of the bosses when designing the matrix to make the wax prototype. Since a layout line would have been engraved on the matrix, it would be seen standing in relief on the wax and hence on the bronze.

Bubbles on the Surfaces

In lost-wax investment casting, generally the investment step is not perfect. When the investment is applied to the wax, it entraps air, and these air bubbles can be reproduced as tiny spheres on the metal surface. These are seen often on modern lost-wax castings but also occasionally on ancient ones. If the hypothesis that these mirrors are lost-wax castings is true, we might be able to find some reproduced air bubbles somewhere on the surface. In fact, one spherical bubble was found on the Freer mirror (fig. 15).

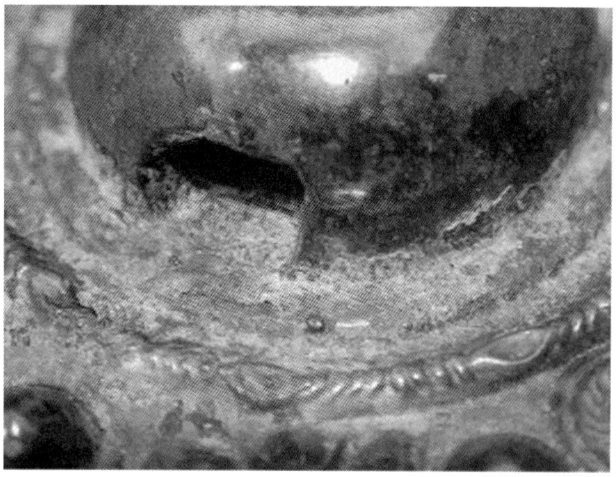

Figure 15. *One spherical bubble on the Freer mirror*

Core in the Knob and Lost-Wax Casting

As described above, the bar of mold material forming the hole within the boss must be placed individually within the wax for each casting. It is not a part of the original mold, and there is no reason why the positions of the holes within the bosses must be the same on otherwise identical mirrors made by the lost-wax method. In fact, to ensure that the hole was in an identical position from one mirror to the next would require a more complex matrix to be made. Such an effort would seem to be without worthwhile advantage for the final product.

Conclusions

The hypothesis that these identical mirrors were made by the lost-wax method appears to be strongly supported by the physical evidence observed on the pair of mirrors in question. This hypothesis, of course, should be tested further on other sets of identical mirrors. In the Freer Gallery, there are two more triangle-rim bronze mirrors, which were excavated in Japan. For one of these, an identical mirror exists in Japan; for the other, an identical mirror is in the collection of the Berlin Museum (Umehara 1931). Further study of these other pairs of mirrors will, it is hoped, definitively determine the methods used in the making of this class of objects.

Acknowledgments

This research was done under a collaboration that started in 1990 between the Smithsonian Institution and the Bunkacho, the Japanese Agency of Cultural Affairs.

References

Chase, W. T. 1975. *Bronze disease and its treatment.* Bangkok: Bangkok National Museum, Department of Fine Arts.

Institute of Archaeology of Shanxi Province. 1996. *Art of the Houma foundry.* Princeton: Princeton University Press.

Kondo, Takaichi. 1988. *Triangular-rim mirrors.* Tokyo: Tokyo University Press.

Lawton, Thomas, and Linda Merrill. 1993. *Freer: A legacy of art.* Washington, D.C.: Freer Gallery of Art.

Murakami, Ryu, and Masaaki Sawada. 1996. Scientific comparison of the bronze mirrors discovered at the Yukinoyama tumulus and in the Freer collection at Smithsonian Institution. In *Studies of the Yukinoyama tumulus,* ed. Hiroshi Tsude, Shinya Fukunaga, and Takeshi Sugii. Yōkaichi-shi: Yōkaichi-shi Kyōiku Iinkai.

Murakami, Ryu, Takayasu Koezuka, and Masaaki Sawada. 1996. Scientific approach to the bronze mirrors discovered at the Yukinoyama tumulus. In *Studies of the Yukinoyama tumulus,* ed. Hiroshi Tsude, Shinya Fukunaga, and Takeshi Sugii. Yōkaichi-shi: Yōkaichi-shi Kyōiku Iinkai.

Okamoto, Hidenori. 1999. *The period of the triangular-rim mirrors.* Tokyo: Yoshikawa-kobunkan.

Pearson, Richard. 1992. *Ancient Japan.* Washington, D.C.: Arthur M. Sackler Gallery.

Tsude, Hiroshi, Shinya Fukunaga, and Takeshi Sugii, eds. 1996. *Studies of the Yukinoyama tumulus.* Yōkaichi-shi: Yōkaichi-shi Kyōiku Iinkai.

Umehara, Sueji. 1931. Bronze mirrors transported to Europe and America, excavated in Japan. *Shigaku* 9(1).

13 A Millennium of Khmer Bronze Metallurgy: Analytical Studies of Bronze Artifacts from the Musée Guimet and the Phnom Penh National Museum

David Bourgarit, Benoit Mille, Thierry Borel, Pierre Baptiste, and Thierry Zéphir

ABSTRACT Combined bulk and surface analyses as well as radiographic examinations were performed on seventy-five metallic artifacts from both the Musée Guimet and the Phnom Penh National Museum, taking advantage of the opportunity of the international exhibition *Sculpture of Angkor and Ancient Cambodia: Millennium of glory* which brought to France major art pieces from the Phnom Penh National Museum. This extensive analytical study has brought to light a very specific metallurgy, characterized by both diversity and conservatism. Diversity is shown in the coeval use of two kinds of alloys, leaded and unleaded bronzes, the use of which does not depend on any apparent technological reason. On the other hand, the metal choice reflects an astonishing continuity throughout the ten centuries of Khmer civilization, during which time bronzes were never replaced by the more common brass alloys found in use in neighboring civilizations. Gold-rich bronze alloys have also been found, which brings to mind the famous *samrit* known in the literature. This study, in revealing the original aspects of Khmer metallurgy, exhibits also some of its mysteries.

Introduction

The chronology of ancient Cambodian art is divided roughly into three periods: the pre-Angkorian period, running from the first centuries of the common era to the end of the eighth century; the Angkorian period, from the consecration of King Jayavarman II, the founder of the Angkorian royal and religious institutions (ca. 802–35 or later), to the traditionally accepted year 1431, in which Angkor would have been abandoned as a capital under the military pressure of Siamese armies; and the post-Angkorian period, from the fifteenth century onward. Leaving apart the magnificent brick and sandstone monuments of Angkor and numerous other sites in Cambodia, Khmer art is mainly known by its stone statuary (Dupont 1955; Boisselier 1955, etc.). References to bronze works of art exist in the comprehensive studies on Khmer sculpture, and a handful of articles are specifically devoted to bronze sculpture or ritual objects (Groslier 1921; Boisselier 1967; Le Bonheur 1972; Bunker 1971–72). However, they mostly refer to iconography, style, or history of very important pieces or groups of pieces. They give little, if any, information on technological aspects of Khmer metallurgy. Up until now, the only analytical studies concerning Khmer bronze sculpture are, first, the presentation of George Coedès (1923)—still useful but now obsolete—and, more recently, Woodward's study on the sacred sculpture of Thailand (Woodward 1997). This last work is also remarkable, with an in-depth technological study of the statues. Nevertheless, as mentioned by Strahan in the second chapter, the laboratory investigations following each entry of the catalogue only represent a starting point on the subject. The different assessments concerning the Khmer metallurgy were made on only a few pieces, rather minor ones related to the Khmer domination period in what is today central and northeastern Thailand (fourteen artifacts were analyzed, mainly statues), so further investigations are necessary.

Our own investigations were initiated in 1997 in conjunction with the international exhibition *Sculpture of Angkor and Ancient Cambodia: Millennium of glory* (Paris; Washington, D.C.; Tokyo; and Osaka). Most of the major bronzes from the Phnom Penh National Museum were then brought to France for conservation and study. A collaborative scientific research program on Khmer metallurgy supported by the Cambodian Ministry of Culture, and with the help of our Cambodian colleagues, was initiated, with the partners being the Center for Research and Restoration of the French Museums (C2RMF) and the Musée National des Arts Asiatiques—Guimet, Paris. As the first results were quite promising, it was decided to add to the corpus under study the Khmer bronze collection from the Musée Guimet. Bulk and surface analysis and radiographic examinations were performed on these artifacts. To follow is a presentation of the corpus, the experimental procedures used, and a discussion of the results of the three investigative methods employed and what those methods revealed. First, the radiographic examination and what radiography revealed about the casting processes used are discussed; second, the different alloy compositions as well as the impurity patterns in the metals are presented and discussed; and finally, the surface analysis of the artifacts is considered.

Presentation of the Corpus

Seventy-five metallic artifacts preserved in both the Phnom Penh National Museum and the Musée Guimet were thoroughly studied (table 1, fig. 1). Most of them are religious statues or votive images. Significant sets of prestige artifacts were also studied, including ritual objects as well as other artifacts, like palanquin hooks, and so forth.

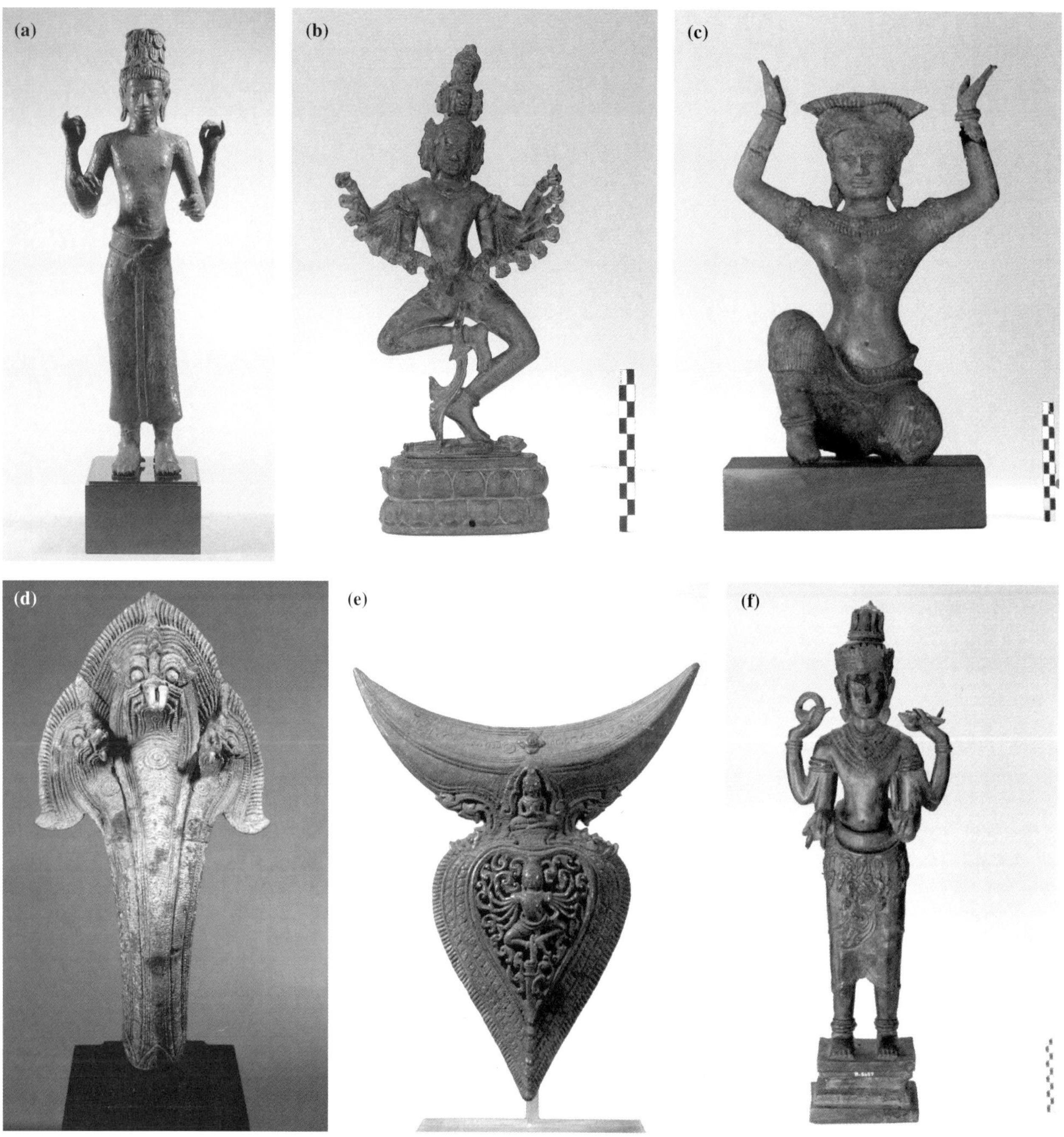

Figure 1. *Photographs of some of the Khmer bronze artifacts studied*

(a) Maitreya (pre-Angkorian, 8th century, inv. MA 3321) (copyright Musée Guimet)

(b) Hevajra (10th–11th century, 2nd half 11th century, inv. 5331) (copyright C2RMF D. Bagault)

(c) Support for a mirror. Kneeling female figure (Angkor Vat, 1st quarter 12th cent., inv. 5476) (copyright C2RMF D. Bagault)

(d) Finial for a shaft in the shape of a three-headed Naga (1st quarter 12th century, Angkor Vat, inv. MG 18889) (copyright Musée Guimet)

(e) Mirror handle (Bayon, 12th–13th century, inv. EO801) (copyright Musée Guimet)

(f) Visnu (post-Angkorian, 16th century, inv. 5457) (copyright C2RMF D. Bagault)

Table 1. *The seventy-five Khmer artifacts studied*

	Designation	Museum	Inv. no.	Height (cm)	Style	Date	Provenance	Discovery location	Aesthetic quality	Radiography	Metal anal.	Surface anal.
Pre-Angkorian	Maitreya	G	MA3321	46	Pre-Angkorian	8th	Northeast Thailand	"cache of 300"	Very good	Y	Y	Y
	Bodhisattva Lokesvara	G	MA4985	64	Pre-Angkorian	8th	Northeast Thailand	"cache of 300"	Very good	Y	Y	N
	Male figure	G	MG18318	23	Pre-Angkorian	8th–9th ?	Vietnam	Tra Vinh	Minor	Y	Y	Y
	Standing Buddha	G	MA3785	19	Dvâravâti	8th–9th	Northeast Thailand	?	Minor	Y	Y	N
	Avalokitesvara	P	3548	24	Srivijaya	7th–8th	Malaysia ?	Phnom Ta Kream	Good	Y	Y	N
	Lokesvara	P	5330	18	Pre-Angkorian	7th–8th	South Cambodia	Vat Kompong Luong	Very good	Y	Y	N
	Buddha	P	5405	28	Pre-Angkorian	7th–8th	Angkor	West Baray	Very good	Y	Y	N
	Standing Buddha	P	5412	28	Pre-Angkorian	7th–8th	Cambodia ?	Vat Banon	Very good	Y	N	Y
10th–11th cent.	Perfume censer or oil lamp —kneeling male figure	G	MA1067	19	Bakheng/Koh Ker	1st half of 10th	Cambodia ?	?	Minor	Y	Y	N
	Female divinity	G	MA2239	19	Koh Ker	2nd half of 10th	Cambodia	?	Good	N	Y	N
	Hevajra	P	5331	30	Baphuon	2nd half of 11th	Angkor	Banteay Kdei Temple	Very good	Y	Y	Y
	Visnu	G	MA1339	87	Baphuon	2nd half of 11th	Cambodia	?	Good	Y	Y	Y
	Linga	P	3557	12	?	10th–11th ?	Central Cambodia	?	Minor	Y	Y	N
Angkor Vat	Visnu astride Garuda	G	MA2678	18	Baphuon-Angkor Vat	end of 11th	Cambodia ?	?	Minor	Y	Y	N
	Visnu-Vasudeva-Narayana	P	5291	40	Angkor Vat	1st half of 12th	Angkor	Kapilapura Temple	Very good	Y	Y	N
	Adorned Buddha	P	2081	80	Angkor Vat	1st half of 12th	Angkor	Angkor Vat Temple	Major	Y	Y	N
	Support for a mirror—kneeling female figure	P	5476	35	Angkor Vat	1st half of 12th	Angkor	Bayon Temple	Very good	Y	Y	N
	Finial for a shaft (three-headed Naga)	G	MG18889	34	Angkor Vat	1st half of 12th	Angkor	Banteay Srei Temple	Very good	Y	Y	N
	Adorned Buddha Protected by the Naga	P	3296	23	Angkor Vat	1st half of 12th	Champa (South Vietnam)	Silver Towers	Good	Y	Y	Y
	Male divinity	G	MA5960	63	Angkor Vat	1st half of 12th	Cambodia or Thailand	?	Very good	Y	Y	N
	Dancer	G	C6500	21	Angkor Vat/Bayon	12th	Cambodia	?	Good	Y	Y	N
	Visnu Astride Garuda	G	MA539	14	Angkor Vat/Bayon	12th	Cambodia	?	Minor	N	Y	N
	"Chevet de statuette"	G	MG18175	20	Angkor Vat/Bayon	12th	Cambodia	?	Good	N	Y	N
	Sadasiva	P	2682	34	Angkor Vat, Bayon, Post-bayon	12th–13th	Angkor	?	Good	Y	Y	N
	Dagger	G	MA5986	34	Angkor Vat/Bayon	12th–13th	Cambodia or neighb.	?	Major	N	N	N
	Dagger	G	MA5987	35	Angkor Vat/Bayon	12th–13th	Cambodia or neighb.	?	Major	N	N	N
Bayon	Mirror handle	G	EO801	21	Bayon/Angkor Vat	12th–13th	Laos	Vat Phu Temple	Major	N	Y	N
	Perfume censer	G	MA5984	25	Bayon/Angkor Vat	12th–13th	Cambodia or neighb.	?	Very good	N	Y	N
	Lokesvara	P	5340	42	Bayon	end of 12th–beginning of 13th	Angkor	North Khleang Temple	Good	Y	Y	Y
	Buddhistic Triad	P	2424	30	Bayon	end of 12th–beginning of 13th	Angkor	Roluos	Major excavation	Y	Y	Y
	Miniature shrine of Hevajra	P	2494	7	Bayon	end of 12th–beginning of 13th	Angkor	Roluos	Major excavation	Y	Y	Y
	Hevajra	P	4170	22	Bayon	end of 12th–beginning of 13th	Angkor	Royal Palace	Very good	Y	Y	Y
	Brahmanic Triad	P	2283	21	Bayon	end of 12th–beginning of 13th	Angkor	?	Good	Y	Y	Y

(Table 1 continued)

	Designation	Museum	Inv. no.	Height (cm)	Style	Date	Provenance	Discovery location	Aesthetic quality	Radiography	Metal anal.	Surface anal.
	Kneeling female figure	P	5467	17	Bayon	end of 12th–beginning of 13th	Cambodia	Danrun	Good	Y	Y	N
	Prajnaparamita with eleven heads	P	5333	14	Bayon	end of 12th–beginning of 13th	Cambodia	?	Good	Y	Y	N
	Hevajra	G	MA849	15	Bayon	end of 12th–beginning of 13th	Cambodia	?	Minor	Y	Y	Y
	Prajnaparamita	G	MA508	12	Bayon	end of 12th–beginning of 13th	Cambodia	?	Minor	N	Y	N
	Lokesvara	G	MA509	8	Bayon	end of 12th–beginning of 13th	Cambodia	?	Minor	N	Y	N
	Yogini	G	MA540	9	Bayon	end of 12th–beginning of 13th	Cambodia	?	Minor	N	Y	N
	Siva	G	MA2110	18	Bayon	end of 12th–beginning of 13th	Cambodia	?	Good	N	Y	N
	Ganesa	G	MA1829	9	Bayon	end of 12th–beginning of 13th	Cambodia	?	Good	N	Y	N
	Cosmic Lokesvara with radiating arms	G	MA5940	39	Bayon/Lopburi	end of 12th–beginning of 13th	Thailand ?	?	Very good	Y	Y	N
	Seated Buddha	G	MG9807	9	Bayon/Lopburi	end of 12th–beginning of 13th	Thailand ?	?	Minor	N	Y	N
	Finial-Adorned Buddha Protected by the Naga	P	5593	14	Bayon	end of 12th–beginning of 13th	Cambodia	Kas Prag	Good	Y	Y	N
	Finial with Garuda	G	MA102	19	Bayon	end of 12th–beginning of 13th	Cambodia?	?	Minor	Y	Y	Y
	Damaru	G	MA5988	13	Bayon	end of 12th–beginning of 13th	Cambodia or neighb.	?	Good	Y	Y	N
	Vajra	P	5483	12	Bayon	end of 12th–beginning of 13th	Cambodia	Mongkol Borei	Good	Y	Y	N
	Vajra	G	MA1617	14	Bayon	end of 12th–beginning of 13th	Cambodia	?	Good	N	Y	N
	Ritual conch	P	5484–5485	29	Bayon	end of 12th–beginning of 13th	Angkor	Tonle Sap	Major	Y	Y	N
	Ritual conch	G	MA3564	25	Bayon	end of 12th–beginning of 13th	Cambodia	?	Good	N	Y	N
	Tripod	G	MA3565	9	Bayon	end of 12th–beginning of 13th	Cambodia	?	Minor	N	Y	N
	Tripod	G	MA541	12	Bayon	end of 12th–beginning of 13th	Cambodia	?	Minor	N	Y	N
	Naga	G	MG18888	29	Bayon	end of 12th–beginning of 13th	Angkor	West Mebon?	Good	Y	Y	N
	Bed "ring"	G	MA3563b	17	Bayon	end of 12th–beginning of 13th	Cambodia	?	Good	N	Y	N
	Bed "hook"	G	MA5989	19	Bayon	end of 12th–beginning of 13th	Cambodia	?	Very good	N	Y	N
	Bed "ring"	G	MG18851	19	Bayon	end of 12th–beginning of 13th	Cambodia	?	Good	N	Y	N
	Bell	G	MG18850	22	Bayon	end of 12th–beginning of 13th	Cambodia	?	Very good	N	Y	N
	Buddhistic triad plaque	G	MG18204	8	Bayon	end of 12th–beginning of 13th	Cambodia	?	Minor	N	Y	N
	Ganesa	P	2320	26	Bayon	1st half of 13th	Angkor	?	Good	Y	Y	N
Post-Angkorian	Buddha	P	2057	86	Post-Angkorian	14th	Cambodia	Kong Pisei	Major	Y	Y	Y
	Visnu	P	5457	64	Post-Angkorian	16th	Angkor	North Khleany Temple	Major	Y	Y	N
	Lion	G	MA2341	24	Post-Angkorian	16th	Cambodia or Thailand	?	Minor	Y	Y	Y
	Lion	G	MA2342	24	Post-Angkorian	16th	Cambodia or Thailand	?	Minor	Y	Y	Y
Uncertain dating	Linga	G	MG18187	13	?	10th ?	Cambodia	?	Minor	N	Y	N
	Hand of big statue	G	C6501	12	?	10th ?	Cambodia	?	Good	N	Y	N
	Garuda	G	MA2592	7	Angkor Vat/Bayon	12th ?	Cambodia	?	Minor	N	Y	N
	Female divinity	G	MA3789	21	Bayon	end of 12th–beginning of 13th	Cambodia or Thailand	?	Minor	Y	Y	Y
	Visnu	G	C6496	19	Bayon ?	end of 12th–beginning of 13th	Cambodia?	?	Minor	N	Y	N
	Prajnaparamita	G	MA2677	29	Bayon ?	end of 12th–beginning of 13th	Cambodia?	?	Good	N	Y	N
	Seated Buddha	G	MA3788	15	Bayon/Lopburi ?	end of 12th–beginning of 13th	Thailand?	?	Minor	Y	Y	Y
	Standing Buddha	G	MA3786	27	Bayon/Lopburi ?	end of 12th–beginning of 13th	Thailand?	?	Minor	Y	Y	N
	Female divinity	G	MG18333	12	Bayon or late ?	?	Cambodia?	?	Minor	N	Y	N
	Standing Buddha	G	MA3787	24	Post-Angkorian ?	?	?	?	Minor	Y	Y	N
	Cup	G	MG24921	16	?	?	?	?	Good	N	Y	N

Several types of data were recorded for each artifact: style, date (derived from the style), provenance (associated with the style or the place of discovery), and location of discovery. Our study is based on a chronological grouping of the objects:

1. Pre-Angkorian group, seventh to ninth century (eight artifacts). This group contains only a few pieces and is characterized by a great diversity in provenance.
2. Tenth-to-eleventh-century group (four artifacts). This group contains only a few pieces and also is characterized by a great diversity in provenance.
3. Angkor Vat-style objects and sculptures, circa first three quarters of the twelfth century (thirteen artifacts).
4. Bayon-style objects and sculptures, circa last quarter of the twelfth to the early thirteenth century (thirty-three artifacts). This group is perhaps the most important to our study, because it is the largest and very well characterized from a stylistic point of view.
5. Post-Bayon/post-Angkorian group, fourteenth to sixteenth century (four artifacts). This group contains only a few artifacts that vary widely in date.

The objects are also classified according to their aesthetic quality in order to determine a possible social attribution. Of course, this topic may be most controversial; however, it is a useful approach for the study. This classification includes four aesthetic groups:

1. "Minor" artifacts. These are rather crude pieces and are presumed to be a popular production.
2. "Good" artifacts. The pieces are of good quality and suggest a middle-level patronage.
3. "Very good" artifacts. The pieces are of high skill and suggest a high-level patronage.
4. "Major" artifacts. These are masterpieces of very high skill. They indicate a very high-level patronage (perhaps dignitaries and/or kings).

Experimental

Radiographic Procedure

The radiographic exposure conditions mainly depended upon the mean atomic number, the density, the thickness, and the shape of the object to be examined. Copper or lead filters and intensifying lead screens in contact with both sides of the film often were used. They allowed an optimal view of the whole object, since they prevented the image from blurring due to scattered radiation. In the case of the Khmer statues, the most common exposure conditions were 300 kV, 24 mA-min, source-to-film distance of 1.3 m, with a 12 mm copper filter and two 0.1 mm lead screens, and AX or AA400 Kodak film. This allowed us to evaluate the wall thickness of the hollow-cast statues and the presence of armatures, casting defects, and assemblies. A longer distance (2 m) was used for the tall statues, with 60 mA min exposure, to avoid distortion in the image. For thick-walled objects such as the two post-Angkorian lions, a higher voltage was necessary—up to 420 kV—with 4 mm lead filter and 60 mA-min exposure. For thinner objects, such as arms, wings, and pedestals, a lower voltage around 120 kV was sufficient, with a 2 mm copper filter, and 5 mA-min exposure.

Bulk-Metal Analysis

In order to determine the chemical compositions of the bronzes, analyses were performed by atomic emission spectrometry (AES) at the C2RMF. Circa 20 mg of metal was removed by drilling (1 mm diameter, 10 mm deep) after eliminating most surface corrosion products; the drillings were also carefully inspected under stereobinocular magnification to insure that no corrosion products or dust were contained in the samples. In the case of artifacts made of separately cast sections, all of the sections of interest were sampled. Circa 10 mg of the drillings in each sample were precisely weighed and dissolved in 5 ml aqua regia solution (hydrochloric and nitric acids). The solution was then analyzed using AES.

The objects from the two museums were not analyzed using the same instrumentation. The artifacts from Phnom Penh were analyzed using a direct current plasma spectrometer (DCP-AES), and twelve elements were measured. The artifacts from Musée Guimet were analyzed using an inductively coupled plasma atomic emission spectrometer (ICP-AES), with twenty-six chemical elements characterized (appendix 2). (For more details on the operating conditions, see Mille and Bourgarit 2000.) To make sure that the two sets of analyses are compatible, some samples were analyzed using both instruments (appendix 1). In these cases, comparable results were achieved.

Surface Analysis

The chemical compositions of surface layers were obtained either by proton-induced x-ray emission (PIXE) on the ion beam facility AGLAE (3 MeV, proton beam extracted in air with a spot size of 100 μm; see Calligaro et al. 1998) or by x-ray fluorescence spectrometry (XRF).

Results

Radiographic Examinations

Forty-five bronze artifacts of the corpus were radiographed (table 2). Examples of the radiographs are shown in figures 3–5. The three following general observations can be made. First, most of the artifacts have a rather homogeneous radiographic image. Few porous metals and no casting defects are observed, indicating that the castings are generally of good quality, even for the minor artifacts. Second, the casting method used seems rather logical: small statues are solid, whereas the largest ones are hollow cast, with an armature

Table 2. Data derived from the radiographic examinations of forty-five Khmer bronze artifacts

	Designation	Inv. no.	Height (cm)	Element	Kind of casting[a]	Solid parts[b]	Wall thickness[c]	Wall evenness[d]	Assembly[e]	Assembly technique[f]	Lost-wax casting process[g]
Pre-Angkorian	Maitreya	MA3321	46		HCA	Fa	t	E	—	—	D or ip
	Bodhisattva Lokesvara	MA4985	64		HCA	?	t	E	—	—	D or ip
	Male figure	MG18318	23		HCA	A,L	TH	U	—	—	D
	Standing Buddha	MA3785	19		S	—	—	—	—	—	D
	Avalokitesvara	3548	24		HA	Fa,F	t	U	O (aureole)	?	D or ip
	Lokesvara	5330	18		S	—	—	—	O (crown)	SW	D
	Buddha	5405	28		HA	H	t	U	—	—	D
	Standing Buddha	5412	28		HCA	H	t	U	—	—	D or ip
10th–11th cent.	Perfume censer or oil lamp—kneeling male figure	MA1067	19		H	A	t	U	—	—	D
	Hevajra	5331	30		H	A,S	TH	U	A	SW, arm in shoulder	D or i1
	Visnu	MA1339	87		HCA	n	t	E	O (bracelet)	SW?	D or ip
Angkor Vat	Visnu astride Garuda	MA2678	18		S	—	—	—	O (feet in shoulders)	Tangs	D
	Visnu-Vasudeva-Narayana	5291	40		HA	A,L	t	E	A	Pin, shoulder in arm	D or ip
	Adorned Buddha	2081	80		HCA	H	t	E	H?O (earrings)?	?	D or ip
	Support for a mirror—kneeling female figure	5476	35		HA	H	t	U	A	?	i1
	Finial for a shaft (three-headed Naga)	MG18889	34		H	n	t	U	H	Clip	D
	Adorned Buddha Protected by the Naga	3296	23		H	Fa	TH	U	O (Naga)	Tangs	D or i1
	Male divinity	MA5960	63		HA	H	TH	U	A	Pin, shoulder in arm	ip
	Dancer	C6500	21		HA	A,F	t	U	—	—	D
	Sadasiva	2682	34		H	A,F	TH	U	O (crown)?	SW?	D

*(**Table 2** continued)*

	Designation	Inv. no.	Height (cm)	Element	Kind of casting[a]	Solid parts[b]	Wall thickness[c]	Wall evenness[d]	Assembly[e]	Assembly technique[f]	lost-wax casting process[g]
Bayon	Lokesvara	5340	42		H	H	TH	U	—	—	i1
	Buddhistic Triad	2424	30	Buddha	HA	A	TH	U	O (Naga)	Tangs	D
			30	Lokesvara	HA	A	TH	U	—	—	D
			30	Prajnaparamita	HA	A,L	TH	U	—	—	D
	Miniature shrine of Hevajra	7192	7	Yogini	S	—	—	U	—	—	D
			20	Hevajra	H	n?	TH	U	—	—	D
	Hevajra	4170	22		HA	A,S	TH	U	A?	?	D or ip
	Brahmanic Triad	2283	21	Siva	S	—	—	—	—	—	D
			21	Devi	S	—	—	—	—	—	D
			21	Visnu	S	—	—	—	—	—	D
	Kneeling female figure	5467	17		HA	A,F	t	E	O (lotus)	SW	D
	Prajnaparamita with eleven heads	5333	14		H	A	TH	U	—	—	i1
	Hevajra	MA849	15		S	—	—	—	—	—	D
	Cosmic Lokesvara with radiating arms	MA5940	39		H	Fa	TH	E	A	Pin, shoulder in arm	i1
	Finial-Adorned Buddha Protected by the Naga	5593	14	Buddha	H	A	TH	U	—	—	D
	Finial with Garuda	MA102	19		H	A	TH	U	—	—	D
	Damaru	MA5988n	8	Central figure	H	A	t	E	—	—	D
			5	Little figures	S	—	—	—	O	SW?	D
	Naga	MG18888	29		CS		TH	U	—	N	D
	Ganesa	2320	26		H	Fa	TH	U	A	SW, arm in shoulder	i1
Post-Angkorian	Buddha	2057	86		HA	F	TH	U	—	—	D
	Visnu	5457	64		HA	H	TH	U	A+O (crown)	SW, arm in shoulder	ip
	Lion	MA2341	24		H	L	TH	U	—	—	D
	Lion	MA2342	24		H	L	TH	E	—	—	D
Uncertain dating	Female divinity	MA3789	21		H	A,F	TH	U	—	—	D
	Seated Buddha	MA3788	15		H	A,L	TH	U	—	—	i1
	Standing Buddha	MA3786	27		HA	A,F	t	E	—	—	D
	Standing Buddha	MA3787	24		HCA	H,F	TH	U	—	—	i1

a) S = solid, H = hollow, HA = H+armat., HCA = H+complex armat. b) n = none, H = hands, Fa = forearms, A = arms, F = feet, S = shine, L = legs c) TH = thick, t = thin d) E = even, U = uneven
e) B = body, A = Arms, H = head, O = other f) SW = soldering/welding, N = no g) D = direct, i1 = indirect/one-piece mold, ip = indirect/ piece mold

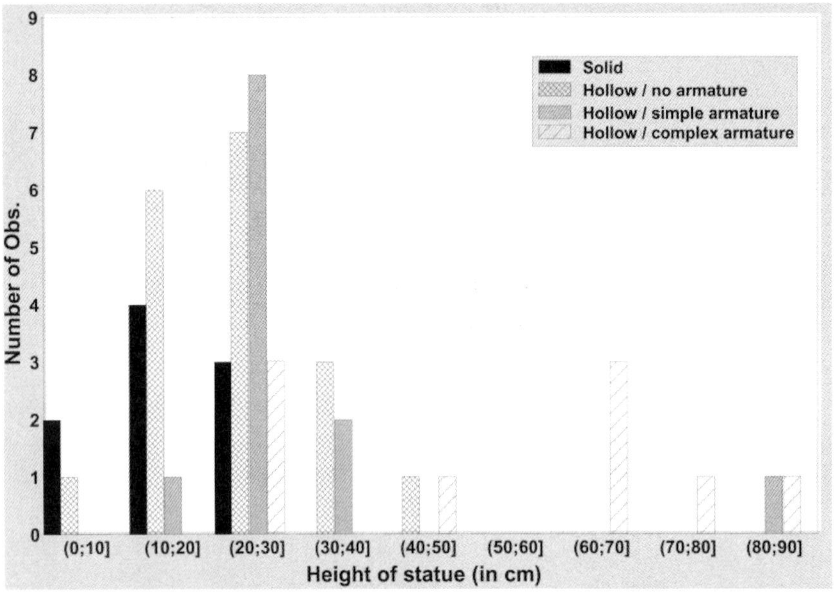

Figure 2. *Graph showing the differences in the height of the Khmer statues according to the type of casting. Casting complexity increases with height: introduction of a core, then simple armature, then complex armature*

growing in complexity with the statue size (fig. 2). The statues are seldom entirely hollow cast; almost systematically, some members are solid. Statues with solid principal members are all below circa 30 cm in height, with one exception—the 40-cm Visnu-Vasudeva-Narayana (inv. 5291)—which has solid arms. Third, several features, such as the complexity of the general designs, the numerous undercut details, and the frequent use of wax appliqués, indicate that the lost-wax process was widely used instead of sand-mold casting.

However, apart from these three aspects common to all objects, quite a large diversity is observed in casting choices. This diversity appears through several features, and most often there does not appear to be a technical explanation, nor can it be related to any chronological or cultural parameters. The use of separately cast parts may represent the best example. There are no obvious reasons why particular statues were cast in sections—the most common separation occurs at the level of the arms—while others were cast in one piece (table 3, fig. 3). Furthermore, there were other separately cast parts, such as the crowns on several statues or the bracelets on the eleventh-century Visnu (inv. MA1339), for which no reason for separate castings could be determined. Similarly, the joining techniques vary in an unexplained manner, as seen by the two techniques used to join the arms (fig. 3). Finally, the kind of lost-wax process used also shows discrepancy. If the direct process[1] seemed to be the most common process, the indirect process was also used at least for nine out of forty-five artifacts[2] (table 2; for more details on the distinction between the two processes, see also the radiographs on figs 4 and 5).

Table 3. *Descriptions of the method of joining the arms for standing images taller than 30 cm.* Arms separately cast: *Arms cast separately and joined to the torso of the figures.* Arms integrally cast: *No arm assembly.*

	Arm assembly					**No assembly**			
	Designation	Inv. no.	Height (cm)	Aesthetic quality	Joining technique	Designation	Inv. no.	Height (cm)	Aesthetic quality
Pre-Angkorian						Maitreya	ma3321	46	Very good
						Bodhisattva Lokesvara	ma4985	64	Very good
10th–11th	Hevajra	5331	30	Very good	SW, arm in shoulder	Visnu	ma1339	87	Good
Angkor Vat	Visnu-Vasudeva-Narayana	5291	40	Very good	Pin, shoulder in arm	Adorned Buddha	2081	80	Major
	Support for a mirror—kneeling figure	5476	35	Very good	?	Sadasiva	2682	34	Good
	Male divinity	ma5960	63	Very good	Pin, shoulder in arm				
Bayon	Cosmic Lokesvara with radiating arms	ma5940	39	Very good	Pin, shoulder in arm	Lokesvara	5340	42	Good
	Ganesa	2320	26	Good	SW, arm in shoulder	Buddhistic Triad	2424	30	Major
Post-Angkorian	Visnu	5457	64	Major	Pin, shoulder in arm				

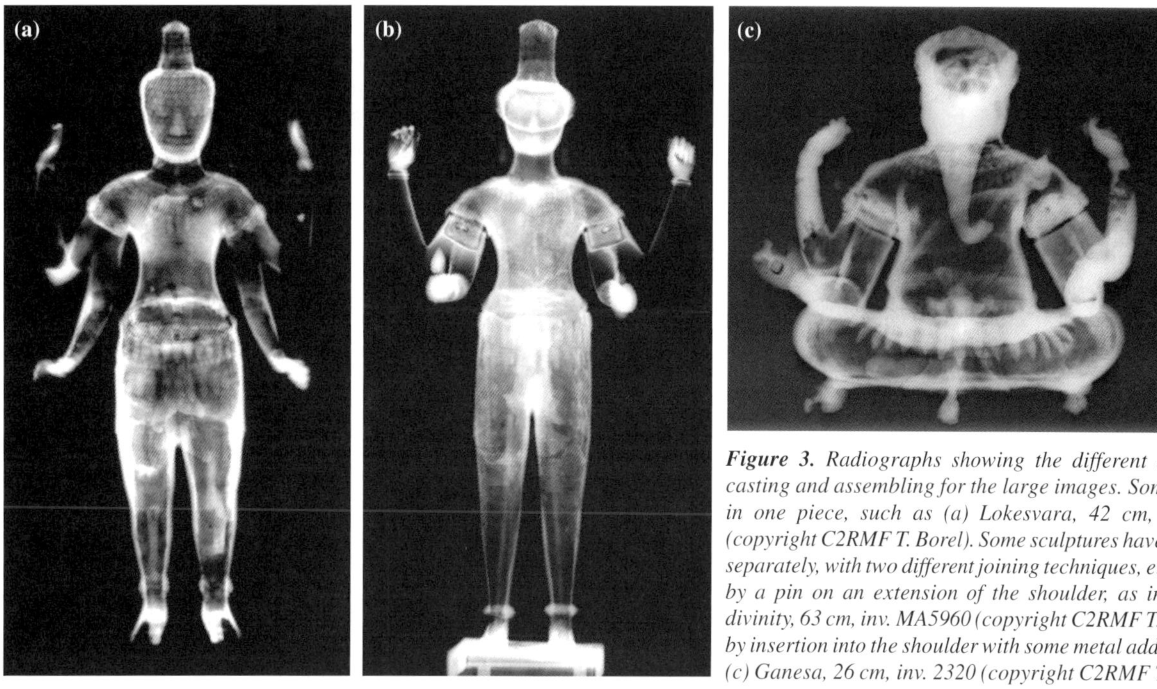

Figure 3. Radiographs showing the different choices of casting and assembling for the large images. Some are cast in one piece, such as (a) Lokesvara, 42 cm, inv. 5340 (copyright C2RMF T. Borel). Some sculptures have arms cast separately, with two different joining techniques, either fixing by a pin on an extension of the shoulder, as in (b) male divinity, 63 cm, inv. MA5960 (copyright C2RMF T. Borel), or by insertion into the shoulder with some metal addition, as in (c) Ganesa, 26 cm, inv. 2320 (copyright C2RMF T. Borel)

Figure 4. Radiograph of the Maitreya (46 cm, inv. MA3321) showing the thinness and evenness of the metal wall thickness (copyright C2RMF T. Borel)

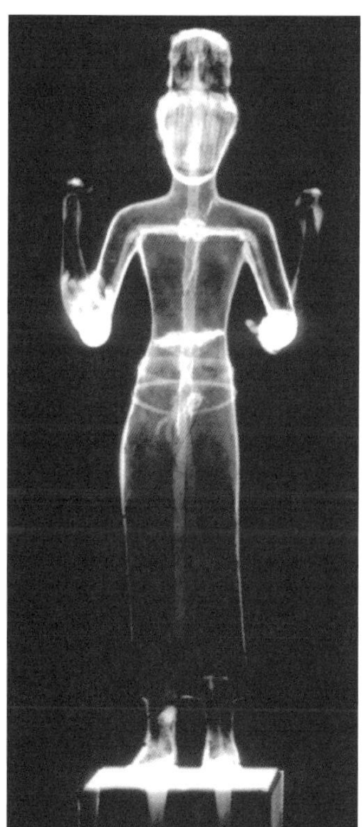

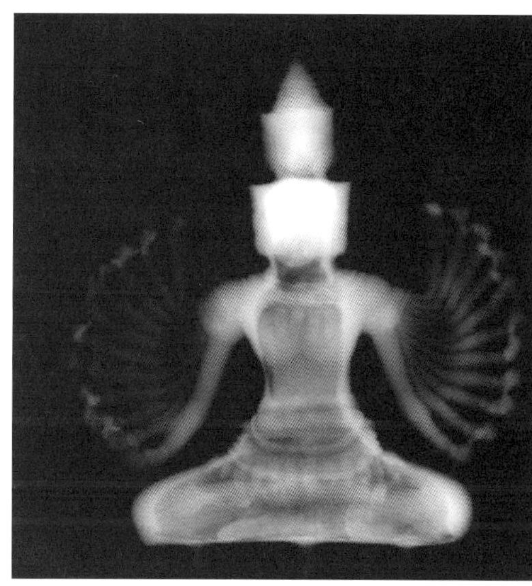

Figure 5. Radiograph of the Prajnaparamita with eleven heads (14 cm, Bayon, inv. 5333) showing the coarse aspect of the metal inner walls where wax was removed mechanically from the mold after it had set. The feature indicates an indirect lost-wax casting process in a one-piece mold (copyright C2RMF T. Borel)

Bulk-Metal Analyses

One hundred and twelve samples from sixty-nine artifacts were analyzed by AES, either DCP or ICP-AES.[3] The results are reported in appendixes 2 and 3. To follow are summaries of the analyses.

Alloys

Most of the objects analyzed were bronzes or leaded bronzes (sixty-two out of sixty-nine artifacts).[4] The other seven artifacts are mostly pure copper or a zinc-tin-copper alloy with some lead. (The Angkor Vat-related Adorned Buddha [inv. 2081] exhibits almost 2 wt. % zinc, but it is counted here as a bronze.) Two remarkable alloys are observed: the pre-Angkorian Lokesvara (inv. 5330) shows only tin and lead, while the Koh Ker Female divinity (inv. MA2239) is a silver-tin-copper alloy. The Hevajra from the miniature shrine (inv. 7192) is also such an unusual alloy.[5] The seven artifacts associated with these particular alloys are left out of the discussion of composition below, as are as the artifacts that were not considered authentic with absolute confidence (see table 1). The summary comments to follow refer to a total of fifty-six artifacts.

The compositions of the bronzes show large discrepancies in their tin and lead contents, ranging from a low-tin alloy (2 wt. % Sn) to high-tin bronze (15 wt. % Sn).[6] Some of the bronzes contained virtually no lead, while some were found to have more than 20 wt. % lead. However, there appear to be some general tendencies among objects of similar composition. For example, the highest tin contents are usually seen in the thin-walled castings (fig. 6), which may correspond to a technological choice.[7] Also of note is the fact that the prestige artifacts show less diversity in the alloy recipes (fig. 7).[8] However, as a general rule, the compositions of the bronzes varied widely, and no explanation can be found for this. The chronological or cultural context of objects does not lead to any convincing sorting (fig. 8).[9] Nor does the aesthetic quality (fig. 9): the tendency observed in the tenth-to-eleventh-century group and the Bayon groups, where the best statues are unleaded bronzes, must be considered with precaution given the few artifacts concerned.

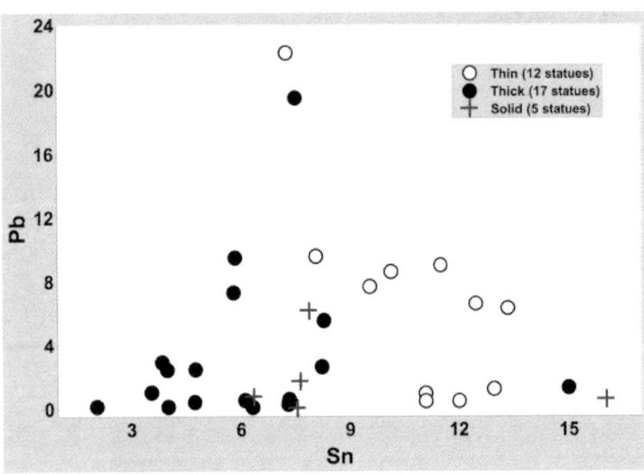

Figure 6. Sn and Pb content (in wt. %) in the statues *according to the thickness of the metal wall. (The "solid" group refers to the solid-cast statues.) The "thin" hollow castings show the highest Sn contents (only bronzes and leaded bronzes; redundant analyses and data from poorly related statues were not included)*

Thus, the hypothesis that lead was introduced in order to replace the tin—which may probably have been a more expensive metal—cannot be proved. Even after a highly selective grouping, the variability of bronze recipes remains large (for example, the Bayon-related statues from Cambodia, seen in fig. 10).

Impurity

Here, as for the alloying elements, some particular aspects must be considered prior to undertaking the statistical analysis of the copper impurity patterns (appendixes 2 and 3).[10]

Of particular note is the high gold content observed in numerous Khmer bronze objects (fig. 11): twenty out of fifty-five artifacts reveal more than 150 wt. ppm gold. Gold content even reaches more than 450 wt. ppm, as seen in six statues. Gold at such levels cannot come from the copper ore. Its presence is due to another reason: either it originates from the remelting of gilded bronzes or it was deliberately added. To test the first hypothesis, a small calculation can be made: the remelting of a bronze image together with its gilding layer would lead to a few tenths of ppm of gold in the bronze.[11] The smallest values are compatible with some of our analyses. Nevertheless, the highest values (more than 500 ppm) are very improbable, since they imply the remelting of an artifact with a gold layer thicker than 50 μm on a bronze wall thinner than 1 mm. Thus, at least for the six statues with a high gold content, the deliberate addition of gold is more probable. This point will be further discussed below. Note that three ambiguous Bayon-style statues also have gold contents of more than 500 ppm: the Female Divinity (inv. MA3789), the Prajnaparamita (inv. MA2677), and the Seated Buddha (inv. MA3788).)

The other impurities show few variations within the corpus studied (fig. 12), with a very narrow content distribution turning around the same mean-contents for all elements (0.05–0.1 wt. %). Only a few extremes stand out: two pre-Angkorian images, the Avalokitesvara (inv. 3548), and the Buddha (inv. 5405) show high antimony contents;[12] the Bayon Prajnaparamita (inv. 5333) has a very high cobalt content, and the Bayon Ganesa (inv. 2320) and the Cosmic Lokesvara (inv. MA5940) are quite high in arsenic and iron.

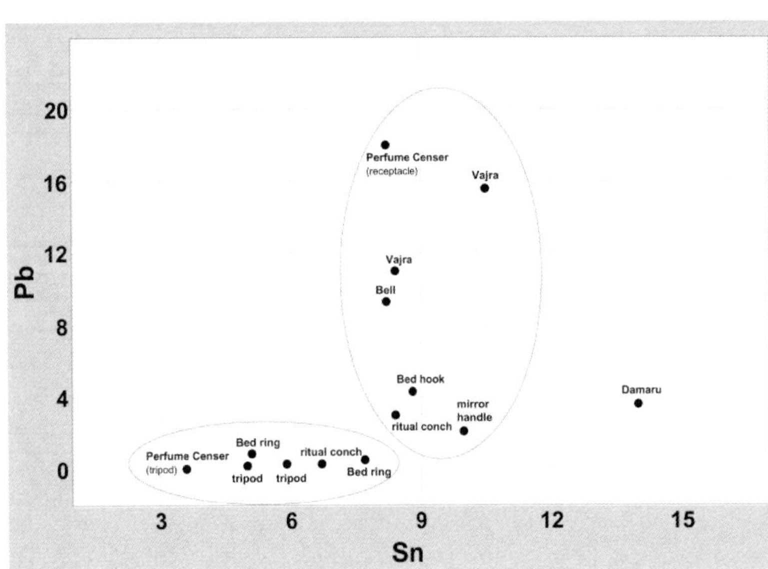

***Figure 7.** Sn and Pb content (in wt. %) in the* prestige artifacts. *Two groups stand out: bronzes without Pb showing dispersed low Sn content (3–8 wt. % Sn) and leaded bronzes (2–18 wt. % Pb) showing a narrow Sn content (8–10 wt % Sn) in fourteen analyses representing thirteen artifacts, with the same restrictions as in figure 6; all of the artifacts analyzed belong to the Bayon group*

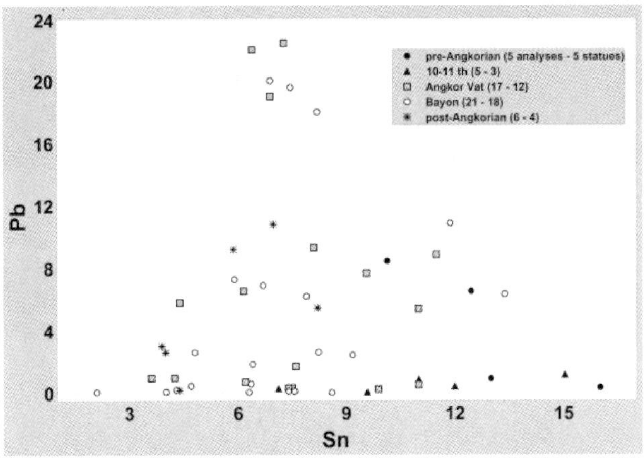

Figure 8. *Sn and Pb content (in wt. %) in the* statues *according to their chronological or cultural assignations. Large discrepancies in the alloy compositions remain within each of the five stylistic groups. Nevertheless, and although only small populations are being considered, three tendencies can be seen: (1) the high Sn content of the pre-Angkorian statues (four statues); (2) the absence of Pb in the bronzes from the tenth–eleventh group (four statues); and (3) the high Pb contents restricted to Angkor Vat and Bayon styles (six statues). There are the same restrictions on the population represented as in figure 6*

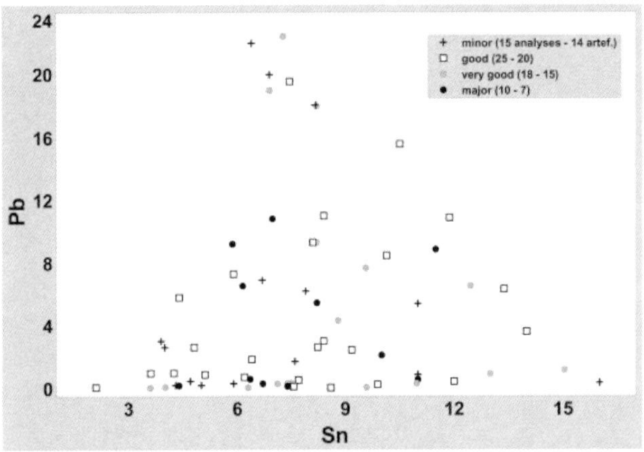

Figure 9. *Sn and Pb content (in wt. %) in* statues and prestige artifacts *according to their aesthetic quality. The alloy composition and especially the lead content cannot be correlated to aesthetic quality. There are the same restrictions on the population represented as in figure 6*

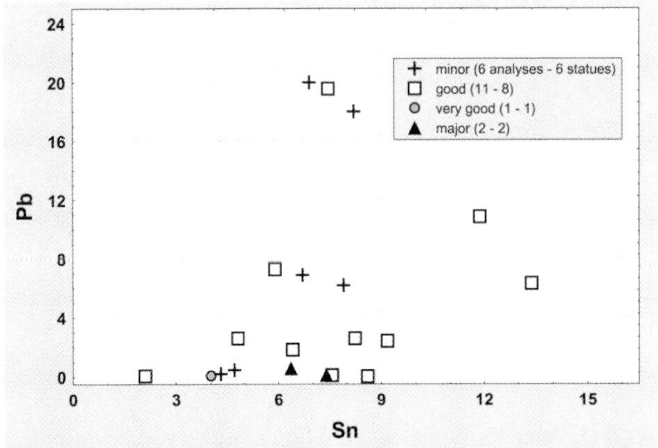

Figure 10. *Sn and Pb content (in wt. %) in the* Bayon statues from Cambodia, *according to aesthetic quality. Even in such a highly selective grouping, the variation in alloy compositions remains. For statistical reasons, aesthetic quality cannot be considered here as a relevant parameter. There are the same restrictions on the population represented as in figure 6*

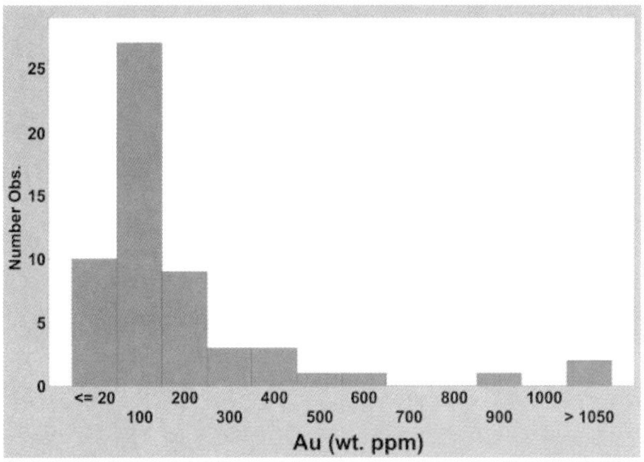

Figure 11. *Distribution of the Au contents. (For each column the central value is indicated instead of the borders.) Most of the bronzes contain more than 100 ppm Au, and in some statues the content can reach more than 500 ppm. (Fifty-seven analyses representing fifty-five artifacts. Only bronzes and leaded bronzes were included; redundant analyses and the Maitreya figure MA3321 were not used.)*

We have here a quite unusual situation, where the copper circulating in the Khmer empire exhibits a uniform fingerprint that remains unchanged for almost one thousand years.

Assemblies

The results derived from those objects composed of multiple parts were studied using statistical methods to see if trends or groupings could be made. Principal component analysis was performed using the results for seven elements: Zn, As, Ag, Fe, Co, Ni, and Sb (fig. 13). Then by looking at the alloying element compositions, tin and lead, we sorted the multipiece castings into three different groups according to their homogeneity in composition (table 4).

1. In the homogeneous assemblies group, each of the seven multipiece castings shows a remarkable metal composition homogeneity, where the alloying elements (tin and sometimes lead) as well as the impurities are present at the same levels in all the analyzed parts with less than 20% relative variation. Note that the base of the Bayon-related Ganesa is made of a different metal when compared with the image itself (arms and body); the tripod and the Perfume Censer are clearly not associated (inv. MA5984).
2. In the alloy variation group, four artifacts show variations in the alloying elements, ranging from more than 20% up to 50%, whereas the impurity patterns remain homogeneous.
3. In the heterogeneous group, nine artifacts show large variations in the alloying compositions as well as in their impurity patterns.

The differences in the groups may reflect workshop or production habits. The artifacts belonging to the homogeneous group, for example, are similar in that each sculpture's various parts are very close in composition to one another, thus suggesting that each sculpture was the product of a unique pour. Also, the relatively small size of the artifacts under concern is compatible with such a process. For the

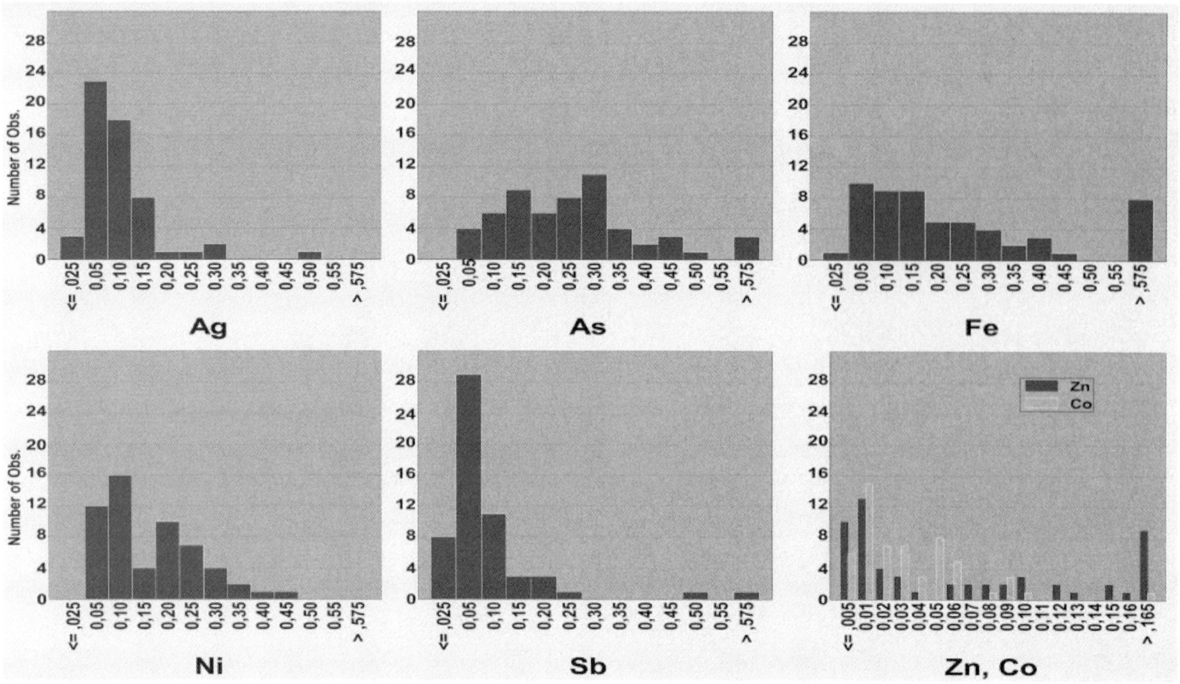

Figure 12. *Distribution of impurity contents (wt. %; for each column the central value is indicated instead of the borders). Most of the impurities (Ag, Sb, Co, Ni) show a very narrow range of content distribution centered on 0.05–0.10 wt %; Zn and Fe are more widely distributed. This uniform and unchanging copper-impurity pattern during the entire Khmer period is puzzling. (Same population as for the previous graph.)*

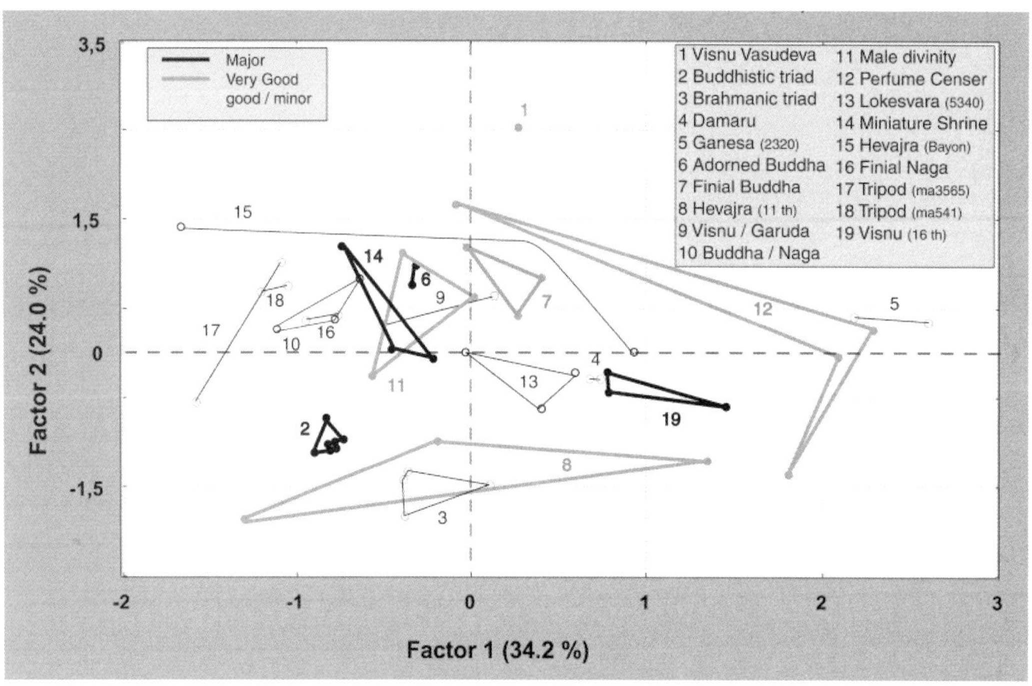

Figure 13. *Graph of the principal component analysis of the impurity contents in the bronze assemblies. The different sections of each assembly are joined, and the space between the points gives an idea of the relative homogeneity or heterogeneity of the impurities. The more distant the points, the more different the impurity patterns. The inverse is not necessarily true, since this two-dimensional graph is a projection of 34 + 24 = 58% of the total information contained in the seven-dimensional space considered (i.e., the content values in the seven elements Zn, Fe, Co, Ni, Ag, As, and Sb). Thus, in some cases, such as for the tripod, no. 18, the differences may be masked by the projection*

sculptures belonging to the alloy variation group, the different parts are made of a copper that has exactly the same impurity patterns, but the alloy compositions vary. This clearly shows that the various parts of the sculptures were made from separate pours. The artifacts belonging to the heterogeneous group are clearly made of different metals. In several cases the different parts were not initially intended to be associated. This is probable for the two Bayon tripods (inv. MA3565 and MA541), and this would not be surprising for the Bayon Hevajra and its base (inv. MA849). Similarly, the Yogini of the shrine of Hevajra (inv. 7192) shows a bronze very different from the ones of the cage and the base (and the Hevajra is a silver-copper alloy). For other artifacts of the heterogeneous group, casting was probably performed in the same workshop but using different alloys. The three different methods of production do not depend on the aesthetics: major artifacts as well as "good" or minor ones are to be found in the three groups (table 4).

Surface Analysis

Surface analysis was employed mainly for characterizing the vestiges of gilding visible on some statues. Traces of polychromy were only looked at for the artifacts of the Phnom Penh National Museum, and only the pre-Angkorian Avalokitesvara (inv. 3548) and the post-Angkorian Buddha (inv. 2057) showed significant remains of paint. The red pigment on the face of the Buddha has been identified as cinnabar.[13] Also, one figure, the Bodhisattva Maitreya (inv. MA3321), exhibited a strange silvery surface; it was examined, and the results for it will be discussed separately.

Gilding

Many of the statues exhibited some minute areas of gold on their surfaces. Unfortunately, all the areas were not accessible for surface analysis without sampling. Nevertheless, the surfaces of sixteen gilded statues were analyzed (table 5). The gold was generally found associated with mercury, which surely comes from the gilding layer.[14] For the Adorned Buddha protected by the Naga (inv. 3296), metallographic examination of a sample taken from the base shows the typical porous and granular structure of mercury-amalgam gold layer (Arc-Antique pers. comm.). For most surfaces with mercury, traces of silver were also detected. The silver was not contained in the substrate. The surface Ag/Cu ratios are far higher than those measured in the metal, and the much-too-low silver content in the metal is not compatible with segregation processes that occur during casting or corrosion. This means that the gold used for the gilding contains silver. As an example, on the Finial with Garuda (inv. MA102), gold appeared to contain almost 2 wt. % of silver. The silver content strongly indicates that gold leaf was not used for gilding. While gold leaf can be applied using mercury (Lins and Oddy 1975), leaf cannot contain such a high level of silver or other metals (Anheuser 1995). Thus, these artifacts were fire gilded. Note that the absence of mercury—that is, a content under the detection limits—does not exclude fire gilding, since mercury can be removed to a large extent during (and after) the gilding process. Unfortunately, most statues showing no mercury have not undergone any careful surface examination to allow us to determine with confidence which process may have been employed. Only three artifacts (inv. MA3785, 3788, and 3789) were observed under light with a binocular lens during a previous study (Drilhon, Francaix, and Lahanier 1977), revealing only for the Female Divinity (inv. MA3789) the presence of gold foil. But in most of those cases where traces of gilding remain, the sculptures were fire gilded.

The "Silvered" Maitreya

The pre-Angkorian Maitreya (inv. MA3321) exhibits a silverlike surface that was analyzed using PIXE. Two locations were chosen for analysis because of their particularly

Table 4. *The three different types of bronze assemblies according to the chemical composition variation between the different sections*

	Designation	Inv. no.	Sections analyzed	Aesthetic quality
Homogeneous assemblies	Brahmanic Triad	2283	Siva = Devi = Visnu = base	Good
	Damaru	MA5988	Drum = little figures	
	Ganesa	2320	Body = arm = base	
	Visnu-Vasudeva-Narayana	5291	Arm = body	Very good
	Perfume censer	MA5984	Receptacle = lid/tripod	
	Bell	MA18850	Bell = ball	
	Buddhistic Triad	2424	Naga (body, head) = Buddha (body, base) = Lokesvara = Prajnaparamita	Major
Alloy variations	Lokesvara	5340	Head/body (upper part/ lower part)/arm	Good
	Finial-Adorned Buddha Protected by the Naga	5593	Finial/Buddha/Naga	
	Finial for a shaft (three-headed Naga)	MG18889	Central head Naga/side head Naga	Very good
	Adorned Buddha	2081	Head/body	Major
Alloy and impurity variations	Visnu Astride Garuda	MA2678	Visnu/Garuda	Minor
	Hevajra	MA849	Hevajra/base	
	Tripod	MA3565	Tripod/bassin	
	Tripod	MA541	Tripod/bassin	
	Adorned Buddha Protected by the Naga	3296	Buddha/Naga (body/head)	Good
	Hevajra	5331	Body/arm/base	Very good
	Male divinity	MA5960	Body/arm (right = left)	
	Miniature shrine of Hevajra	7192	Yogini/cage = base	Major
	Visnu	5457	Head = base/arm/crown	

Homogeneous assemblies: Same contents in all analyzed sections in alloying elements (Sn and Pb) as well as in impurities (less than 20% relative variation). *Alloy variations:* Same impurity patterns, but variations on alloying elements (20–50%). *Alloy and impurity variations:* Large internal variations on alloying elements and impurities

Table 5. *Results of surface analyses: Quantitative when obtained by PIXE (wt. %), qualitative when obtained by XRF (O = more than 5 wt. %; tr = between 5 and 1 wt. %; nd = less than 1 wt. %; — = not measured)*

	Designation	Inv. no.	Analyt. Technic	Localization	Au	Hg	Ag	Cu	Sn	Pb
Pre-Angkorian	Maitreya	MA3321	PIXE	Internal right forearm	—	—	—	81.7	19.3	—
				External right arm	—	—	—	77.0	22.5	—
	Male figure	MG18318	PIXE	Chest	72.2	6.4	2.3	18.3	< 0,4	< 0,2
	Standing Buddha	5412	XRF	Back, belt, dress	O	O	tr	—	—	—
10th–11th cent.	Hevajra	5331	XRF	Cage	tr	nd	nd	—	—	—
	Visnu	MA1339	PIXE	Left shin	89.0	< 0,7	0.3	7.1	3.0	< 0,3
Angkor Vat	Adorned Buddha Protected by Naga	3296	XRF	Naga head	O	O	tr	—	—	—
				Naga body	O	O	tr	—	—	—
				Buddha	O	O	tr	—	—	—
Bayon	Lokesvara	5340	XRF	Left buttock	O	nd	tr	—	—	—
	Buddhistic Triad	2424	XRF	Naga head	O	O	tr	—	—	—
				Lokesvara	O	O	tr	—	—	—
	Miniature shrine of Hevajra	7192	XRF	Cage	tr	nd	nd	—	—	—
	Hevajra	4170	XRF	Leg	tr	nd	nd	—	—	—
	Brahmanic Triad	2283	XRF	Siva	O	O	tr	—	—	—
				Devi	O	O	tr	—	—	—
				Visnu	O	O	tr	—	—	—
	Hevajra	MA849	PIXE	Chest	6.6	< 0,2	< 0,5	40.7	6.6	45.5
				Left arm	8.6	< 0,3	< 0,2	64.4	13.3	12.7
	Finial with Garuda	MA102	PIXE	Foot	63.3	11.3	1.0	22.1	2.0	< 0,2
Post-Angkorian	Buddha	2057-1	XRF	Right thigh	nd	—	—	—	—	—
	Lion	MA2341	PIXE	Back left thigh	35.5	< 0,5	0.4	58.7	2.7	1.4
	Lion	MA2342	PIXE	Back left thigh	10.8	< 0,3	< 0,3	82.3	2.5	1.7
Uncertain dating	Female divinity	MA3789	XRF	?	O	nd	—	O	—	—
	Seated Buddha	MA3788	XRF	?	O	nd	tr	O	tr	—

shiny surface: the internal right forearm and the external right arm. The results (table 5) are to be compared with the bulk analysis (appendix 2).

Previous assumptions regarding this statue were that the element responsible for this particular surface feature was not tin but rather silver. However, analyses showed clear evidence of tin surface enrichment; there was approximately 20 wt. % of tin on the surface, as opposed to 11 wt. % in the alloy. Unfortunately, without cross-sections to study, it is difficult to distinguish between the different processes possibly involved (Meeks 1986, 1993). The first question would be to determine if this surface aspect was deliberate or the consequence of a natural selective corrosion. If deliberate, the tin-enriched surface may have been produced by many methods, from "tin sweat," or inverse segregation, during casting to a variety of different tinning processes possibly used after casting.

Interpretation: Toward an Original Metallurgy?

From the scientific investigations performed on the seventy-five Khmer metallic artifacts, several features of the Khmer devotional bronze metallurgy can be considered distinctive. These features will be discussed hereafter, grouping them into three main aspects: the metal supply, the casting, and the finishing.

The Metal Supply

The first distinctive feature of this metallurgy is the omnipresence of tin in the copper-based alloys used during the entire Khmer period. Sixty-two of the sixty-nine artifacts analyzed are bronzes, so this omnipresence surely lasts at least up to the end of the Bayon period (mid-thirteenth century) and probably much later. The most recent objects from the corpus are dated to the sixteenth century (the Visnu and the two lions), and those are bronzes as well. Thus, the Khmer metallurgists did not employ other alloys, such as brass, which was used in some surrounding areas in early periods.[15] This confirms the general tendency shown first by Werner (1972). The analyses of the sacred sculptures of Thailand from the Griswold collection also give a late date for the appearance of zinc in the region (Woodward 1997). The first high-zinc-containing images in Thailand (more than 10 wt. % Zn) date to the fifteenth century (cat. nos. 69–73). As remarked by Strahan (1997), the omnipresence of tin may not be surprising given (1) the proximity of the tin-ore belt running from central Burma down to Malaysia (Bronson 1992, 78), which is part of the very rich deposits of tin located in Southeast Asia, and (2) the difficulty of making zinc alloys (although zinc ores might have been available from the area near what is now the Thai/Burmese border). As an extreme extension of this omnipresence of tin, the analysis of the pre-Angkorian Lokesvara (inv. 5330) has revealed a tin-based alloy; this analysis brings to mind the Oc-èo culture in southern Vietnam, which had developed tin metallurgy between the first and the fifth century C.E. (Malleret 1948).

Nevertheless, two authentic Bayon statues, which were made using quaternary alloy compositions including zinc, tin, and lead (Seated Buddha [inv. MG9807] and Ganesa [inv. MA1829]), shed some color on this monochromatic situation. Strahan also observed such an unusual composition in the Dvaravati Standing Buddha dated to the ninth century (Woodward 1997, cat. no. 14). Thus, the "tin hegemony" may have allowed some imports of zinc-containing copper alloys or recipes. Incidentally, quite high zinc contents as an impurity are observed in many cases (more than 0.6 wt. % zinc). Several Cambodian statues from different periods are so: the Dancer of an Angkor Vat/Bayon style (inv. C6500), the Bayon Damaru (inv. MA5988), the Bayon Ritual Conch (inv. MA3564). So are a few images from Thailand associated with the Khmer domination period in the Griswold collection: the Visnu Astride Garuda (cat. no. 17; 0.7 wt. % zinc) and the Seated Buddha in meditation (cat. no. 32; 1 wt. % zinc). The Prakon Chai Avalokitesvara of the Musée Guimet, a high-tin bronze, also contains almost 1 wt. % zinc (inv. MA4985; see Lahanier 1974). These high-zinc contents may have several origins. First, they may come from remelting of zinc-containing copper alloys. Second, zinc may also be an impurity coming from high-zinc-bearing copper or lead ores. We must indeed bear in mind that important lead-zinc deposits are located in Burma, particularly at the Thai/Burmese border (Bronson 1992, 78). As a matter of fact, the artifacts supposed to come from Thailand show such a tendency, where zinc content is proportional to lead content (fig. 14).

Concerning the copper ores, a remarkable feature of our corpus deserves attention: almost all the artifacts analyzed present a similar main impurity pattern (Ag, As, Co, Ni, Sb). Such similarity in the copper-impurity patterns may indicate the use of a unique source of copper. We know that copper and, more generally, nonferrous deposits were rare in the Khmer empire (Dupouy 1913, cited by Malleret 1954a; Bronson 1992, 98), so that most of nonferrous metals probably had to be imported. Does it mean that the copper used throughout an entire millennium came from a single place? Unfortunately, the chemical elements that happen to be relevant in this study (Ag, As, Co, Fe, Ni, Sb) are rather ubiquitous geochemical fingerprints (Ixer 1999). The more specific chemical markers determined for the artifacts of the Musée Guimet also show an unchanged pattern, but this is not sufficient to ascertain and identify a unique copper ore source. At least this rarity of copper deposits might have led to extensive recycling. Recycling is a common practice in modern foundries, as observed in Hanoi (Malleret 1954b), where old vessels as well as Chinese tea boxes often enter the melt. In our corpus, evidence of remelting is strong, but it is unfortunately not conclusive. As an example, the compositional heterogeneities observed within many assemblies may be a consequence of remelting, yet the metals used for the separate casts, obviously different, can originate from recycling as well as from different "fresh" metal supplies.

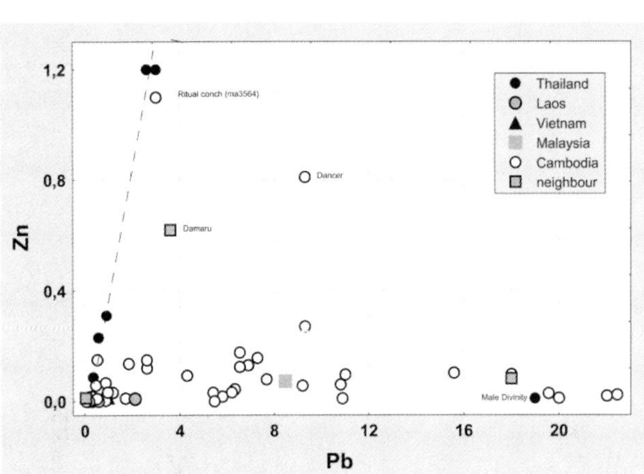

Figure 14. *Zn vs. Pb contents according to provenance. The artifacts supposed to come from Thailand show a linear correlation, suggesting that for these artifacts, Pb and Zn are associated to the original ore. Two artifacts whose origin is uncertain show the same relationship: the ritual conch and the Damaru. The male divinity, whose Thai origin is uncertain, does not follow the tendency. (Results in wt %; same population as for the graph of figure 11.)*

In the Foundry

While the lasting use of tin-and-copper-based alloys seems to constitute an original feature of Khmer metallurgy, it is paired with another distinctive trait: the apparent absence of any standardized bronze recipe. Tin contents vary drastically within our corpus, a phenomenon even more conspicuous given the use of both leaded and unleaded bronzes. Our attempt to correlate these variations and the use of one or the other alloy to any parameter has proved unsuccessful; none of a number of traits—chronology, cultural context, aesthetic quality, or technical aspects—could be associated in a convincing manner with any particular alloy composition. The provenance of the artifacts, although not clearly determined for several pieces, does not seem to affect the alloy composition as first suggested by Strahan (1998). She noticed that the tin and lead contents of the bronzes decrease as we proceed toward the east, away from the Burmese/Thai border where tin and lead ores are localized. We may argue that the related populations studied here are too small to show conclusively any tendency and that the composition discrepancies reflect the diversity of our corpus. Nevertheless, when a grouping happened to be sufficiently numerous to allow relevant statistical treatment, the results still showed an absence of correlation. As an example, the grouping of all of the statues dated from the Bayon period and supposed to come from Cambodia has not led to any conclusive tendency regarding the variation of the alloy recipes.

One further aspect may prove that the compositions do not conform to any uniform standard. This is the heterogeneity in alloy composition often found within the multipiece objects, where the variation in composition between the different parts can reach 50%.

The explanations for this nonstandardization of the bronze recipes may come from recycling as well as from devotional offerings (see below). In some cases, such as for the Adorned Buddha or the Finial (table 4), alloying was obviously made on the basis of a single copper supply in the foundry prior to casting, so that the variations are clearly due to a lack of alloy control. As a matter of fact, there may be only a few controls, the alloy composition not being a key parameter for the Khmer metallurgists, at least in the range of compositions measured. One tendency nevertheless deserves further study: in our corpus, the highest tin contents are almost exclusively found in the castings with the thinnest walls.

As with the bronze compositions, the present study leaves the impression of an absence of normalized casting processes. The lost-wax method was used with remarkable skill. Some very thin and regular hollow-cast statues were produced. We have also noticed that well-cast pieces are typical; porosity and casting defects are rare. However, some casting choices remain unexplained and thus appear randomly made. The example of the arms is demonstrative, where equivalent images show either a one-piece casting or a separate casting of the arms. These differences may principally be associated with the coexistence in the Khmer empire of the two lost-wax processes, the direct and the indirect, although it seems that most of the images in our corpus were cast by the direct lost-wax process. Yet the issue still remains: Why was one process used instead of another? The question of reproducibility is out of the question, since even major artifacts, supposed as such to be without any doubt unique, one-of-a-kind artifacts, can be cast using an indirect process. Variations of working capacity between the different foundries may explain the different processes, especially the use of separate castings. Its principal advantages are (1) easier molding, (2) easier building of the mold and especially the core, (3) easier handling, (4) use of a smaller furnace, and (5) the availability of less metal at one time for the casting (see Rama 1988). On the other side, they need an assembly and good finishing to hide the junctions. However, it may be also a question of local habits and traditions, which probably is also the main explanation for the modern version of this phenomenon in Thailand: in one modern foundry, a Seated Buddha of some 30 cm height was cast with a direct process (Strahan 1997), whereas in another workshop a similar image was executed using the indirect process (Chase 1975). The tradition would then explain also the a priori unnecessarily separately cast crowns or bracelets encountered several times. It would also explain the coexistence of different joining processes for the arms, wherein either the arm was inserted into the shoulder and fixed by metal or the opposite arrangement, where an extension of the shoulder fits into a hollow in the arm, the join being mechanical. And again tradition may explain why in some cases the different molds of an assembly have been poured from a unique melt, whereas in other cases several melts have been used. At least it does not depend on the social attribution as deduced from aesthetic qualities, since major images as well as minor ones are made using the same processes.

Surface and Special Treatments

Within this plurality of workshop habits, the gilding process appears much more consistent. Fire gilding is indeed encountered very often, regardless of the period. This may be characteristic of the Khmer civilization, at least in comparison to what can be found in the non-Khmer Thailand (Strahan 1998). Let us note also the fact that the gold used for fire gilding is not pure and systematically reveals traces of silver.

Another special treatment may have been performed on the Koh Ker Female Divinity (inv. MA2239) and on the Bayon Hevajra found at the Royal Palace at Angkor (inv. 4170), which are made of silver-copper alloy (20 wt. % silver, with about 5% tin). This kind of alloy is better known on Greek and Roman coinage (Aristophanes fifth century B.C.E.; Cope 1972) and well known in Japan as the famous *shibuichi* (25% silver; see Murakami 1993). It has also been identified on central Andean metalwork, from the first millennium B.C.E. until the Spanish invasion (Lechtman 1988), and most recently on a death mask from the Liao dynasty (tenth to twelfth century C.E., northeastern China–Inner Mongolia; Jett and Logan in press). The purpose of such a composition is to get a special surface aspect, which is quite easy through silver surface enrichment by inverse segregation (see experimentation noted in Jett and Logan in press). The two examples of our corpus cited above suggest that this coloring technique was known quite early by the Khmer metallurgists as well.

Tin surface enrichment is another way to give a silver appearance to the surface of a bronze, as observed on the pre-Angkorian Bodhisattva Maitreya and also on several images from the Griswold collection dated before the tenth century, including Standing Bodhisattva (cat. no. 12), Standing Buddha (cat. no. 15), and Naga Protected Buddha (cat. no. 16). The number of images is still too small to state with confidence that tin surface enrichment was exclusive to the early periods of the Khmer civilization. Moreover, without any metallographic examinations, it cannot be concluded whether or not these tin surface enrichments were the result of natural corrosion or of a deliberate treatment. Further study on this particular coloring technique would be quite interesting and would provide useful information. In particular, it would aid in determining the authenticity of the big Maitreya stylistically dated to the sixth to seventh century and analyzed by A. Bennett (1999) that shows tin surface enrichment.

Clearly deliberate is the addition of gold in the bronze melts of at least six statues of our corpus. This phenomenon may also appear in several statues examined by Strahan, since she observes gold amounts larger than trace by XRF surface analysis where no gilding was visible (Woodward 1997). We are far from established recipes such as those for objects made from the famous *samrit* that are said to contain more gold than copper (Coedès 1923). A much more realistic explanation might be the habit of jewelry offerings added to the melt, as attested in modern foundries in Vietnam (Malleret 1954a) and Thailand (Strahan 1997). In our corpus, this might further be confirmed by the fact that in some cases, the only impurity varying from one section of an assembly to another is gold (see the Angkor Vat-related Adorned Buddha protected by the Naga [inv. 3296]). This feature is seen throughout the whole Khmer period from the early pre-Angkorian times, as seen in the Buddha dated from the seventh to eighth century, to the latest ones, as shown by the Visnu from the sixteenth century. Whether or not such a gold content is a criterion of authenticity is beyond our concern: we note that in addition to several authentic objects, three ambiguous images of the corpus have revealed fairly high gold contents.

Conclusions

Extensive technological study of major metal art images and objects belonging to the entire millennium of the Khmer civilization revealed a very well developed bronze devotional metallurgy. These highly skilled creations also appear to have employed quite innovative processes, such as copper-silver alloying and perhaps also tin surface enrichment, in order to obtain particular surface colors. As a general rule, the images were well designed and executed with great care. Beautifully thin and even casting on quite tall images was typical, and very few casting defects or porosity were detected in all of the artifacts studied, be they major or minor pieces.

Some surprisingly persistent features distinguish this metallurgy. Gilding for the large majority of objects consisted of fire gilding, also known as amalgam gilding, throughout the ten centuries of the Khmer civilization. Bronze was never replaced by brass, even in the late periods. Also surprisingly constant is the copper impurity pattern, a feature for which there is no easy explanation. Is this feature a consequence of the well-known rarity of copper supplies in the Khmer empire, which led to the use of a unique supplier and, at the same time, to extensive recycling? At least the problem of metal supply may explain the second original feature of this metallurgy: the absence of any standardized alloy recipe, where low-tin, high-tin, and lead-tin bronze alloys coexist without any apparent logical reason. Since this absence of standardization not only affected the alloy compositions but also the casting processes, metal supply may not be the only explanation for this characteristic phenomenon. Thus, the idea of workshops or even schools working independently, with their own traditions and habits, may be raised. The confirmation and the characterization of these localized production centers requires further study, as does the metal supply issue. Finally, the legend of the *samrit* alloy continues to keep its secrets, although the unusual amounts of gold found in numerous images from all periods may be evidence of devotional offerings of jewelry and thus part of the mysterious recipe of this metal.

Acknowledgments

Many thanks to Nicolas Borghi who, as a student of the Ecole Superieure de Physique-Chimie Industrielle de Paris, assisted us during sampling and analyzing of the artifacts from the Musée Guimet collections.

Notes

1. We can distinguish between two main lost-wax processes, the direct and the indirect. In the direct process, the model is worked in wax, either by carving a wax block or by applying a layer of wax on a core (by painting on several coats or applying solid wax slabs) and carving it; the wax model is then encased in a refractory mold and melted out, and the metal is poured. The indirect lost-wax process first makes a mold of the original model. A wax layer is then applied to coat the mold surface, either by filling it with molten wax and pouring most of it out before it sets (one-piece mold), or by applying the wax by hand (piece mold). A core can then be added to fill the hollow space, and so forth.
2. To distinguish between the two processes, we used four features deduced from the radiographic examinations.
   (1) The best way to rule out the use of the direct process is to observe that the casting was performed in separate sections, at least when the separate sections are hollow. If the direct process had been used, it would have entailed either the cutting apart of the wax model with its inner core (and armature!) or the modeling of the sections separately to avoid discrepancies between the separate parts. Both possibilities are quite unlikely (see Haynes 1992, 38). When the separate section is solid, the cutting of the wax before molding may be far more secure (see arms of the Visnu-Vasudeva-Narayana [inv. 5291]).
   (2) The thickness and evenness of the metal walls may be another indication, at least partly. Thick, uneven metal walls indicate that wax was pressed on to the core according to a direct lost-wax process instead of having been applied in the internal walls of a mold. Unfortunately, the opposite situation is not that discriminating; to achieve evenness and thinness, the indirect lost-wax process can be employed as well as the direct one, provided that great care is taken in sizing and shaping the core.
   (3) The presence of an armature is also instructive, since its role is to ensure the integrity of the core (and the wax around it) in the casting process. During indirect lost-wax casting in a one-piece mold, an armature is hard to set before pouring the core. On the contrary, when a hollow-cast statue of a relatively big size shows no armature, it has been surely cast that way, as in Sadasiva (inv. 2682) and Lokesvara (inv. 5340), but the effective lack of armature may be difficult to demonstrate. An iron armature that is severely corroded can be invisible by radiographic examinations; so is a wooden armature, as observed by D. Strahan on modern Thai castings (1997).
   (4) A fourth feature that can be indicative of either direct or indirect casting is the nature of the cavity in hollow castings. In some statues the coarse aspect of the inner metal walls clearly shows that wax was removed mechanically after it set, during an indirect process using a one-piece mold, as in the Seating Buddha (inv. MA3788), the main body of the Prajnaparamita with eleven heads, the legs of Hevajra (inv. 5331), and the arms and legs of Lokesvara (inv. 5340). In the two last statues, the pouring out of the wax was incomplete in one arm, leading to a solid elbow.
3. The artifacts cast in separate sections generate several analyses, which may constitute redundant information. Depending on the aspect looked at and the composition homogeneity, we will consider for one such artifact either all of the analyses of all the different sections, or part of them, or only one analysis.
4. We use the term *bronze* for any composition with more than 2 wt. % Sn. The distinction between bronze and leaded bronze is more difficult, as one is often unable to distinguish between a metal impurity and deliberate addition. We know neither at which quantity lead starts to be effective (in lowering the melting point, enhancing fluidity, avoiding porosity, facilitating chiseling, etc.) nor what knowledge the Khmers had on this subject. We have chosen to differentiate between those bronzes with lead contents more than or less than 1.3 wt. %, based on the statistical distribution of the analytical results for our corpus.

5. A small section was cut from the base, and the bulk metal was observed by SEM and analyzed by SEM-EDX. The analytical results were 17 wt. % Ag and 15 wt. % Sn (Arc-Antique pers. comm.).
6. According to Scott (1991), we consider the usual limit of solubility of tin in copper to be about 14 wt. % (the theoretical limit is around 17%), and thus, the tin-content frontier between low-tin and high-tin alloys.
7. This property of tin seems to be well known by the modern bronze casters (Rama 1988, 280); the explanations nevertheless are far from clear. The ability of a molten metal to flow depends on the effective fluidity of the melt, which is not so much dependent on the tin content as on the presence or not of a deoxidizing agent such as phosphorous (Smithells 1967; Hanson and Pell-Walpole 1951). However, it also depends on the freezing range and thus directly on the tin composition. Nevertheless, according to Hanson and Pell-Walpole (1951), the largest freezing range in chill-cast conditions is reached at 6 wt. % Sn, whereas the thin-walled castings contain more than 10 wt. % Sn. Another important parameter for fluid-flow property is the gas absorption, but in the range 5–12 wt. % Sn, the tin content has little influence on gassing; above 12% it becomes deleterious (at least in considering the steam reaction in chill casting; see Hanson and Pell-Walpole 1951, 135). Incidentally, let us note that lead does not play any significant role in our corpus.
8. However, the casting complexity does not seem to play any role in the bipolar distribution of the prestige artifacts. For example, the tripod of the Perfume Censer (inv. MA5984), a rather complex object, has low tin and lead contents, whereas the simple-shaped Linga (inv. MA18187) and Vajra (inv. 5483) are high in tin and lead.
9. There do appear to be three general tendencies, but the small number of samples does not permit conclusions based on the analyses: (1) all four pre-Angkorian statues exhibit the highest tin contents (more than 10 wt. % Sn); (2) all statues of the tenth to eleventh century have less than 1 wt. % Pb; and (3) a particular composition exhibiting very high lead contents (more than 15 wt. % Pb) and a rather steady tin content (8 ± 2 wt. % Sn) appears only during the time period of Angkor Vat to Bayon.
10. First, only the chemical elements analyzed for both sets of measurements (DCP and ICP) will be taken into account in the general treatment (appendix 2). Nevertheless, the other elements prove to be useful for discriminating, since they allow distinguishing between several unusual compositions (appendix 3). Thus, the Bayon-related Siva (inv. MA2110) shows a relatively high Te content. The lid of the Perfume Censer (inv. MA5984) exhibits high values for Cd and Cr. The central Naga head of the Finial (inv. MG18889) also contains an unusually high amount of Cr. The most unusual composition remains that of the pre-Angkorian Maitreya (inv. MA3321), where In, P, Se, and V contents are much higher than for the other artifacts. For the latter, its composition is so remarkable it will not be considered in the following statistical treatment. Second, we will not take into account the Mn contents, which are very close to the detection or quantification limits for most artifacts, or the Bi, which was in several cases due to the lead used (linear correlation factor Pb/Bi = 0.77). Finally, we will focus on the bronzes and leaded bronzes. Therefore, eighty-one analyses associated with fifty-five artifacts were compared based on their Ag, As, Au, Co, Fe, Ni, Sb, and Zn contents.
11. We assume a gilding layer thickness varying between 5 and 50 μm and a bronze wall thickness ranging from 1 to 5 mm. Due to porosity, we assume a loss of density of 2 for the mercury-gold layer, which leads to a density ratio between pure gold and pure copper of (19/2) / 9 1.
12. The Avalokitesvara is distinctive, since it is associated with the Srivijaya culture in Malaysia. The high antimony content may be as such some fingerprint, and finding this fingerprint in another image, the Buddha, raises some questions.
13. A micro-fragment was sampled behind one ear. Some was powdered and characterized by XRD, and some was cross-sectioned and observed under light optical microscopy. A black layer (75 μm thick) was found under the red cinnabar one (40 μm). The infrared absorption spectra are very close to those of natural resins (thanks to Ms. P. Hugon from the French Laboratory of Research on the Historical Monuments).
14. We cannot ensure for every statue that cinnabar (HgS) was not used as a pigment in the area of the proton or x-ray beam spot, but we know for sure that the size of the gilded area analyzed was systematically larger than the spot size. Therefore, the only way for the mercury detected not to come from the gilding is that a cinnabar layer was present under the gilding layer.
15. The copper alloys used in the Himalayan devotional images seldom exhibit significant amounts of tin, zinc being the favored alloying element as early as the eleventh century (Beguin and Liszak-Hours 1982; see also the few brass statues of western Tibet dated from the twelfth to the fifteenth century in Craddock 1981). Von Schroeder reaches the same conclusions and extends this phenomenon to the entire northwestern part of India (Von Schroeder 1981).

References

Anheuser, K. 1995. Antike und neuere Techniken zur Vergoldung von Metallen und Nichtmetallen. *Berliner beiträge zur archäometrie* 13:87–97.

Béguin, G., and J. Liszak-Hours. 1982. Objets himâlayens en métal du musée Guimet: étude en laboratoire. *Annales du Laboratoire de Recherche des Musées de France* 28–82.

Bennett, A. 1999. Scientific examination of a Maitreya bronze. *Arts of Asia* 29(4):91–96.

Boisselier, J. 1955. Sculpture khmere et son evolution. *Publication École Française d'Extrême Orient* XXXVII(1–2).

——. 1967. Notes sur l'art du bronze dans l'ancien Cambodge. *Artibus Asiae* XXIX(4):275–334.

Bronson, B. 1992. Patterns in the early Southeast Asian metals trade. In *Early metallurgy, trade, and urban centers in Thailand and Southeast Asia,* ed. I. Glover, P. Suchitta, and J. Villiers, 63–114. Bangkok: White Lotus.

Bunker, E. C. 1971–72. Pre-Angkor period bronzes from Prakon Chai. *Archives of Asian art* XXV:67–76.

Calligaro, T., J. C. Dran, H. Hamon, et al. 1998. An external milli-beam for archaeometric applications on the AGLAE IBA facility of the Louvre Museum. *Nuclear instruments and methods B* 136–38:339–43.

Chase, W. T. 1975. *Bronze disease and its treatment.* Bangkok: National Museum.

Coedès, G. 1923. Bronzes Khmèrs. *Ars Asiatica* 5.

Cope, L. H. 1972. The metallurgical analysis of Roman imperial silver and Aes coinage. In *Methods of chemical and metallurgical investigation of ancient coinage,* vol. 8, ed. E. T. Hall and D. M. Metcalf, 3–47. London: Royal Numismatic Society.

Craddock, P. T. 1981. The copper alloys of Tibet and their background. In *Aspects of Tibetan metallurgy,* ed. W. A. Oddy and W. Zwalf, 1–31. British Occasional Paper no. 15. London.

Drilhon, F., J. Francaix, and C. Lahanier. 1977. Rapport d'analyse de huit statuettes appartenant au Musée Guimet. *C2RMF internal report*, no. 682.

Dupont, P. 1955. La statuaire préangkorienne. *Artibus Asiae* supp. XV.

Dupouy, G. 1913. *Etudes minéralogiques sur l'Indochine française*. Paris.

Groslier, G. 1921. L'art du Bronze au Cambodge. In *Art et archéologie Khmers,* vol 1, 413–23. Paris.

Hanson, D., and W. T. Pell-Walpole. 1951. *Chill-cast tin bronzes*. London: Edward Arnold and Co.

Haynes, D. 1992. *The technique of Greek bronze statuary,* ed. Philipp von Zabern. Mainz/Rhein.

Ixer, R. A. 1999. The role of ore geology and ores in the archeological provenancing of metals. In *Metals in antiquity,* ed. S. M. M. Young, P. A. M. Pollard, P. Budd, and R. A. Ixer, 43–52. BAR International Series 792. Oxford: Archaeopress.

Jessup, H. I., and T. Zéphyr, eds. *Sculpture of Angkor and ancient Cambodia: Millennium of glory*. Washington, D.C.: National Gallery of Art.

Jett, P., and A. Logan. In press. A metallographic examination of two death masks from the Liao dynasty China. Presented at the Founders and Platers Conference, Sept. 1999, Oxford.

Le Bonheur, A. 1972. Un bronze d'époque préangkorienne representant Maitreya. *Arts Asiatiques* XXV:155–98.

Lahanier, C. 1974. Examen d'un torse de Bodhisattva Avalokitesvara. *C2RMF internal report*, no. 439.

Lechtman, H. 1988. Traditions and styles in central Andean metalworking. In *The beginning of the use of metals and alloys,* ed. R. Maddin, 344–78.

Lins, P. A., and W.A. Oddy. 1975. The origins of mercury gilding. *Journal of archaeological science* 2:365–73.

Malleret, L. 1948. L'art et la métallurgie de l'étain dans la culture d'Oc-èo. *Artibus Asiae* XI.

——. 1954a. A propos d'analyses de bronzes archéologiques. *Bulletin de la Société des Etudes Indochinoises* XXIX(4):1–11.

——. 1954b. Fonte d'une statue du Buddha à Hanoi. *Bulletin de l'École Française d'Extrême-Orient* XLVI(2).

Meeks, N. D. 1986. Tin-rich surfaces on bronze: Some experimental and archaeological considerations. *Archaeometry* 28(2):133–62.

——. 1993. Surface characterization of tinned bronze, high-tin bronze, tinned iron, and arsenical bronze. In *Metal plating and patination,* ed. S. La Niece and P. Craddock, 247–75. London: Butterworth Heinemann.

Mille, B., and D. Bourgarit. 2000. L'analyse des alliages anciens à base de cuivre: état des connaissances et développement d'un protocole d'analyse par ICP-AES. *Revue d'archéométrie* 24:13–26.

Murakami, R. 1993. Japanese traditional alloys. In *Metal plating and patination,* ed. S. La Niece and P. Craddock, 85–94. London: Butterworth Heinemann.

Rama, J.-P. 1988. *Le bronze d'art et ses techniques*. Dourdan: H. Vial.

Scott, D. A. 1991. *Metallography and microstructure of ancient and historic metals*. Los Angeles: The Getty Conservation Institute.

Smithells, C. J. 1967. *Metals reference book*. 4th ed. London: Butterworth.

Strahan, D. K. 1997. Bronze casting in Thailand. In *The sacred sculpture of Thailand,* H. W. Woodward Jr., 27–41. Baltimore: Walters Art Gallery.

——. 1998. A technical study of 11–12th century Khmer bronze sculpture. In *The beginning of the use of metals and alloys,* 97–102. BUMA-IV. Sendai City, Japan: Japan Institute of Metals.

Von Schroeder, U. 1981. Geographical distribution of the basic alloys. In *Indo-Tibetan bronzes*. Hong Kong: Visual Dharma Publications.

Werner, O. 1972. *Spektralanalytische und metallurgische untersuchungen um Indischen bronzen*. Leiden: Brill.

Woodward, H. W., Jr.. 1997. *The sacred sculpture of Thailand*. Baltimore: Walters Art Gallery.

Appendix 1. *Comparison of the analyses performed by direct current plasma atomic emission spectrometry (DCP-AES) and by inductively coupled plasma atomic emission spectrometry (ICP-AES)*

Lab no.			Inv. no.		technique	Ag	As	Au	Bi	Co	Fe	Ni	Pb	Sb	Sn	Zn
23577	Pre-Angkorian	Buddha	5405		ICP	0.22	0.38	0.063	0.03	0.01	0.40	0.08	7.0	0.65	13.3	0.14
					DCP	0.16	0.42	0.062	0.02	0.02	0.04	0.08	6.5	0.63	12.5	0.12
23559-1	Angkor Vat	Visnu-Vasudeva-Narayana	5291	Arm	ICP	0.06	0.30	0.001	0.07	0.01	0.22	0.03	19.9	0.20	7.5	0.02
					DCP	0.06	0.41	0.003	0.09	0.01	0.24	0.04	22.4	0.24	7.3	0.02
23580-2	Bayon	Miniature shrine of Hevajra	7192	Base	ICP	0.06	0.15	0.003	0.03	0.02	0.18	0.20	0.8	0.04	8.0	0.20
					DCP	0.06	0.17	0.005	0.02	0.02	0.15	0.22	0.9	0.04	7.0	0.20
23570-1	Bayon	Buddhistic Triad	2424	Base	ICP	0.15	0.05	0.017	0.01	0.03	0.02	0.15	0.04	0.004	7.4	0.003
					DCP	0.14	0.05	0.017	< 0.01	0.05	0.04	0.13	0.17	< 0.01	7.4	< 0.005

Appendix 2. *The analytical results for the sixty-nine Khmer artifacts (112 samples). The objects of the Phnom Penh National Museum (inv. number without any letter) were analyzed by DCP-AES; the objects from the Musée Guimet by ICP-AES. For the latter, results for additional elements were obtained. The corresponding contents are reported in appendix 3. All results are in wt. % except for the chemical elements followed by * (wt ppm). The figures followed by ° correspond to values under the quantitative limit (three times the detection limit). The results for copper were not quantified and have not been reported.*

	Designation	Inv. no.	Section	Sample	Sn	Pb	Zn	Ag	As	Fe	Ni	Sb	Au*	Bi*	Co*	Mn*
Pre-Angkorian	Maitreya	MA3321		Head	11	0.6	0.23	0.03°	0.09	0.52	0.05	0.02	< 200	< 140	8600	< 12
	Bodhisattva Lokesvara	MA4985		Arm	13	0.9	0.31	0.05	0.19	0.75	0.06	0.05	36	160	870	< 4
	Male figure	MG18318		Arm	0.8	0.6	0.02	0.28	0.08	0.02	0.04	0.05	200	390	8	< 4
	Standing Buddha	MA3785		Foot	16	0.4	0.09	0.05	0.12	0.66	0.09	0.03	29	140	650	5
	Avalokitesvara	3548		Right foot	10.1	8.5	0.07	0.12	0.51	0.17	0.07	0.48	240	240	260	< 10
	Lokesvara	5330		Left foreleg	83.8	8.5	< 0.005	0.01	0.03	0.02	0.00	0.01	200	< 100	190	< 10
	Buddha	5405		Dress bottom	12.5	6.5	0.12	0.16	0.42	0.04	0.08	0.63	620	220	240	< 10
10th–11th cent.	Perfume censer or oil lamp – kneeling male figure	MA1067		Hair	11	0.9	0.07	0.49	0.24	0.12	0.12	0.07	120	120	480	< 4
	Female divinity	MA2239		Base	4.6	0.3	0.24	19	1.10	0.08	0.07	0.06	890	220	34	21
	Hevajra	5331	Base	Base	15.0	1.2	0.03	0.15	0.25	0.79	0.20	< 0.02	240	< 200	820	< 10
			Body	Foot	9.6	0.05	< 0.005	0.28	0.08	0.02	0.25	< 0.01	30	< 100	< 50	< 10
			Arm	Internal right elbow	7.1	0.3	0.02	0.11	0.18	0.11	0.34	0.01	50	110	190	< 10
	Visnu	MA1339		Arm	12	0.4	0.01	0.10	0.17	0.10	0.21	0.03	180	75	260	< 4
Angkor Vat	Visnu astride Garuda	MA2678	Visnu	Foot	7.6	1.7	0.01	0.07	0.14	0.19	0.12	0.04	75	120	130	< 4
			Garuda	Base	11	5.4	0.03	0.10	0.29	0.37	0.09	0.09	110	310	230	< 4
	Visnu-Vasudeva-Narayana	5291	Arm	Internal right elbow	7.3	22.4	0.02	0.06	0.41	0.24	0.04	0.24	30	900	90	< 10
			Body	Left ear	7.3	22.4	0.02	0.06	0.41	0.24	0.04	0.24	30	900	90	< 11
	Adorned Buddha	2081	Head	Near right ear	11.5	8.9	1.6	0.06	0.26	0.09	0.08	0.11	250	340	250	< 10
			Body	Dress, top left	6.2	6.5	0.18	0.09	0.26	0.10	0.07	0.10	210	280	290	< 10
	Support for a mirror – kneeling female figure	5476		Belt	9.6	7.7	0.08	0.12	0.45	0.13	0.09	0.19	110	820	280	< 10
	Finial for a shaft (three-headed Naga)	MG18889	Central head Naga		11	0.5	0.01°	0.05	0.15	0.14	0.08	0.02	16	50°	59	6°
			Side head Naga		7.5	0.3	0.004°	0.04	0.11	0.06	0.09	0.02	16	47	64	< 4
	Adorned Buddha Protected by the Naga	3296	Naga head		4.2	1.0	0.02	0.10	0.16	0.18	0.06	0.10	1790	590	70	< 10
			Naga body		9.9	0.2	0.01	0.08	0.10	0.13	0.11	0.06	110	100	60	< 10
			Buddha		6.2	0.7	0.01	0.15	0.11	0.06	0.06	0.08	1570	590	70	< 10
	Male divinity	MA5960	Body	Base	7.4	0.3	0.007	0.05	0.10	0.04	0.25	0.02	16	110	130	9
			Right arm	Thumb	6.9	19	0.01	0.06	0.25	0.12	0.07	0.11	23	760	140	< 4
			Left arm	Thumb	6.9	11	0.01	0.06	0.24	0.26	0.15	0.08	38	480	190	6
	Dancer	C6500		Foot	8.1	9.3	0.81	0.08	0.26	0.18	0.09	0.09	60	430	220	< 4
	Visnu astride Garuda	MA539		Arm of Garuda	6.4	22	0.02	0.05	0.29	0.24	0.03	0.12	40°	1200	94	< 4
	Chevet of statuette	MG18175			4.4	5.8	0.02	0.04	0.23	0.07	0.21	0.05	23	80	150	< 4
	Sadasiva	2682		Left internal elbow	3.6	1.0	0.03	0.18	0.22	0.03	0.15	0.07	350	170	130	< 10

	Designation	Inv. no.	Section	Sample	Sn	Pb	Zn	Ag	As	Fe	Ni	Sb	Au*	Bi*	Co*	Mn*
Bayon	Mirror handle	EO801			10	2.1	0.009	0.11	0.20	0.17	0.10	0.07	180	340	210	< 4
	Perfume censer	MA5984	Receptacle		8.2	18	0.08	0.07	0.31	0.22	0.06	0.16	27	720	100	< 4
			Lid		7.9	16	0.03	0.08	0.29	0.63	0.13	0.17	88	640	1600	< 5
			Tripod	Foot	3.6	0.03	0.01	0.23	0.86	0.36	0.39	0.06	9	32	520	5
			Tripod	Plateau	3.4	0.03	0.01	0.03	0.82	0.34	0.38	0.05	< 10	32	500	< 4
	Lokesvara	5340	Body, upper part	Left foot	4.8	2.6	0.15	0.11	0.26	0.29	0.23	0.04	300	90	530	< 10
			Arm	External right elbow	6.4	1.8	0.14	0.05	0.28	0.18	0.18	0.02	100	380	240	< 10
			Head	Left ear	7.0	2.4	0.11	0.06	0.35	0.23	0.22	0.04	110	310	560	< 10
			Body, lower part	Belt	9.2	2.4	0.20	0.12	0.27	0.50	0.19	0.04	210	90	570	< 10
	Buddhistic Triad	2424	Base		7.4	0.2	< 0.005	0.14	0.05	0.04	0.13	< 0.01	170	< 100	540	< 10
			Buddha body	Left elbow	5.9	0.1	< 0.005	0.11	0.04	0.06	0.14	< 0.01	300	< 100	380	< 10
			Buddha base	Base	8.1	0.1	< 0.005	0.13	0.07	0.05	0.15	< 0.01	200	< 100	450	< 10
			Naga body	Snake (junction)	7.4	0.1	< 0.005	0.15	0.06	0.04	0.14	< 0.01	190	< 100	460	< 10
			Naga head		7.0	0.1	< 0.005	0.14	0.06	0.03	0.14	< 0.01	190	< 100	450	< 10
			Lokesvara	Left elbow	7.9	0.1	< 0.005	0.14	0.06	0.07	0.15	< 0.01	200	< 100	400	< 10
			Prajnaparamita	Left elbow	7.1	0.05	< 0.005	0.16	0.06	0.04	0.16	< 0.01	490	< 100	370	< 10
	Miniature shrine of Hevajra	7192	Yogini		6.4	0.6	< 0.010	0.08	0.15	0.09	0.08	0.13	20	780	< 100	< 10
			Base	Underneath	7.0	0.9	0.20	0.06	0.17	0.15	0.22	0.04	50	210	230	< 10
			Cage		5.2	0.2	0.51	0.04	0.07	0.15	0.20	0.03	30	100	110	< 10
	Hevajra	4170		Principal left elbow	4.0	0.1	< 0.005	0.02	0.47	0.47	0.47	0.02	10	< 100	980	< 10
	Brahmanic Triad	2283	Siva	Left ear	7.6	0.1	< 0.005	0.15	0.07	0.10	0.25	0.02	230	< 100	510	< 10
			Devi	Left ear	7.9	0.1	< 0.005	0.17	0.08	0.09	0.26	0.02	420	< 100	500	< 10
			Visnu	Left ear	7.6	0.1	< 0.005	0.17	0.10	0.32	0.26	0.01	360	< 100	580	< 10
			Base	Underneath	8.9	0.1	< 0.005	0.18	0.07	0.07	0.27	< 0.01	230	< 100	600	< 10
	Kneeling female figure	5467		Left forearm	13.4	6.3	0.04	0.07	0.33	0.35	0.22	0.05	350	< 100	650	< 10
	Prajnaparamita with eleven heads	5333		Internal left arm	8.2	2.6	0.12	0.07	0.28	1.1	0.26	0.05	80	< 100	4890	< 10
	Hevajra	MA849	Hevajra	Base	7.9	6.2	0.03	0.06	0.37	0.42	0.21	0.05	150	140	510	< 4
			Base		5.3	7.2	7.7	0.05	0.06	0.17	0.03	0.02	< 20	77	< 10	< 4
	Prajnaparamita	MA508		Base	6.7	6.9	0.13	0.06	0.18	0.41	0.11	0.04	38	140	200	< 4
	Lokesvara	MA509		Base	8.2	18	0.10	0.18	0.19	0.16	0.03	0.18	26	810	40	< 4
	Yogini	MA540		Base	4.3	0.2	0.02	0.04	0.04	0.04	0.30	0.05	8	130	80	5
	Siva	MA2110		Base	8.6	0.03	0.01	0.14	0.04	0.12	0.13	0.00	130	160	250	< 4
	Ganesa	MA1829		Base	6.8	2.7	8.5	0.03	0.10	0.36	0.10	0.19	20°	170	14	5
	Cosmic Lokesvara with radiating arms	MA5940		Arm	6.3	0.05	0.01	0.02	0.65	0.97	0.37	0.05	< 10	37	920	< 4
	Seated Buddha	MG9807		Arm	3.9	1.0	14	0.11	0.17	0.24	0.10	0.02	72	280°	240	12°
	Finial-Adorned Buddha	5593	Finial	Bottom	11.9	10.9	0.01	0.08	0.27	0.30	0.09	0.08	100	280	480	< 10
	Protected by the Naga		Buddha	Left elbow	7.5	19.6	0.03	0.08	0.29	0.27	0.04	0.12	50	640	250	< 10
			Naga	Junction	9.4	19.1	0.01	0.07	0.33	0.30	0.11	0.11	60	480	400	< 10
	Finial with Garuda	MA102		Tail	4.7	0.5	0.06	0.07	0.29	1.1	0.31	0.04	460	110	650	< 4
	Damaru	MA5988	Drum		14	3.6	0.62	0.05	0.28	0.30	0.26	0.04	62	130	620	< 4
			Little figures	Base	14	3.7	0.57	0.05	0.29	0.25	0.26	0.04	61	130	620	< 4
	Vajra	5483			10.5	15.6	0.10	0.11	0.36	0.20	0.08	0.15	90	690	290	< 10
	Vajra	MA1617			8.4	11	0.10	0.14	0.28	0.24	0.21	0.10	400	440	940	13
	Ritual conch	5484–5485	Tripod		6.7	0.3	< 0.005	0.04	0.17	0.03	0.26	0.08	40	< 100	300	< 10

	Designation	Inv. no.	Section	Sample	Sn	Pb	Zn	Ag	As	Fe	Ni	Sb	Au*	Bi*	Co*	Mn*
	Ritual conch	MA3564	Conch		8.4	3.0	1.10	0.10	0.29	0.63	0.20	0.07	360	170	390	< 4
	Tripod	MA3565	Tripod	Foot	5.9	0.3	0.01°	0.28	0.09	0.06	0.04	0.07	23°	370	< 50	7°
			Bassin	Bassin	7.9	0.5	< 0.004	0.092	0.10	0.03	0.04	0.11	20	800	< 50	6°
	Tripod	MA541	Tripod	Foot	5.0	0.2	0.005	0.07	0.08	0.04	0.03	0.05	20	240	< 50	7
			Basin	Basin	7.0	0.6	0.01	0.10	0.10	0.11	0.03	0.08	54	760	< 50	10
	Naga	MG18888		Head	2.1	0.1	0.00	0.02	0.32	0.06	0.28	0.04	36	32	83	< 4
	Bed "ring"	MA3563b			5.1	0.9	0.003°	0.10	0.15	0.27	0.09	0.09	51	510	120	< 4
	Bed "hook"	MA5989			8.8	4.3	0.09	0.13	0.26	0.17	0.15	0.06	360	200	200	< 4
	Bed "ring"	MG18851			7.7	0.5	0.15	0.09	0.32	0.39	0.08	0.19	19	130	350	< 4
	Bell	MA18850	Bell		8.2	9.3	0.27	0.05	0.32	0.27	0.25	0.06	170	140	460	5
			Ball		8.0	9.4	0.27	0.05	0.32	0.29	0.25	0.06	170	130	450	< 4
	Buddhistic Triad plaque	MG18204			6.9	20	0.01	0.07	0.28	0.10	0.03	0.12	7	870	62	< 4
	Ganesa	2320	Body	Underneath	5.9	7.3	0.16	0.07	0.65	0.88	0.29	0.10	1780	300	610	< 10
			Base	Left foot	6.8	6.5	0.07	0.31	0.55	1.6	0.32	0.06	1300	200	480	< 10
			Arm	Right external elbow	7.3	8.6	0.16	0.08	0.61	0.77	0.24	0.10	1390	200	590	< 10
Post-Angkorian	Buddha	2057-1	Lower part	Dress, bottom left	5.9	9.2	0.06	0.08	0.37	0.31	0.18	0.08	880	330	380	< 10
			Upper part	Dress, middle right	7.0	10.8	0.06	0.09	0.44	0.35	0.21	0.11	710	570	360	< 10
	Visnu	5457	Arm	Left external forearm	8.2	5.5	< 0.005	0.05	0.43	0.14	0.33	0.06	230	110	430	< 10
			Head	Top	4.4	0.2	< 0.005	0.03	0.64	0.17	0.54	0.06	920	< 100	950	< 10
			Base	Base	4.8	0.1	< 0.005	0.03	0.60	0.08	0.53	0.06	510	< 100	380	< 10
			Crown	Crown	5.9	2.1	< 0.005	0.06	0.41	0.16	0.34	0.05	300	< 100	440	< 10
	Lion	MA2341		Leg	4.0	2.6	1.2	0.12	0.17	0.12	0.09	0.05	22	110	52	< 4
	Lion	MA2342		Leg	3.9	3.0	1.2	0.12	0.18	0.09	0.08	0.06	40	120	44	< 4
Uncertain	Linga	MA18187		Underneath	8.0	14	0.28	0.09	0.28	0.09	0.08	0.14	200	590	100	< 4
dating	Linga	3557		Bottom	6.2	0.2	< 0.005	0.07	0.08	0.10	0.06	0.05	50	370	< 50	< 10
	Hand of big statue	C6501			7.2	0.2	0.007	0.11	0.14	0.16	0.09	0.02	23	40	150	< 4
	Garuda	MA2592		Leg	6.0	0.5	15	0.01	0.03	0.13	0.02	0.03	< 10	110	20	4
	Female divinity	MA3789		Base	4.3	0.1	0.01	0.07	0.30	0.67	0.30	0.05	2500	66	480	< 4
	Visnu	C6496		Arm	8.9	19	0.03	0.08	0.26	0.20	0.04	0.10	4	740	80	< 4
	Prajnaparamita	MA2677		Foot	13	5.0	3.1	0.04	0.29	0.23	0.19	0.05	610	200	360	< 4
	Seated Buddha	MA3788		Lotus	4.2	0.3	0.03	0.20	0.26	0.38	0.20	0.04	1000	150	190	19
	Standing Buddha	MA3786		Base	12	9.2	0.36	0.07	0.20	0.89	0.15	0.05	51	210	450	9
	Female divinity	MG18333		Arm	8.1	1.4	0.03	0.02	0.72	0.24	0.17	0.08	7	81	240	< 4
	Standing Buddha	MA3787		Base	10.0	1.4	0.32	0.23	0.48	0.26	0.11	0.07	88	120	380	< 4
	Cup	MG24921			7.8	7.2	10	0.08	0.16	0.24	0.33	0.05	5	150	20	< 4

Appendix 3. *Additional results obtained for the artifacts from the Musée Guimet using ICP-AES. All results are in wt ppm except for P and S (wt %). The figures followed by ° correspond to values under the quantitative limit (three times the detection limit)*

	Designation	Inv. no.	Section	Sample	Ba*	Cd*	Cr*	Ge*	In*	Mg*	Mo*	P	S	Se*	Te*	Ti*	U*	V*
Pre-Angkorian	Maitreya	MA3321		Head	130	< 5	< 28	< 97	480	160	< 15	0.21°	< 0.25	770	< 93	40	< 30	93
	Bodhisattva Lokesvara	MA4985		Arm	< 10	< 5	< 5	< 5	47	< 5	< 0.8	< 0.005	0.051	< 15	43	1.2	5	< 2
	Male figure	MG18318		Arm	9	< 5	< 5	< 6	46	< 5	< 0.9	< 0.006	0.028°	15	42	1.3	< 1.7	< 2
	Standing Buddha	MA3785		Foot	< 10	< 5	< 5	< 6	29	9	< 0.9	< 0.007	0.21	33	48	3.4	3.7°	< 2
10th–11th cent.	Perfume censer or oil lamp – kneeling male figure	MA1067		Hair	< 10	< 5	< 5	< 6	24	5	< 0.9	< 0.006	0.058	18	57	0.8°	< 1.7	< 2
	Female divinity	MA2239		Base	22	< 5	17°	1000	35	20	< 4	< 0.027	< 0.065	80	66°	13	< 8	< 2
	Visnu	MA1339		Arm	< 10	< 5	< 5	< 6	42	6	< 0.9	< 0.006	0.015°	24	58	1.0°	< 1.7	< 2
Angkor Vat	Visnu astride Garuda	MA2678	Visnu	Foot	< 10	< 5	< 5	< 6	24	< 5	0.9°	< 0.006	0.028°	40	41	< 0.4	< 1.8	< 2
			Garuda	Base	11	< 5	< 5	< 6	23	< 5	1.0°	< 0.007	0.11	23	44	0.7°	< 1.8	< 2
	Finial for a shaft (three-headed Naga)	MG18889	Central head Naga		51	< 5	99	< 33	51	17	< 5	0.039°	< 0.082	86	74°	7.8	< 10	< 2
			Side head Naga		10	< 5	< 5	< 7	51	< 5	< 1.1	0.017°	< 0.018	63	43	0.7°	< 2.1	< 2
	Male divinity	MA5960	Body	Base	< 10	< 5	< 5	< 5	19	7	< 0.7	< 0.005	< 0.012	17	41	2.2	< 1.4	< 2
			Right arm	Thumb	< 10	< 5	< 5	< 7	38	< 5	< 1.0	0.069°	0.35	< 15	50	0.5°	< 1.9	< 2
			Left arm	Thumb	< 10	< 5	< 5	< 6	30	< 5	< 0.9	0.065°	0.27	< 15	50	0.9°	< 1.8	< 2
	Dancer	c6500		Foot	< 10	< 5	< 5	< 6	26	< 5	< 0.8	< 0.006	0.15	< 15	37	0.6°	< 1.6	< 2
	Visnu astride Garuda	MA539		Arm of Garuda	< 10	< 5	< 5	< 6	33	< 5	< 0.9	< 0.007	0.41	< 15	42	4.5	< 1.8	< 2
	Chevet of statuette	MG18175			< 10	< 5	< 5	< 6	42	< 5	< 0.9	< 0.006	< 0.015	20	38	1.1°	< 1.7	< 2
Bayon	Mirror handle	eo801			< 10	< 5	< 5	< 6	24	< 5	< 0.9	< 0.006	0.038°	25	52	1.7	< 1.7	< 2
	Perfume censer	MA5984	Receptacle		10	< 5	< 5	< 9	37	48	< 1.4	< 0.010	0.34	< 15	52	4.5	< 3	< 2
			Lid		20	21	83	< 35	56	15	< 6	< 0.006	0.35	< 28	63°	8	< 11	< 2
			Tripod	Foot	< 10	< 5	< 5	39	27	< 5	< 0.9	< 0.007	0.41	70	85	1.3	< 1.8	< 2
			Tripod	Plateau	< 10	< 5	< 5	32	25	< 5	< 0.9	< 0.037	0.23°	61	72	1.0°	< 1.7	< 2
	Hevajra	MA849	Hevajra	Base	< 10	9	< 5	11°	23	< 5	< 0.9	< 0.006	0.25	16	47	2.3	< 1.7	< 2
			Base		< 10	< 5	< 5	< 6	20	8	< 0.9	< 0.021	0.042°	18	15°	1.3	< 1.8	< 2
	Prajnaparamita	MA508		Base	< 10	< 5	< 5	< 6	23	8	< 0.8	< 0.006	0.06	< 15	36	2.1	< 1.6	< 2
	Lokesvara	MA509		Base	31	< 5	< 5	< 6	28	< 5	< 0.9	< 0.006	0.30	15	60	1.4	< 1.7	< 2
	Yogini	MA540		Base	20	< 5	6.2°	< 9	59	7	< 1.4	< 0.010	< 0.022	< 15	53	9.9	< 3	< 2
	Siva	MA2110		Base	< 10	< 5	< 5	< 7	48	< 5	< 1.0	< 0.007	0.041°	58	220	3.5	< 1.9	< 2
	Ganesa	MA1829		Base	41	22	9.2	< 6	24	6	< 0.9	0.012°	0.017°	< 15	23	0.6°	< 1.8	< 2
	Cosmic Lokesvara with radiating arms	MA5940		Arm	< 10	< 5	< 5	38	43	< 5	< 0.8	0.006°	0.37	< 15	49	0.9°	< 1.5	< 2
	Seated Buddha	MG9807		Arm	130	< 5	< 5	< 83	110	46	< 13	< 0.089	< 0.22	< 67	< 80	20	< 26	< 5
	Finial with Garuda	MA102		Tail	27	< 5	< 5	< 11	42	8	< 1.7	< 0.012	0.10	19°	56	1.0°	< 4	< 2
	Damaru	MA5988	Drum		< 10	< 5	< 5	25	22	8	< 1.0	< 0.007	0.11	< 15	37	3.3	< 2.0	< 2
			Little figures	Base	12	< 5	< 5	26	21	< 5	< 0.9	< 0.006	0.078	< 15	43	1.3	< 1.7	< 2
	Vajra	MA1617			< 10	< 5	< 5	< 9	31	12	< 1.3	< 0.011	0.17	< 15	58	12	< 2.5	< 2

	Designation	Inv. no.	Section	Sample	Ba*	Cd*	Cr*	Ge*	In*	Mg*	Mo*	P	S	Se*	Te*	Ti*	U*	V*
	Ritual conch	MA3564	Conch		14	< 5	< 5	< 11	69	5	< 1.6	< 0.009	0.14	< 15	50	1.9°	< 3	< 2
	Tripod	MA3565	Tripod	Foot	130	14	15°	< 35	50	26	< 6	< 0.037	< 0.087	< 28	61°	12	< 11	< 2
			Basin	Basin	29	< 5	< 6.8	< 24	43	5	< 3.6	0.035°	< 0.059	< 19	80	< 1.4	< 7	< 2
	Tripod	MA541	Tripod	Foot	< 10	< 5	< 5	< 6	46	10	< 0.9	< 0.007	< 0.015	< 15	40	0.7°	< 1.8	< 2
			Basin	Basin	< 10	< 5	11°	< 16	52	8	< 2.4	< 0.017	< 0.039	< 15	72	7.2	< 5	< 2
	Naga	MG18888		Head	< 10	< 5	< 5	< 6	22	< 5	< 0.8	< 0.006	< 0.014	27	56	1.3	< 1.6	< 2
	Bed “ring”	MA3563b			< 10	< 5	< 5	< 7	24	< 5	< 1.0	0.012°	0.025°	15°	49	1.0°	< 1.9	< 2
	Bed “hook”	MA5989			< 10	< 5	< 5	< 6	21	< 5	< 1.0	< 0.007	0.093	15°	67	1.4	< 1.9	< 2
	Bed “ring”	MG18851			< 10	< 5	< 5	< 5	25	< 5	< 0.7	< 0.005	0.041	< 15	35	0.2°	1.5°	< 2
	Bell	MA18850	Bell		< 10	< 5	30	14°	22	< 5	< 0.9	< 0.007	0.15	< 15	42	0.9°	< 1.8	< 2
			Ball		< 10	< 5	< 5	15°	21	12	< 0.9	< 0.006	0.15	16	44	1.3	< 1.7	< 2
	Buddhistic Triad plaque	MG18204			< 10	< 5	< 5	< 6	34	6	< 0.9	< 0.006	0.38	< 15	50	1.9	< 1.7	< 2
Post-Angkorian	Lion	MA2341		Leg	< 10	< 5	< 5	< 6	31	< 5	< 0.9	< 0.007	< 0.015	< 15	31	0.5°	< 1.8	< 2
	Lion	MA2342		Leg	< 10	< 5	< 5	< 6	34	< 5	< 0.9	< 0.007	0.026°	15	36	0.7°	< 1.8	< 2
Uncertain dating	Linga	MA18187		Underneath	23	< 5	< 5	< 16	98	8	< 2.4	0.026°	0.21	< 15	53	1.1°	< 5	< 2
	Hand of big statue	c6501			< 10	< 5	< 5	< 6	23	< 5	< 0.9	0.013°	< 0.015	18	64	0.6°	< 1.7	< 2
	Garuda	MA2592		Leg	10	6	< 5	< 6	24	7	< 0.9	0.009°	< 0.015	< 15	12°	2	< 1.8	< 2
	Female divinity	MA3789		Base	< 10	< 5	< 5	< 6	49	< 5	< 0.9	< 0.007	0.13	27	87	2.7	< 1.8	< 2
	Visnu	c6496		Arm	< 10	< 5	< 5	< 6	47	< 5	< 0.9	< 0.007	0.34	15°	54	0.9°	< 1.8	< 2
	Prajnaparamita	MA2677		Foot	< 10	< 5	< 5	27	30	< 5	< 1.0	< 0.007	0.11	18	64	< 0.4	< 20	< 2
	Seated Buddha	MA3788		Lotus	25	9	< 5	< 8	26	31	< 1.2	0.043	0.14	16°	68	3.7	< 2.4	< 2
	Standing Buddha	MA3786		Base	21	< 5	< 5	9°	24	35	< 0.9	0.010°	0.39	24	77	10	4°	< 2
	Female divinity	MG18333		Arm	< 10	< 5	< 5	< 6	23	< 5	< 0.9	< 0.007	0.22	15	41	1.7	< 1.8	< 2
	Standing Buddha	MA3787		Base	< 10	< 5	< 5	< 6	22	6	< 1.0	< 0.007	0.064	15°	40	1.0°	< 1.8	< 2
	Cup	MG24921			< 10	< 5	< 5	< 6	43	< 5	< 0.9	< 0.007	0.018°	36	9°	0.6°	< 1.8	< 2

14 Trace Elements in North Indian and Tibetan Statuettes Composed of Copper Alloys

Josef Riederer

ABSTRACT Statuettes from Tibet, Nepal, and the northern parts of India have been cast from copper and various copper alloys that are often characteristic of their place and date of origin. Besides differences in their main component, such as tin, lead, and zinc, there are considerable differences in the trace-element concentrations due to the use of copper from different ore deposits. This permits the classification of the analyses into groups of similar trace-element patterns. These groups are well defined, but they do not correspond with regional groups based on iconographic observations. In a certain area, ores from different deposits were probably used. Comparison of trace-element patterns of statuettes from northern India and the Himalayas with their supposed origin permits conclusions on the transfer of materials along with the transfer of technologies in the course of the spread of Buddhism.

The Copper Alloys of Statuettes from Tibet, Nepal, and Northern India

At the Rathgen Research Laboratory in Berlin, about four hundred statuettes from northern India, Nepal, and Tibet in various museums and private collections, have been analyzed by atomic absorption spectroscopy. The main elements tin, lead, and zinc, as well as the principal trace elements iron, nickel, silver, antimony, arsenic, bismuth, cobalt, gold, and cadmium, have been analyzed quantitatively. The concentration of copper has been calculated as the difference between the sum of the concentrations of the quantitatively determined elements and 100%. Through the results of this research (Riederer 1977, 1979, 1984, 1989, 1991; Schroeder 1981), and those by other authors (Craddock 1981; Werner 1972) it becomes obvious that for the manufacture of statuettes in this region, a variety of technically feasible copper alloys have been used. The principal alloys found by the analysis of the statuettes—copper, bronze, and brass—could be subdivided based on the large variations in the concentration of the main elements tin, lead, and zinc. The common term *Indian bronzes* is too general and usually incorrect, since many more statuettes are made from copper or brass than from bronze. Terms such as *tin-lead-bronze with medium amounts of tin and high amounts of lead* or *tin-brass with small amounts of tin and very high amounts of zinc* were introduced with definitions of the ranges of concentrations of small (0–5%), medium (5–10%), high (10–20%), and very high (> 20%) amounts. It became clear that more precise definitions of the alloys are needed.

In this study, eight main material groups are defined as follows: copper (A), tin-bronze (B), lead-bronze (C), tin-lead-bronze (D), brass (E), tin-brass (F), lead-brass (G), and tin-lead-brass (H). These groups are subdivided further according to the concentrations of tin, lead, and zinc, which results in twenty-six subgroups of correctly defined copper alloys. This subdivision is necessary for the discussion of the use of these copper alloy types from various regions and periods and is the basis of a scheme to place statuettes in a context of material use during the early periods in Tibet, Nepal, and northern India. Various types had been preferred in these areas, which were gradually replaced in southern Tibet and Nepal by copper and different types of brass in other regions.

In central Tibet, the use of bronze alloys predominated until the eleventh century, when the use of brasses became more common. Through the early periods, there is a gradual increase in zinc contents of the bronzes, and later, brasses with lower zinc contents occur together with brasses with high amounts of zinc, as shown in table 1. The parallel use of different types of brass can be an argument for the existence of several production centers, or it could be a sign of the spread of technologies from one center to other areas.

The alloys of statuettes of western Tibet are similar to those of central Tibet, although it appears that the zinc contents increase slightly later in western Tibet. Bronzes in the early centuries from western Tibet are not as common, and confirmation of this trend will have to await more analysis of these materials.

From the thirteenth century onward, gilt statuettes from southern Tibet and Nepal are predominantly composed of copper, as shown in table 2. Almost no change in this composition is seen over seven centuries.

In northeast India, a peculiar alloy of brass with medium to high amounts of zinc and high amounts of lead has been used since the tenth century, as shown in table 3. This alloy is quite typical for this area, and it may be possible to investigate the provenance of brass statuettes rich in lead with respect to this alloy common in northeast India.

From Rajastan-Gujarat, Swat, Kashmir, and Ladakh, the number of analyses is not yet sufficient to define precisely

Table 1. *Examples of statuettes from brass used in central Tibet from the eleventh to nineteenth century*

Century	Cu	Sn	Pb	Zn	Fe	Ni	Ag	Sb	As	Bi	Co	Au	Cd
11	81.97	4.20	5.01	7.81	0.35	0.12	0.07	0.06	0.40	< 0.025	0.015	< 0.01	< 0.001
14	77.32	8.17	7.95	5.55	0.44	0.07	0.05	0.23	0.20	< 0.025	< 0.005	< 0.01	< 0.001
15	82.19	2.87	1.76	12.46	0.14	0.05	0.05	0.15	0.18	< 0.025	0.005	0.14	<0.001
16	79.53	6.15	2.91	10.91	0.15	0.09	0.05	0.07	0.14	< 0.025	<0.005	< 0.01	< 0.001
17	83.82	0.92	2.05	12.77	0.34	0.04	0.01	0.02	< 0.05	< 0.025	0.018	< 0.01	< 0.001
18	83.13	< 0.25	0.24	15.85	0.15	0.05	0.17	0.03	0.39	< 0.025	< 0.005	< 0.01	0.001
19	60.87	0.38	1.82	36.27	0.47	0.09	0.04	0.03	0.05	< 0.025	0.005	< 0.01	0.003

Table 2. *Composition of statuettes of copper from Nepal from the ninth to nineteenth century*

Century	Cu	Sn	Pb	Zn	Fe	Ni	Ag	Sb	As	Bi	Co	Au	Cd
9	99.16	< 0.02	< 0.025	0.01	0.70	0.06	0.01	—	—	—	—	—	—
11	98.87	0.13	0.04	0.01	0.02	0.09	0.01	—	—	—	—	—	—
13	98.55	< 0.2	0.18	0.01	0.01	1.05	0.05	0.11	—	—	—	—	—
14	96.50	0.60	1.36	0.90	0.30	0.15	0.06	0.02	0.08	< 0.025	< 0.005	0.03	< 0.001
15	98.50	< 0.20	0.15	0.73	0.21	0.10	0.17	0.01	0.08	< 0.025	< 0.005	0.05	< 0.001
16	98.65	< 0.20	0.09	0.48	0.09	0.04	0.05	0.09	0.51	< 0.025	< 0.005	< 0.01	< 0.001
17	99.36	< 0.2	0.06	0.01	0.01	0.16	0.16	—	—	—	—	—	—
18	99.32	< 0.20	< 0.025	0.01	0.55	0.04	0.02	0.02	< 0.05	< 0.025	< 0.005	0.05	< 0.001
19	99.22	< 0.20	0.01	0.02	0.03	0.05	0.01	0.02	< 0.05	< 0.025	0.005	0.03	0.003

Table 3. *Composition of statuettes of brass from northeastern India in the tenth and twelfth centuries*

Century	Cu	Sn	Pb	Zn	Fe	Ni	Ag	Sb	As	Bi	Co	Au	Cd
10	77.95	5.67	9.51	5.62	0.32	0.11	0.09	0.07	0.56	0.071	0.015	0.01	< 0.001
12	80.42	< 0.20	5.84	12.89	0.10	0.31	0.01	< 0.01	0.43	< 0.025	< 0.005	< 0.01	< 0.001

the variety of copper alloys used for statuettes. Brasses predominate in all of these regions, and it appears as though they underwent an alloy development similar to the neighboring regions of the Himalayan area.

Trace Elements

The determination of trace elements such as iron, nickel, bismuth, cobalt, antimony, arsenic, silver, gold, and cadmium by atomic absorption spectrometry can be accomplished with the necessary accuracy to deduce the type of copper ores that had been used. Most of these trace elements are present as impurities in the copper ores. The common tin ore, cassiterite, does not contribute considerably to the trace-elemental composition of an alloy. The zinc ore, calamine, may contain higher amounts of iron and cadmium than are likely to be found in brass alloys. In the following discussion of trace elements, gold and cadmium are not considered because gold may originate from a gilded surface and cadmium usually occurs in concentrations below the detection limit of the analytical system of 0.001%.

Iron

In statuettes from Tibet, Nepal, and northern India, iron was found in concentrations between < 0.01 and 3%. The higher values up to 3% are not uncommon, and this composition does not seem to relate to a particular region or period of origin. The lowest concentrations of iron occur in statuettes from Nepal and southern Tibet. Also, there is a slight tendency toward lower iron values in the eighteenth and nineteenth centuries, but this trend cannot be generalized.

If the iron concentrations of different groups of origin are considered, two general groups of objects emerge: one group with low iron (0.01–0.05%), and another group with higher iron (0.5 or 1%). Unfortunately, these groups are not separated, and there are as many values between 0.1 and 0.5% as in both groups themselves. The extremely high iron values of the Swat statuettes may be due to a statistical insecurity of a low number of analyses, but it is remarkable that the Swat statuettes have iron concentrations of about 1–2%, and similar high values occur in early post-Gandhara as well as Kashmir statuettes. Nine analyses of statuettes from southern Tibet with these low, intermediate, and high iron values show how large these differences are (table 4).

Nickel

Nickel also shows a considerable variety of concentrations between < 0.005 and 1.60%. Again, no gradual trends in concentrations with time or region of origin were seen. The single exception where the nickel concentrations are considerably higher are the statuettes from western Tibet, where nickel concentrations of more than 0.5% are quite common. Two statuettes from western Tibet with low nickel content suggest that different types of copper ores may have been used. One could argue that since nickel concentrations are generally lower in other areas, those statuettes rich in nickel

Table 4. *Comparison of iron concentrations in statuettes from southern Tibet from different periods*

Century	Cu	Sn	Pb	Zn	**Fe**	Ni	Ag	Sb	As	Bi	Co	Au	Cd
14	98.98	< 0.20	0.12	0.71	**0.01**	0.04	0.01	< 0.01	< 0.05	< 0.025	< 0.005	0.12	< 0.001
19	98.65	< 0.20	< 0.025	0.03	**0.01**	0.04	0.06	0.03	< 0.05	< 0.025	0.018	1.16	< 0.001
16	97.50	1.30	0.18	0.67	**0.01**	0.02	0.05	0.01	< 0.05	< 0.025	< 0.005	0.26	< 0.001
15	98.49	< 0.2	0.22	0.50	**0.53**	0.09	0.02	0.03	0.12	—	—	—	—
13	98.37	0.30	0.07	0.30	**0.54**	0.10	0.02	0.02	0.28	—	—	—	—
18	99.15	< 0.20	0.07	0.15	**0.57**	0.03	0.19	0.01	< 0.05	< 0.025	< 0.005	< 0.01	< 0.001
18	98.05	< 0.20	0.38	0.41	**1.01**	0.04	0.04	0.03	< 0.05	< 0.025	< 0.005	0.05	< 0001
15	89.39	< 0.2	1.18	7.41	**1.44**	0.12	0.04	0.09	0.33	—	—	—	—
15	92.64	< 0.2	0.49	4.86	**1.55**	0.20	0.03	0.03	0.20	—	—	—	—

Table 5. *Concentrations of cobalt in statuettes from central and western Tibet from different periods*

	Century	Cu	Sn	Pb	Zn	Fe	Ni	Ag	Sb	As	Bi	**Co**	Au	Cd
CT	14	97.80	1.58	0.17	0.08	< 0.01	0.05	0.03	0.05	< 0.05	< 0.025	**< 0.005**	0.24	< 0.001
CT	17	86.64	< 0.20	0.30	9.10	2.80	0.39	0.05	0.08	0.48	0.117	**0.035**	< 0.01	< 0.001
WT	12	89.83	< 0.20	4.52	0.08	1.07	1.60	0.12	<0.01	2.52	< 0.025	**0.262**	< 0.01	< 0.001

CT = Central Tibet. WT = Western Tibet.

Table 6. *Comparison of bismuth concentrations in statuettes from central and western Tibet*

	Century	Cu	Sn	Pb	Zn	Fe	Ni	Ag	Sb	As	**Bi**	Co	Au	Cd
CT	14	97.80	1.58	0.17	0.08	< 0.01	0.05	0.03	0.05	< 0.05	**< 0.025**	< 0.005	0.24	< 0.001
CT	17	84.43	1.49	2.47	9.61	0.87	0.04	0.10	0.04	0.84	**0.112**	0.005	< 0.01	< 0.001
WT	16	75.43	< 0.25	0.61	22.55	0.42	0.12	0.09	0.08	0.54	**0.145**	< 0.005	< 0.01	< 0.001

CT = Central Tibet. WT = Western Tibet.

are related to those of western Tibet. These statuettes rich in nickel can be defined as a reference group to allow comparison of this group of objects with statuettes from central Tibet and northern India.

Silver

The silver concentrations are very low, usually around 0.01%. One group of central Tibetan statuettes is similar in their concentrations to the statuettes from western Tibet.

Antimony

A group of statuettes in western Tibet clearly contains unusually low amounts of antimony along with varying amounts of nickel and arsenic. Statuettes from central Tibet and northeastern India are generally characterized by relatively high amounts of antimony, arsenic, iron, and cobalt but low amounts of nickel. In both cases, precisely defined groups of identical trace-element patterns can be formed.

Arsenic

The compositional differences between statuettes from western Tibet and those from central Tibet become more obvious with regard to their arsenic content. The amounts of arsenic in statuettes from western Tibet reach 2%, and nickel and cobalt occur in relatively high concentrations. In central Tibetan statuettes, the concentrations of arsenic and other trace elements tend to be extremely low. Statuettes from northeastern India generally are rich in arsenic and other trace elements.

Cobalt

Statuettes from western Tibet, which can be characterized by their higher nickel concentrations, are also extremely high in cobalt (up to 0.300%). By contrast, the majority of statuettes from other regions have cobalt concentrations below the detection limit of the analytical method of 0.005%. Some statuettes from central Tibet are quite similar in their high trace-element composition to those of western Tibet. The comparison of a statuette from western Tibet with two statuettes from central Tibet in table 5 clearly shows this difference. Most other statuettes have extremely low cobalt contents that rarely go up to 0.01%. Values between 0.01 and 0.03% are exceptional.

Bismuth

Bismuth usually is found in quite low concentrations of less than 0.025% in statuettes from northern India and the region of the Himalayas, as represented by the first analysis of table 6. In statuettes from western Tibet, the concentrations of bismuth are high, but again, there is a group of statuettes from central Tibet that is quite close in the trace-element pattern to the statuettes of western Tibet.

Trace-Element Groups

The large range of trace-element concentrations demands a classification based on these concentrations. Since the main difference of copper deposits is the existence of ores from the fahlerz group, containing various arsenates and antimoniates of copper and other metals, characterized by high amounts of arsenic and antimony, the analysis has been subdivided into ten groups of increasing arsenic content: I–X. In this series, the odd-numbered groups I, III, V, VII, and IX have low amounts of antimony; the even-numbered groups II, IV, VI, VIII, and X have high amounts of antimony. Each of these groups is further subdivided into two groups: A and B. Subgroup A has high concentrations of iron, and subgroup B has low concentrations of iron. These subgroups are again subdivided, into groups 1 and 2—1 with low concentrations of nickel and 2 with high concentrations of nickel. By this scheme all analyses of statuettes fall into groups with very similar trace-element compositions.

Some examples illustrating the principle of these groups are shown below. In fact, the nearly four hundred statuette analyses in the databank of the Rathgen Research Laboratory fall into these groups and convincingly show that each object has a well-defined position within these groups. Some compositions of statuettes belonging to group IA1 are shown in table 7.

I Arsenic below 0.005%, antimony below 0.05%
IA Arsenic below 0.005%, antimony below 0.05%, iron below 0.1%
IA1 Arsenic below 0.005%, antimony below 0.05%, iron below 0.1%, nickel below 0.05%

The majority of objects in this group are gilt copper statuettes from Tibet and Nepal from the nineteenth century. A smaller number from the same period are brasses with varying amounts of zinc. Very pure copper and copper alloys are characteristic for objects of a more recent origin. This group also contains much earlier objects—for instance, a considerably large group of brass statuettes from Tibet from the fifteenth and sixteenth centuries. Since the amount of zinc in some of these objects is quite high (up to 25%), we cannot eliminate the possibility that some of these objects are not as old as indicated by their stylistic features.

A few objects are similar in their trace-element pattern to that of group IA1 but with unusually high concentrations of nickel, as shown in table 8. These objects, usually made of a low-zinc brass, are relatively late and originate from Tibet and Nepal. They may have the same or similar cultural origin. Some objects of this group have high concentrations of cobalt.

If the whole large series of analyses is regarded, it becomes obvious that there are objects with high concentrations of iron. Compositions of some of these objects are shown in table 9. These statuettes may represent a peculiar group of related objects. As in group IA, statuettes belonging to group IB tend to be from the eighteenth and nineteenth centuries, underlining the fact that objects of this time period were manufactured using pure metals.

Groups II–VIII

The examples of group I show the principle of the subdivision of analyses with very low concentrations of trace elements into subgroups with higher amounts of iron and nickel. In the same way, the remaining analyses of groups II–VIII can be subdivided first into their concentrations of

Table 7. *Some statuettes of group IA1 (very low concentrations of trace elements)*

	Century	Cu	Sn	Pb	Zn	Fe	Ni	Ag	Sb	As	Bi	Co
T	14	98.98	< 0.20	0.12	0.71	0.01	0.04	0.01	< 0.01	< 0.05	< 0.025	< 0.005
T	16	75.74	< 0.20	2.76	21.38	0.02	0.05	0.04	0.02	< 0.05	< 0.025	< 0.005
T	18	98.66	< 0.20	< 0.025	1.19	0.01	0.08	0.04	< 0.01	< 0.05	< 0.025	< 0.005
N	19	96.29	< 0.20	0.06	3.48	0.04	0.04	0.03	0.01	< 0.05	< 0.025	< 0.005
T	19	99.49	< 0.25	0.03	0.02	< 0.01	0.09	0.11	< 0.01	< 0.05	< 0.025	< 0.005

T = Tibet. N = Nepal.

Table 8. *Some statuettes of group IA2 (very low concentrations of trace elements but higher amounts of nickel)*

	Century	Cu	Sn	Pb	Zn	Fe	Ni	Ag	Sb	As	Bi	Co
T	18	99.32	< 0.20	0.02	0.31	0.06	0.21	0.02	0.01	< 0.05	< 0.025	0.011
N	18	96.63	< 0.20	< 0.025	3.00	0.05	0.22	0.07	0.02	< 0.05	< 0.025	< 0.005

T = Tibet. N = Nepal.

Table 9. *Some statuettes of group IB1 (very low concentrations of trace elements and high concentrations of iron)*

	Century	Cu	Sn	Pb	Zn	Fe	Ni	Ag	Sb	As	Bi	Co
N	16	99.33	< 0.20	< 0.025	0.01	0.64	0.02	0.01	< 0.01	< 0.05	< 0.025	< 0.005
T	16	98.50	< 0.20	0.27	0.63	0.46	0.09	0.03	0.02	< 0.05	< 0.025	< 0.005
ET	19	71.36	< 0.20	0.13	28.31	0.13	0.04	0.01	< 0.01	< 0.05	< 0.025	< 0.005
N	19	82.28	< 0.20	0.64	16.76	0.21	0.06	0.02	< 0.01	< 0.05	< 0.025	< 0.005
T	19	90.44	< 0.20	0.06	9.14	0.23	0.06	0.04	0.01	< 0.05	< 0.025	< 0.005

N = Nepal. T = Tibet. ET = Eastern Tibet.

Table 10. *Some statuettes of group IB2 (IBI with very low concentrations of trace elements but higher concentrations of iron and nickel)*

	Century	Cu	Sn	Pb	Zn	Fe	Ni	Ag	Sb	As	Bi	Co
T	18	98.64	< 0.20	0.54	0.05	0.65	0.10	0.02	< 0.01	< 0.05	< 0.025	0.007
T	18	98.99	< 0.20	0.07	0.25	0.26	0.37	0.06	< 0.01	< 0.05	< 0.025	0.008

T = Tibet.

Table 11. *Some statuettes of group IX and X with high concentrations of trace elements*

Prov.	Group	Century	Cu	Sn	Pb	Zn	Fe	Ni	Ag	Sb	As	Bi	Co
WT	IX	14	87.35	< 0.2	2.07	8.80	0.06	< 0.005	0.09	< 0.01	1.63	< 0.025	< 0.005
NEI	IXA1	10	92.37	< 0.2	5.17	0.07	0.01	0.51	0.03	0.05	1.79	< 0.025	< 0.005
WT	IXA1	12	91.40	< 0.20	5.62	0.09	0.07	0.70	0.16	< 0.01	1.90	< 0.025	< 0.005
WT	IXB1	12	79.44	< 0.20	5.78	12.65	0.28	0.02	0.08	0.04	1.67	0.036	< 0.005
T	IXB1	15	85.43	< 0.20	0.11	13.01	0.17	0.04	0.14	0.02	1.00	< 0.025	< 0.005
WT	IXB2	12	89.83	< 0.20	4.52	0.08	1.07	1.60	0.12	< 0.01	2.52	< 0.025	0.262
L	X	16	78.37	< 0.20	18.59	0.07	0.50	0.08	0.13	0.40	1.17	0.673	0.026

WT = Western Tibet. NEI = Northeast India. WT = Western Tibet. T = Tibet. L = Ladakh.

Table 12. *Some statuettes with high concentrations of nickel and arsenic*

	Century	Cu	Sn	Pb	Zn	Fe	Ni	Ag	Sb	As
NEI	10	92.37	< 0.2	5.17	0.07	0.01	**0.51**	0.03	0.05	**1.79**
WT	12	91.40	< 0.20	5.62	0.09	0.07	**0.70**	0.16	< 0.01	**1.90**

IXA2: as IXA1 but higher concentrations of nickel
NEI = Northeast India. WT = Western Tibet.

Table 13. *Some statuettes with high concentrations of iron and arsenic*

	Century	Cu	Sn	Pb	Zn	Fe	Ni	Ag	Sb	As	Bi	Co
WT	12	79.44	< 0.20	5.78	12.65	**0.28**	0.02	0.08	0.04	**1.67**	**0.036**	< 0.005
T	15	85.43	< 0.20	0.11	13.01	**0.17**	0.04	0.14	0.02	**1.00**	< 0.025	< 0.005

IXB1: as IXA1 but higher concentrations of iron
WT = Western Tibet. T = Tibet.

Table 14. *A statuette with high concentrations of iron nickel and arsenic*

	Century	Cu	Sn	Pb	Zn	Fe	Ni	Ag	Sb	As	Bi	Co
WT	12	89.83	< 0.20	4.52	0.08	**1.07**	**1.60**	0.12	< 0.01	**2.52**	< 0.025	**0.262**

IXB2: as IXA1 but higher concentrations of iron and nickel
WT = Western Tibet.

Table 15. *Some statuettes with high concentrations of all trace elements*

	Century	Cu	Sn	Pb	Zn	Fe	Ni	Ag	Sb	As	Bi	Co
L	16	78.37	< 0.20	18.59	0.07	**0.50**	**0.08**	**0.13**	**0.40**	**1.17**	**0.673**	**0.026**

X. Arsenic over 1%, antimony over 0.05%. XB. Arsenic over 1%, antimony over 0.05%, iron over 0.1%. XB1. Arsenic over 1%, antimony over 0.05%, iron over 0.1%, nickel below 0.1%
L = Ladakh.

arsenic and antimony. Further subdivision of the groups is based on the concentrations of iron into subgroups A and B and, again, concentrations of nickel into further subgroups 1 and 2. All of these groups and subgroups are filled with a statistically representative number of analyses so that these groups are relevant for objects of varying cultural origin. Compositions of some of the statuettes belonging to these groups are given in tables 10–15.

IX Arsenic over 1%, antimony below 0.05%
IXA Arsenic over 1%, antimony below 0.05%, low concentration of iron
IXA1 Arsenic over 1%, antimony below 0.05%, low concentration of iron and nickel
X High concentrations of all trace elements

These statuettes with high relatively amounts of arsenic must have been made using copper of different origin from those of group I, which contain extremely low concentrations of arsenic and other trace elements. Objects from western Tibet as well as northeastern India consist of alloys that are rich in trace elements. Based on this type of composition, objects identified as Tibetan by their iconographic properties certainly have a connection with western Tibet.

Conclusions

The conclusions of this research are as follows:

1. Due to the limited scope of this paper, the statistical basis is not as obvious as by a presentation of the total of about four hundred analyses.
2. A broad range of trace-element compositions was found in statuettes from northern India, Nepal, and Tibet.
3. For regions such as northeastern India and western Tibet, the trace-element concentrations are characteristic, while in other areas various groups of trace-element patterns were found.
4. The classification of statuettes by stylistic and iconographic features, such as those from Nepal or Kashmir, does not necessarily correspond to the classification by trace-element groups. In these cases, it appears that different regional stylistic groups developed, but the same or geologically similar ores may have been used.
5. Groups of statuettes with characteristic compositions may be defined as reference groups to evaluate other compositional groups.
6. In some defined regions, the trace element concentrations are similar over long periods, indicating that the same types of copper ores have been used over long periods. Some problems exist in precisely dating statuettes, so such conclusions need to be carefully considered.
7. Statuettes from the nineteenth and twentieth centuries have very low concentrations of trace elements in the whole region of northern India and the Himalayas. This indicates the use of a separate type of copper as compared to earlier statuettes.
8. Since trace-element group classifications are quite distinct, each statuette from northern India and the region of the Himalayas has its own place in the system. It may, therefore, be valuable to first place statuettes into their trace-element groups and subsequently consider their stylistic properties and relationships.

References

Craddock, P. T. 1981. The copper alloys of Tibet and their background. In *Aspects of Tibetan metallurgy,* Occasional Paper no. 15, ed. W. A. Oddy and W. Zwalf, 1–31. London: British Museum (Research Laboratory and Department of Oriental Antiques).

Riederer, J. 1977. Metallanalysen chinesischer Spiegel. *Berliner beiträge zur archäometrie* 2:6–16.

——. 1979. Zur materialanalyse der statuetten. In *Das bild des Buddha*, ed. Kunstamt Berlin-Tempelhof, 63–68. Berlin: Safari-Verlag.

——. 1984. Materialanalysen an bronze- und mesingstatuetten des Staatlichen Museums für Völkerkunde in München. In *Skulpturen aus Indien*, ed. C. Mallebrein, 231–36. München: Staatliches Museum für Völkerkunde.

——. 1989. Zur metallanalyse der kultstatuetten. In *Die götter des Himalaya*, ed. G. W. Essen and T. T. Thingo, 292–96. München: Prestel-Verlag.

——. 1991. The relation between the composition of north Indian statuettes of copper alloys and the region and date of their origin. *Proceedings of the first international colloquium on the role of chemistry in archaeology, 15–18 November, Hyderabad,* ed. M. C. Ganorkar and N. Rama Rao, 95–102. Hyderabad: Birla Institute of Scientific Research.

Schroeder, U. V. 1981. *Indo-Tibetan bronzes.* Hong Kong: Visual Dharma Publications.

Werner, O. 1972. *Spektralanalytische und metallurgische Untersuchungen an indischen Bronzen.* Leiden: E. J. Brill.

15 New Evidence for the Historical Context of Buddhist Bronzes from Swat Valley, Northern Pakistan

Chandra L. Reedy

ABSTRACT In most art historical discussions of medieval northwest Indian sculptural traditions (sixth to thirteenth century), bronzes from the Kashmir Valley receive much attention as major works of art, while those from the Swat Valley of northern Pakistan are often ignored or seen as products of a minor center whose craftsmen were greatly influenced by those of Kashmir. However, new information concerning Swat bronzes has recently emerged. We can now begin to reevaluate ideas about the artistic prominence of medieval Buddhist Swat and its relationship to Kashmiri art.

Carbon-14 test results date some medieval Swat pieces to around the sixth century, earlier than previously thought (eighth to ninth century). Art historical evidence relates dated pieces and other similar bronzes to early stone and terracotta objects from excavations in Swat and to rock-cut images. Historical, archaeological, and technical evidence considered together now points to Swat Valley being a major medieval religious and artistic center before and during the emergence of medieval Kashmir as a major center, and Swat artists probably had a strong impact on the casting centers of Kashmir.

The Swat Valley as a Medieval Buddhist Center

The Swat Valley is located near eastern Afghanistan, just north of the Khyber Pass and the northern Pakistani city of Peshawar. Early Chinese Buddhist pilgrims visited Swat during the fifth to eighth century and left detailed written accounts. These describe a rich area with fertile soil, a pleasant climate, and an irrigation system that helped to produce abundant crops (Beal 1869, 188–96).

Swat clearly had the resources needed for a flourishing settlement and cultural center in the medieval period. However, northern Pakistan has few metal-ore sources. Resource studies have shown that the most likely sources for copper, tin, zinc, and lead used in bronze production in Swat were a variety of mining sites in Afghanistan (Reedy 1997, 87–91). In the early medieval period, there were certainly close cultural contacts between Buddhist Swat and the major Buddhist pilgrimage and trade sites of Afghanistan, such as Bamiyan.

As in Afghanistan, the Buddhist religion and culture flourished in Swat during the early medieval period. Chinese pilgrims reported, for example, the presence in Swat of five hundred Theravadin monasteries; at one temple there were sixty full-length golden figures (probably gilded stone images) (Beal 1869, 26–28, 188–96; Giles 1923, 11; Yung-hsi 1957, 25; Legge 1965, 28–29).

In the seventh century, the pilgrim Xuanzang found that Buddhism was no longer existing as easily as before, mainly because of a series of outside attacks on Swat. Buddhism was still predominant in the region, but many of the fourteen hundred monasteries he claimed to have seen were in ruins (Beal 1906, 120–21).

We know that Buddhism persisted in Swat in the face of Islamic suppression up until at least the second half of the thirteenth century, when the Tibetan pilgrim Orgyan-pa visited the region. What he found was a form of Buddhism that incorporated magic practices and witchcraft. The most prominent images were tantric (fig. 1). However, by the time of the travels of the Tibetan pilgrim sTag-ts'ang ras-pa in the early seventeenth century, Islam had completely replaced Buddhism throughout Swat. He found no living monasteries—only ruins (Tucci 1940, 39–40).

Swat Bronzes

In 1962, Douglas Barrett was the first scholar to describe stylistic features that he believed distinguish a Swat Valley school of sculpture from Kashmiri objects. His core group included bronzes in the collection of the Wali of Swat, all said to have been recovered in the valley. Some of the characteristic features of bronzes in that collection can be seen in figure 1. These include a base supported by animals, tassels at the sides of the base, flower ornaments at the arms and ears, a single pearl-like strand for a necklace, and a tall crown with side extensions and ribbons falling behind the shoulders.

Art historians have always had some difficulty agreeing on whether or not individual bronzes might have been made in Swat or in Kashmir, due to many stylistic and iconographic overlaps between the two regions. The total number of bronzes attributed to Swat is small, especially in comparison with the number of pieces attributed to the nearby very lush and wealthy region of Kashmir. Medieval Kashmiri chronicles mentioning the importance of the bronze-casting centers and their products have survived (Stein 1961), while the written record for Swat is more scattered and sketchy. It has been assumed that Swat Valley bronzes were made a bit later and were heavily influenced, both technologically and stylistically, by the major casting centers of Kashmir. In fact, Swat has often been seen as a "provincial" area of Kashmiri art and culture.

Recent Technical Studies

A recent technical study of Himalayan bronzes, including a focus on differentiating bronzes of Swat from those of Kashmir, has helped to address the attribution problem. The technical program incorporates a wide range of art-historical and technical data in a multivariate statistical analysis to identify, when possible, the most likely region of manufacture for each sculpture. The original study also provides a technical database useful for attributing additional bronzes that undergo a similar technical study. Details of the method and results are reported elsewhere (Reedy 1997). Here I summarize materials and methods used by artists of Swat and discuss a few interesting pieces in greater detail.

Most medieval Swat bronzes were hollow cast around a clay core. The figures themselves were always cast in one piece, although a base and/or backplate were sometimes separately cast. Separate backplates most often became detached from their figure and are now lost. However, most Swat bronzes have a tang on the back, as in figure 2, that was probably used to attach a backplate. A characteristic feature of Swat bronzes is a small, round hole on the back of the head that is found on most of the sculptures (fig. 2). This hole probably indicates where a chaplet or rod was used to hold the mold in place during casting and was then removed.

Many Swat sculptures have at least one iron armature present, straight up the center of the piece (fig. 3). Most also have chaplets. Often these were removed after casting, leaving square or rectangular holes on the surface of the metal, visible in the x-radiograph in figure 3, that were usually covered with a small copper patch. Most of the bronzes also have repair patches placed over casting flaws. Almost all bronzes were decorated with silver inlays, usually placed in the eyes and *urna* (third eye). Copper inlay, usually in the lips, appears on most of the sculptures.

Metal analysis by inductively coupled plasma emission spectrometry (Reedy 1997, 79–80) has shown that 95% of the bronzes from Swat were apparently produced with a deliberate copper-tin alloy (with tin present at levels of 2% or more). Fewer than one-fourth of bronzes from Kashmir have tin at a high enough level to indicate a possible deliberate addition. Many of the Swat bronzes also have unusually high tin concentrations, up to 23% by weight. High levels of tin would create a very brittle metal, which is why many Swat bronzes are severely cracked and broken.

About half of all Swat pieces also contain zinc, and about a quarter have lead additions. In contrast, 100% of sculptures from Kashmir have additions of zinc, and most of those also have lead at levels indicating it might be an intentional addition to the copper. None has the unusually high tin concentrations sometimes found in Swat.

Clay cores in Swat are coarse and sandy (Reedy 1991). They usually have a dark appearance from carbonization of organic matter, probably straw and dung, that was mixed in. Most often, carbon comprises 30–40% of the core material. This carbonized material is suitable for carbon-14 dating (fig. 4).

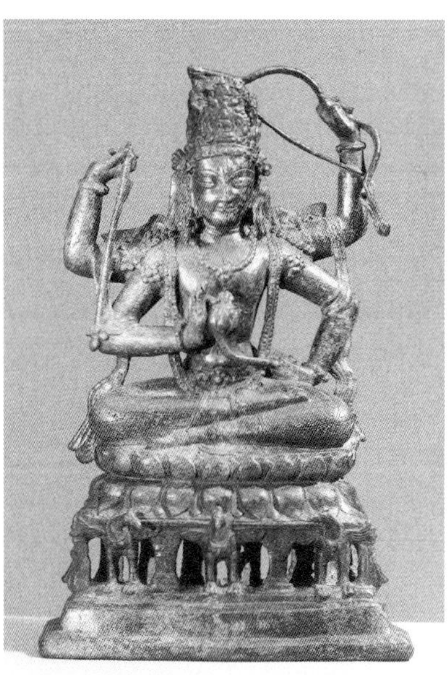

Figure 1. *Vajradharma Lokeshvara, early Swat Valley tantric image, The Metropolitan Museum of Art, gift of Abby Aldrich Rockefeller, 1942 (42.25.20)*

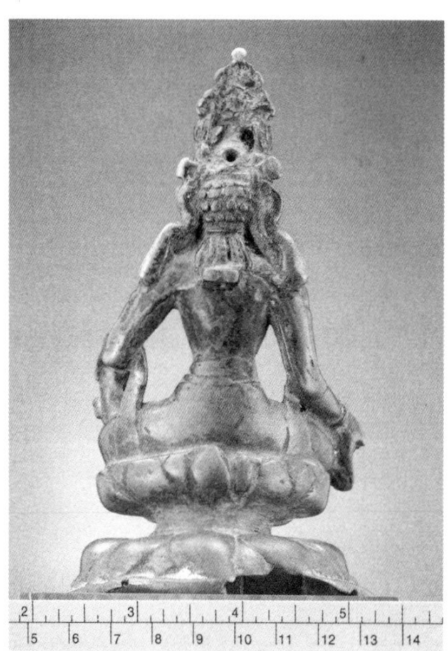

Figure 2. *Bodhisattva Maitreya, reverse side, showing tang on back and hole at back of head typical of Swat Valley bronzes, Robert H. Ellsworth, Ltd.*

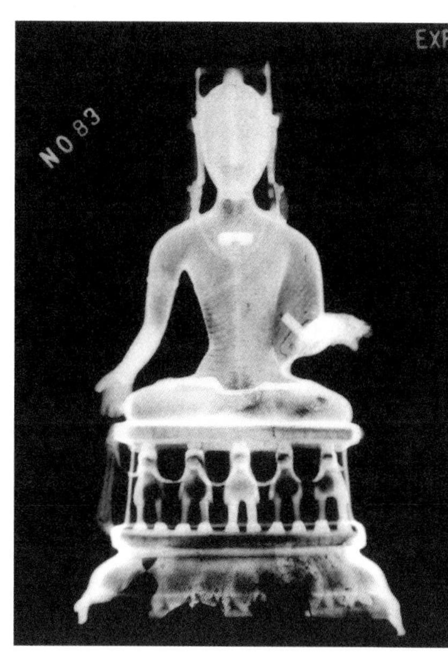

Figure 3. *X-radiograph of Buddha in figure 9, showing typical Swat Valley hollow casting, one iron armature, and rectangular chaplet holes*

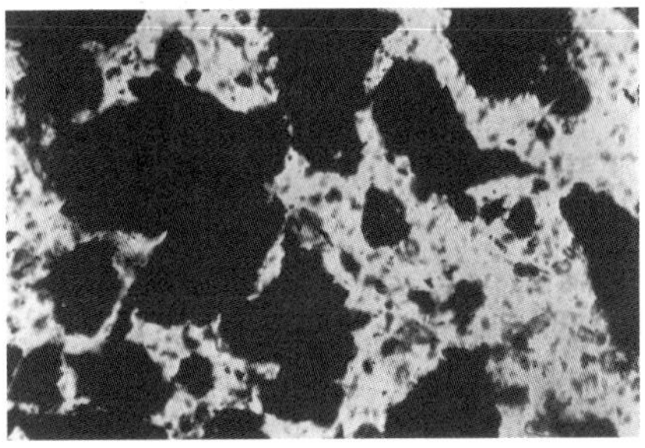

Figure 4. *Carbonized organic matter typical of Swat Valley casting core materials, plane polarized light. ×16*

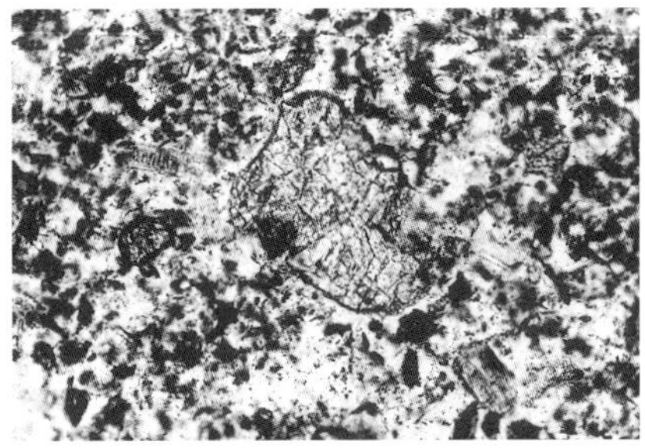

Figure 5. *Clinozoisite mineral, typical of many Swat Valley clay core materials, plane polarized light. ×80*

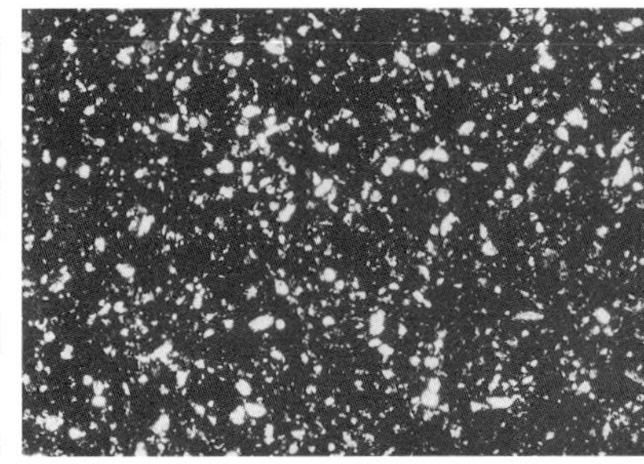

Figure 6. *Sandy clay core material typical for Swat Valley bronzes, crossed polarized light. ×16*

A mineral often characteristic of Swat bronzes is clinozoisite (fig. 5), a form of epidote that occurs in almost all Swat bronze cores yet is rarely found in clay cores of bronzes from other Himalayan regions, including Kashmir. There is no evidence that the sand or any minerals were added as a temper. Instead, they display wear patterns characteristic of minerals transported by water during deposition in a river environment, indicating they were naturally mixed with clay deposits in the stream beds where the clay was selected for collection (fig. 6).

Style and Iconography of Swat Bronzes

Bronzes identified in the technical study as being in the Swat group show that no one stylistic feature is absolutely distinctive of all Swat bronzes. For example, a tasseled textile over a throne, discussed by Barrett (1962), is restricted to Swat objects and does not occur on Kashmir bronzes at all; however, fewer than half of all Swat bronzes have this feature. More common in Swat is the ornament style of a pearl band holding a single flower on the center of the upper arm, often with a single strand of pearls at the neck, as seen in figure 7. Such ornaments are extremely rare in Kashmir.

The use of animals to support a throne is also distinctive. Although animals occasionally are used on a base in Kashmir, there they are either two lions, one at each side of the base, or there is a combination of griffins or lions at the sides and a *yaksha* figure in the center. In Swat, the animals (lions, elephants, or birds) can be located all along beneath the Buddha figure or just to the sides; but in every case there are tassels at their sides, as in figure 8; no *yaksha* figures appear.

Carbon-14 Dating and Evidence for Early Swat Bronzes

Accelerator mass spectrometry (AMS) carbon-14 dating is a technique applicable to all organic materials; with direct measurement of the carbon-14 isotope, only very small amounts of carbon are needed. Clay-core samples from two Swat bronzes were used for AMS carbon-14 dating by D. J. Donahue at the University of Arizona (Reedy 1997, 303).

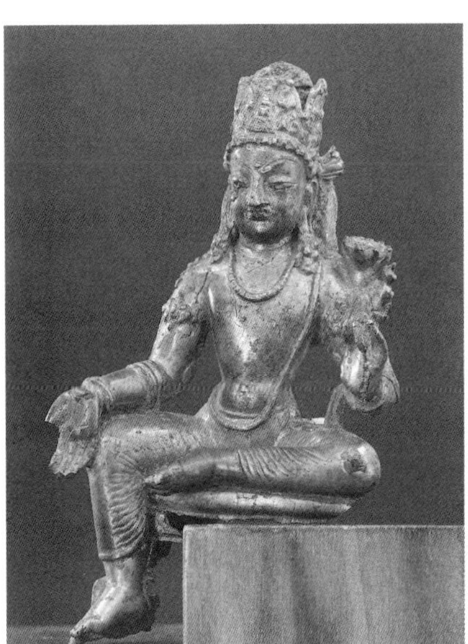

Figure 7. *Bodhisattva Maitreya, with ornaments typical of Swat Valley bronzes, Robert H. Ellsworth, Ltd.*

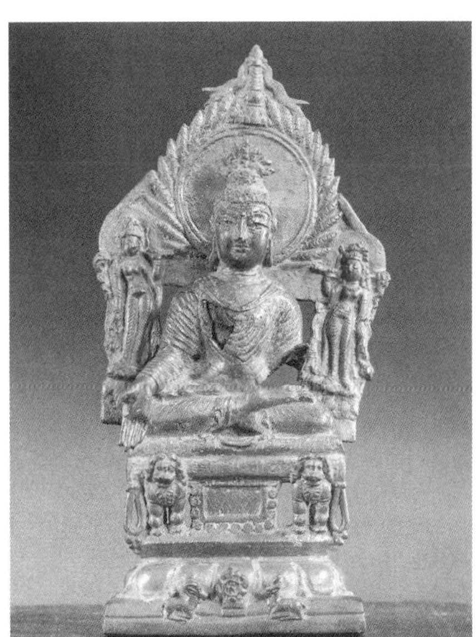

Figure 8. *Buddha Shakyamuni, Robert H. Ellsworth, Ltd.*

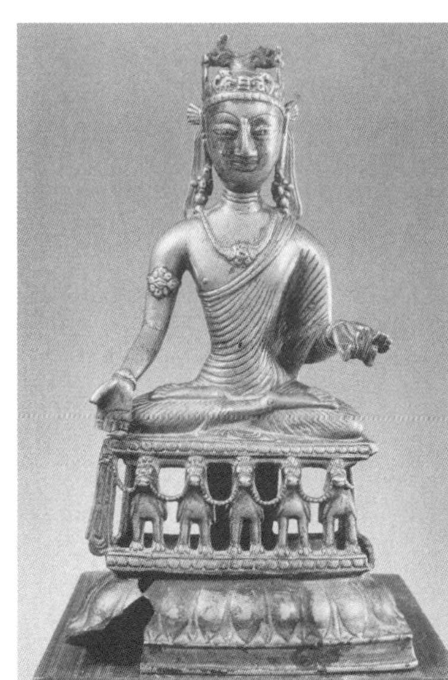

Figure 9. *Seated Buddha, Pacific Asia Museum, gift of Robert and Nancy King*

One of these is a Swat Valley bronze now in the Pacific Asia Museum, Pasadena (fig. 9). It represents a transcendent or celestial Buddha; however, we do not yet know the exact identification, as the iconography is rather unusual. There are a crown and ornaments, along with a monastic robe. The hand gestures are similar to those seen with both Shakyamuni Buddha and Ratnasambhava, and the animals supporting the base are birds, the vehicle of Amoghasiddhi.

The technical study clearly placed the piece within the Swat group and excluded an origin elsewhere in the Himalayan regions, including Kashmir (Reedy 1997, fig. P14). There are many breaks and repairs. Obviously, the alloy that was used to cast the piece is very brittle, causing the piece to be quite fragile. Metal analysis shows that this piece is a very high-tin bronze, with about 20% tin added to the copper, which results in a silvery appearance that is probably what the artists were trying to achieve.

Carbon comprises 40–50% of the core material in this case. AMS carbon-14 analysis gave a radiocarbon age of 1,600 BP ± ninety years (ca. fourth to sixth century). Comparative art-historical evidence supporting a sixth-century date is provided by the work of P. G. Paul (1979). He worked with terra cottas from Swat, convincingly placing them to the sixth to seventh century; many of the pieces have clear stylistic relationships to Swat bronzes.

Paul hypothesized that at the end of the flourishing period of artistic production in Gandhara in the fifth century, the center of Buddhist art in northwest India shifted from the Indus basin up to Swat, as evidenced by numerous rock-cut reliefs, terra-cotta plaques and molds, and bronze sculptures. Coupled with the results of the carbon-14 date for the Pacific Asia Museum piece, it is likely that the later (eighth to ninth century) dates generally assigned to Swat bronzes in the art-historical literature may be later than the actual dates of manufacture.

A bronze Vairochana image (fig. 10) in *bodhyangi mudra* has been attributed by art historians to Swat in the past, or elsewhere in northern Pakistan (von Schroeder 1981, fig. 11G; Pal 1983, 49; Lerner and Kossak 1991, 119–20), an attribution supported by the technical study (Reedy 1997, fig. P29). The image most similar to this piece is another Vairochana in the collection of the Wali of Swat and attributed by Barrett (1962, fig. 12) to the Swat Valley of the eighth to ninth century. The thrones of both pieces are supported by a row of lions atop a double lotus, both objects have a textile placed over the seat with large tassels hanging down at each side, and both display the same *bodhyangi mudra*.

AMS carbon-14 dating places the piece in the fifth to sixth century, roughly contemporaneous with the Pacific Asia Museum piece. The Vairochana, too, is made from a very brittle alloy that imparts a silvery color (a high-tin bronze). Metal analysis shows that the tin content is 18.2% by weight. Like the Pacific Asia Museum piece, this sculpture has many breaks and missing areas because of the brittleness of the alloy.

Other Possible Early Swat Bronzes

We can hypothesize that other bronzes might have been produced in the sixth to seventh century or earlier based upon comparison to the two for which carbon-14 dates are available, to the excavated terra-cotta plaques discussed by Paul (1979), and to excavated stone sculptures. For example, the Buddha image of figure 11, found in the technical study to be in the Swat group (Reedy 1997, fig. P13), has been published with a date as early as the sixth or seventh century, perhaps eighth (Pal 1975, no. 83; von Schroeder 1981, fig. 5F; Klimburg-Salter 1982, plate 5; Lerner and Kossak 1991, 117–18). Rather than being a high-tin bronze, this is a brass sculpture (15.5% zinc) with copper inlay in the lips and silver inlay in the eyes and *urna*. The closest parallel for

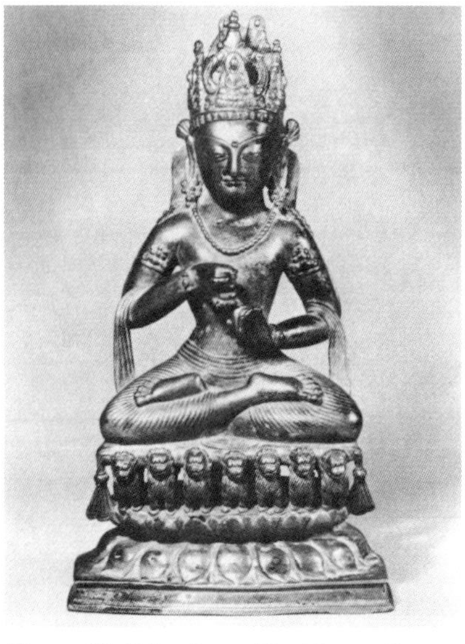

Figure 10. *Vairochana, The Metropolitan Museum of Art, Samuel Eilenberg Collection, Ex Coll.: Columbia University, Purchase, Louis V. Bell and Fletcher Funds, 1987 (1987.218.7)*

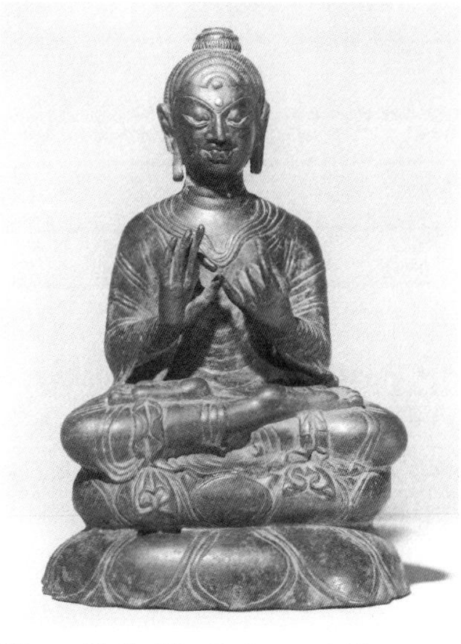

Figure 11. *Buddha Shakyamuni, The Metropolitan Museum of Art, Samuel Eilenberg Collection, Ex Coll.: Columbia University, Purchase, Lita Annenberg Hazen Charitable Trust Gift, 1987 (1987.218.3)*

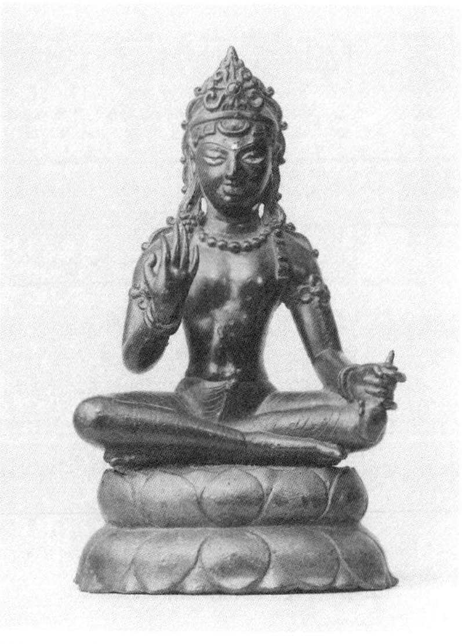

Figure 12. *Bodhisattva Maitreya, The Metropolitan Museum of Art, gift of Mr. and Mrs. A. Richard Benedek, 1978 (1978.536)*

some of the peculiar stylistic features of the piece can be seen on a fragment of a schist sculpture excavated at the Swat Valley site of Butkara (Facenna 1981, plates 85a–c). That object is a seated figure, probably a Buddha. It, like the brass sculpture, is in the *padmasana* posture. The method of depicting the folds of the robe is the same, with the design consisting of well-separated groups of three parallel lines. The lotus petals in both objects are formed by a double line rather than the more usual singular line. The Butkara stone sculpture has been dated by the excavators of the site to the fourth-to-sixth-century stratigraphic level. Thus, a date of the sixth century or even earlier for the brass sculpture is not unrealistic.

The bodhisattva Maitreya in figure 12, also assigned to the Swat group by the technical study (Reedy 1997, fig. P16), was previously been published with a sixth-century (von Schroeder 1981, fig. 5G) or seventh-century (Klimburg-Salter 1982, plate 14) date. Like the Buddha image discussed above, it has very wide and rounded lotus petals although without the double outlining. The material is a copper-zinc-tin alloy (8.8% zinc and 3.6% tin added to the copper).

And, finally, the standing Tara of figure 13, now in the Metropolitan Museum of Art, is very similar to a Tara image on a terra cotta excavated at the same site of Butkara (Paul 1979, fig. 15). Similarities include the use of a simple lotus base, the scarf ends flanking the image, and the method of modeling the figure with a slim waist and somewhat stocky lower body with unusually long legs. The Tara is a leaded brass sculpture (11% zinc and 3% lead added to the copper) and has the usual copper inlay in the lips and silver inlay in the eyes. Since the similar terra-cotta image can be dated stratigraphically to the sixth to seventh century, that is probably the most likely date of manufacture for the sculpture as well. The technical study placed this piece clearly in the Swat group rather than in Kashmir (Reedy 1997, fig. P15).

Swat and Kashmir

Although comparative dating of many Swat and Kashmiri statues is problematic, there are some specific images and stylistic conventions for which evidence exists pointing to Swat as the originator and Kashmir as the later adopter.

Contemplating bodhisattva images, consisting of a seated Padmapani or Avalokiteshvara with one leg extended and the right hand placed near the face with the elbow resting on top of the knee, are very common in Swat. These images occur as stone carvings and rock-cut images throughout the valley (Stein 1929, fig. 66; Tucci 1958, figs. 1–4, 10, 13, 19, 22, 23). Some copper-alloy statues of this same image have appeared, and they have usually been attributed by art historians to Kashmir, although Swat has occasionally been considered a possible region of production (Lee 1975, no. 9; Lerner 1975, plate 4; Pal 1975, nos. 45, 46a–b; Harle 1979, 127–34; von Schroeder 1981, figs. 6G and 6I; Klimburg-Salter 1982, plate 28). In technical studies, although some of these images do indeed fit the Kashmir group, others best fit the Swat Valley group of objects.

An example of a Swat piece is the seated Padmapani in figure 14, which very closely resembles some of the stone carvings of that image in Swat. Similar features are the stylized rock base with the patron at its bottom corner, the very shallow lotus seat at the top of the base, and the build and proportions of the figure. The piece also displays a small image of a deer at the bottom of the base, across from the kneeling patron figure. The sculpture is made of a leaded copper-zinc-tin alloy (with 5.6% zinc, 5% tin, and 4.5% lead added to the copper). The lips are inlaid with copper, and the eyes and *urna* with silver.

A second Swat example, seen in figure 15, differs in its garment style, lotus form, base (a stylized wicker throne atop one row of sharp and downturned lotus petals), hairstyle and facial features; the eyes are half-closed rather than fully open, and there is no third eye. However, the metal here

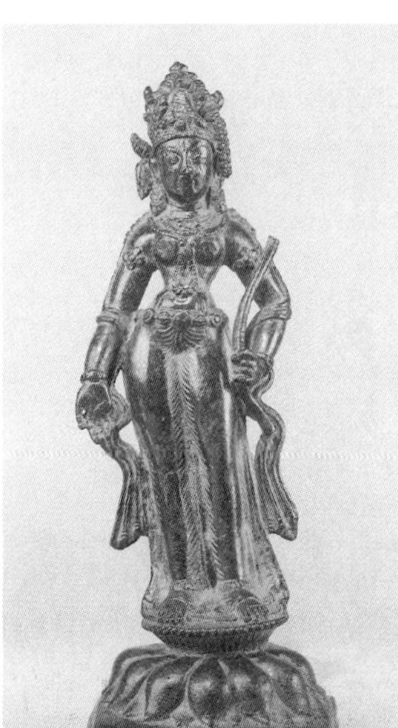

Figure 13. *Tara, The Metropolitan Museum of Art, gift of Dr. and Mrs. Gary Katz, 1984 (1984.484.2)*

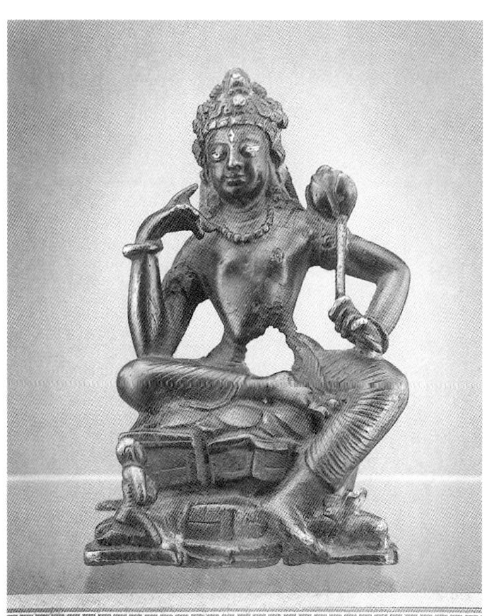

Figure 14. *Padmapani, private collection*

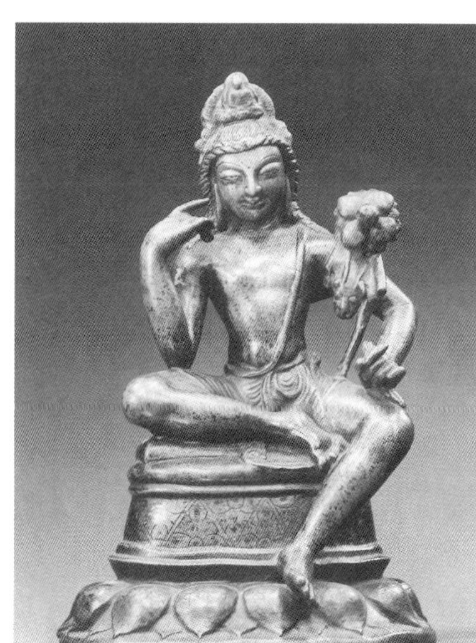

Figure 15. *Avalokiteshvara, The Metropolitan Museum of Art, Harris Brisbane Dick Fund and Fletcher Funds, 1974 (1974.273)*

is very similar to the Padmapani shown in figure 14 (5.3% zinc, 6% tin, and 2.5% lead in addition to the copper).

A Kashmiri version of this image, seen in figure 16, differs only slightly in style (mainly with the garments). It has a more rounded appearance, but this is mainly because it is much more worn than the Swat pieces. There is a difference in material, however. The Kashmiri piece is a leaded copper-zinc-tin alloy, too, but is much higher in zinc than the two Swat pieces (14.8% zinc, 3% tin, 1.8% lead). The result is that the Kashmiri piece has more of a golden glow.

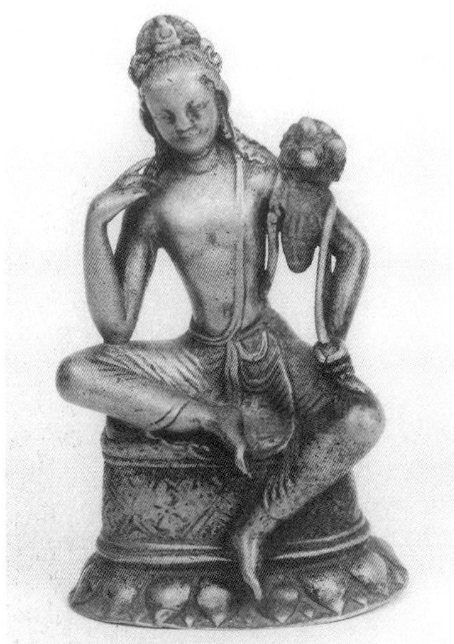

Figure 16. *Avalokteshvara, Kashmir version of Swat Valley form, private collection, photograph by Jim Strong, Hempstead, New York*

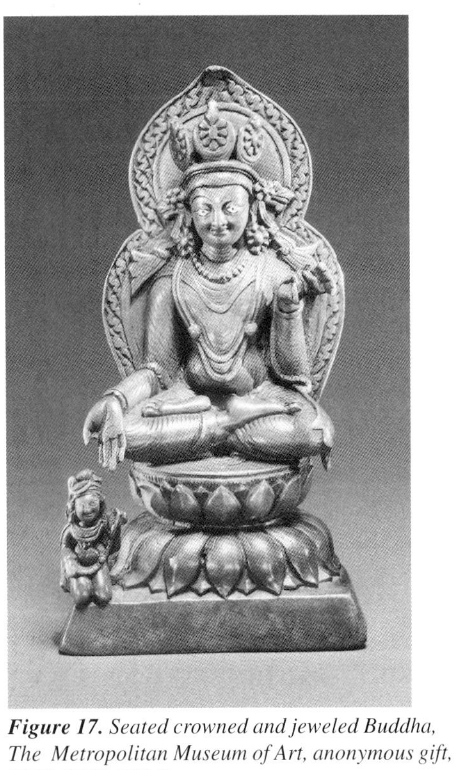

Figure 17. *Seated crowned and jeweled Buddha, The Metropolitan Museum of Art, anonymous gift, 1983 (1983.555.4)*

Although many rock-cut examples of the contemplating bodhisattva have been photographed in situ in Swat, none has been published for Kashmir. Since technical studies support the idea that this image was cast in metal in both regions, it is reasonable to hypothesize that this image type was cast first in Swat, probably during the period when rock-cut depictions of the deity were being made.

There are other types of images pointing to a Swat influence on later medieval art of Kashmir. For example, there is a Vairochana image in the Sri Pratap Singh Museum in Srinagar that is iconographically related to the early dated Vairochana image in figure 10 and the Vairochana sculpture in the collection of the Wali of Swat (Barrett 1962, fig. 12), discussed above. The crowned and jeweled Vairochana in Srinagar is also in *bodhyangi mudra*, an unusual hand gesture for Indian art since Vairochana is most often depicted in *dharmachakra mudra*. The Kashmiri piece is attributed by art historians (Pal 1975, no. 37) to a much later date (eleventh century) than the two Swat examples previously discussed. It is therefore possible that this particular image is another example of the movement of an iconographic type from Swat into Kashmir.

Finally, a seated Buddha image (fig. 17) now in the Metropolitan Museum of Art is somewhat different in style from other bronzes attributed to Swat yet was clearly assigned to that group in the technical study, and an origin in Kashmir was excluded (Reedy 1997, fig. P31). The backplate follows the basic flame motif of Kashmir, but the actual pattern used to depict the flames is unusual and not seen in this exact form in Kashmiri art. However, the same method of depicting the flames (as thick, slanted squiggles) is seen on a sixth-to-seventh-century terra-cotta plaque excavated in Swat by a team from Peshawar University and on a bronze frame of a Buddhist image found in Swat (Paul 1979, figs. 7, 21). The costume and turban style worn by the small patron figure are consistent with what one would expect from either northern Pakistan or Afghanistan.

Many aspects of the style of this piece represent features of later Kashmiri art and provide more indications of the inspirational role that Swat may have played in the development of Kashmiri art. Examples are the rounded facial features, the tripartite crown with each section composed of a crescent encircling a floret, and the piece of cloth coming over the left knee.

Conclusions

The role of earlier Gandhara art from Afghanistan and Pakistan in the development of Kashmiri art of the eighth century and later is widely acknowledged (Pal 1975, 36; Huntington 1985, 353, 357; Harle 1986, 189–92). Evidence for continuous artistic production on a large scale exists in the archaeological remains of Swat. Although the medieval Swat bronzes are stylistically only superficially related to earlier Gandhara objects, there is no reason to assume that any technological break ever occurred, with bronze casting disappearing at the end of the Gandhara period and suddenly reappearing in the eighth to ninth century under Kashmiri influence.

In fact, we can see an evolution in casting technology from the early Gandhara to post-Gandhara periods, moving from lesser to greater complexity (Reedy 1992). The earlier (pre-third-century) Gandhara works are generally small and are cast solid, either in the round or in an open mold, with no inlay work. Some of the later Gandhara works are larger and more sophisticated sculptures, and some are hollow cast with iron armatures as we frequently find with the medieval Swat bronzes. With a continuous casting technology present in northern Pakistan, the production of fifth-to-seventh-century Buddhist bronzes in Swat as complex as those produced in less isolated Kashmir during the early medieval period is not surprising.

We should, however, never expect tremendous numbers of such sculptures to come to light due to the scarcity of ore resources in the region. Such works must always have been items of great luxury in Swat. However, I would argue that many of these pieces are probably much older than most art historians currently credit them as being; many are highly creative in subject matter, visual style, and technological style; and these works clearly may have had an impact on certain avenues in the development of medieval copper-based sculptural styles in Kashmir.

Acknowledgments

Dr. Pieter Meyers of the Los Angeles County Museum of Art was instrumental in helping me initiate technical studies of Himalayan bronzes. He freely shared his extensive expertise, and I owe him a large debt of gratitude. At the Freer Gallery of Art, W. Thomas Chase was always helpful to me, sharing his infectious interest and deep knowledge of bronze-casting techniques.

References

Barrett, D. 1962. Bronzes from northwest India and western Pakistan. *Lalit Kala* 11:35–44.

Beal, S. 1869. *Travels of Fah-hian and Sung Yun*. London: Trubner.

——. 1906. *Si-yu-ki*. London: Kegan Paul, Trench, Trubner, and Co.

Facenna, D. 1981. *Sculptures from the sacred area of Butkara (Swat, Pakistan)*. Reports and memoirs of ISMEO 2(2). Rome: Istituto Italian per il Medio ed Estremo Oriente.

Giles, H. A. 1923. *The travels of Fa-hsien*. Cambridge: Cambridge University Press.

Harle, J. C. 1979. An early brass image of a bodhisattva with Kashmiri or Swat Valley affinities. In *South Asian archaeology*, ed. J. F. Van Lohuizen-de Leeuw, 127–34. Leiden: Brill.

——. 1986. *The art and architecture of the Indian subcontinent*. Harmondsworth and New York: Penguin Books.

Huntington, S. 1985. *The art of ancient India*. New York and Tokyo: Weatherhill.

Klimburg-Salter, D. 1982. *The Silk Route and the diamond path*. Los Angeles: UCLA Art Council.

Lee, S. 1975. *Asian art: Selections from the collection of Mr. And Mrs. John D. Rockefeller, part 2*. New York: Asia Society.

Legge, J. 1965. *A record of Buddhist kingdoms*. New York: Paragon and Dover.

Lerner, M. 1975. *Bronze sculptures from Asia*. New York: Metropolitan Museum of Art.

Lerner, M., and S. Kossak. 1991. *The lotus transcendent: Indian and southeast Asian art from the Samuel Eilenberg collection*. New York: Metropolitan Museum of Art.

Pal, P. 1975. *Bronzes of Kashmir*. Graz: Akademische Druck-U. Verlagsanstalt.

——. 1983. Off the Silk Route and the diamond path. *Art international* 26(2):44–65.

Paul, P. G. 1979. Some terra-cotta plaques from the Swat-Indus region: A little-known phase of the post-Gandhara art of Pakistan. In *South Asian archaeology*, ed. H. Hartel, 411–28. Berlin: Dietrich Reimer Verlag.

Reedy, C. 1991. Petrographic analysis of casting core materials for provenance studies of copper alloy sculptures. *Archeomaterials* 6(2):121–63.

——. 1992. The materials and technology of Gandhara and related copper-based sculptures. In *The crossroads of Asia: Transformation in image and symbol*, ed. E. Errington and J. Cribb, 241–55. Cambridge: Ancient India and Iran Trust.

——. 1997. *Himalayan bronzes: Technology, style, and choices*. Newark and London: University of Delaware and Associated University Presses.

Stein, M. A. 1929. *On Alexander's track to the Indus*. London: Macmillan.

——. 1961. *Kalhana's Rajatarangini*. 2 Vols. Delhi: Motilal Banarsidass.

Tucci, G. 1940. *Travels of Tibetan pilgrims in Swat Valley*. Calcutta: Greater India Society.

——. 1958. Preliminary report on an archaeological survey in Swat. *East and West* 9:279–328.

von Schroeder, U. 1981. *Indo-Tibetan bronzes*. Hong Kong: Visual Dharma.

Yung-hsi, L. 1957. *A record of the Buddhist countries*. Beijing: Chinese Buddhist Association.

16 Scientific Investigations of an Enthroned Buddha from Kashmir

John Twilley

ABSTRACT A technical and scientific examination of an important early-eighth-century enthroned Buddha and Adorants on the Cosmic Mountain (F.1972.48.2.s) from Kashmir in the collection of the Norton Simon Museum has been undertaken for the purpose of characterizing its materials and understanding the working methods of the sculptor. The sculpture is noted for the exceptional quality of its workmanship, the variety of its decorative elements, and its excellent state of preservation. It is one of three or four in Western collections that best exemplify the high aesthetic quality of Kashmiri art of this time period. While other investigations have often sought to characterize the metallurgical technologies by which Himalayan sculpture has been created, the objective of this study was to go beyond that step to identify all of the sculptor's materials and the techniques by which they were applied insofar as they could be determined from surviving original remnants. The condition of this work is especially conducive to accurate noninvasive quantitative analyses of the brass. As such, it represents a study case in the application of quantitative x-ray fluorescence analysis to the study of ancient casting materials. The results demonstrate a very high level of workmanship, incorporating both cast and cold-worked foundry repairs in different alloys. The original surface decoration of the sculpture was found to be more complex than previously supposed. Beneath modern retouchings with gold paint, remnants of pigmented resin coatings survive that simulate fabric garments. Elsewhere, surface inlay work was discovered to consist both of real metallic inlays of copper and silver and of visually indistinguishable simulated inlays in nonmetallic materials. The results allow a more complete picture of the original appearance of the Buddha and a broader understanding of the decorative techniques of the Kashmiri artist. They suggest that the original appearance of the sculpture was considerably more colorful.

Observations on the Construction Techniques

The figure of the Buddha, weighing 2463.5 g, has been hollow cast by the lost-wax process in a single pour (figs. 1 and 2). A strong response to a magnet at several locations is indicative of the presence of an iron armature inside. Such a response may be found near the top of the head and particularly along the left forearm. X-radiography of the sculpture reveals that this armature is still in place and that it consists of a central shaft, terminating near the top of the head, to which separate arm wires are attached (fig. 3). The arm wires extend at least as far as the forearms. Where they attach to the central shaft they have been flattened and turned up to follow the shaft as far as the base of the neck.

The legs also contain wire armatures that curve gracefully well inside the knees. The way in which the leg armatures connect to the central shaft is hidden by areas of radiographic density in the hands, feet, and abdomen. The cause of this feature in the abdomen is unclear. Whether it represents penetration of molten metal into internal fractures in the clay core or the introduction of some consecrated object is not clear from the radiograph. However, there was no external evidence of access to the casting interior through the base. The foundry repairs that could be identified are smaller in area than the opacity in the radiograph, suggesting that they could not have served as access for the introduction of such an article.

Numerous repairs were required to correct minor defects in the casting. These were expertly carried out by chiseling rectangular recesses into the casting and inserting patches. These patches have provided the opportunity to measure the composition of separate bits of metal from the artisan's workshop.

Repair patches that are nearly invisible may be found on some of the most prominent parts of the sculpture, including the right cheek (fig. 4), the back of the right hand (fig. 5), and the heel of the right foot (fig. 6). Four patches exist on the back of the Buddha, of which two overlap. The smallest finger of the right hand also appears to be a repair piece but it is not clear whether this is a repaired casting defect or a repair of subsequent damage. Although a suggestive line exists on the bottom of the Buddha along the right side, it does not appear that the bottom of the figure has been inset.

One of the most curious details of the Buddha is an incised drapery pattern on the right-side tang extending into the throne. While the drapery has been completed across the full underside of the figure as a system of more or less parallel folds (fig. 7), the sharply curved folds of this tang appear to bear no relationship to those on the figure itself (fig. 8). One possible explanation for this is that the three tangs were fragments of metal inserted into the wax model and enclosed within the investment so as to be *cast on* when the figure was poured. The fragment with the apparently

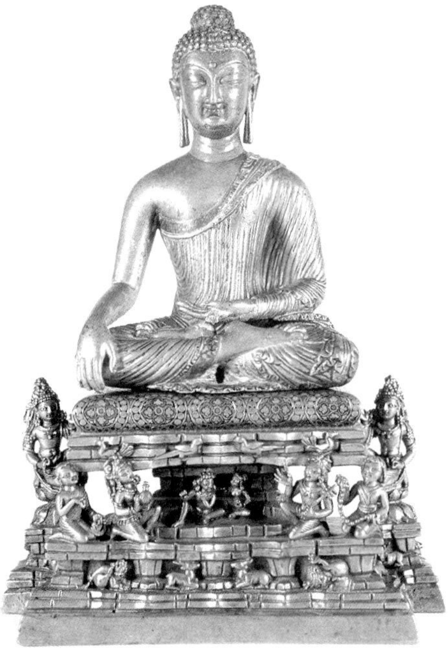

Figure 1. *Seated Buddha on separately cast throne, Kashmir, eighth century, brass, inlaid with copper and silver with added polychrome, Norton Simon Museum of Art, F.197248.2.S, 33.6 × 24.1 × 12.1 cm (PAL 2003)*

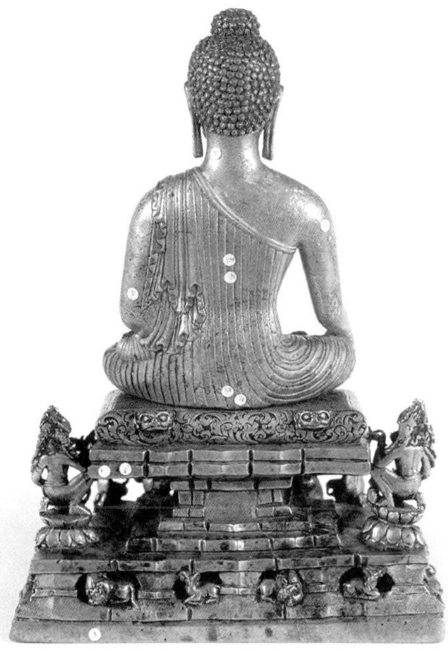

Figure 2. *Enthroned Buddha, rear. Dots temporarily mark analysis points*

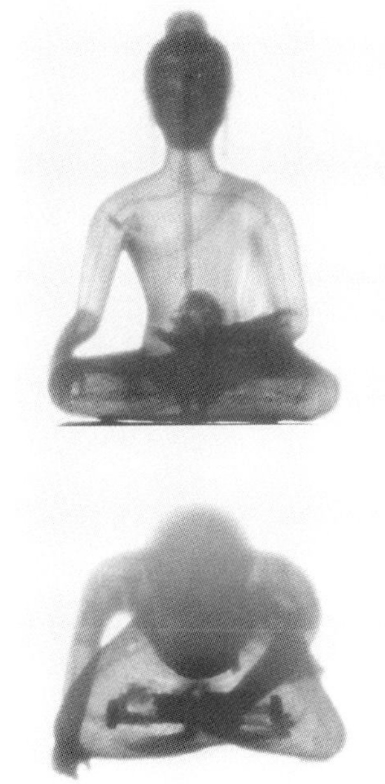

Figure 3. *X-radiographs of Buddha with inverted grayscale. Top: front view; bottom: top view. Conditions: 200 kV, 8 mA, 4 min, source to object distance 1.5 m*

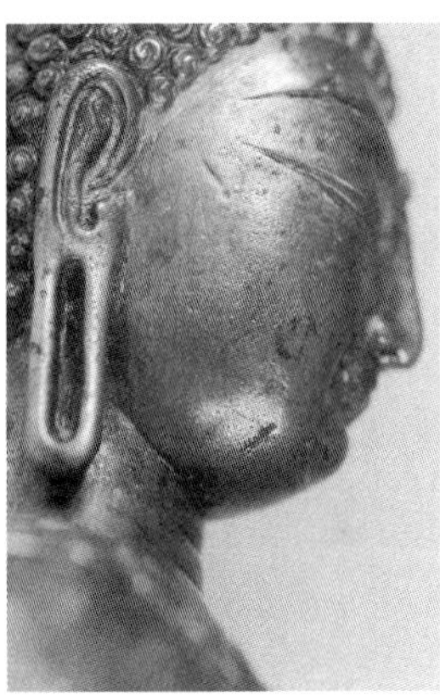

Figure 4. *Right cheek with finely wrought foundry patch*

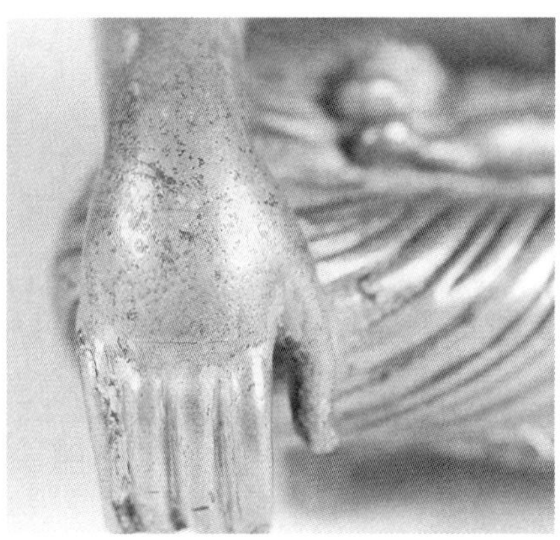

Figure 5. *Right hand with finely wrought foundry patch*

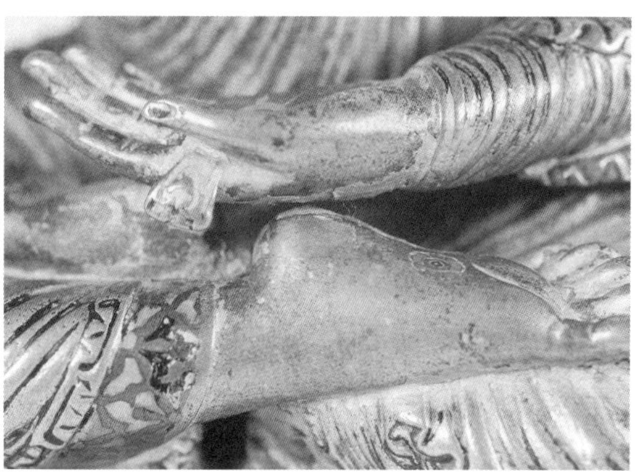

Figure 6. *Right heel with finely wrought foundry patch. Copper and silver embroidery inlays along ankle contain losses that reveal working methods and underlayer of black resin*

unrelated drapery lines may have been cut from an earlier miscast statue. No other elements of the figure appear to have been made separately and then cast on.

The throne is considerably more complex in that it contains numerous highly detailed figures on a smaller scale. It weighs 3718.3 g. The composition originally included a third element—a gateway, or *torana*—the existence of which is demonstrated by the two brackets on the rear of the throne. In the case of the throne, there are subtle indications that some of the figural elements may have been separately made and fixed in the investment for the main casting. In addition, unlike the Buddha, the throne contains repaired foundry defects that were made by subsequently casting minor pours of additional brass.

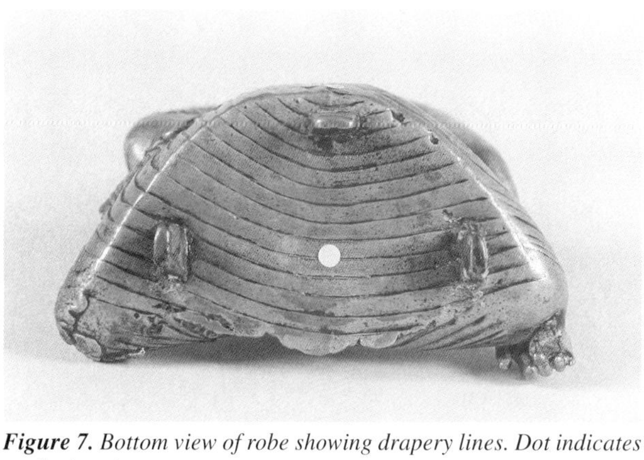

Figure 7. *Bottom view of robe showing drapery lines. Dot indicates analysis point*

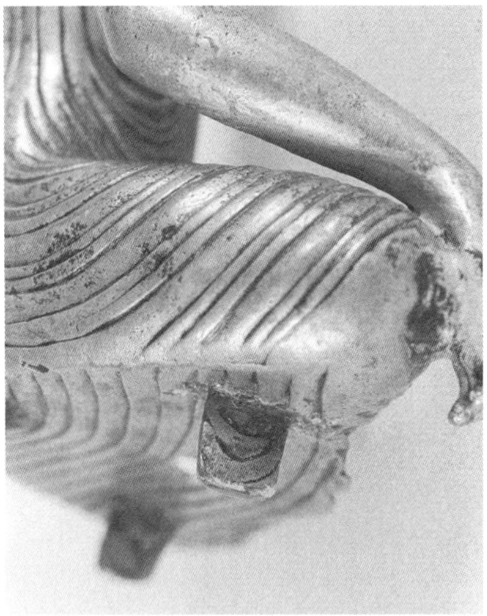

Figure 8. *Side view of right tang, showing sharply folded drapery lines unrelated to those of the robe*

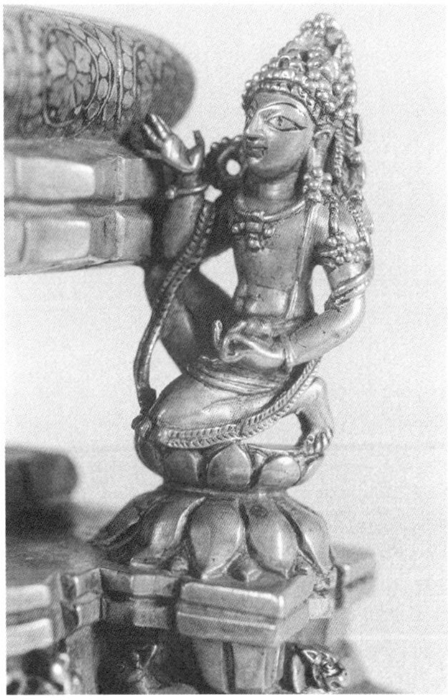

Figure 9. *One of two flanking attendants that may have been incorporated into the throne assembly by casting on*

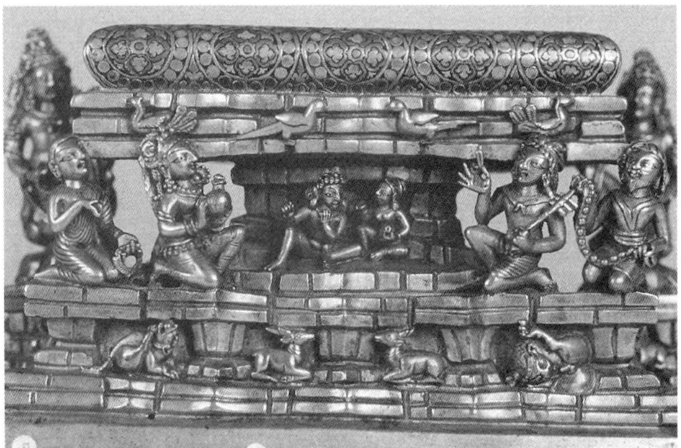

Figure 10. *The central pair of figures beneath the throne, although obstructed at the rear, are fully finished in the round, suggesting that they were incorporated by casting on. Dots indicate analysis points*

The large flanking figures seated on lotuses (fig. 9) have a series of indentations that surround each of their locations on the underside of the base. These figures may have been cast separately, joined to the wax model for the base, covered with investment and "cast on." The indentations would seem to follow the junction between the precast figure and the base. However, it is also possible that these indentations are simply the result of joining separately formed wax models of the flanking figures to the wax model of the base and casting them simultaneously. They may, therefore, be vestigial features from the joining of the wax parts that have been replicated in the bronze. No features could be found that could completely settle this question.

The central pair of figures beneath the "mountain" (fig. 10) likewise may have been separately cast and inserted into the investment for the base. The evidence for this is more compelling. Within the interior of the base, an irregular thickening exists behind these figures. In addition, they are highly finished, in the round, despite the inaccessibility of their reverse sides. It appears, therefore, that they were finished prior to being situated in this place.

The left rear corner of the base failed to fill properly with metal during casting, resulting in a large void. Subsequently, new metal was poured into the space. Such a repair would be made by creating a localized covering of investment material around the gap and pouring additional metal after the original investment had been removed and the problem assessed.

Other areas of incomplete filling of the metal were left unrepaired if they were out of view and resulted in no weakening of the base. The underside, horizontal surface of the throne cushion, which meets the upraised knees of the two flanking figures, is extremely thin. Where gaps exist in this surface, they were left alone.

All of the visible surfaces of the base are highly finished. An idea of the as-cast surface condition of the metal can be derived from the inspection of the base interior. Finishing subsequent to casting included the creation or sharpening of the fine detail by punch and chisel work and burnishing. Nearly all of the surfaces of the figure and the throne have been so worked to create or to sharpen details too fine to be cast.

The surface conditions of the Buddha and throne are consistent with artifacts that have been protected from the elements and handled regularly over generations. Their surfaces are free of corrosion and brightly polished from this handling.

Surface Decoration with Inlay Metals

Copper and silver have been extensively used to provide contrast to details of both components. In the Buddha the drapery folds and embroideries have been accentuated by copper and silver (fig. 11). The manner of their addition may be observed at a couple of small losses (fig. 6). Chisel cuts were made in the casting and the surface carved away in between them. Bits of the inlay metal were then hammered into place, spreading their edges to lock against the sides of

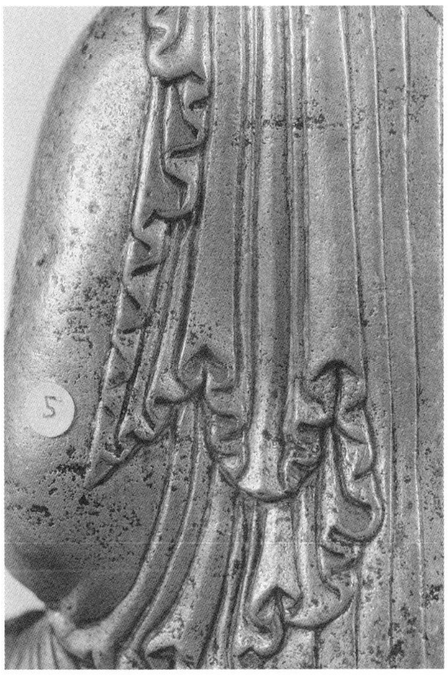

Figure 11. Copper inlay work on the hem of the robe behind the left shoulder

the cut. A black material, identical in appearance to that used in its own right to accentuate drapery and as a base layer for paint in the hair, has been placed beneath such inlays to serve as a compliant foundation upon which to get a snug fit. It does not appear to be an adhesive.

On the throne cushion, two different approaches to the same embroidery pattern have been used, depending upon the degree of visibility on the final piece. On the front, the circular patterns of silver dots have been created by individual inlays of silver within raised circles of brass. The intervening brass surface between the raised circles has been chased down and covered with black-pigmented resin (plate 16). On the sides of the cushion, however, the same design has been created with greater economy of labor by inlaying a large open circle of silver onto which dots have been formed by chasing down both the intervening silver and brass and covering the surrounding areas with black-pigmented resin (plate 17). The incised lines used to strengthen the outlines of the inlaid metal designs have been created last and may be observed to be cut into the inlay metals as well as the brass (plate 18).

An important detail of the tool handling in the inlaid design may be seen in the form of a serrated edge that exists along the inlay metal. This feature is also apparent in the textile pattern on the top of the cushion, where a pigmented resin layer was applied to accentuate the design. This type of microfeature has been frequently encountered on later Indian and Islamic metalwork. It is the result of using the corner of a flat chisel in a series of closely spaced indentations to create a curving line.

The same manner of handling a tool may be observed in the making of a much fainter "sketch" to plan out the design. It remains visible where the final pattern fails to precisely follow this line in the three preceding images (plates 16, 17, and 18).

By contrast, certain other features were created by another technique. The lips of the Buddha and tiny details of figures on the base are applied in copper or silver metal that appears to have been hammered *onto* a surface roughened to receive it rather than in a chased recess.

Decorative Coatings and Surface Residues

The most obvious coating is a superficial layer of gold powder paint that not only covers earlier paint layers but even obscures some of the fine silver and copper inlay work. This is of relatively recent origin. Paint remains of several different kinds persist in the hair of the Buddha. A clear stratigraphic relationship exists between three of them. The lowest is a black, tarlike material with occasional bubbles. This is similar to the black material applied in other areas to accentuate the pattern on the throne cushion and the folds in the robe. Atop the black is a tan-colored ground layer followed by a medium blue shade (plate 19). Remnants of a separate application of blue paint remain situated variously atop the black, the ground, and the original blue. These remnants are clearly from a later addition and have a deep blue color (plate 20).

The robe contains a surprising application of brown paint to the folds—surprising because even under the microscope, it is virtually indistinguishable from the copper inlay. In plate 21 this paint appears on the left and the copper inlay on the right. (Traces of more recent gold powder paint that has not completely worn away clutter the edges of the areas of interest in the photo.) One could speculate that this brown paint is a later repair intended to replace copper metal inlay that has been lost. However, there is a lack of proof that the metal is original and the paint restoration due to the impossibility of discovering what lies beneath each.

Beneath the right hand there is a thick layer of red, lacquerlike material with a pebbly, variegated appearance. This appears to be the remnant of a more extensive layer. The edges of this red material are surrounded by shellac that has probably been applied as an adhesive to prevent further flaking of the red. The pattern of the throne cushion was formerly accentuated by red- and black-colored resin in the chiseled fabric designs (plate 18).

Surface deposits exist in at least three categories that are sometimes difficult to distinguish: remnants of ancient decorative coatings, incense residue, and grime. The protected or hidden surfaces of the throne are heavily covered with a tarry residue that seems partially to be the product of incense burning. The underside of the base has a black patina. The tarry residue has a peripheral fringe of transparent green corrosion. When the residue itself is sampled, it is found to be green in color when viewed by transmitted light. There are scattered green whiskers composed of soft single crystals of an organic corrosion product that have grown, undisturbed, over many years.

Analytical Investigation of the Metals

The objective of the quantitative analysis of the metals was to learn the composition of the alloys involved in the cast-

Table 1. *Nonsampling quantitative elemental analysis by energy dispersive x-ray fluorescence spectrometry (normalized weight %)*

Element	Cu	σ	Zn	σ	Sn	σ	Pb	σ	Fe	σ	Ni	σ	Mean Analytical Totals	σ	No. of analyses
Buddha	74.7%	1.1	24.5%	1.0	0.06%	0.05	0.16%	0.06	0.47%	0.06	0.078%	0.003	95.1%	4.0	(6)
Throne	74.2	0.5	24.3	0.6	0.12	0.07	0.57	0.07	0.69	0.04	0.074	0.006	100.1	1.0	(5)
Throne welds	76.8	0.2	21.2	0.1	0.20	0.11	0.80	0.08	0.92	0.05	0.072	0.004	93.4	5.6	(2)
Buddha patches	74.0	0.7	25.2	0.8	0.03	0.05	0.12	0.10	0.49	0.05	0.086	0.005	94.1	3.2	(6)
Throne patch	76.0		21.8		0.22		0.91		0.92		0.082			n.a.	(1)

ings, the inset repairs, and the cast-on repairs. At the time of this work, a *torana* of similar size and shape to the one that must have been a part of this sculpture was being evaluated as potentially being the missing component. Ultimately, compositional and technical differences coupled with a dimensional incompatibility with the throne led to the conclusion that the *torana* was not part of the original composition.

In total, twenty alloy analyses were conducted. Six locations on the initial Buddha casting were analyzed. Six of its repairs, all cold-worked patches, were analyzed. Five locations on the initial throne casting were analyzed. Only three repairs to the throne were accessible and large enough for analysis. One of these was a cold-worked patch, and two were cast-in-place welds. Normalized mean compositions and their corresponding standard deviations are reported in table 1. The analysis protocols and a critique of the data are presented in the appendix.

Most important, it can be seen that the average contents of the principal alloy metal, zinc, in the throne, the Buddha, and all the patches on the Buddha lie within a single standard deviation of one another. The welded repairs to the throne lie significantly lower in zinc content, and the one patch repair in the throne that was deemed large enough to be analyzed independently of the surrounding metal lies close to the weld composition. There may be a sound technological reason for using a brass of lower zinc content for the welded repairs. Perhaps the slightly lower melting point of the casting to be repaired relative to that of the repair material promoted fusion between the metals when they came into contact. Reference to the alloy phase diagram for the copper-zinc binary system provides a liquidus temperature of approximately 980ºC for this composition, which would not be significantly altered by disregarding the remaining alloy constituents.

The *torana* examined at the same time as the Buddha and throne possessed a zinc content significantly higher than all of the components of the statue. Its mean value of 30.8% lies above a critical technological value of 28% that is believed to be the highest value normally achievable by cementation methods. The cementation method, which depends on the absorption of zinc vapor driven out of its ore into copper metal in a closed retort, avoids the technological difficulty of isolating pure zinc metal for subsequent melting with copper. While the technological developments necessary for the isolation of zinc metal seem to have been achieved in India long before most other areas of the world, the values found in these artworks are quite high. The region of Kashmir and its immediate environs has produced the earliest high-zinc brasses, but the reliably reported examples of those that exceed 28%, prior to the eighteenth century, are very few. Reedy reported four such examples in her dissertation work on Himalayan bronzes (1986, 1997). However, for two of them, difficulties in the analysis, probably due to the state of preservation of the metal, resulted in extraordinarily high analytical totals of 117 and 139%. Clearly, any example of brass with zinc in excess of 28%, stylistically datable to the eighth century, should be cautiously scrutinized. The second notable deviation in the composition of the *torana* from those of all components of the sculpture was its low level of iron—fully an order of magnitude lower and comparable to modern cast brass.

In order to allow further comparisons with other sets of data, inductively coupled plasma emission spectroscopy (ICPES) was carried out on a single sample drilled from each component of the statue. The sample from the Buddha was drilled from the right proper tang with the fold lines and that from the throne base, from the underside of the right proper, largest figure. Samples were analyzed along with standards of similar compositions using a dissolution regimen utilized for the largest body of data of this type acquired to date (Reedy 1986, 1997).

Table 2. *Quantitative major and trace-element analysis by inductively coupled plasma emission spectroscopy (un-normalized)*

Element	Buddha	Throne
Cu	69.6	75.0
Zn	23.9	22.8
Sn	.225	.363
Pb	.545	1.51
Fe	.377	.644
Ni	.071	.085
As	.020	.052
Co	.005	.021
Sb	.079	.096
Ag	.017	.031
Bi	.010	.012
Se	.030	.032
In	.029	.040
Al	.008	.009
B	.003	.003
Ba	.001	.001
Ca	.006	.006
Cd	.001	.001
Cr	.001	.002
Mg	.001	.001
Mn	.002	.003
Mo	.002	.001
P	.012	.015
Si	.006	.006
V	.001	.001
Total	95.2	100.7

The results are shown in table 2 (Holguin pers. comm. 1996). When normalized these were in good agreement with the quantitative energy dispersive x-ray fluorescence spectrometry (EDXRF) data for those elements in the base. The difference in the values for the Buddha is believed to derive from the fact that the ICPES sample came from the engraved tang that has subsequently been postulated to be a cast-on fragment from another sculpture in the same workshop. Its position could not be approached from the proper geometry for quantitative EDXRF to be carried out, so a direct comparison of results from the two methods on that area was not possible. Analytical totals for the ICPES analyses ranged from 94.95 to 100.74, with one of the lowest analytical totals being that of a standard included with the artwork samples. Therefore, variations from 100% in the ICPES method are due to factors other than the hypothetical presence of corrosion products in the ancient samples.

Decorative Colorants

The colored materials on the surface of the sculpture were analyzed in order to determine the composition of their pigments and to learn the identity of the medium or binder used to apply them. The combined results of polarized light microscopy (PLM), x-ray diffraction (XRD), and Fourier transform infrared spectroscopy (FTIR) were used to identify the pigments, while the latter was used alone to categorize the media.

The original blue color of the hair was achieved with azurite in which traces of malachite may be found. FTIR analysis shows that this was applied in a glue tempera medium. Strong peaks at 1640 cm^{-1} and 1540 cm^{-1} superimposed upon those of the azurite confirm the proteinaceous nature of the medium. The later restoration blue that is much more intensely colored may be seen by PLM to be synthetic ultramarine, a pigment available only since the Industrial Revolution. The strongest peak of the spectrum, near 1100 cm^{-1}, is due to silicate bonds in the ultramarine. Peaks at 1640 cm^{-1} and 1540 cm^{-1} indicate that the medium is again proteinaceous matter. The original "ground" layer of tan paint beneath the azurite contains a mixture of colorless mineral matter and the orange pigment realgar, a rare and unstable sulfide of arsenic occasionally encountered in Himalayan painting. XRD (by the Gandolfi camera method) and PLM conducted sequentially on the same, very small sample reveal the presence of quartz, the sheet silicate rectorite, and mixed clays consisting of illite and kaolinite-montmorillonite (fig. 12). The thick black layer beneath appears, on the basis of PLM, to contain very fine anisotropic mineral matter consistent with its identification as bone black; however, confirming hydroxylapatite absorptions in the FTIR spectrum, if present, could not be resolved from interfering substances.

The brown paint that simulates inlaid copper so accurately on the Buddha is a complex mixture of colorless clay with a filamentary morphology, the yellow arsenic sulfide pigment orpiment (finely ground), some charcoal, an isotropic yellow ochre, a trace of cinnabar, and red ochre. Its FTIR spectrum contains absorptions confirming the presence of clay, calcite, silicate minerals, and an oil or resin medium, while XRD of this material confirms the presence of calcite, quartz, and illite clay (fig. 13). The brown paint is unlikely to be a modern restoration due to the presence of orpiment, a pigment that is not essential to the appearance of the layer and is not used today due to the toxicity of arsenic.

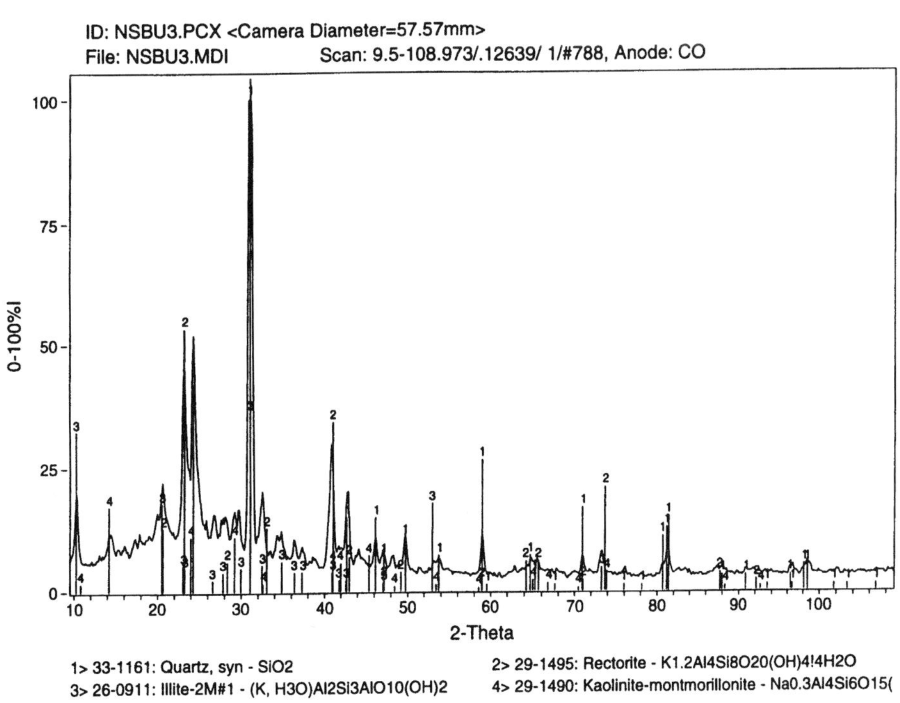

Figure 12. The original tan ground layer of the hair analyzed by x-ray diffraction via the Gandolfi method, showing the correlation to reference patterns for its constituent minerals

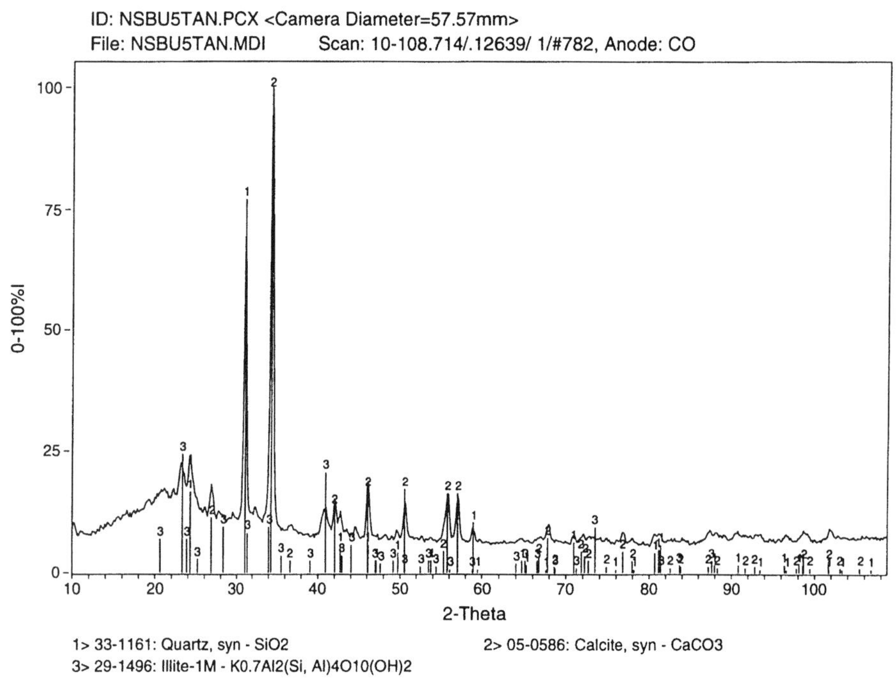

Figure 13. The copper-simulating pigmented inlay of the drapery analyzed by x-ray diffraction via the Gandolfi method, showing the correlation to reference patterns for its constituent minerals

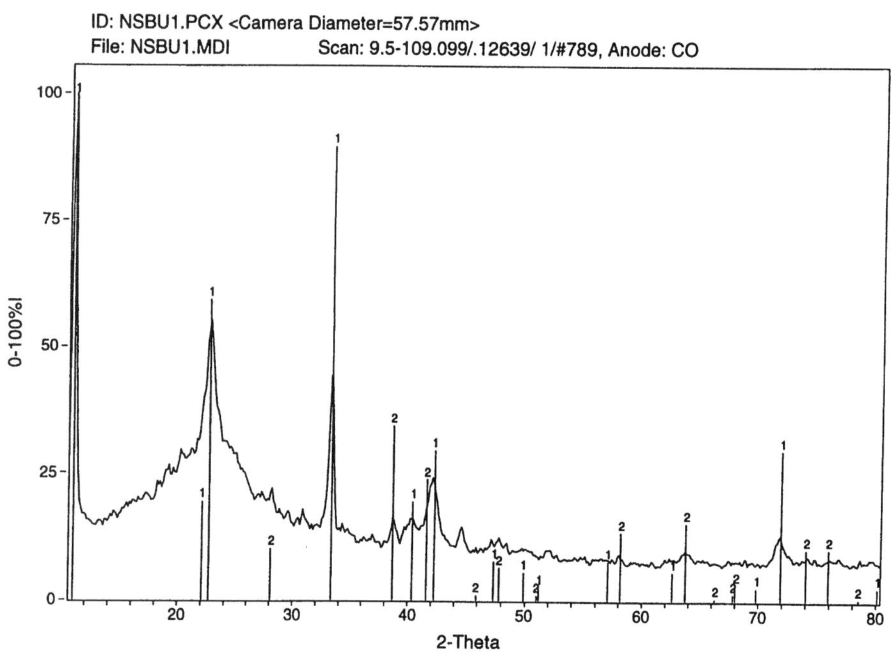

Figure 14. The residual red polychrome layer that survives behind the hand analyzed by x-ray diffraction via the Gandolfi method, showing the correlation to reference patterns for its constituent minerals

The red material that makes up the thick mass beneath the proper right hand is quite different from that on the throne and is pigmented with fine hematite mixed with highly fibrous talc. XRD confirms these identifications (fig. 14), and FTIR demonstrates that the medium in which they are applied is a natural resin. A sharp peak at about 3675 cm^{-1} and the strongest one at 1100 cm^{-1} are from the talc, while a group of peaks around 2900 cm^{-1} and a closely spaced pair at 1737 cm^{-1} and 1715 cm^{-1} are due to a resin. This is a material distinct from both the water-soluble paints that were used in the hair of the Buddha and from oriental lacquer, a material that this superficially resembles. Along the margin of the remaining red there is a clear orange resin. Its FTIR spectrum contains shoulders on the short wavelength side of each of the two strong peaks near 2900 cm^{-1} that are absent in the main mass of colored material. This unpigmented material with a spectrum similar to that of shellac may have been added later as an adhesive in an

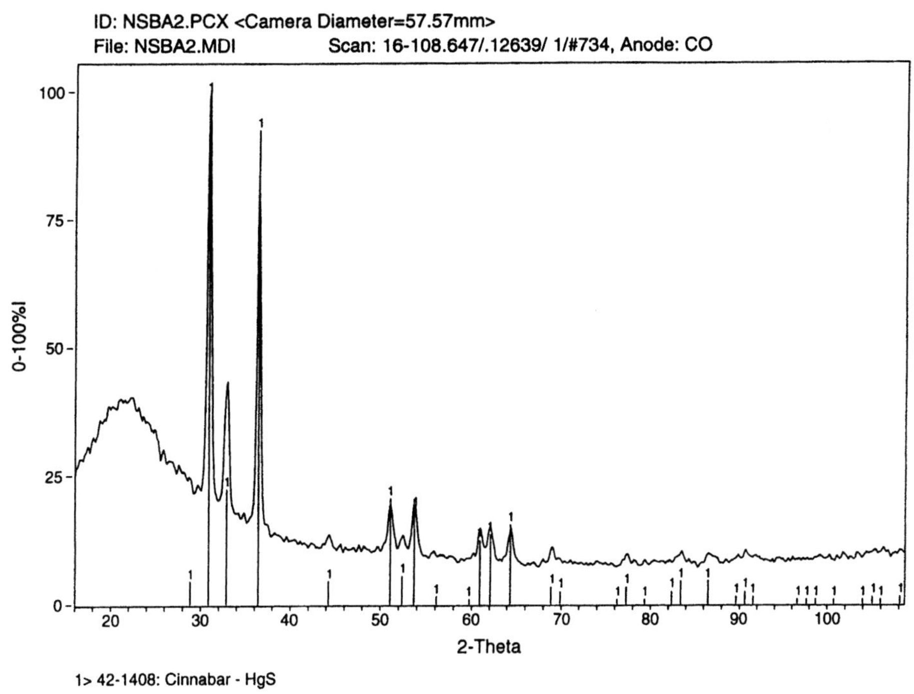

Figure 15. *The residual red polychrome layer that survives in the incised decoration of the throne cushion analyzed by x-ray diffraction via the Gandolfi method, showing the correlation to the reference pattern for cinnabar*

attempt to prevent the further loss of the colored layer or may have segregated and migrated out of a resin mixture used in the layer itself.

The red used to accent the pattern of the throne cushion is pigmented with an odd mixture that contains both coarse and very fine cinnabar with a size difference of nearly 100 to 1. The XRD pattern is dominated by the lines attributable to this mineral (fig. 15). In addition, PLM examination revealed that it contains charcoal, lead white, and a strongly discolored medium that is intimately mixed in places with a blue material that is believed to be copper corrosion from the underlying metal. Much of the cinnabar is badly discolored to dark brown or black, a phenomenon that is believed to be partially responsible for the variegated color of the surviving red part of the throne. Direct subtraction of the FTIR spectrum of the throne red from the red beneath the hand yields a difference spectrum that, apart from minor variations in the proportions of the ingredients, is the spectrum of talc. (As a sulfide, cinnabar does not contribute to the infrared absorptions.) The two colored coatings are, therefore, very similar in their composition of organic material.

The black material filling some of the incised decorations of the throne cushion is distinguished from the other decorative materials by its content of tiny bubbles. Its FTIR spectrum is that of a natural resin.

Conclusions

All visible surfaces of the enthroned Buddha were given a highly skilled level of finishing by either metalworking or color application. The evidence supports the idea that the sculpture was brilliantly colored by both inlay metals and polychromy when first completed and that the polychrome was restored when damaged. This contrasts with the present appearance of the sculpture, which, though polychromatic, is predominantly the color of the brass of which the casting is made. The highly complex mixture of pigments that simulates inlaid copper can be argued to represent an early restoration of a missing patch of copper. Since the color used matches that of the adjacent thoroughly oxidized copper, it is clear that the pigment application followed the inlay work by some years, or else it would not have initially matched the copper appearance. The skill of this simulation and the choice of simulating the inlay in paint (rather than replacing the copper) point toward the involvement of a painter in the restorations and suggest that access to a suitably skilled metalworker was not available. An alternate social or religious interpretation is that the act of retouching the loss with paint was deemed more consistent with the care of a consecrated object than a return to the atelier of a metalworker. Gold paint, added in the latest redecoration prior to its incorporation in the museum collection, imposed a uniformity of appearance on the Buddha as an aesthetic priority that was not that of the original artist nor of his immediate successors. It has now worn away to an extent that reveals part of the prior decorative scheme. The present state of the sculpture's decoration, therefore, represents a historical chronology of these changing tastes.

Acknowledgments

Mike Cowell of the Research Laboratory of the British Museum provided brass standards that were essential to the quantitative analysis reported here. Brent Holguin of UCLA performed ICPES analyses using the protocols and instrumentation employed for much of Reedy's work, thereby reducing the potential for interlaboratory variation between the results. Pratapaditya Pal, Fellow for Research at the Norton Simon Museum at the time of this undertaking, recognized the potential for technical analysis and suggested this project. I am grateful to Sarah Campbell and the trustees of the Norton Simon Museum for facilitating the investigation and for granting permission to publish the results. I have taken all of the photographs.

Appendix: Quantitative Alloy Analysis

The quantitative alloy analyses for this sculpture provide two unusual opportunities. First, the multiple analyses allow for a reduction in experimental error, leading to greater confidence that the differences identified between separately formed parts of the sculpture are real and not the effect of random experimental effects or localized alloy segregation. Second, the relatively homogenous alloy type allows an assessment of the experimental repeatability that can be achieved by nonsampling EDXRF on a subject that is appropriate to this method.

Quantitative x-ray fluorescence spectrometry has the potential for very high accuracy and high repeatability on suitable classes of material (Twilley 1980; Lutz and Pernicka 1996). The ease of nonsampling XRF analysis has induced some researchers in archaeological materials to attempt quantitative analysis on inappropriate surfaces with a disregard for the physical properties of x-ray absorption and emission. This lack of attention to experimental detail has often led to correspondingly poor results. Some of the preconditions necessary for accurate quantitative results include the use of standards that are not only similar in gross composition to the unknowns but also similar in microstructure so that *matrix* effects are consistent for all samples. Single-phase alloys such as those on the copper-rich side of the copper-zinc system are very good candidates in this regard as long as minor secondary phases are uniformly distributed and of similar size in all samples and standards. For *typical* archaeological bronzes of the copper-tin-lead system, leaded brasses, or quaternary alloys of these four metals, nonsampling XRF analysis is often less accurate than alternate methods involving the destruction of a sample where the textural and grain-size effects of the metal are eliminated. Objects wherein surface patination or subsurface corrosion has proceeded are similarly inappropriate for quantitation.

On the other hand, unlike XRF, many alternate quantitative methods presume the ability to maintain complex mixtures of metal and semimetal salts in solution without prior knowledge of the chemistries involved, leading to occasional systematic errors related to sample preparation. Emission spectroscopy by older spark and flame excitation methods has been applied to the investigation of Himalayan metalwork (Werner 1972) but suffers from problems of interelement interference and nonlinearities in both excitation and detection.

The uncorroded surfaces of this enthroned Buddha make it an excellent candidate for quantitative analysis of the main alloy constituents by nonsampling EDXRF. We therefore had the luxury of conducting numerous analyses and arriving at average values for the statue and throne that overcome the minor variations that one encounters from point to point on any large casting. (Thereafter, trace-element compositions for each of the three castings were obtained by the use of ICPES on a single sample drilling taken from each.)

EDXRF analysis was conducted using the characteristic excitation radiation from secondary targets of molybdenum and gadolinium. Brass standards used for obtaining calibration coefficients for each of the elements were produced by the BNF Metals Technology Group and supplied by the British Museum Research Laboratory under the designations C42x01 and C42x03. Results for the elements copper, nickel, iron, zinc, and lead were obtained through excitation with the molybdenum target and combined with those of tin obtained with the gadolinium target using the *Quantex* algorithm of Kevex Corp. Standard counting times were sixty and one hundred and eighty seconds, with a tungsten x-ray tube operating at either 40 kV, 0.6 mA; or 60 kV, 1.2 mA for the molybdenum and gadolinium runs, respectively.

The minimum and maximum analytical totals for individual analyses are reported for each category to convey a sense of geometric effects on the normalization. The variation in analytical total for this sculpture is largely dependent upon geometric effects, such as surface curvature. The analytical implications of these variations that are induced by nonplanar geometry are quite different from those arising from a copper alloy that has undergone corrosion in burial, where corrosion products occupy part of the metal volume. In the present case dominated by geometric effects, the normalization of the total to 100% is more valid than in the corroded case, where vastly nonuniform matrix effects (corrosion compounds vs. metal) and the selective loss of alloy constituents dominate the experimental uncertainties.

The throne presented more flat surfaces that could be analyzed and has a correspondingly narrower *range* in the individual analytical totals and a higher *average* analytical total near 100%, further reinforcing the conclusion that these variations are geometric in origin.

References

Lutz, J., and E. Pernicka. 1996. Energy dispersive x-ray fluorescence analysis of ancient copper alloys: Empirical values for precision and accuracy. *Archaeometry* 38(2):313–23.

Pal, P. 2003. *Himalayas: An Aesthetic Adventure*. Chicago: Art Institute of Chicago. 104–5.

Reedy, C. L. 1986. Technical analysis of medieval Himalayan copper alloy statues for provenance determination. Ph.D. diss., University of California at Los Angeles.

——. 1997. *Himalayan bronzes: Technology, style, and choices*. Delaware and London: Associated University Presses.

Twilley, J. 1980. Compositional analysis of Colombian tumbaga by x-ray fluorescence spectrometry. In *Actes du XXeme symposium international d'archeometrie, Vol. III,* ed. L. Langouet, 303–12. Rennes: Université de Rennes.

Werner, O. 1972. *Spektralanalytische und metallurgische untersuchungen an Indischen bronzen*. Leiden: E. H. Brill.

Paintings

17 Pigments on Japanese Ukiyo-e Paintings in the Freer Gallery of Art

Elisabeth West FitzHugh

ABSTRACT The Japanese paintings on silk and paper in the Freer Gallery of Art include over five hundred of the ukiyo-e school, dating from the late sixteenth to the late nineteenth century. Pigment analysis of five hundred paintings on which color was used was carried out between 1968 and 1989. The intent was to characterize the pigments used at these dates in Japan and to determine any chronological differences in pigment use during the period. Analytical methods used were polarized light microscopy (PLM), chemical microscopy, and x-ray powder diffraction (XRD). Scanning electron microscopy with energy dispersive x-ray analysis (SEM-EDXA) was used to a limited extent. Seventeen major pigments were identified. The eleven pigments used throughout the period were shell white, vermilion, red lead, organic red, red and brown earths, yellow earth, organic yellow, malachite, azurite, indigo, and organic brown. The other six, generally not used early in the period, were smalt, copper greens, orpiment, Prussian blue, lead white, and ultramarine.

Introduction

The collection of the Freer Gallery includes over six hundred Japanese paintings, drawings, and sketches of the ukiyo-e school, dating from the late sixteenth to the late nineteenth century. Pigment analysis of five hundred of these works was carried out between 1968 and 1989 and has been the subject of a preliminary report (FitzHugh 1979). The project was first conceived by Rutherford J. Gettens, then head curator of the Freer Technical Laboratory, with the aim of characterizing pigments used and to determine any chronological differences in pigment use. This work forms a basic database on pigment use in Japanese paintings of this period and has facilitated further work by others (Leona and Winter 2001; Winter, Giaccai, and Leona this volume).

The investigation of carbon inks and pigments was beyond the scope of this project. Some identifications were made of metal pigments, but they were not systematically searched for. Mica, identified as muscovite, was found on five paintings.

New coloring materials from China became available in Japan when Buddhism reached there from China and Korea in the sixth century. The range of pigments in Japan before this time was limited chiefly to earth colors and clays (Yamasaki and Emoto 1979). China is often mentioned as the best source of some of these pigments, or as the best source of information about them, so some references to pigment use in China are included.

Paintings in the ukiyo-e style depicted the everyday life of the ordinary people of Japan. An example can be seen in plate 22. They were executed on silk and paper supports and are in the form of hanging scrolls *(kakemono)*, handscrolls *(makimono)*, folding screens *(byōbu)*, and drawings or sketches.

Methods of Analysis

Identifications were carried out by traditional methods: PLM, chemical microscopy, and XRD. SEM-EDXA was used in a few cases. The range of colors on each painting was noted by visual inspection and with a binocular microscope at low magnification. Every color was sampled, within the limits of subjective judgments that had to be made. Pigments on nineteenth-century drawings and sketches were often applied too thinly to permit analysis by the methods used, but with these exceptions, the majority of the pigments were characterized. Samples were taken from areas showing no obvious repaint, although a few repainted areas were sampled when questions arose.

PLM is an excellent and simple way to begin an examination of a pigment sample. In addition to revealing color and optical properties, it is a reliable way to determine how homogeneous or heterogeneous the sample is, to see if it is a single pigment or a mixture, to determine if it may be organic or inorganic, and to detect traces of other materials that might be overlooked. PLM can then provide a firm identification of the inorganic pigments in most cases. The sampling and mounting techniques have been described (Gettens 1976), and the application of PLM to pigment identification has been summarized (Feller and Bayard 1986). Detailed descriptions of the optical properties of these pigments, with the exception of the iron earths and the copper greens, can be found in volumes 1, 2, and 3 of *Artists' pigments: A handbook of their history and characteristics*, cited in the sections on each pigment. Every pigment sampled in this series was examined by PLM, and the microscope slides are permanently filed.

Chemical microscopy was carried out on an unmounted sample under the microscope to determine solubility and to test for the presence of major elements. The application of chemical microscopy to pigment identification was pioneered by Joyce Plesters (1956) and Walter McCrone (1982) and is covered in the chapters in *Artists' pigments*. These methods are now rarely used, but at the time they served their purpose admirably. Chemical microscopy was used on selected samples of all pigments except smalt; the tests used, with a few exceptions, are not detailed here. The small amount of animal-glue medium present did not interfere with chemical microscopy.

XRD was used on selected samples to confirm the identification of crystalline materials, such as minerals and inorganic pigments. SEM-EDXA was used for element determination in the copper greens and was carried out by Melanie Feather, Smithsonian Center for Research and Materials Education.

Major Pigments of the Ukiyo-e Period

Eleven pigments were used throughout the ukiyo-e period, as summarized in table 1.

Table 1. *Pigments used throughout the ukiyo-e period (late sixteenth to mid-nineteenth century)*

Shell white	Organic yellow
Vermilion	Malachite
Red lead	Azurite
Red and brown earths	Indigo
Yellow earths	Organic brown
Organic red	

Shell White

Shell white, *gofun*, is a form of calcite, calcium carbonate ($CaCO_3$), made of crushed and ground seashells, usually oyster shells (Gettens, FitzHugh, and Feller 1993). It is the only pigment in this series unique to Japan as well as the only white pigment found, except for a few occurrences of lead white, on some late paintings. It is also an important constituent of pigment mixtures designed to obtain lighter color tones. Shell white gradually replaced clay as the common white pigment in Japan during the fifteenth to sixteenth century (Yamasaki and Emoto 1979).

The characteristic microscopic appearance of shell white by PLM was generally accepted as sufficient for identification. The particles are highly birefringent, vary in size, and are elongated, flat, and lathlike, with a striated appearance as seen in plate 23. The particles reflect the way the calcium carbonate was deposited in layers at the edge of the growing mollusk shell (Gettens, FitzHugh, and Feller 1993).

Vermilion

Vermilion, *shu* or *shin-sha*, red mercuric sulfide (α-HgS) appears on the paintings as a bright clear red. It was sampled from the painting proper, not from the seals. The naturally occurring mineral cinnabar is ground to produce the pigment. In addition, synthetic vermilion can be made by the dry process and the wet process (Gettens, Feller, and Chase 1993). Wet-process vermilion, first prepared in Germany in 1687, is characterized by fine, uniform particles, typical of a chemically precipitated material. It does not appear on these paintings.

Dry-process vermilion is microscopically identical to natural vermilion, and there is no way to differentiate between them. Dry-process vermilion was made by the eighth century in the West and was known to the Chinese in antiquity (Silbergeld and McNair 1988). It has been suggested that knowledge of the process may have been transmitted from China to the West (Gettens, Feller, and Chase 1993). In the dry process, mercury and sulfur are heated together to form black mercuric sulfide, which in turn is heated, causing the red mercuric sulfide to form by sublimation. The mineral that was not pure enough to be used as a pigment was used as a source of mercury for the dry process (Sung 1966). The distillation of mercury from cinnabar was known as early as the third century B.C.E. in China (Needham 1976).

Deposits of cinnabar occur throughout China (Golas 1999), and it was found there in graves dated to 4000 B.C.E. (Wang and Wang 1999). It was exported from China to Japan during the Edo period (Oguchi 1969). Mineral cinnabar occurs in Japan (Takamatsu 1878). Vermilion was used as a pigment in Japan from the eighth century on, and dry-process vermilion was made there from the beginning of the seventeenth century (Yamasaki and Emoto 1979).

By PLM ground cinnabar and dry-process vermilion both show variation in particle size, some particles being quite large, with a fractured look and sometimes a hexagonal shape. Large particles are often difficult to distinguish from hematite by PLM; by chemical microscopy the solubility of vermilion in hydriodic acid serves to differentiate it from hematite (Gettens, Feller, and Chase 1993).

Red Lead

Red lead, *entan* or *tan*, artificial lead tetroxide (Pb_3O_4) appears on the paintings as a bright orange-red and sometimes a dull orange. It is made in two ways: by heating lead white or by heating metallic lead in a two-step process, with the yellow compound litharge (PbO) being an intermediate product (FitzHugh 1986). This two-step process was used in Japan in the eighth century (Yamasaki 1959), and red lead has been in continuous use there since then.

In China, a fifth-century B.C.E. manuscript appears to describe the manufacture of red lead from the metal (Needham 1976). Red lead was used on wall paintings in western China as early as the fourth century C.E. (Xu, Zhou, and Li 1983).

By PLM red lead appears as tiny, irregularly rounded grains, red to orange-brown in transmitted light. The highly characteristic blue-green interference colors seen between crossed polars, as shown in plate 24, were considered sufficient for identification (FitzHugh 1986).

Organic Red

Organic red supplied a pink or purplish-pink color on these paintings. *Enji* is the common term used during the Edo period for an organic red pigment (Oguchi 1969). The organic reds that could have been available in Japan at this period are madder (Schweppe and Winter 1997), the two anthraquinone derivatives carmine and lac dye (Schweppe and Roosen-Runge 1986), and safflower red.

The organic red on these paintings shows by PLM as a red stain with no particulate structure. Limited chemical microscopy suggests that more than one pigment was in use. Work has begun in the Freer on the identification of these organic reds (Winter, Giaccai, and Leona this volume).

Red and Brown Earths

Red and brown earths, *shido, taisha,* and *bengara,* are related in composition. The coloring agents of both are various iron oxides, mixed with clay minerals, silicates, and other materials (Gettens and Stout 1966). The red mineral hematite, the anhydrous iron oxide, α-Fe_2O_3, supplies the color of red earths and is also found in brown earths. The hydrated iron oxides, generally termed limonite, include goethite, $HFeO_2$, which can be yellow to yellowish-brown, and lepidocrocite, γ-FeO(OH), which can be deep red to reddish-brown (Roberts, Rapp, and Weber 1974). Lepidocrocite may be a source of color in the brown earths.

Red and brown earth pigments were used early worldwide. Iron oxide minerals are found in Japan and China (Golas 1999). In Japan earth pigments were known from before the third century B.C.E. and have been used at all periods (Yamasaki and Emoto 1979).

By PLM hematite is a clear transparent red with high refractive indices, close to that of vermilion. Finely divided red earth occurs in a range of uniformly small particle size, as seen in plate 25. Sometimes it is so fine that it appears only as a rosy stain. Identification by PLM was supported by a positive test for iron by chemical microscopy. As noted above, large hematite particles can be distinguished from vermilion by chemical microscopy.

Yellow Earth

Yellow earth, *ōdo,* which is generally assumed to derive its principal color from goethite, was used rarely but consistently in Japan from the sixth century (Yamasaki and Emoto 1979). In China yellow earth has been found on wall paintings as early as the fifth century (Piqué 1997). It was found on only six of these paintings.

By PLM yellow earth is usually heterogeneous in composition, a mixture of silica with particles that, as observed on these paintings, can be yellow, brown, or orange red. A positive test for iron by chemical microscopy confirms the presence of an iron-containing pigment.

Organic Yellow

Organic yellow on these paintings may be gamboge, *tō-ō,* or *gamboji.* Gamboge is a gum resin obtained from trees of the genus *Garcinia* throughout south and southeast Asia (Winter 1997). It was said to be imported from China into Japan (Conder 1911). An organic yellow assumed to be gamboge was widely used in Japan from the eighth century on (Yamasaki and Emoto 1979).

By PLM the organic yellow on these paintings shows as a yellow stain. The formation of a deep orange color was noted with the application of 5% NaOH by chemical microscopy. This phenomenon is characteristic of gamboge (Winter 1997) and strongly suggests its possible presence on these paintings. In fact, four of these organic yellows were confirmed to be gamboge by thin-layer chromatography (Winter 1987). Gamboge has been identified on a late-seventeenth-century scroll by reflectance spectroscopy (Taguchi and Taguchi 1977).

Malachite

Malachite, *rokushō,* is the green basic copper carbonate mineral, $CuCO_3.Cu(OH)_2$, which was crushed and ground to produce the pigment. It has been used in Japan continuously from the seventh century on (Yamasaki and Emoto 1979). Several localities in Japan supply good-quality malachite (Geerts 1883).

Malachite is found widely in China (Golas 1999) and was used on third-century wall paintings at the Mogao Grottoes at Dunhuang (Xu, Zhou, and Li 1983). It is less widely used on paintings on paper and silk from China proper.

Artificial malachite, green verditer, has been reported on nineteenth-century Chinese paintings (Wise and Wise 1998) but is not known on Japanese paintings. It was mixed with barite in a repainted area on one painting in this series.

The appearance of malachite by PLM was generally considered sufficient for identification. The particles are pale green, almost colorless when small or thin, with squarelike fracture, sometimes fibrous, and highly birefringent with high refractive indices (Gettens and FitzHugh 1993b).

Azurite

Azurite, *gunjō,* the natural blue mineral, a basic copper carbonate, $2CuCO_3.Cu(OH)_2$, is ground to produce the pigment (Gettens and FitzHugh 1993a). Blue verditer, the artificial form, was not found here and has not been reported on East Asian paintings. The color of the mineral pigment can vary from dark to bright and is sometime a pale blue. It was used in Japan from the seventh century on (Yamasaki and Emoto 1979).

Azurite occurs with malachite—but less abundantly—in Japan and China (Geerts 1878). It was known to the Chinese by the late fourth century B.C.E. (Needham 1976) and was identified on wall paintings at the Mogao Grottoes at that date (Xu, Zhou, and Li 1983).

The appearance of azurite by PLM was usually considered sufficient for identification. Large particles are highly birefringent, a deep blue with greenish undertone, while small or thin particles are pale blue, as seen in plate 26. Particles are irregular and fractured in appearance, sometimes with conchoidal fracture (Gettens and FitzHugh 1993a).

Indigo

Indigo, *ai,* varies in color on these paintings from bright to dull blue, or greenish-blue when thinly applied. Indigotin, $C_{16}H_{10}O_2N_2$, the blue constituent, comes from several species of plants. The most important ones are available in Asia, but the plant sources of the indigo for these paintings are unknown. Synthetic indigo was first available commercially in 1897 (Schweppe 1997), so it would not be expected here.

Indigo was known in Japan at least as early as the eighth century (Yamasaki and Emoto 1979). It has been identified by spectral reflectance on nineteenth-century Japanese prints (Feller, Curran, and Bailie 1984), and on a late-seventeenth-century scroll by reflectance spectroscopy (Taguchi and Taguchi 1977). Indigo was known in China at least as early as the third century B.C.E., and its cultivation, preparation, and use as a dye there was described in the fifth century (Sung 1966). It is the common blue on these paintings.

By PLM indigo appears as a blue stain, with tiny opaque particles sometimes visible. It is not distinguishable from Prussian blue by PLM, but it is possible to distinguish the two by chemical microscopy as discussed below. The characteristic behavior of indigo by chemical microscopy with alkali and acid was accepted as indication of its presence; it is unaffected by 5% NaOH, whereas the blue color is destroyed by dilute HNO_3, sometimes with the formation of yellow isatin (Schweppe 1997). UV-visible reflectance spectroscopy has been used to identify indigo and Prussian blue on these paintings (Leona and Winter 2001; Winter, Giaccai, and Leona this volume).

Organic Brown

Organic brown appears as brown or reddish-brown on these paintings. By PLM it shows as a nonparticulate brown, sometimes reddish, stain. Tests by chemical microscopy indicate that more than one material was in use. Identification of these complex browns is proceeding at the Freer Gallery (Winter, Giaccai, and Leona this volume).

Pigments Used Late in the Ukiyo-e Period

Six pigments were used for part of the ukiyo-e period, after the mid-seventeenth century, as summarized in table 2.

Table 2. *Pigments used in the later part of the ukiyo-e period*

Late seventeenth century on	Smalt
	Copper greens
Eighteenth century on	Orpiment
Late eighteenth century on	Prussian blue
	Lead white
Nineteenth century only	Ultramarine

Smalt

Smalt, *hana-konjō,* is an artificial pigment, a coarsely ground, potassium-rich glass, colored blue by the addition of cobalt oxide. On these paintings it is usually a gray-blue, or occasionally a bright blue, often with a granular, glittering appearance. Smalt was long thought to be a sixteenth-century European invention, but we now know of isolated occurrences earlier on the Asian mainland, in China on Tang-dynasty iron sculpture (Larson 1988), and on eleventh- to twelfth-century wall paintings in Central Asia (Mühlethaler and Thissen 1993). A single occurrence is known on a sixteenth- to early-seventeenth-century Japanese print (Magurn 1942). No occurrences of smalt earlier than the seventeenth century were found on East Asian paintings in the Freer Gallery, and it was not used before the late seventeenth century on these ukiyo-e paintings. An artificial bluish glass pigment colored with cobalt oxide was known in Japan in the early eighteenth century (Oguchi 1969), and its preparation was described in the late nineteenth century (Takamatsu 1878).

The appearance of smalt by PLM was considered sufficient for identification (Mühlethaler and Thissen 1993). Particles vary in color from pale blue, often purplish in tone, to a bright blue similar in color to azurite. They are isotropic and transparent, with a low refractive index, and show square and angular corners, or thin flat edges sometimes with conchoidal fracture. Tiny entrapped air bubbles are occasionally present.

Copper Greens

Copper greens, as viewed on the paintings, often look superficially like coarse granular malachite, sometimes more yellow in tone. They are found after the late seventeenth century on these paintings.

A number of copper greens were identified by XRD as atacamite and paratacamite, green dimorphic basic copper chlorides with the composition $Cu_2Cl(OH)_3$, which occur in nature as minerals and are also found as corrosion products on bronze and copper. The pigments consist of extremely finely divided material made up of rounded particles with every appearance of being artificially prepared, as seen in plate 27. Atacamite and other basic copper chloride are known as pigments in the West, although they may sometimes be alteration products (Scott 2000). Atacamite has been found on eighth-century Japanese objects in the Shōsōin (Naruse 1996), and it has been identified on a number of early Chinese wall paintings (Duang et al. 1987; Xu, Zhou, and Li 1983; Piqué 1997).

The other copper green pigments have yet to be characterized. They are complex mixtures, which include glassy particles, quartz, calcite, and starch particles as well as unusual forms of malachite and basic copper chlorides. In addition to the problems of identification, there is confusion in naming. The term *verdigris,* which is copper acetate, has been applied incorrectly to both malachite and emerald green (Oguchi 1969; Geerts 1883). Emerald green, a synthetic copper aceto-arsenite, appears here on a single nineteenth-century painting. It was invented in Germany in the first decade of the nineteenth century (Fiedler and Bayard 1997) and may have been brought to Japan by the Dutch.

Criteria for the presence of a copper green were the appearance by PLM of a green material that is not typical malachite, coupled with a positive test for copper by chemical microscopy or SEM-EDXA. Paratacamite or atacamite were identified by XRD and the presence of chloride ions determined by SEM-EDXA.

Orpiment

Orpiment, *sekiō,* is the yellow arsenic sulfide mineral, As_2S_3, which is ground to produce the pigment. Its micaceous character makes it difficult to grind finely and contributes to the glittery appearance on the painting. Artificial orpiment has the same chemical composition as the mineral. Realgar, *yuō,* a red to orange arsenic sulfide mineral, As_4S_4, occurs with orpiment. It had limited use as a pigment in the West and has been found on Central Asian wall paintings but not on paintings in Japan or China.

Orpiment is known from several locations in Japan (Palache, Berman, and Frondel 1944), but better sources of orpiment and realgar are available in China (Geerts 1878), including the world's largest realgar deposit (Silbergeld and McNair 1988). Both minerals were important in Chinese medicine and metallurgy as early as the fourth century B.C.E. (Needham 1974), although there was no direct mention at that date of their use as pigments. Orpiment has been identified on nineteenth-century Chinese paintings (Wise and Wise 1998).

Orpiment was referred to in Japanese texts as early as the tenth century (Schafer 1955). It is said to be a standard painting material in Japan (Oguchi 1969), but it is not often mentioned in lists of traditional Japanese pigments. One author says it was "rarely used by painters" (Anderson 1886). It was widely used on nineteenth-century works in this series and on twelve earlier paintings, none dating before the eighteenth century. It has not been found on other Japanese paintings in the Freer Gallery.

Artificial orpiment is made by the dry and wet processes (Wallert 1984). Wet-process orpiment, produced by precipitation methods from solution, has never been used as a pigment (FitzHugh 1997). Dry-process material is prepared by heating sulfur with an arsenic compound, usually arsenic dioxide, to produce the pigment by sublimation. In addition, the natural mineral can be sublimed to get a more uniform product (FitzHugh 1997), a method described in mid-nineteenth-century Japan (Takamatsu 1878). Evidence from an 1837 text suggests that dry and wet methods were learned from the Dutch (Kitano and Koezuka 2000).

By PLM orpiment particles are usually coarse and of limited transparency, with a foliated, micaceous structure. Especially characteristic are the distinctive anomalous green to blue polarization colors (FitzHugh 1997), which were generally considered sufficient for identification. The natural mineral can include large and small particles, while the dry-process material tends to be more uniform. The orpiment used here is almost certainly the natural mineral, as the particles show great variety in size.

Prussian Blue

Prussian blue, *bero, bero-ai, konjō, berensu* is an artificial pigment, a ferric ferrocyanide, with the basic formula $Fe_4(Fe(CN)_6)_3.xH_2O$, first made in Germany in 1704 (Berrie 1997). It was imported into Japan by Dutch and Chinese traders at least as early as 1763 and appears never to have been manufactured there (Smith in press). Prussian blue is used on Japanese prints (Feller, Curran, and Bailie 1984), but there are no other reports of its use in Japan or China.

Prussian blue was used extensively by nineteenth-century artists on these paintings. It occurs on eight earlier paintings. On three it was applied as repaint mixed with kaolin. The others date within the time frame suggested by Smith (in press).

Prussian blue is not distinguishable from indigo by PLM. It is so finely dispersed that it appears as a nonparticulate greenish-blue stain even at high magnifications. By chemical microscopy it is unaffected by dilute HNO_3, but the color is destroyed by 5% NaOH, sometimes turning a reddish-brown color due to the formation of ferric hydroxide. If enough pigment is present, the application of dilute HCl will cause re-formation of the blue. This characteristic chemical behavior provided a clear indication of the presence of Prussian blue rather than indigo, which has the opposite chemical behavior with acid and alkali. When an apparent mixture of the two was present, no firm identification of either pigment was possible. In a single case, identification was made by XRD. Recent studies at the Freer Gallery have expanded our knowledge of the use and identification of Prussian blue and indigo on these paintings (Leona and Winter 2001; Winter, Giaccai, and Leona this volume).

Lead White

Lead white, *empaku,* artificially made basic lead carbonate, $2PbCO_3.Pb(OH)_2$, was known by the seventh century in Japan (Yamasaki and Emoto 1979). In the "Dutch process," the common preparation method in the West, lead was exposed to acetic acid vapors from vinegar (Gettens, Kühn, and Chase 1993). This method was well established in Japan in the late nineteenth century (Takamatsu 1878).

The Dutch process was described in fifth-century B.C.E. China (Needham 1976). Lead white has generally been found in use persistently but not commonly on Chinese works. However, it was found to have been used extensively on fifth-century wall paintings at the Tiantishan Grottoes (Zhou, Zhang, and Cheng 1997).

Lead white has been found on a number of pre-ukiyo-e

Japanese paintings in the Freer Gallery (Winter 1981), but it is generally accepted that lead white was not used after the seventeenth century in Japan (Yamasaki and Emoto 1979). However, lead white was found on nine paintings in this study. Two artists did these paintings, and they date from the late eighteenth century on. Thus, lead white was evidently occasionally in use after the seventeenth century.

By PLM lead white appears as finely divided, rounded, birefringent particles of uniform size. Anomalous interference colors are present but are usually not visible because of the finely divided nature of the material (Gettens, Kühn, and Chase 1993).

Ultramarine

Ultramarine, *ruri,* in its natural form is the powdered blue mineral lazurite (lapis lazuli), a sulfur-containing sodium aluminum silicate. Artificial ultramarine was first made in France in 1828 (Plesters 1993); it was probably imported into Japan from Europe. It was found on six nineteenth-century paintings in this series. No ultramarine, either artificial or natural, has been found on other Japanese paintings in the Freer Gallery.

By PLM ultramarine has a low refractive index and is deep blue in transmitted light. Artificial ultramarine shows finely divided particles, uniform, somewhat rounded, and opaque, as seen in plate 28. Particles of the natural material are irregular and angular in shape, sometimes with calcite and silicate mineral impurities (Plesters 1993).

Conclusions

The dual purpose of this survey was to identify the pigments used by painters of the ukiyo-e school in Japan and to determine when they were in use in the period between the late sixteenth and the nineteenth centuries. The analytical methods have inherent limitations, but they served to make possible firm identification of a large majority of the pigments. The exact characterization of some of them, notably the organic pigments and the copper greens, remains to be accomplished. However, we now know which pigments were used throughout the period and which were not used early in the period, and by extension we know which were not used at all.

The nonuse of certain pigments early in the ukiyo-e period can generally be accounted for. Prussian blue and artificial ultramarine were invented at known dates. Smalt was evidently not available in Japan before the sixteenth century. The lead white occurrences are too few to warrant any conclusions. Orpiment was known early for alchemical and metallurgical purposes but does not appear to have been a standard pigment in Japan. Further research on the copper greens is necessary before any conclusions can be drawn about their dates of use.

This study, and further research arising from it, should assist scholars in the field of ukiyo-e painting with problems of attribution and dating. In addition, the accumulation of data supplies information for the history of technology in East Asia.

References

Anderson, W. 1886. *The pictorial arts of Japan.* London: Sampson, Low, Marston, Searle and Rivington.

Berrie, B. H. 1997. Prussian blue. In *Artists' pigments: A handbook of their history and characteristics,* vol. 3, ed. E. W. FitzHugh, 191–217. New York: Oxford University Press; Washington, D.C.: National Gallery of Art.

Conder, J. 1911. *Paintings and studies of Kawanabe Kyōsai.* Tokyo: Maruzen.

Duang, S., J. Miyata, N. Kumagai, and R. Sugishita. 1987. Analysis of pigments and plasters from wall paintings of Buddhist temples in northwest China. *Kobunkazai no Kagaku* 32:13–20.

Feller, R. L., and M. Bayard. 1986. Terminology and procedures used in the systematic examination of pigment particles with the polarizing microscope. In *Artists' pigments: A handbook of their history and characteristics,* vol. 1, ed. R. L. Feller, 295–98. New York: Oxford University Press; Washington, D.C.: National Gallery of Art.

Feller, R. L., M. Curran, and C. Bailie. 1984. Identification of traditional organic colorants employed in Japanese prints and determination of their rates of fading. In *Japanese woodblock prints: A catalogue of the Mary A. Ainsworth Collection,* ed. R. S. Keyes, 253–66. Oberlin, Ohio: Allen Memorial Art Museum, Oberlin College.

Fiedler, I., and M. Bayard. 1997. Emerald green and Scheele's green. In *Artists' pigments: A handbook of their history and characteristics,* vol. 3, ed. E. W. FitzHugh, 219–71. New York: Oxford University Press; Washington, D.C.: National Gallery of Art.

FitzHugh, E. W. 1979. A pigment census of ukiyo-e paintings in the Freer Gallery of Art. *Ars Orientalis* 11:27–38.

——. 1986. Red lead and minium. In *Artists' pigments: A handbook of their history and characteristics,* vol.1, ed. R. L. Feller, 109–39. New York: Oxford University Press; Washington, D.C.: National Gallery of Art.

——. 1997. Orpiment and realgar. In *Artists' pigments: A handbook of their history and characteristics,* vol. 3, ed. E. W. FitzHugh, 47–79. New York: Oxford University Press; Washington, D.C.: National Gallery of Art.

Geerts, A. J. C. 1878. *Les produits de la nature Japonaise et Chinoise (Partie inorganique et minéralogique).* Vol. 1. Yokohama: C. Lévey.

——. 1883. *Les produits de la nature Japonaise et Chinoise (Partie inorganique et minéralogique).* Vol. 2. Yokohama: C. Lévey and S. Salabelle.

Gettens, R. J. 1976. Japanese paintings: Technical studies and conservation at the Freer Gallery of Art. In *Conservation and restoration of pictorial art,* ed. N. Brommelle and P. Smith, 241–52. London: Butterworths.

Gettens, R. J., and E. W. FitzHugh. 1993a. Azurite and blue verditer. In *Artists' pigments: A handbook of their history and characteristics,* vol. 2, ed. A. Roy, 23–35. New York: Oxford University Press; Washington, D.C.: National Gallery of Art.

——. 1993b. Malachite and green verditer. In *Artists' pigments: A handbook of their history and characteristics,* vol. 2, ed. A. Roy, 183–202. New York: Oxford University Press; Washington, D.C.: National Gallery of Art.

Gettens, R. J., and G. L. Stout. 1966. *Painting materials.* New York: Dover.

Gettens, R. J., R. L. Feller, and W. T. Chase. 1993. Vermilion and cinnabar. In *Artists' pigments: A handbook of their history and characteristics,* vol. 2, ed. A. Roy, 159–82. New York: Oxford University Press; Washington, D.C.: National Gallery of Art.

Gettens, R. J., E. W. FitzHugh, and R. L. Feller. 1993. Calcium carbonate whites. In *Artists' pigments: A handbook of their history and characteristics,* vol. 2, ed. A. Roy, 203–26. New

York: Oxford University Press; Washington, D.C.: National Gallery of Art.

Gettens, R. J., H. Kühn, and W. T. Chase. 1993. Lead white. In *Artists' pigments: A handbook of their history and characteristics,* vol. 2, ed. A. Roy, 67–81. New York: Oxford University Press; Washington, D.C.: National Gallery of Art.

Golas, P. J. 1999. *Science and civilization in China.* Vol. 5: Chemistry and chemical technology. Part 13: Mining. Cambridge: Cambridge University Press.

Kitano, N. and T. Koezuka. 2000. Fundamental research on *sekiō* (orpiment) pigment contained in excavated *urushi* objects of the Edo period (I). *Bunkazai Hozon Shūfuku Gakkai shi* 44:70–79.

Larson, J. 1988. A treatment and examination of polychrome Chinese sculpture at the Victoria and Albert Museum. In *The conservation of Far Eastern art,* ed. J. S. Mills, P. Smith, and K. Yamasaki, 120–25. London: International Institute for Conservation of Historic and Artistic Works.

Leona, M., and J. Winter. 2001. Fiber optics reflectance spectroscopy: A unique tool for the investigation of Japanese paintings. *Studies in conservation* 46:153–62.

Magurn, B. 1942. Daiitoku Myō-ō: A Japanese Buddhist deity. *Bulletin of the Fogg Museum of Art* 10(1):14–23.

McCrone, W. C. 1982. The microscopical identification of artists' pigments. *Journal of IIC-Canadian group* 7, nos. 1 and 2:11–34.

Mühlethaler, B., and J. Thissen. 1993. Smalt. In *Artists' pigments: A handbook of their history and characteristics,* vol. 2, ed. A. Roy, 113–30. New York: Oxford University Press; Washington, D.C.: National Gallery of Art.

Naruse, M. 1996. Investigation of inorganic pigments used on the Shōsōin objects by x-ray analyses. In *International symposium on the conservation and restoration of cultural property: Spectrometric examination in conservation,* ed. Chie Sano, 168–76. Tokyo: Tokyo National Research Institute for Cultural Properties.

Needham, J. 1974. *Science and civilisation in China.* Vol. 5: Chemistry and chemical technology. Part 2: Spagyrical discovery and invention: Magisteries of gold and immortality. Cambridge: Cambridge University Press.

——. 1976. *Science and civilisation in China.* Vol. 5: Chemistry and chemical technology. Part 3: Spagyrical discovery and invention: Historical survey from cinnabar elixirs to synthetic insulin. Cambridge: Cambridge University Press.

Oguchi, H. 1969. A scientific study on the coloring materials in Japanese paintings. *Bulletin of the Faculty of Fine Arts (Tokyo)* 5:27–82.

Palache, C., H. Berman, and C. Frondel. 1944. *The system of mineralogy of James Dwight Dana and Edward Salisbury Dana.* 7th ed. Vol. 1. New York: John Wiley and Sons.

Piqué, F. 1997. Scientific examination of the sculptural polychromy of cave 6 at Yungang. In *Conservation of ancient sites on the Silk Road,* ed. N. Agnew, 348–61. Los Angeles: Getty Conservation Institute.

Plesters, J. 1956. Cross-sections and chemical analysis of paint samples. *Studies in conservation* 2:110–57.

——. 1993. Ultramarine blue, natural and artificial. In *Artists' pigments: A handbook of their history and characteristics,* vol. 2, ed. A. Roy, 37–65. New York: Oxford University Press; Washington, D.C.: National Gallery of Art.

Roberts, W. L., G. R. Rapp Jr., and J. Weber. 1974. *Encyclopedia of minerals.* New York: Van Nostrand Rheinhold.

Schafer, E. H. 1955. Orpiment and realgar in Chinese technology and tradition. *Journal of the American Oriental Society* 75(2):73–89.

Schweppe, H. 1997. Indigo and woad. In *Artists' pigments: A handbook of their history and characteristics,* vol. 3, ed. E. W. FitzHugh, 81–107. New York: Oxford University Press; Washington, D.C.: National Gallery of Art.

Schweppe, H., and H. Roosen-Runge. 1986. Carmine: Cochineal carmine and kermes carmine. In *Artists' pigments: A handbook of their history and characteristics,* vol. 1, ed. R. L. Feller, 255–83. New York: Oxford University Press; Washington, D.C.: National Gallery of Art.

Schweppe, H., and J. Winter. 1997. Madder and alizarin. In *Artists' pigments: A handbook of their history and characteristics,* vol. 3, ed. E. W. FitzHugh, 109–42. New York: Oxford University Press; Washington, D.C.: National Gallery of Art.

Scott, D. A. 2000. A review of copper chlorides and related salts in bronze corrosion and as painting pigments. *Studies in conservation* 45:39–53.

Silbergeld, J., and A. McNair. 1988. *Chinese painting colors, studies of their preparation and application in traditional and modern times.* Trans. Yü Feian, *Chung-kuo Hua Yen-se Ti Yen-chiu.* Hong Kong: Hong Kong University Press; Seattle and London: University of Washington Press.

Smith, H. D., II. In press. Hokusai and the blue revolution in Edo prints. In *Hokusai and his age,* ed. G. C. Calza and J. T. Carpenter. Venice: International Hokusai Research Centre, University of Venice.

Sung, Y. 1966. *Chinese technology in the seventeenth century. T'ien-Kung K'ai-wu,* trans. E. Z. Sun and S. Sun. University Park and London: The Pennsylvania State University Press.

Taguchi, E., and M. Taguchi. 1977. Spectrophotometric identification of color materials, especially of organic pigments and combined mixed colors, with reference to the pigments used in the scroll "Genji-monogatari 54-jo emaki" of the Edo period. *Kobunkazai no Kagaku* 20–21:18–37.

Takamatsu, T. 1878. *On Japanese pigments.* Tokyo: Department of Science in Tokio Daigaku.

Wallert, A. 1984. Orpiment and realgar. *Maltechnik-Restauro* 90(4):45–57.

Wang, J., and J. Wang. 1999. Application of vermilion pigments in ancient China. *Wenwu Baohu Yu Kaogu Kexue* 11(1):39–44.

Winter, J. 1981. "Lead white" in Japanese paintings. *Studies in conservation* 26:89–101.

——. 1987. Organic colors in East Asian paintings: The identification of gamboge. In *ICOM Committee for Conservation, eighth triennial meeting,* ed. K. Grimstead, 111–17. Marina de Rey, California: Getty Conservation Institute.

——. 1997. Gamboge. In *Artists' pigments: A handbook of their history and characteristics,* vol. 3, ed. E. W. FitzHugh, 143–55. New York: Oxford University Press; Washington, D.C.: National Gallery of Art.

Winter, J., J. Giaccai, and M. Leona. East Asian painting pigments: Recent progress and remaining problems. This volume, 157–63.

Wise, D., and A. Wise. 1998. Observations on nineteenth-century Chinese pigments with special reference to copper greens. In *IPC conference papers London 1997,* ed. J. Eagan, 125–36. Leigh, Worcester, UK: Institute of Paper Conservation.

Xu, W., G. Zhou, and Y. Li. 1983. X-ray analysis of the inorganic Dunhuang pigments. *Dunhuang Yanjiu* 1:187–97.

Yamasaki, K. 1959. Chemical studies on the eighth-century red lead preserved in the Shōsōin at Nara. *Studies in conservation* 4:1–4.

Yamasaki, K., and Y. Emoto. 1979. Pigments used on Japanese paintings from the protohistoric period through the seventeenth century. *Ars Orientalis* 11:1–14.

Zhou, G., J. Zhang and H. Cheng. 1997. Pigment analysis of polychrome statuary and wall paintings of the Tiantishan Grottoes. In *Conservation of ancient sites on the Silk Road,* ed. N. Agnew, 362–68. Los Angeles: Getty Conservation Institute.

18 East Asian Painting Pigments: Recent Progress and Remaining Problems

John Winter, Jennifer Giaccai, and Marco Leona

ABSTRACT Recent research and many of the remaining problems in East Asian pigment history concern organic colorants. We have recently completed a project on identifying indigo and Prussian blue in Edo-period Japanese paintings, using primarily fiber-optics reflectance spectroscopy (FORS). The results and major conclusions are summarized. Organic red pigments offer a number of possibilities, but the use of insect dyes, such as lac dye and cochineal, has clearly been important, and the progress of introduction of cochineal, a New World product, is of historical interest. Our approaches to identifying these and other red colorants include FORS, UV-visible fluorescence spectroscopy, and high-performance liquid chromatography. A rather puzzling organic brown pigment found on Edo-period Japanese paintings has been studied primarily by infrared spectroscopy but also by other methods, including x-ray photoelectron spectroscopy. The remaining possibilities for the origin of this pigment are summarized.

Introduction

Twenty years ago, East Asian pigment history was thought to be fairly well understood. It consisted largely of Japanese results, the basis having been laid by Yamasaki and his coworkers (Yamasaki and Emoto 1979), and by Gettens and FitzHugh working on the Freer Gallery's ukiyo-e school paintings (FitzHugh 1979). The picture presented was of many of the usual inorganic pigments, often minerals, plus some organic materials like gamboge, indigo, and an unknown red colorant. As for Korea and China, it seemed to be assumed that if anyone should go to the trouble of identifying pigments from their paintings, the situation would look much like Japan.

Since that time, there have been one or two surprises. Here in the Freer Gallery, FitzHugh and Zycherman (1983, 1992) discovered two early and hitherto unsuspected Chinese pigments based on barium copper silicates, naming them Han blue and Han purple, and it was also shown in the Freer laboratory that lead white of earlier periods in Japan was basic lead chloride far more often than it was the conventional basic carbonate (Winter 1981; Winter and Joel 1988; Naruse 1996). We have also been seeing a kind of gradual encroachment of basic copper chlorides on the malachite hegemony in mineral greens. Chinese and Korean results are no longer hypothetical but have been actually appearing (Xu, Zhou, and Li 1983; Winter 1989).

Questions still remain. Not surprisingly, many are in areas where unambiguous identifications are technically difficult, and many of these involve organic colorants. The latter pose marked challenges: since most of them are strong absorbers of light, the superficial density on a painting is nearly always low; most analytical methods are subject to interferences from other organic components; they are frequently mixtures of compounds, the relative proportions of which may vary; and the colorants are liable to have undergone changes, such as fading. Apart from being a technical challenge, organic colorant usage implies a number of interesting historical questions. The present phase of our research includes attacking the "organic colorant problem."

Indigo and Prussian Blue

Indigo is one of the better-known dyestuffs, but it has also been used as a pigment in East Asia and indeed elsewhere (March 1935; Sze 1977, 584; Silbergeld and McNair 1988). It is obtained from a number of different plant species, the insoluble blue indigotin (fig. 1) being formed from the plant material by a process of fermentation and oxidation. Its use in East Asian paintings is well attested in the historical literature and also in Japan by studies of paintings and painted objects of as early as the eighth century. Most commercial indigo now is synthetic, but since synthetic manufacture only became possible late in the nineteenth century, this work is concerned with the natural product. This means that it may vary in purity and may have other colored compounds co-occurring, such as the red isomer indirubin.

Prussian blue is entirely synthetic and was first made in Berlin around 1704 (Berrie 1997). However, it is also

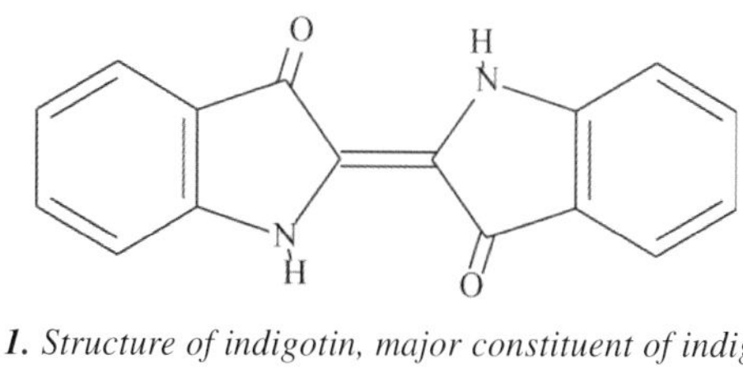

Figure 1. *Structure of indigotin, major constituent of indigo*

subject to some variability arising from differences in manufacturing methods. Two typical formulae are $KFe[Fe(CN)_6]$ and $Fe_4[Fe(CN)_6]_3$, but in the first other cations may replace potassium, and there is usually some degree of hydration of the molecule. These differences can be associated with differences of exact color and of working properties.

There has been some historical interest in the introduction of Prussian blue into Japan and its early use there (Smith 1998; Screech 2000, 203–4). There appear to have been very sporadic references to it during the eighteenth century and records of some importation from the 1780s. The general impression received is that it was known but not widely used at that time. During the first half of the nineteenth century, there were varying degrees of competition between Dutch and Chinese importers. This resulted in generally declining prices and increasing quantities imported from around the 1820s. The increasing availability of Prussian blue seen from trading documents has been associated with the increasing use of a dark blue on Japanese woodblock prints after about 1830—the dark blue in question has usually been assumed to be Prussian blue (Smith 1998).

This was the background to the question of indigo and Prussian blue occurrences on paintings of the eighteenth and nineteenth centuries. They are both dark blue, finely divided pigments and a priori their use for similar design elements might be expected. They are not identical in color, Prussian blue being usually brighter and of a higher chroma, but both can vary in color for the reasons mentioned. It appeared to be of interest to see when Prussian blue began to appear and whether it would be preferred by a painter when both were available.

However, distinguishing the two in a painting has frequently not been easy. It is well known that optical microscopy is unreliable for this purpose and other methods are frequently frustrated by the low superficial density in either case (both pigments have a high tinting strength). Our main method was FORS in the visible and near-infrared region. The method is fully described in a recent paper (Leona and Winter 2001). It provides a diffuse reflectance

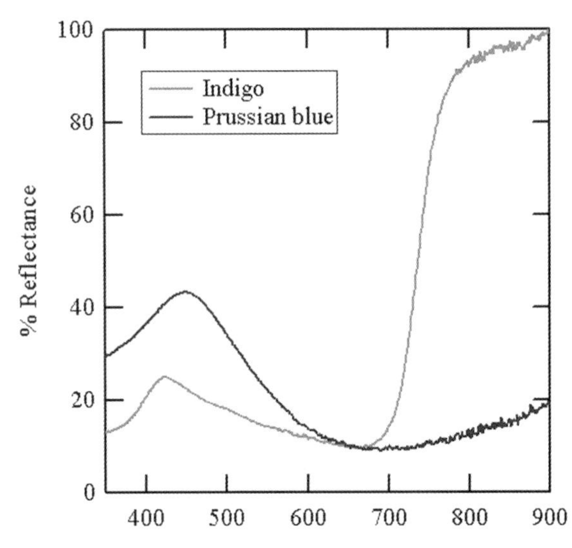

Figure 2. *Reflectance spectra for indigo and Prussian blue*

Table 1. *Overall statistics, Indigo and Prussian blue, 130 paintings*

One pigment only as blue		96
Indigo only as blue[1]	85	
Prussian blue only as blue[2]	11	
Both pigments used as blue		34
Used in separate design elements	27	
Used as mixtures	13	
Both situations on same painting	6	

1. Eighteen paintings had indigo also in a green mixture
2. Three paintings had indigo in a green mixture

spectrum from the near-infrared into the ultraviolet regions from a spot on the painting about 0.7 mm in diameter. Curves obtained for indigo and Prussian blue are shown in figure 2. A characteristic feature of the indigo spectrum is the unusually low absorption in the near-infrared region, which not only allows the ready identification of this pigment but was also useful in imaging work. Mixtures of these pigments with each other and with other colorants are identifiable, if necessary, using Kubelka-Munk transforms to match the spectra against references.

Other methods used were x-ray fluorescence spectroscopy (directly on the painting surface) to check for the presence of iron in Prussian blue areas, and, where sample sizes permitted, infrared spectroscopy using an FTIR microscope.

Altogether, 139 paintings of the eighteenth to nineteenth century proved to have either indigo or Prussian blue (or both). Nine of these had the blue only in a mixture with another color (typically a yellow), leaving the 130 paintings with blue features in indigo, Prussian blue, or a mixture of the two (table 1).

Many of the paintings examined have not been firmly dated, but it is clear that the majority are from the nineteenth century. It was possible to select thirty-four paintings that are confidently attributed to some time after about 1830, at which point Prussian blue was available at a relatively low price, as noted above. Prussian blue was identified on eighteen and indigo on thirty-one of these works; Prussian blue appears as the *sole* blue on three and indigo as the *sole* blue on sixteen of them. It seems clear that indigo continued to be used in paintings far more often than Prussian blue, notwithstanding the availability of the latter and its claimed importance in color prints.

The occurrence of mixtures of the two pigments may arise from a fine adjustment of the blue color by at least some painters but may also represent adulteration of the imported Prussian blue by the (presumably cheaper) domestic indigo.

Some specific instances are interesting. The earliest painting (by its current attribution) on which Prussian blue, in addition to indigo, was found was a landscape by Ike no Taiga (Freer accession number F1964.11). This artist's life dates were 1723–1776, and stylistic arguments place the painting probably in the 1760s. However, the upper part of this work has major repairs, leading to the possibility that the Prussian blue, present on the lower part, is a later addition. Another case of both pigments was on a painting from Nagasaki by the painter Nanreisai, dated 1817 (Leonhart collection, on loan). Since imports of this kind came in through Nagasaki and since the date is at the very beginning of serious imports and its use as an artists' material, the finding is fairly credible.

One or two paintings are interesting in that indigo and Prussian blue were used for different design features, clearly as a deliberate choice. An album leaf by Hokuga (nineteenth century) seen in plate 29 depicts a piece of blue and white porcelain, seen in close-up in plate 30. Both indigo and Prussian blue have been used and were readily distinguished by the FORS technique. They can also be separated by infrared imaging. Since indigo has (for a blue pigment) unusually low absorption in the infrared (fig. 2), it has low density in an infrared image, so that indigo-painted areas are largely dropped out, as seen in figure 3. Digital subtraction of this image from a black and white visible-light image taken to register with it removes the Prussian blue and leaves the indigo, as shown in figure 4. Thus, figure 3 shows those features painted in Prussian blue, and figure 4 shows those features painted in indigo.

A rather similar case is a painting by Katsushika Hokusai (1760–1849), *The Lute and White Snake of Benten* (plate 31). A close-up of the mouth of the bag containing Benten's lute (plate 32) is seen to have two shades of blue, a darker shade overpainting a lighter. It was demonstrated spectroscopically that the lighter, duller blue was indigo and that Prussian blue was present in the darker area. An infrared image (fig. 5) confirmed this, as the indigo-painted areas have a low density.

Apart from the use of these pigments for blue design features, sixty paintings had one or the other mixed with either yellow for a green color, or with red for a purple, as seen in table 2. The relatively large number of cases where blue-yellow mixtures have been used for a green color is interesting. Apparently, indigo was much preferred in such applications, especially with red. No mixtures of Prussian blue with a red have so far been encountered.

***Table 2.** Greens and purples*

Mixtures with other colors, total		60
Indigo + yellow	48	
Prussian blue + yellow	4	
Indigo + Prussian blue + yellow	1	
Indigo + red	13	

Organic Red Pigments

Organic red colors have long been noted on paintings and cited in historical documents from Japan and China (see, for example, Sze 1977; Silbergeld and McNair 1988, and references below). Such colors are usually known collectively as *enji* in Japanese and *yan zhi* in Chinese. Dictionaries often translate this as *cochineal,* but in fact it is a rather vague term signifying a darkish red color often used in cosmetics.

Organic red colors have been recorded a number of times on Chinese and Japanese paintings. Two early examples, of the Tang period, are from the northwestern desert areas of China (Oguchi and Hayashi 1981; Xu, Zhou, and Li 1983). In Japan, examples are paintings in the eighth-century Shōsōin repository (Yamasaki 1968), a twelfth-century handscroll of *The Tale of Genji* (Yamasaki and Nakayama 1954), a hand-colored print of the sixteenth to seventeenth

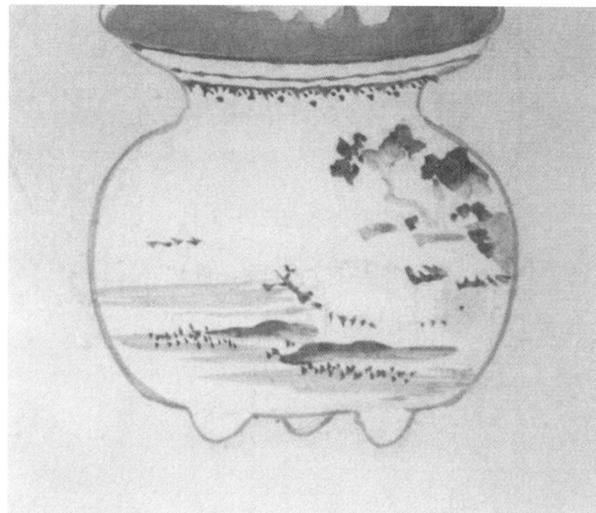

***Figure 3.** Infrared image of detail in plate 30; indigo suppressed*

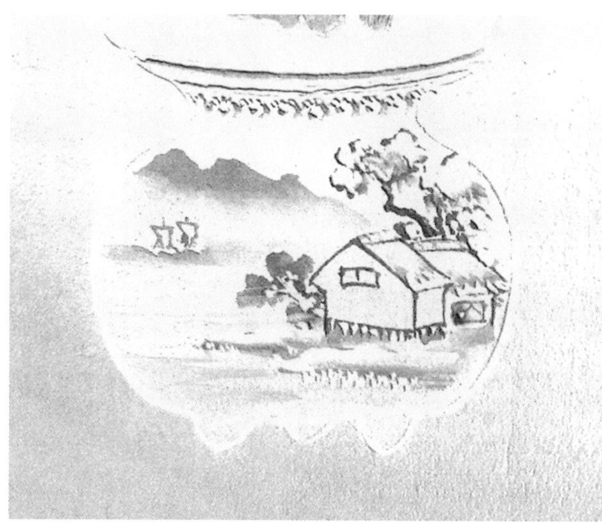

***Figure 4.** Subtraction image of detail in plate 30; Prussian blue areas suppressed*

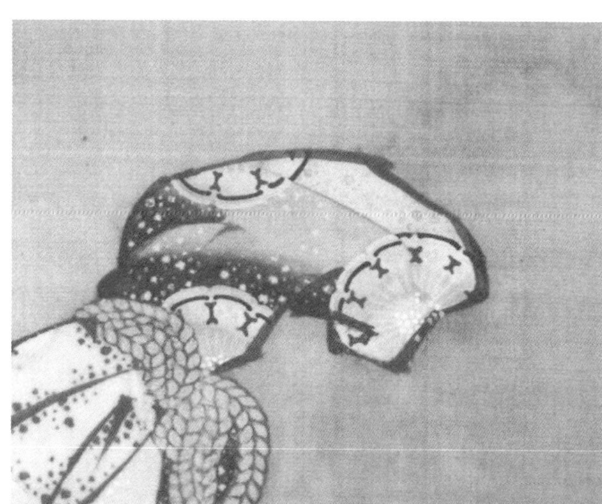

***Figure 5.** Infrared image of plate 32; indigo areas suppressed*

century (Magurn 1942), and a seventeenth-century Kano school handscroll (Taguchi and Taguchi 1977). Elisabeth FitzHugh in her work on the Freer ukiyo-e paintings used optical and chemical microscopy to show the presence of an organic red on around 120 such paintings. The color of the materials used was not altogether constant, more than one shade of red being found.

When it comes to what red organic materials might have been used, there are a number of possibilities. The two most often cited by Japanese authors are lac dye and cochineal. Both are products of scale insects, and both have a long history as dyestuffs. The colored constituents are heavily substituted anthraquinones typified by the structures shown in figure 6. Carminic acid is the major constituent of cochineal. Lac dye has been found to be a mixture of up to five laccaic acids and other compounds, with laccaic acid A, shown here, usually being the major one. The heavy substitution patterns on the anthraquinone nucleus are typical of scale insect products. The other material secreted by the lac insect is, of course, the resin shellac, which became commercially important from the late nineteenth century onward.

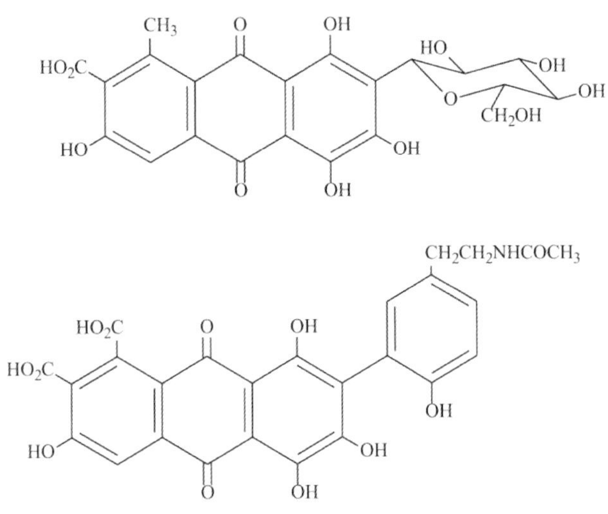

Figure 6. Structures of carminic acid and laccaic acid A

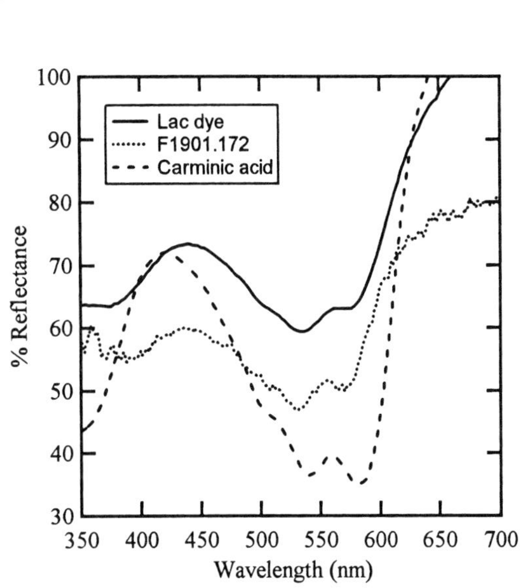

Figure 7. Reflectance spectra for lac dye and carminic acid

Cochineal is a New World product, the insect species occurring in Central and South America. It is generally believed that the earliest that such a product could be found in East Asia would be about the sixteenth century, when the Jesuits were active in that part of the world, although most references to it seem to be later. Lac certainly goes back much further. There is a specimen of stick lac in the eighth-century Shōsōin (Asahina 1955), and there are historical accounts of it in Tang-period China (Schafer 1957). The course of the introduction of cochineal into an application already served by the older lac dye may therefore be of interest in much the same way as with Prussian blue versus indigo.

There are many other possibilities, and in order to find some degree of focus amidst this plethora, we began with the insect-dye problem in Japanese paintings. This gave us a concrete historical question to start out with, and a suite of paintings that we knew had some version of organic red—the ukiyo-e works where this had been described by FitzHugh. The immediate questions were: were these organic reds in fact insect dyes, and, if so, can we identify them as either lac dye or cochineal? It is reasonable to assume that anything that is not a highly substituted anthraquinone characteristic of insect products or not an anthraquinone at all will rapidly become obvious from the spectroscopic properties. Thus far, a great deal of the time has been taken up with developing techniques.

The experimental protocol included checking for elements heavier than potassium with x-ray fluorescence spectroscopy in order to identify iron oxide-based pigments, which can also appear in thin red washes. That done, the same FORS technique used with the blues was applied. Typical reflectance spectra are shown in figure 7. These are references for cochineal and lac dye plus the curve obtained from a painting. It has become clear that reflectance spectroscopy will distinguish the heavily substituted insect anthraquinones from other classes of colorants, including plant anthraquinones, such as alizarin or purpurin, but will not distinguish lac dye from cochineal. Thus, the painting here has a detail in an organic red that may be characterized as an insect dye. The more specific question, however, remains unresolved.

We have experimented with UV-visible fluorescence spectroscopy, also using a fiber optic probe. This method has been applied by others, either for samples in solution (Wallert 1986, 1991) or using a similar probe to investigate colors in Japanese woodcut prints (Shimoyama, Noda, and Katsuhara 1998). Plate 33 shows three-dimensional fluorescence spectra for lac dye and for carminic acid as paint-outs on paper. Fluorescence spectroscopy is a complex technique; emission intensities, for example, are heavily influenced by the molecular environment. In particular, it has been found that fluorescence of these compounds is only stimulated to a usable degree when a metal ion able to form coordination complexes is present. In the present case the paper had been sized with an alum preparation, providing Al^{3+} ions for coordination. Under these conditions, lac dye and cochineal gave distinguishable fluorescence spectra that might, in principle, enable them to be identified.

In our hands the method suffers from a number of problems. The presence of back-scattered (nonfluorescent) radiation from the painted surface is a major nuisance, and the relative inefficiency of fiber optics leads to poor signal/noise ratios. It may be usable for highly fluorescent compounds, in which cases it may also be possible to work with very small samples in solution, which permits more experimental control.

More recently we have looked at chromatographic methods, in particular high-performance liquid chromatography (HPLC). Figure 8 shows the separation and relative retention times of lac dye constituents and carminic acid using HPLC with diode-array detection. Clearly, good separations are possible; the question therefore becomes whether detection limits are adequate for the very small samples available from paintings. With the diode array detector monitoring at 275 nm and specifying a signal/noise ratio of 5, we found limits of detection of 2.5 ng for carminic acid and 10 ng for lac dye. These are considered usable for heavy applications of paint, although probably not for light washes. Our current concerns are with improving the sensitivity of detection.

The FORS technique enabled red design colors to be identified as insect dyes on over fifty Edo-period Japanese paintings. Five were checked further by HPLC. Although samples from three of these were too small to yield valid results, the other two (see table 3) were shown to contain carminic acid, indicating the presence of cochineal. Figure 8 shows the HPLC scan from one of the paintings.

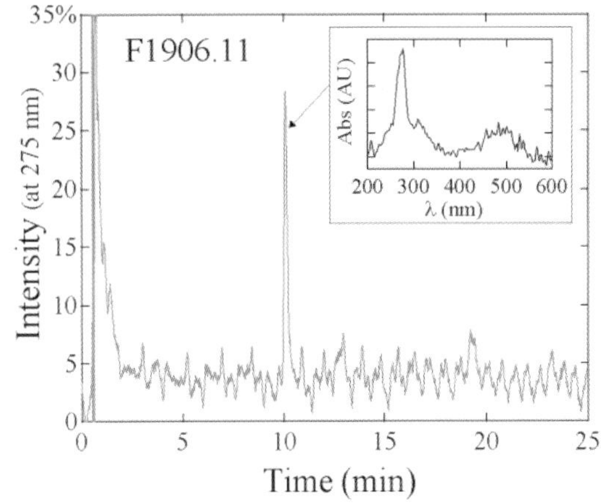

Figure 8. *HPLC scan of sample of red from* Yang Guifei *hanging scroll*

Table 3. *Paintings with identified cochineal*

Accession #	Artist	Description
F1906.11	Eishi (style of)	Hanging scroll on silk, 19th century, Yang Guifei
F1907.369	Hokusai (attrib.)	Hanging scroll on paper, 19th century, Ono no Komachi

Organic Brown

FitzHugh (1979) described an organic brown pigment on twenty Japanese ukiyo-e school paintings in the Freer collection; the total number is currently about twenty-five (FitzHugh pers. comm.). It was commonest from the mid-seventeenth to the mid-eighteenth century and appeared to have behavior similar to that of van Dyke brown (Feller and Johnston-Feller 1997). So far as is presently known, the material is not historically described. It is not clear whether it is to be found in East Asia other than on Japanese Edo-period paintings.

An identification of this pigment in terms of its source or detailed chemical constitution has so far proved elusive. The color shows some variation, some examples showing a purplish tint, but the evidence at the moment seems to support a basically similar origin for all with some variability that would be likely in a mixture of complex compounds. It is chemically rather intractable, and of course only microscopic samples are available from the paintings. Consequently, a number of approaches have proved fruitless, including gas chromatography-mass spectrometry with variations on pyrolytic introduction, UV-visible absorption spectroscopy, UV-visible fluorescence spectroscopy, and HPLC. The most usable evidence at present comes from the infrared absorption spectra. Figure 9 shows four examples; glue binder was removed by hydrolysis with 6M aqueous HCl at 90°C for a minimum of three hours.

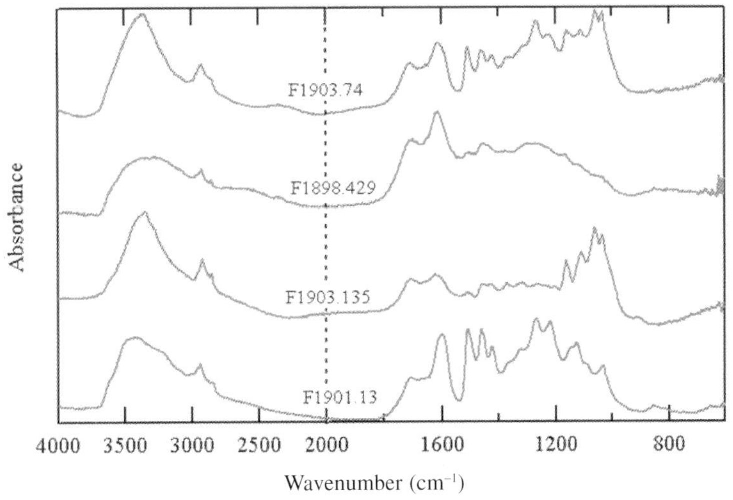

Figure 9. *Infrared spectra of acid-extracted samples of organic brown*

Table 4. XPS analyses; minor elements omitted

Sample	C (%)	N (%)	O (%)	Total
F1901.13, acid extracted[1]	67.76	(n.d.)	31.31	99.07
F1901.166, acid extracted[1]	77.05	2.23	18.53	97.81
F1901.166, water extracted[2]	75.31	2.24	21.41	98.96
F1903.63, acid extracted[1]	75.19	2.08	21.70	98.97
F1898.114, acid extracted[1]	65.89	2.06	28.09	96.04
Sepia, untreated[3]	67.31	8.35	23.35	99.01
Sepia melanin[3]	69.29	7.06	21.12	97.47
Van Dyke brown, untreated[3]	74.95	0.00	22.73	97.68
Van Dyke brown, acid extracted[1,3]	81.83	0.97	16.27	99.07
Animal glue	71.14	8.84	17.85	98.43

1. Sample treated with 6M aqueous HCL at 90°C for twenty-four hours
2. Sample treated three times with water at 90°C for two hours
3. Mean of duplicate measurements

n.d. = not detected

Numbers preceded by "F" are accession numbers for paintings in the Freer Gallery of Art collection

A hypothesis briefly pursued was that this pigment was related to sepia, the dark brown secretion in the ink sacs of certain cephalopods, which has been used as an ink and watercolor paint in the West. Despite some similarities in the infrared spectra, this idea was eventually dropped on the basis of results from x-ray photoelectron spectroscopy, which showed nitrogen levels consistently too low for a material of this type. Results for samples of four paintings are compared with those for melanins, van Dyke brown, and animal glue in table 4. In addition, the presence, as noted below, of woody cellular fragments is inconsistent with this hypothesis.

In any event, the best fit of the infrared spectra has been with combinations of humic acids and cellulosic materials. Absorption maxima from 1200 to 900 cm^{-1} can be attributed to cellulosic materials, while those from 1700 to 1200 cm^{-1} are attributed to humic acids (see fig. 9). Microscopic fragments having the appearance of woody cellular material have been noted in several samples. The following possibilities appear to be available, although none can be said to be definitively proved.

(1) One or more naturally occurring humic deposits analogous to van Dyke brown or Cassel earth form the source. There would have to be some color variation, and at least some of the deposits would have to have the remains of a cellulosic and probably woody origin.
(2) Partly decomposed woody material has been used, available either coincidentally or by deliberate preparation.
(3) A brown dyewood (several are available) has been extracted to yield the desired pigment. A vigorous hot aqueous extraction could well bring in minute cellular fragments, and the various cellulosic, hemicellulosic, tannin-like materials and lignins would be likely to have an infrared spectrum similar to the overall spectra actually observed.

The organic brown problem remains so far unresolved, although we have narrowed down the field of candidates. At this stage, it seems difficult to do more with examples of use on paintings, given the intractability of the material and the realities of sampling, although we continue to be alert to other possible methods. One may hope for more information from the historical record or perhaps for the discovery of a bulk sample that would permit more extensive investigation.

Conclusions and Further Progress

Work on the materials of East Asian paintings continues, concentrating now on those that are technically difficult to identify and hitherto intractable. We are pursuing especially cases that have some historical or conservation interest. The fields of yellow organics and of support dyes, in several colors, will be of interest. We expect the largely unstudied areas of Chinese and Korean paintings to feature more largely in future research.

Acknowledgments

This work was supported by a grant from the Andrew W. Mellon Foundation. Dr. Rene de la Rie and Dr. Suzanne Lomax of the Department of Conservation Science, National Gallery of Art, Washington, D.C., are thanked for making available to us the HP 1090 HPLC unit in that department and for assistance in its use. The XPS work was performed in the Surface Laboratory of the Bausch and Lomb Corporation, Rochester, N.Y. Dr. Dan Hook and his staff in that laboratory are thanked for performing the analyses. Dr. Richard W. Nowak of the Surface and Thin Films Laboratory, University of Buffalo, Buffalo, N.Y., is thanked for arranging for these analyses and for many valuable discussions on the results. Professor Henry Smith, Columbia University, is thanked for providing a copy of the English version of his paper in advance of publication (see Smith 1998).

Experimental

Reference Materials

Indigo: synthetic indigo, Fisher Chemical Co.

Prussian blue: ferric ferrocyanide, $Fe_4[Fe(CN)_6]_3$, Fisher Chemical Co.

Lac dye: Kremer Pigments, 228 Elizabeth Street, New York, NY 10012

Carminic acid: Sigma-Aldrich Company, Milwaukee, Wisc.

Reflectance Spectroscopy

This was performed using a Varian Associates Cary 50 UV-Visible Spectrophotometer fitted with a 3 m fiber optic probe by CIC Photonics, Albuquerque, N.M. The method is described in Leona and Winter (2001).

Fluorescence Spectroscopy

Performed using a Perkin Elmer LS50B Luminescence Spectrophotometer fitted with a similar fiber optic probe as above, also by CIC Photonics.

High-Performance Liquid Chromatography

Performed using an HP 1090M HPLC unit equipped with diode array detector, from the Hewlett Packard Corporation (now Agilent Technologies). Conditions used followed those of Halpine (1996).

X-Ray Photoelectron Spectroscopy (XPS)

Performed on a PHI Quantum 2000 Scanning ESCA microscope, from Physical Electronics, Eden Prairie, Minn.

References

Asahina, Y. 1955. *Shōsōin Yakubutsu (Shōsōin Medicinals).* Osaka: Shokubutsu Bunken Kankō-kai. 42 plates.

Berrie, B. 1997. Prussian blue. Chap. 7 in *Artists' pigments: A handbook of their history and characteristics,* vol. 3, ed. E. W. FitzHugh, 191–217. New York: Oxford University Press; Washington, D.C.: National Gallery of Art.

Feller, R. L., and R. M. Johnston-Feller. 1997. Vandyke brown, Cassel earth, Cologne earth. Chap. 6 in *Artists' pigments: A handbook of their history and characteristics,* vol. 3, ed. E. W. FitzHugh, 157–90. New York: Oxford University Press; Washington, D.C.: National Gallery of Art.

FitzHugh, E. W. 1979. A pigment census of Japanese ukiyo-e paintings in the Freer Gallery of Art. *Ars Orientalis* 11:27–38.

FitzHugh, E. W., and L. A. Zycherman. 1983. An early man-made blue pigment from China—barium copper silicate. *Studies in conservation* 28:15–23.

——. 1992. A purple barium copper silicate pigment from early China. *Studies in conservation* 37:145–54.

Leona, M., and J. Winter. 2001. Fiber optics reflectance spectroscopy: A unique tool for the investigation of Japanese paintings. *Studies in conservation* 46:153–62.

Magurn, B. W. 1942. Daiitoku Myō-ō, a Japanese Buddhist deity. *Bulletin of the Fogg Art Museum* 10:14–23.

March, B. 1935. *Some technical terms of Chinese painting.* Baltimore: Waverly Press. Reprinted, 1969, Paragon Book Corporation.

Naruse, M. 1996. Investigation of inorganic pigments used on the *Shōsōin* objects by x-ray analyses. In *International symposium on the conservation and restoration of cultural property – Spectrometric examination in conservation,* ed. C. Sano, 168–76. Tokyo: Tokyo National Research Institute of Cultural Properties.

Oguchi, H., and I. Hayashi. 1981. On the coloring materials and painting techniques of artifacts unearthed from the tomb of the T'ang dynasty at Turfan. *Tokyo Geijutsu Daigaku Bijutsugakubu Kiyō* 16:1–40.

Schafer, E. H. 1957. Rosewood, dragon's blood, and lac. *Journal of the American Oriental Society* 77:129–36.

Screech, T. 2000. *The shogun's painted culture: Fear and creativity in the Japanese states, 1760–1829.* London: Reaktion Books Ltd.

Shimoyama, S., Y. Noda, and S. Katsuhara. 1998. Non-destructive determination of colorants used for traditional Japanese ukiyo-e woodblock prints by the three-dimensional fluorescence spectrum using fibre optics. *Bunseki kagaku* 47:93–100.

Silbergeld, J., and A. McNair. 1988. *Chinese painting colors: Studies of their preparation and application in traditional and modern times.* Translation of Yu Feian, *Chung-kuo Hua Yen-se Ti Yen-chiu.* Hong Kong, Seattle and London: Hong Kong University Press and University of Washington Press.

Smith, H. 1998. Ukiyoe ni okeru "burū kakumei" [The "blue revolution" in ukiyoe]. *Ukiyoe geijutsu* 128:3–26.

Sze, Mai-mai. 1977. *The mustard seed garden manual of painting.* Trans. of *Chieh Tzu Yüan Hua Chuan,* Nanking, 1679–1701. Princeton: Princeton University Press. Reprint of *The mustard seed garden manual of painting,* 1956 and 1963, New York: Bollingen Foundation.

Taguchi, E., and M. Taguchi. 1977. Spectrophotometric identification of color materials, especially of organic pigments and combined mixed colors—with reference to the pigments used in the scroll "Genji monogatari 54-jo emaki" of Edo period. *Kobunkazai no kagaku* 20–21:18–37.

Wallert, A. 1986. Fluorescent assay of quinone, lichen and redwood dyestuffs. *Studies in conservation* 31:145–55.

——. 1991. "Cimatura de grana": Identification of natural organic colorants and binding media in mediaeval manuscript illumination. *Zeitscrift für kunsttechnologie und konservierung* 5:74–83.

Winter, J. 1981. "Lead white" in Japanese painting. *Studies in conservation* 26:89–101.

——. 1989. Identification of some early Korean pigments. *Misul charyo* 43:1–36.

Winter, J., and E. Joel. 1988. "Lead white" in Japanese paintings (II): Measurement of lead isotope ratios. *Kobunkazai no kagaku* 33:33–44.

Xu Weiye, Zhou Guoxia, and Li Yunhe. 1983. X-ray analysis of the inorganic Dunhuang pigments. *Dunhuang yanjiu* 1:187–97. (Also identified as issue 3.)

Yamasaki, K. 1968. Shōsōin kaiga no gihō to zaishitsu (Techniques and materials of the Shōsōin paintings). In *Shōsōin no Kaiga (Paintings in the Shōsōin),* 155–62. Tokyo: Nihon Keizai Shimbun. Compiled by the Office of the Shōsōin.

Yamasaki, K., and Y. Emoto. 1979. Pigments used in Japanese paintings from the protohistoric period through the nineteenth century. *Ars Orientalis* 11:1–14.

Yamasaki, K., and H. Nakayama. 1954. Studies of the pigments used in the Genji Monogatari scroll paintings. *Bijutsu Kenkyū* 174:229 34. (Eng. summary, p. 2.)

19 *Five Protective Goddesses (Pancaraksha)*: A Study of Color Notations and Pigments

Kate Duffy and Jacki Elgar

ABSTRACT The *Five Protective Goddesses (Pancaraksha)* housed at the Museum of Fine Arts, Boston, is one of several thangkas in the museum's collection of current interest to scholars of Tibetan art. It is believed to date from the mid-fifteenth to the sixteenth century, and its style is thought to be predominantly Nepalese, created by a Nepalese artist living and/or working in southern Tibet. Some art historians have stated that this thangka is an excellent example of the Nepalese-Tibetan style that was developed at the Ngor monastery in Tsang.

In keeping with the earlier studies of thangkas we have presented (Duffy and Elgar 1995, 1999), this thangka was initially examined by infrared reflectography in order to identify color notations. The master painter drew these notations on the ground layer as an instructional guide to himself and the apprentices in his studio. The color notations found are written in Tibetan *Ume* script and were documented throughout the painting. Pigment samples were taken corresponding to the color notations for analysis by Fourier transform infrared spectroscopy (FTIR), polarized light microscopy (PLM), and electron probe microanalysis (EPMA). The results are compared to previous studies with the ongoing goal of producing a body of scientific information that will help scholars to more accurately identify a location and date for a particular thangka.

Introduction

The dating of Tibetan thangkas is extremely difficult because the creators of these religious paintings did not routinely sign and/or date their work. Inscriptions that are found are often mantras or creeds, names of deities or monks, and identifications of scenes, such as scenes from the life of Buddha. Identifying a particular school or monastery from these inscriptions is usually futile. Images of historical figures, such as monks or prominent devotees, usually do not aid in dating because their images are often executed posthumously. Connoisseurship has predominantly been used to date thangkas on the basis of style, religious lineage, and comparison to known dated thangkas. It is because of this inexact process and a dearth of technical information on Tibetan thangkas in Western literature that this study and our previous studies were undertaken.

Five Protective Goddesses (Pancaraksha)

The *Five Protective Goddesses* (fig. 1), referred to throughout this paper as the *Five Goddesses*, is accepted as a meditational thangka depicting the Five Protective Goddesses and fifty-six other deities. In the lower left corner (fig. 2), a wealthy female practitioner is seated before an elaborate altar. She wears a green jacket over red robes and

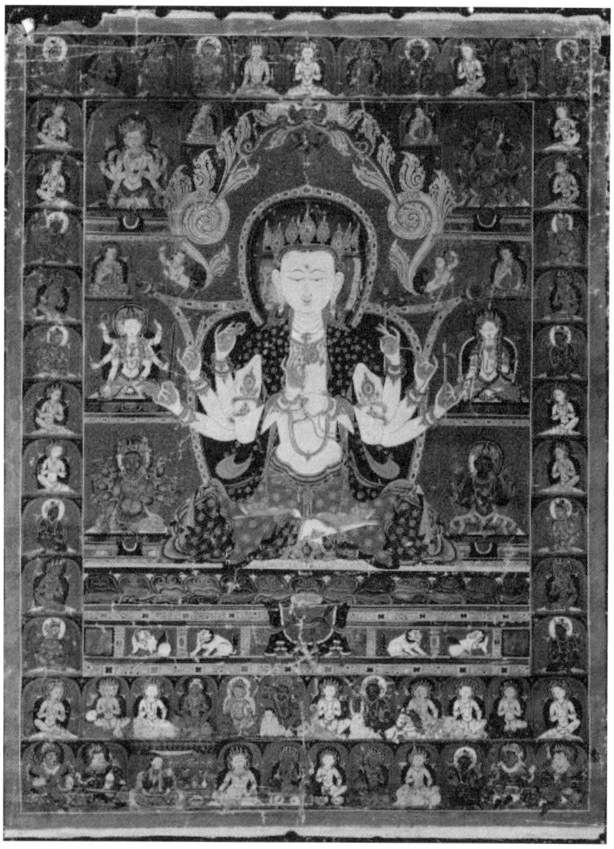

Figure 1. Five Protective Goddesses, *Museum of Fine Arts, Boston. Gift of John Goelet; 67.822*

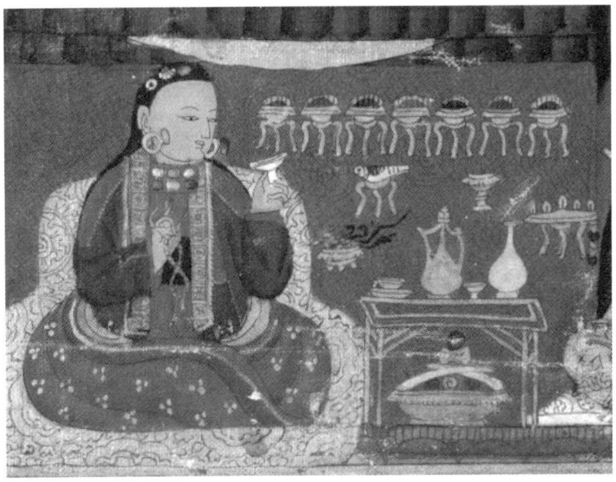

Figure 2. *Detail of female devotee from* Five Protective Goddesses

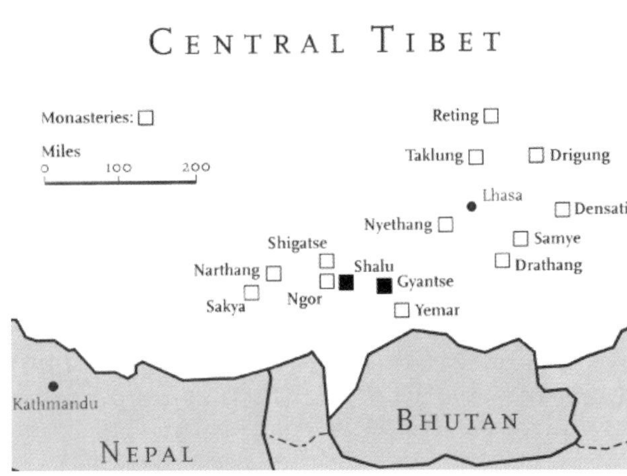

Figure 3. *Map of Central Tibet showing relevant monasteries*

is adorned with earrings and hair ornaments. The altar set with ritual implements is often associated with the painting's consecrating monk, who is usually depicted here and not a practitioner. The female figure may in fact be the patron or donor for this particular thangka and may hold a clue to its date.

The artist of the *Five Goddesses* is unknown, but the painting is presently believed to be from Tibet with an uncertain lineage dating from 1400 to 1499. Earlier scholars have dated the *Five Goddesses* from 1450 to as late as 1699. The style has been and is still thought to be predominantly Nepalese, possibly created by a Nepalese artist living and/or working in southern Central Tibet. The painting was widely accepted as an excellent example of the Nepalese-Tibetan style that was developed at the Ngor monastery in Tsang (southern Central Tibet). The Ngor monastery, an important center for the Buddhist Sakya order, was founded in 1429 by Ngor-chen Kun-dga'-bzang-po (1382–1456), who intensively patronized Nepalese painters (Jackson 1996, 77). In the mid- to late twentieth century, the Ngor style was indiscriminately used to describe all Sakya-pa thangkas, many of which were commissioned elsewhere in Tibet (Jackson 1996, 82). The *Five Goddesses* is no longer given the blanket term Ngor style; although the Nepalese influences are prominent, they are not the only elements currently under scholarly discussion. As stated above, in the lower left corner of the painting there is a female devotee whose clothing and accessories identify her as Tibetan, substantiating the belief that the painting was created for a Tibetan client.

Scholars' Current Observations

The study of Tibetan murals and their influence on thangkas may prove helpful in dating these anonymous paintings. Monasteries such as Shalu (another important Sakya center in southern central Tibet) and Gyantse, whose construction and decorative murals are well documented, provide helpful information. Murals from Shalu and Gyantse (fig. 3) are studied today for their high aesthetic quality and their art-historical importance as the beginning of the Tibetan national style (Jackson 1996, 82).

Scholars have been looking closely at changes in the depiction of garments or costumes as an aid in dating thangkas from Central Tibet in the late thirteenth through the fifteenth century. Kossak (2000) recently wrote about the *Five Goddesses* and compared the drapery of the central figure, the white goddess Pratisara, with the drapery of a Shakyamuni Buddha in a mural painting dated to after 1306 at Shalu monastery. Possibly the Shakyamuni and other murals at Shalu served as models for thangkas. Kossak stated that the garment worn by the Shalu Shakyamuni is traditional, but the ways in which it is worn and the material from which it is made are unprecedented in Tibetan art. He then stated that the garment worn by the central deity in the *Five Goddesses* is a later version of the Shalu garment. He continues, "[T]he main deity wears a floral shawl that covers both shoulders, forms free floating arcs below [the] arms, and is attached to [the] overskirt at [the] hips. The mannered treatment of the shawl with its gyrating curves and sweeping scallops ... is a motif that does not become prominent in Tibet until the mid-fifteenth century."

Kossak (2000) also notes that by the early fourteenth century, garments depicting Chinese brocades had replaced the traditional Indian fabric motifs in many thangkas. The central deity of the *Five Goddesses* wears such brocades, suggesting that the thangka cannot predate the fourteenth century and is feasibly dated to the fifteenth century.

In order to understand the significance of the style depicted in the Shalu murals, some background information about their construction and the political atmosphere at the time of their creation must be explained. The construction of Shalu is well documented, and the building phase that is relevant to the *Five Goddesses* occurred during the period 1290–1335. The local prince of Shalu, Drakpa Gyeltshen, traveled to the Yuan Court in 1306 to be officially declared ruler of Shalu. He returned in 1307 with a number of artists trained at the Yuan Court under the tutelage of Aniko, a renowned Nepalese artist who flourished in China until his death in 1306. Many of the murals at Shalu exhibit what some scholars have termed the Nepalese-Yuan style. Others call this style a Nepalese variant with Bengali and Central

Asian elements (Kossak and Singer 1998). The prevalent color scheme is more subdued and subtle than the colors found in what is accepted as a purely Nepalese style. The Nepalese-Yuan styled murals are believed to be the results of this entourage of artists brought to Shalu from China. The group most likely consisted of Tibetan, Chinese, and Nepalese artists. However, the name of only one remains at the site: a Tibetan artist, Chimpa Sonambum (mchims.pa bsod nam 'bum) (Kreijger 1997).

Although the murals at Shalu predate the *Five Goddesses* by a hundred years or so, they introduce a new style, Nepalese-Yuan, from which the *Five Goddesses* seems to be derived. The head of the central figure in the *Five Goddesses* has large slit eyes, a downward-turned nose, a small mouth, and a heart-shaped face (fig. 4). Her facial type is quite similar to those of numerous deities found at Shalu.

Even more compelling is the comparison between the facial features of the central figure in the *Five Goddesses* and deities depicted in the Gyantse murals. The Gyantse murals decorate two great monuments in the religious enclosure known as the Pelkhor Chöde, which dates from 1427 to 1442. Unlike Shalu, the Gyantse murals are inscribed with the names of twenty-nine artists, all Tibetan. These painters came from various regions of Central Tibet, indicating the presence of numerous Tibetan ateliers (Ricca 1997, 197). Ricca (1997, 199) describes the facial features of the peaceful deities depicted in the Gyantse murals as follows: "The forehead is bounded by hair, which emerges from below the horizontal band of the tiara in a regular array of precisely designed curls with a small indentation at the center to hint at the parting. The mouth is gently curved upwards, ... the outlines of the lips are traced with thin lines and only the lower lip ... is partially coloured in red. The chin is suggested by a wavy line, which almost touches the lower contour of the face. The sinuous edges of the eyelids (the upper one black and the lower one red) frame half-closed, exquisitely rendered eyes ... The eyebrows are drawn with thin, decidedly curved lines, while a vertical stroke marks the ridge of the nose from the bridge to the tip." This description easily applies to the central deity in the *Five Goddesses*.

Furthermore, many of the deities depicted in the Gyantse murals wear garments similar to those worn by the central deity in the *Five Goddesses*. Floral shawls flow freely around and below the arms, covering both shoulders. Ricca (1997, 198) summarizes by saying, "Indian, Newar, and Chinese influences appear to have been mastered and interwoven equally by all the painters working at Gyantse [and] ... are the product of a fully mature Tibetan art ... Gyantse [was] a powerful centre for the dissemination of solutions to problems of style. The results appear in a number of thangkas from various monasteries in Tsang."

Based on stylistic similarities between the central deity in the *Five Goddesses* and the Gyantse murals as well as Kossak's garment theory, the *Five Goddesses* seems to date to the fifteenth century. The question of for whom this thangka was made seems to be clearly answered in the Tibetan female practitioner, shown in the lower left corner of the painting in front of an altar. Analyses of color notations and correlating pigments may provide more information that is useful in determining who painted the *Five Goddesses* as well as contributing to the body of knowledge that we have been and continue to compile on thangkas.

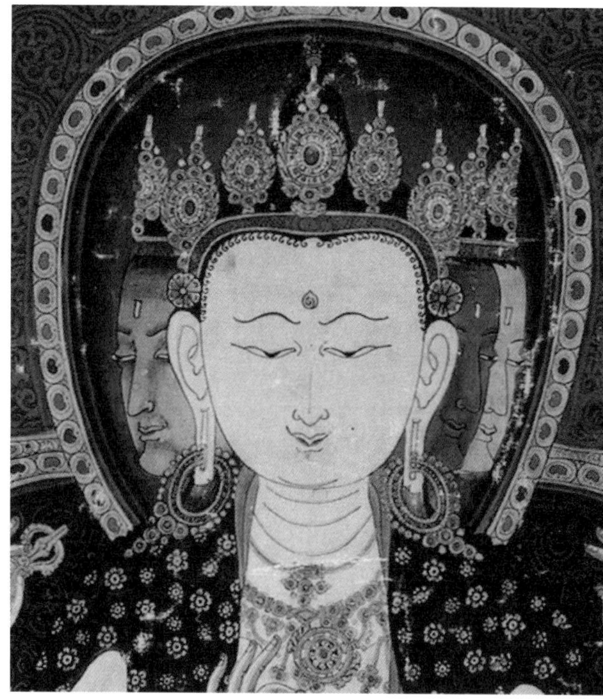

Figure 4. *Detail of the facial features of the central goddess*

Examination and Analysis

The *Five Goddesses* was initially surveyed by infrared reflectography (IRR), a nondestructive technique that permits visualization and photographic documentation of preliminary underlying compositions as well as instructions for colors inscribed on the ground layer. Previous studies have documented handwritten Tibetan *Ume* script, printed Tibetan *Uchen* script, and handwritten Chinese characters. The color notations found throughout the *Five Goddesses* are written in Tibetan *Ume* script. Pigment samples were taken from areas corresponding to the color notations and were analyzed by FTIR, PLM, and EPMA. Table 1 summarizes the results and compares them to results from earlier studies (Duffy and Elgar 1995, 1999).

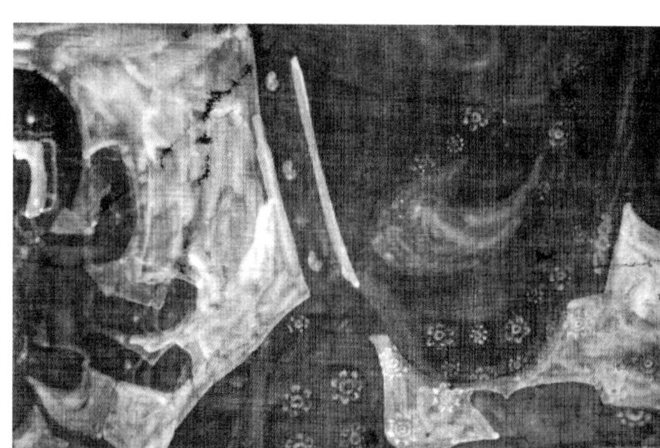

Figure 5. *X-radiograph, detail just to the left of the central deity's proper right knee*

Table 1.

COLOR	NOTATION	TRANSLITERATION	COLOR NAME	LANGUAGE	PIGMENT IDENTIFIED	STUDY
white	ཀ	*ka*	*dkar po*	Tibetan	kaolin	1,3
	白	*bai*	*bai*	Chinese	lead white	2
blue	ཐ	*tha*	*mthing*	Tibetan	azurite	1
		tha	*mthing*	Tibetan	indigo	3
	ར	*ra*	*rams*	Tibetan	indigo	3
	七	*qi*	*shi ging*	Chinese	azurite	2
light blue	ངོ	*ngo*	*sngo*	Tibetan	azurite, kaolin	1
	ངོ་ཀྱ	*ngo kya*	*snog skya*	Tibetan	azurite, kaolin	1
	三七	*san qi*	*san shi ging*	Chinese	azurite	2
green	པ	*pa*	*spang po*	Tibetan	malachite	1
	ཇ	*ja*	*sbyar ljang*	Tibetan	analyses pending	3
	六	*liu*	*shi lu*	Chinese	malachite	2
light green	二六	*er liu*	*er shi lu*	Chinese	malachite	2
	[illegible]	none	*cao lu*	Chinese	malachite and atacamite	2
blue/green	ཇང	*jang*	*sbyar ljang*	Tibetan	orpiment/realgar, indigo, kaolin	1
	ཇང	*jang*	*sbyar ljang*	Tibetan	orpiment/realgar, indigo, kaolin	1
light blue/green	ཇང་ཀྱ	*jang kya*	*ljang skya*	Tibetan	sample not taken	1
dark blue/green	ཇང་ནག	*jang nag*	ljang nag	Tibetan	sample not taken	1
yellow	སེ	*se*	*ser po*	Tibetan	orpiment, kaolin, red lead	1
	སེར	*ser*	*ser po*	Tibetan	sample not taken	1
	སར	*sare*	*ser po*	Tibetan	orpiment, kaolin, red lead	1
	ཨ	*a*	unknown	Tibetan	orpiment	3
	[illegible]	*ba*	*huang*	Chinese	sample not taken	2
orange	ལ	*la*	*li khri*	Tibetan	red lead, calcite	1
	丹	*dan*	*dan*	Chinese	sample not taken	2
red	ག	*ga*	*rgya mtshal*	Tibetan	vermilion, kaolin	1
	མ	*ma*	*dmar po*	Tibetan	vermilion, lac dye, kaolin	1
		ma	*dmar po*	Tibetan	vermilion	3
dark red	མུ	*mu*	*smug po*	Tibetan	vermilion, indigo	3
	[illegible]	none	*hong*	Chinese	vermilion, traces of red lead	2
pink	ན	*na*	*na ros*	Tibetan	lac dye	1,3
	ན་ཀར	*na kar*	*na ros+dkar po*	Tibetan	lac dye, kaolin	1
red flesh	肉	*rou*	*rou*	Chinese	red lead, orpiment	2
beige flesh	米	*mi*	*mi she*	Chinese	orpiment, red lead	2
gray	瓦	*wa*	*ran*	Chinese	sample not taken	2
	ར་ཁ	*ra kha*	unknown	Tibetan	analyses pending	3

1. Duffy and Elgar 1995. 2. Duffy and Elgar 1999. 3. *Five Protective Goddesses*

X-radiographs of the *Five Goddesses* showed brushstrokes invisible to the human eye (fig. 5). X-rays may also be useful in revealing hidden mantras or creeds, which are sometimes written on the front of the painting under its surface, although such writings were not found on this painting.

Radiographs reveal a low x-ray density over the entire painting surface, suggesting either a thin sparing application of mineral pigments, which, applied thickly, would give a more opaque image, or a liberal use of dyestuffs, which have little x-ray density (fig. 6).

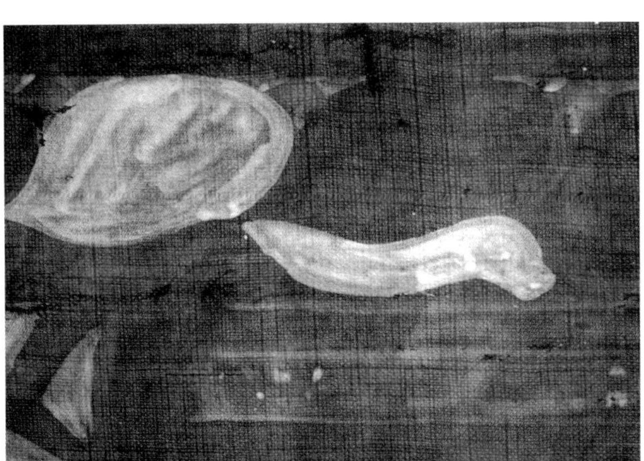

Figure 6. *X-radiograph, detail of right section of throne with lotus petals*

Construction of a Thangka

The support of the *Five Goddesses* is cotton, identified by PLM. Although other supports, such as leather, silk, and linen, have been documented, only cotton has been found in thangkas we examined. In preparation for painting, the support is stitched to four small, pliable sticks, which are then attached to a larger frame, leaving a space of several inches between the inner and outer frame. The space is crucial for adjusting tension. The stretcher must be large enough to provide a taut support for the painter. In preparing a thangka, the support is sized by applying an aqueous glue solution to either side and allowing it to dry. Once the sized support has dried in the sun, the ground layer is applied and burnished with a stone or conch shell until the ground and cloth become a flexible entity, as is necessary for a rolled painting. FTIR and EPMA identified the ground layers as a mixture of kaolin and calcium carbonate.

FTIR spectra of the samples taken from the thangka indicated protein. The results are unexpected when compared with our earlier analyses, wherein the thangkas were thinly bound with a low binder-to-pigment ratio. The results of the *Five Goddesses* are believed to represent subsequent restoration or conservation treatments rather than the paint medium.

Once the ground was prepared, the painter executed the preliminary composition in one of three ways: by hand, by a pouncing technique through a stencil, or by printing with a woodblock. Often when the preliminary composition is rendered by hand, horizontal and vertical lines are applied to divide the space upon which the figures are drawn. The painter of the *Five Goddesses* used such vertical axis lines for the central deity as well as many of the smaller figures. When the preliminary composition is finalized, color notations are drawn on the ground layer as an instructional guide for the master painter as well as apprentices. A number of examples exist of drawings or sketches with color notations or instructions in Newari, Chinese, and Tibetan script.

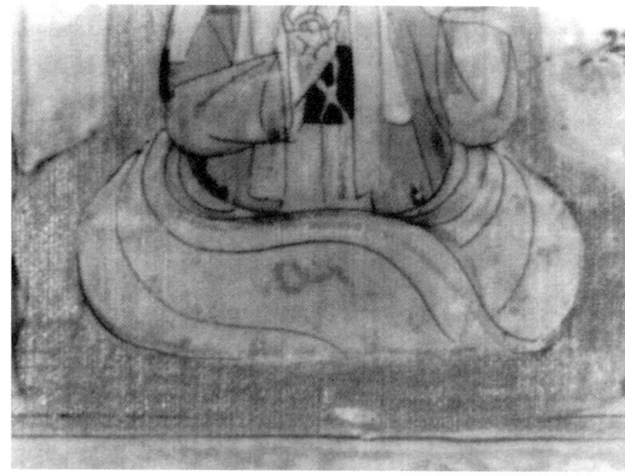

Figure 7. *Color notation* ma (dmar po), *meaning red color*

Color Notations

White

The color notation used for white is *ka (dkar po).* This notation was found throughout the small deities surrounding the central goddess. FTIR and EPMA identified kaolin as the white pigment, similar to the kaolin identified in the ground layer.

Reds

IRR examination revealed the *Ume* character *ma (dmar po),* meaning red color (fig. 7). *Ma* is found throughout the painting to depict red deities, lotus petals, stripes on a deity's dhoti, and the skirt of the female practitioner. EPMA detected Hg and S, indicating the presence of mercuric sulfides. Examination of a dispersed pigment slide revealed coarsely ground particles; however, optical microscopy cannot distinguish between cinnabar and dry-process vermilion.

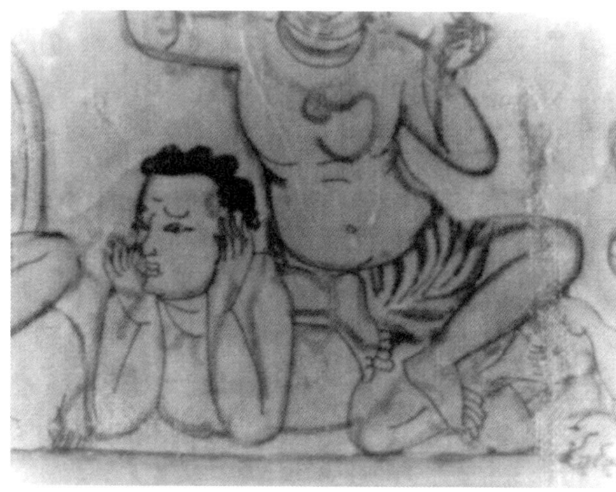

Figure 8. *Color notation* mu (smug po), *meaning dark maroon*

Also analyzed was the dark red coloration of a figure in the lower registers. The color notation is *mu (smug po),* meaning dark maroon (fig. 8). Examination of a dispersed pigment slide revealed a mixture of a red dyestuff and particles of cinnabar or dry-process vermilion. Historical sources as well as previous studies we have done identified lac dye as the most commonly used red dyestuff. Cross-sectional analysis of *mu* did not identify a separate organic layer over the cinnabar/vermilion, as was seen in earlier analyses of this color notation. For the *Five Goddesses,* it is believed that the materials are mixed.

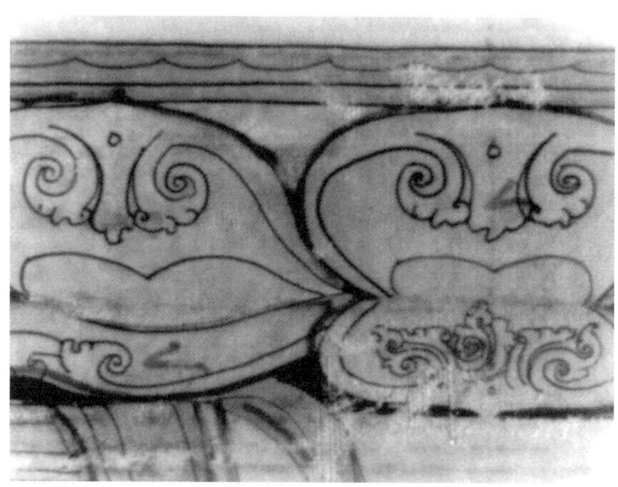

Figure 9. *Color notation* ra (rams), *meaning indigo*

Pink

The color pink is used for lotus petals and parts of garments. Pink is denoted by *na* for *na ros,* which means pink with a maroon tinge. FTIR and PLM showed the presence of a red dyestuff over the kaolin/calcite ground. Historical sources as well as previous studies we have done indicated lac dye.

Blues

Two notations for blue were documented. The first was *ra (rams)*, meaning indigo (fig. 9). This color notation was found on lotus petals and parts of garments. The second character documented was *tha (mthing),* meaning azurite, and was used to depict blue deities. Analysis by FTIR identified a mixture of indigo and kaolin in both notations. The appearance of indigo in both areas is unusual; coarsely ground azurite was identified corresponding to all *tha* notations in earlier studies. The result may indicate the painter's difficulty in obtaining the blue mineral.

Green

The color notation documented for green was *ja.* This notation was used throughout the painting to depict all the green areas (lotus petals, parts of garments, green deities). The notation for basic green, documented on all examined Tibetan thangkas, was *pa (spang),* representing malachite; however, the malachite was not detected on the *Five Goddesses.* Preliminary results of the pigment analysis suggest the use of several colorants for this compound green. Examination of a dispersed pigment slide revealed finely ground particles of orpiment and realgar, mixed with a dark blue/ black dyestuff that is unidentified at the time of this writing.

Yellow

The notation for basic yellow was *a,* which is not translated at this writing. The expected notation, found in earlier studies, in yellow areas was *sa (ser),* meaning yellow. EPMA and PLM were used to confirm the presence of medium-to-coarse particles of orpiment.

The yellow-orange of the central lotus petal was identified by EPMA and PLM as a mixture of orpiment and realgar, accounting for the orange color. Realgar is often associated with orpiment deposits; however, it may also have been a deliberate addition to the pigment mix. The color notation for this petal is not clearly decipherable but is believed to be the same as for basic yellow, *a.*

Gray

A mule mount for a deity, found in the lower register, is painted gray. The color notation documented is *ra kha,* which possibly means indigo gray color. Research is currently ongoing to decipher this notation as well as to identify the notations' corresponding pigment mixture.

Conclusions

Normally, the first colors applied on thangkas are mineral pigments, such as azurite and malachite. The *Five Protective Goddesses* contains neither of these pigments. This is the first time in our studies that neither azurite nor malachite was identified, a distinction that may assist in attribution of this thangka. The main deposit in Tibet for these minerals was Snye-mo-thange in Tsang, which has been worked since the fifteenth century (Jackson and Jackson 1984). The Lhasa government controlled mining and distribution. Artists therefore purchased these minerals directly or indirectly from a government office (Jackson and Jackson 1984).

The use of malachite never seemed to be in question, for the artist never intended to use this mineral pigment. The artist chose only to use a "compound green," and his notations reflect this. The green areas were created with a mixture of pigments, none of which included malachite.

The use of indigo on the blue face of the multiheaded small deity in the upper left is contrary to its color notation, which instructs the painter to use azurite. During the painting of the *Five Protective Goddesses,* the artist may have had difficulty obtaining azurite, using indigo as a substitute. This dyestuff was easily obtainable, since the trade routes between India and the capital of Tibet, Lhasa, traverse Nepal into Tsang (southern Central Tibet).

Since the notations are written in Tibetan script, the artist was probably Tibetan and not Nepalese, as previous scholars had thought. The painting was probably painted for a Tibetan, since the depicted female practitioner is dressed as a Tibetan. Art historians have compared the *Five Protective Goddesses* to the murals at Shalu in style and clothing. The *Five Protective Goddesses* is, however, stylistically closer to the Gyantse murals, which date to the first half of the fifteenth century. Further investigations into the mining operations at Snye-mo-thange, specifically problems in distribution or the difficulty in obtaining azurite during the fifteenth century, may lead to a more exact date for the *Five Protective Goddesses.*

Acknowledgments

We would like to thank the following people at the Museum of Fine Arts, Boston: Joan Cummins, Richard Newman, and George Hagerty. Thanks also go to Janice Carlson of the Winterthur Museum and Gardens for her editorial expertise.

References

Duffy, K., and J. Elgar. 1995. An investigation of palette and color notations used to create a set of Tibetan Thangkas. In *Historical painting techniques, materials, and studio practice,* ed. A. Wallert, E. Hermens, and M. Peek, 78–84. Los Angeles: The Getty Conservation Institute.

——. 1999. Examination of thangkas from Central and Eastern Tibet. In *Sixth international conference on "non-destructive testing and microanalysis for the diagnostics and conservation of the cultural and environmental heritage," Rome 17–20 May 1999,* ed. M. Marabelli, C. Parisi, S. Gagliardi, G. M. Parisi, and G. Torcinaro, vol. 3, 1751–66.

Jackson, D. 1996. *A history of Tibetan painting: The great Tibetan painters and their traditions.* Vienna: Beitrage zur Kultur- und Geisteschichte Asiens.

Jackson, D. P., and J.A. Jackson. 1984. *Tibetan thangka painting: Methods and materials.* London: Serindia Publications. 75.

Kossak, S. 2000. Chronological clues: Changes in sartorial and pictorial styles in the late-thirteenth–early-fifteenth-century Central Tibetan painting. *Oriental art* 46(4):11–19.

Kossak, S., and J. C. Singer. 1998. *Sacred visions: Early paintings from Central Tibet.* New York: The Metropolitan Museum of Art. 43.

Kreijger, H. 1997. Mural styles at Shalu. In *Tibetan art: Towards a definition of style,* ed. J. C. Singer and P. Denwood, 170–77. London: Laurence King Publishing.

Ricca, F. 1997. Stylistic features of the Pelkhor Chöde at Gyantse. In *Tibetan art: Towards a definition of style,* ed. J. C. Singer and P. Denwood, 196–209. London: Laurence King Publishing.

20 Recent Studies on Sri Lankan Mural Painting Technology

B. D. Nandadeva

ABSTRACT Materials and techniques of the three schools of mural painting of Sri Lanka belonging to Late Historical period 2 (1750–1900) were studied. Samples of supports, grounds, primers, paints, and coatings from six randomly selected sites were examined using optical microscopy, scanning electron microscopy (SEM), x-ray spectrometry by energy dispersive microanalysis, x-ray powder diffraction (XRD), Fourier transform infrared spectroscopy (FTIR), ultraviolet-visible absorption spectroscopy, thin-layer chromatography (TLC), and pyrolysis gas chromatography-mass spectrometry (Py-GC-MS). It was found that paintings of all three schools belong to the tempera technique. They were done on a clay and sand ground sometimes mixed with plant fibers and followed by the application of a white primer. Huntite, clay, barium sulfate, lead white, gypsum, calcite, and lithopone were identified as primers or as the white pigment. Cinnabar or dry-process vermilion, orpiment, Prussian blue, manganese blue, and soot were identified as pigments. Binding media were characterized as gums, drying oils, resins, and emulsions of these three types of material.

Introduction

This paper presents the results of a recent study on characterization of materials and techniques of Sri Lankan Buddhist temple paintings belonging to Late Historical period 2. Samples of supports, grounds, primers, and paint layers were examined to elucidate the constituent materials and application methods. Although this is the second attempt to employ scientific analytical techniques to examine Sri Lankan mural paintings, it is the first study to employ scientific research methodologies that involved random site selection, random sampling, sample replication, and statistical methods for data analyses. It is also the first time that a combination of a wide range of modern analytical techniques, such as x-ray diffraction, SEM, energy dispersive x-ray microanalyzing spectrometry, FTIR, and so forth have been used to study any branch of Sri Lankan art.

The most extensive body of traditional Sri Lankan mural-painting sites, comprising over nine hundred temples that survive today as active places of Buddhist worship, belong to Late Historical period 2. These murals have been identified as representing three distinct but related schools on the basis of style, iconography, treatment, and pictorial expression (Gunasinghe 1978; Bandaranayake 1986). The schools are known as the central Kandyan, provincial Kandyan, and southern and western maritime schools. The materials and technology of the paintings had not previously been adequately studied.

The main objective of this study was to examine the hypothesis that visible stylistic and iconographic distinctions between the three schools will be reflected in technological differences between them. Several cultural and art-historical implications of the paintings of the period and deterioration implications in relation to the materials and techniques of the paintings have emerged from the study. Details of those implications are discussed elsewhere (Nandadeva 1997).

Background

Under the patronage of Sinhalese kings, a Buddhist religious revival occurred in the Kandyan kingdom around the mid-eighteenth century and subsequently spread into the southern and western coastal areas. As a result, a large number of new temples were constructed and old ones renovated or refurbished. The term *central Kandyan school* is applied strictly to painters who produced mural paintings in temples in the central districts of Kandy, constructed with royal or feudal patronage during the period between 1750 and 1815 (at which date the Kandyan kingdom fell to the British). As generally the best artists of the day were in the service of the king or feudal chiefs (who patronized such work on behalf of the king), it is expected that the work from this school represents the highest standards of technique.

The term *provincial Kandyan school* applies to substandard artists of provincial centers situated away from the capital. Belonging to the same school are artists who previously worked for the royalty but were dispersed into several localized substylistic groups in the postconquest period as a result of the loss of royal patronage and consequent disintegration of the traditional guild system. One can speculate that the painting technology of the provincial Kandyan school should show some deviation from the central Kandyan school.

The term *southern and western maritime school* was probably composed of artists who had been employed in various forms of folk arts or rituals that required painterly skills and who were hired to decorate the increasing number of new temples. With little or no connection with the central Kandyan school, they developed a unique painting style that is distinct from the former in pictorial expression and artistic creativity. Furthermore, as the Buddhist institutions of the southern and western coastal areas received support and patronage from the native Christian elites and the colonial rulers, these paintings also contain features that either are of European origin or show clear influence of it. Therefore, it is expected that one might find European materials and painting techniques in southern paintings.

Previous Studies

Scientific examination of Sri Lankan mural-painting technology has previously been confined to a doctoral study by de Silva (1962), which was published piecemeal (de Silva 1963, 1965, 1971, 1975, 1990). He used microscopy, chemical tests, infrared spectroscopy, paper chromatography, ion-exchange chromatography, staining tests, and ultraviolet spectroscopy to examine the samples. His study, mostly on sites from the early and middle historical periods, contained only six sites from the central Kandyan school. He did not include the provincial Kandyan and southern maritime schools in his study. He did not explain whether or not randomized methods were used to select sampling sites and sampling spots or whether he used replicate samples, nor did he mention whether confirmatory analytical methods were used.

Much useful information is given by Davy (1821), Davids (1875), Coomaraswamy (1956), and Charles (1990). This information has been derived from careful observation of the paintings and from what was related to them by descendants of traditional painters. Jivan Naide,[1] the only hereditary painter in Sri Lanka today, has preserved as an oral tradition a vast body of information of materials and techniques of wall paintings as practiced by his family.

Sampling Sites

To allow a full program of sampling and analysis to be conducted for each site selected for the study, the number of sampling sites was limited to two from each of the three schools, providing a total of six. The sampling sites were randomly selected from the largest possible population of sites within each school. The six sites selected (with code letters used below) were Degaldoruva Rajamahaviharaya (DD) and Bambaragala Rajamahaviharaya (BB) representing the central Kandyan school, Konduruvapola Purana Tampita Viharaya (KP) and Keselkotuva Sri Sugataramaya (KS) representing the provincial Kandyan school, and Hikkaduve Sri Jananandaramaya (JN) and Vatagedara Sri Sugataramaya (SR) representing the southern maritime school.

Samples were taken mainly from damaged areas of the wall. Random selection methods were also used to choose among suitable sampling spots to ensure representative samples. This was possible only when there were a large number of damaged areas distributed across a large area of the wall surface. If there were fewer or no damaged areas on the painted surface, random sampling was not attempted.

Materials and Methods

A range of analytical methods, including optical microscopy, microchemical analysis, XRD, ultraviolet and visible spectroscopy (UV-vis), FTIR, scanning electron microscopy (SEM), scanning electron microscopy with energy dispersive x-ray microanalysis (SEM-EDXA), thin-layer chromatography (TLC), and Py-GC-MS, was used. The selection of a method was based on its successful use in the identification of art materials by other researchers and on the availability and accessibility of the necessary instruments.

In most cases, replicates of every type of sample were examined under the same conditions to estimate variations in that type of material. Results obtained from one analytical technique were confirmed by a second technique. Sample replicates were not always used in those confirmatory analyses unless contradictory results were found. Information drawn from these analyses was used in combination with information gathered during in situ field examination to make the final interpretations.

Layer Stratigraphy

The samples were embedded in cubes of methyl methacrylate resins, sectioned and polished, and then examined using optical microscopy under reflected light to determine the sequence and nature of the layers. Most cross-sections contained one or more paint layers, a primer, and a part of the ground. The presence of more than one paint layer suggested either a repainting[2] activity of a later date or "twin-layer application."[3] Repainting was identified by the presence of a thin layer of soot or grit between paint layers or a layer of whitewash between the two paint layers.

Supports

Samples of rock and mud from the walls of temples were prepared for optical microscopy as thin sections by being mounted on glass slides, cut, and polished to 30 μm thickness. These were examined using transmitted polarized light for composition and concentration of minerals to identify types of rocks and constituents of mortars.

Grounds

Samples of grounds were also prepared as thin sections for optical microscopy. They were examined using transmitted polarized light to elucidate the composition and concentration of minerals. Such information was used to characterize differences between preparation technologies of grounds for painting.

Samples of grounds from DD, BB, and SR contained plant fibers. The fiber specimens were separated from the samples under a stereo-binocular microscope and cleaned, then mounted and examined by PLM using existing methods (Stoves 1958; Isenberg 1967; McCrone, Draftz, and Delly 1967; Catling and Grayson 1982).

Pigments and Primers

Optical microscopy was the primary analytical technique used to identify the pigments. The samples were prepared as dispersions mounted on microscope slides and examined under transmitted polarized light according to principles and methods described (McCrone 1982; McCrone, McCrone, and Delly 1984). Types of variables considered for identifying pigments were color, isotropy or anisotropy, and refractive index.

Identification of the priming materials by optical microscopy was difficult, as they showed poor resolution due to fine particle size and their tendency to stay as clumps rather than individual particles. Preliminary identification of primers from all six sites was therefore done by XRD using powder diffraction cameras. Diffraction patterns were obtained by the Debye-Scherrer technique using 114.6 mm diameter Gandolfi cameras operated for eighteen to twenty hours, with CuKα radiation (λ = 1.542 Å) at 40 kV and 20 mA. Preliminary results obtained from XRD were compared with qualitative elemental analysis using SEM-EDXA on cross-sections for confirmation.

White Pigments

It was not possible to use optical microscopy or XRD to identify the white pigment due to the difficulty of manually separating the white paint layer from the primer, which is also white. Qualitative elemental analysis by SEM-EDXA using the paint cross-sections mentioned above was chosen as a better technique, since it enables the pigment particles to be examined under high magnification within their layer contexts. The SEM-EDXA instruments were operated at 15 kV voltage and 1.2 mA beam current.

Red Pigments

Red pigment samples from all six sites were identified by optical microscopy as vermilion. These results were confirmed by XRD.

Yellow Pigments

Yellow pigment samples from KS, JN, SR, and the repainted layer of DD were identified by optical microscopy as orpiment and confirmed by XRD. The chloroform extracts of yellow pigment samples from the sites BB, KP, and DD (from the older or original paint layer) produced FTIR spectra (during binding media analysis described below) similar to that of gamboge. To test this proposition, UV-vis was chosen, as previous studies on gambogic acid and its congeners have found four characteristic bands at 217, 280, 291, and 362 nm (Ollis et al. 1965). The samples were treated with dilute HCl, rinsed with water, and dissolved in ethanol, and the solution centrifuged. A sample of gamboge from Sri Lanka was used as a reference standard.

TLC confirmed gamboge initially identified by UV-vis. The development solution was cyclohexane:ethyl acetate: formic acid (75:25:1); formic acid helps curtail tailing of carboxylic acids. The precoated silica gel plates contain an inorganic phosphor that fluoresces bright yellow under short-wave (254 nm) ultraviolet radiation; spots from gamboge appear dark against this background. The plates were then sprayed with 2% N-bromoacetamide in glacial acetic acid (Winter 1987) and examined under long-wave ultraviolet light (365 nm); this produces a whitish fluorescence from gambogic acid and congeners.

Blue Pigments

Examination of the blue pigments by optical microscopy indicated either indigo or Prussian blue, which look very similar. A microchemical method, described by Hofenk de Graaff (1974), was applied to identify the presence of indigo and was negative. This suggested that the pigment may be Prussian blue, which was confirmed by FTIR. Prussian blue is the only common blue pigment that contains the nitrile (C≡N) group, which gives a characteristic absorption band around 2090 cm^{-1}. No special sample-preparation method was required, a small quantity of blue pigment particles being pressed into the diamond cell and analyzed.

Black Pigments

Black pigments were difficult to identify by optical microscopy because of their opacity. Examination of cross-sections by SEM-EDXA suggested that all of them are carbon-based. The particle morphology of black pigment samples from all six sites was examined by SEM imaging, an approach that can distinguish production techniques for carbon-based pigments (Winter 1983). A series of cleaning steps was used to remove organic materials, such as binders and preservative coatings, and inorganic materials, such as carbonates and silicates, associated with the paint layer. To prevent sample contamination by airborne material, all samples were prepared in a modular clean room in the Department of Conservation and Scientific Research of the Freer Gallery using a microsample cleaning method established by that laboratory.

Binding Media

Staining of cross-sections (Wolbers and Landrey 1987) and FTIR gave general information on binding media. The fluorescent dyes used for staining were triphenyltetrazolium chloride (4% solution in methanol) for carbohydrates, fluorescamine (0.05% in acetone) for proteins, Nile red (0.01% in 80:20 ethanol:water) for oils, and antimony pentachloride (4–5% in chloroform) for resins. The ultraviolet light source was a high-pressure mercury lamp; wavelength range for fluorescence excitation was 360–420 nm.

For analysis with FTIR, binders were extracted using chloroform and water successively. Chloroform extracts were heated at circa 50°C for about fifteen minutes and water extracts at circa 70°C for about thirty minutes and left to stand for about twelve hours. The solutions were then centrifuged and transferred onto microscope slides. The dried extracts were used for examination.

Only a few of the chloroform extracts, which FTIR showed to be mixtures, were analyzed using Py-GC-MS. The pyrolysis chamber and GC injector, used in split mode, were kept at 320°C. The DB-1 capillary column (30 m × 0.32 mm) was ramped from 40°C to 200°C at 5 deg/min and from 200°C to 305°C at 10 deg/min, held there for ten minutes, and then ramped from 305°C to 320°C at 5 deg/min and held there for six minutes. The transfer line and iron trap manifold were maintained at 260°C. The helium flow rate was approximately 2 ml/min.

A reference sample of *dorana-tel,* an exudate of the *dorana* tree *(Dipterocarpus glandulosus)* was analyzed with Py-GC-MS. This substance was a popular binding medium and one of the two ingredients of a coating material, particularly in the southern school. De Silva (1962, 150–53) concluded that it had the properties of a drying oil. This contrasted with two later studies using TLC and IR spectroscopy, which suggested that it is a resin (de Silva, Rodrigo, and Wijesekara 1962; Geevananda, Gunawardana, et al. 1980). It was considered necessary to understand the correct chemical properties of the material.

Results and Discussion

This study has produced a substantial body of information on the materials and techniques of Sri Lankan mural paintings of the Late Historical period. Some of the information gathered here either contradicts previous claims or is new. Other information either confirms what was already known or further explains existing knowledge. A summary of the information yielded from the analyses is given in table 1.

The identification of clay and sand as the two basic materials in the grounds of five of the six sites confirms previous information. The use of cotton fibers mixed with clay and sand as identified by de Silva (1962, 88–90) and also related by Jivan Naide was found at only three sites. The claims regarding the use of coconut fibers by Charles (1990) and Jivan Naide, and kapok *(Ceiba pentandra)* by Charles (1990) are not confirmed.

The materials identified here for the primers are very different from those mentioned in the previous literature. A material commonly used for primers known as *makul* was identified by Coomaraswamy (1956, 164) and de Silva (1962, 98–108) as hydrous magnesite. De Silva (1962, 98–108) has also found clay as the primer at some sites. Here the material used for the primer at the two central Kandyan sites and one provincial Kandyan site (KP) was identified as the mineral huntite ($CaMg_3(CO_3)_4$) (fig. 1). This is the first time that huntite has been reported from Sri Lankan temple paintings. Hydrous magnesite was not found at any of the six sites. The materials identified at other sites (lithopone, clay,

Table 1. *Summary of results obtained from analyses of materials*

School and site	Ground	Primer and white pigment	Pigments (red/yellow/black/blue/green)	Binding media (in ground, primer, paint)
Central Kandyan				
DD	Coarse sand (> 50%), clay, cotton fiber	Huntite	Cinnabar, gamboge, flame carbon	Gum Gum, oil Gum, oil
BB	Coarse sand (> 50%), clay, cotton fiber	Huntite	Cinnabar, gamboge, soot	Gum, oil Oil, resin Gum, oil, resin
Provincial Kandyan				
KP	Coarse sand (> 50%), clay, carbonized plant fiber	Huntite	Cinnabar, gamboge, soot, Prussian blue	Gum, oil Gum, resin Gum, resin
KS	Fine sand (> 50%), clay, uncarbonized fiber similar to straw	Lithopone Lithopone, clay Clay	Cinnabar, orpiment, soot, Prussian blue, manganese blue, blue over yellow	Gum, oil Gum, resin (*dorana-tel*) Gum, resin
Southern Maritime				
JN	Coarse sand (< 50%), lime	Clay Clay, lime Clay, barite, lead white	Cinnabar, orpiment, soot, Prussian blue	Gum, oil Oil, resin Gum, oil, resin
SR	Coarse sand (20–30%), clay, natural lime, cotton fiber	Clay Clay, lime	Cinnabar, orpiment, soot, Prussian blue, manganese blue, blue over yellow	Oil, resin Gum, resin Resin *(dorana-tel)*

lime, barium sulfate, gypsum, and lead white as pure material or as mixtures) also were not previously mentioned.

Cinnabar was mentioned in previous literature as the only red pigment of the Late Historical period. The present study confirms this claim (fig. 2), with the reservation that the methods used here do not distinguish cinnabar from dry-process vermilion. Charles (1990) mentions that red earth (red ocher) was used by some painters, as cinnabar was rare and expensive. The present study did not find evidence for the use of red ocher at any of the six sites.

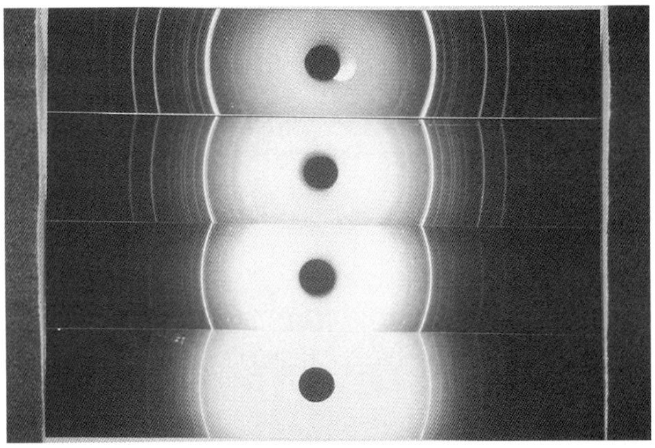

Figure 1. *X-ray diffraction patterns. From top to bottom: reference sample of huntite from West Kimberly, Western Australia; samples of primers from the DD, BB, and KP sites*

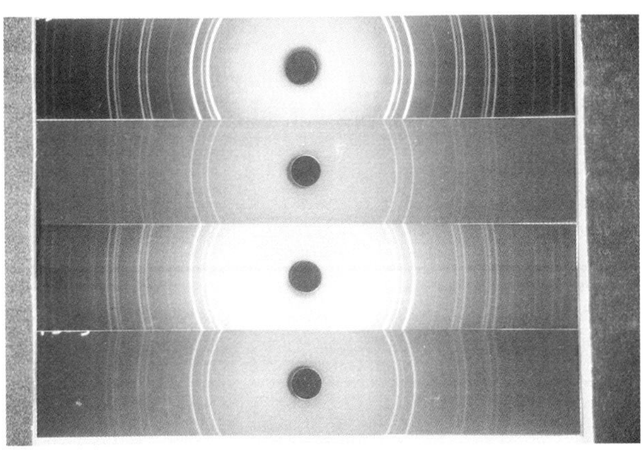

Figure 2. *X-ray diffraction patterns of red pigments. From top to bottom: reference pattern of vermilion (reference collection of the Department of Conservation and Scientific Research, Freer Gallery of Art); red pigment samples from the DD, BB, and KP sites*

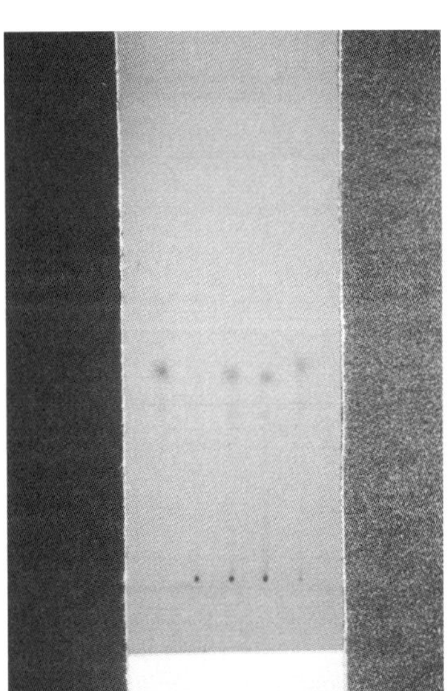

Figure 3. *Thin-layer chromatogram of gamboge. From left to right: gamboge reference sample from Sri Lanka; three yellow paint samples from DD, BB, and KP; reference sample from southeast Asia (reference collection, Department of Conservation and Scientific Research, Freer Gallery of Art)*

Gamboge and orpiment were mentioned in previous literature as yellow pigments (fig. 3), and this is confirmed by the present work. The claim that yellow earth was also used, in addition to orpiment, by some southern painters (Charles 1990) is not supported by this study.

Indigo, mentioned by Coomaraswamy (1956, 164) and Jivan Naide was not found at any of the six sites. Instead, Prussian blue, mentioned only once (Davids 1875), was found in the provincial Kandyan and southern schools. Use of manganese blue mixed with Prussian blue at KS might be a later development.

The identification of the black pigments of all six sites as carbon-based confirmed the general identifications given in previous studies. The characterization of the carbon-based pigments as flame carbon at DD and soot at the other sites is a new addition to the knowledge on painting technology. Pulverized coconut-shell charcoal, claimed to have been used by southern painters (Charles 1990), was not found at any of the sites.

According to Coomaraswamy (1956, 164), a green color was made by mixing blue and yellow. For the central Kandyan school, this statement was found to be true, as two samples from BB showed a mixture of the two. The application of a thin blue layer over a thick yellow layer to produce green was not mentioned in the previous literature, although it is not a particularly new or novel idea. Claims by Jivan Naide that brown was produced by mixing red and yellow and by Coomaraswamy (1956, 164) that gray was produced by mixing black and white were also confirmed.

The information obtained in this study on binding media, while still needing further confirmation, supports some of the previous views. This study found evidence to suggest the use of an oil or an emulsion of gum and oil in the central Kandyan school and *dorana-tel* with or without an oil in the southern school. This partly supports the claim by de Silva (1962, 177) that gum or an emulsion of gum and oil was used in the central Kandyan school. It also partly supports the claims by Coomaraswamy (1956, 165) that gum was the medium (possibly with the use of oil for exterior paintings) used by the central Kandyan painters and that the use of *dorana-tel* was a southern innovation. The present study makes clear that an analytical technique based on a separation method is needed to confirm this preliminary information on media.

This study found that *dorana-tel* is a resin. The Py-GC-MS analysis of fresh and artificially aged samples shows that it does not have the properties of a drying oil as claimed by de Silva (1962, 150–53). This result confirms the findings of the two previous studies of *dorana-tel* using TLC and infrared spectroscopy (de Silva, Rodrigo, and Wijesekera 1962; Geevananda, Gunawardana, et al. 1980).

Conclusions

This research has uncovered a considerable body of new information on the materials and technology of Sri Lankan mural paintings. It has identified materials that had not been reported in previous literature. Those include huntite, lead white, lithopone, calcite, barium sulfate, gypsum, and manganese blue. Also important is the finding of the reintroduction of lime-based grounds, which had been a popular technique during the Early and Middle Historical periods (de Silva 1962, 84–87). Furthermore, it was possible to characterize soot as the black pigment at many of the sites as distinct from a pure flame carbon such as lampblack or carbon black.

As for painting technique, the use of extenders for certain pigments and the application of twin layers, a lighter one to be followed by a brighter one, were revealed as new information in this study. The application of blue over yellow or vice versa to produce green also had not been reported previously.

Comparison of the three schools shows that the stylistic differences between them are reflected in differences in materials and techniques of painting. It is also seen that the degree of control over the materials and techniques within a school vary from one school to another. The closest similarity is between BB and the original paintings at DD. The two southern sites showed some similarities. The provincial Kandyan sites showed the most variability, illustrating that the school had a lesser control over the materials and techniques. Those differences showed that a considerable diversity of mural-painting technology had prevailed during that period of 150 years, distributed across a smaller geographical region.

Several cultural and art-historical implications of Late Historical period 2 have emerged. These relate to the nature of royal patronage for artists, systems of training of artists, state of the arts under British rule, and so forth. A discussion of these implications is beyond the aim of this paper. A detailed examination of such issues is given elsewhere (Nandadeva 1997).

It has also been shown that the problems of deterioration (those not caused by physical or mechanical factors) are related to the materials and techniques of the paintings. Since there was a considerable diversity in painting technology, it can be expected that such material-specific deterioration problems will also be diverse in nature. By illustrating those relationships between diverse painting technologies and complex deterioration problems, this study makes the point that future treatments of the mural paintings should be based on a thorough understanding of the materials and techniques of the paintings of each individual site. The deterioration implications are discussed in detail elsewhere (Nandadeva 1997).

This study has provided preliminary information on binding media and coatings. Although conclusive evidence of the use of gums, acrylic, and polyvinyl acetate resins (from conservation treatments) and some indication of the presence of resin and oil media were obtained through cross-section staining and FTIR, these techniques were insufficient to unravel completely the complex nature of these materials. Future research on media analysis will probably have to rely heavily on GC-MS work.

Fruitful areas for further research include the analysis of samples from additional sites belonging to each of the three schools. It would be interesting to see if the pattern of similarities and differences in this research continues when more sites are studied. The inclusion of more sites in the program will eventually allow more detailed questions and hypotheses to be addressed. This research has provided a solid foundation for future work on Sri Lankan mural paintings of the Late Historical period.

Acknowledgments

This paper was based on a doctoral dissertation submitted to the Department of Art Conservation, University of Delaware, Newark, Delaware, in 1997. Analytical work was carried out at the University of Delaware, the Winterthur Museum, and the Department of Conservation and Scientific Research, Freer Gallery of Art and Arthur M. Sackler Gallery, Smithsonian Institution. Funding was provided by a Kress Foundation Fellowship, a competitive Fellowship from the University of Delaware, and a Forbes Fellowship from the Freer Gallery of Art.

Notes

1. Jivan Naide is a direct descendant of the family of master painters of the Kandyan kingdom and is probably connected to the atelier of Nilagama Patabanda, the principal artist of the Degaldoruva Rajamaha Viharaya (one of the two central Kandyan sites of the present study), dated to 1771–86. I interviewed him twice in 1993, and a summary of information given by him is given elsewhere (Nandadeva 1997).
2. Refurbishment of old temples was considered a duty and responsibility of the ruler throughout the history of Sri Lanka. Stripping of old paintings during such refurbishment was considered unacceptable, perhaps for religious and/or ethical reasons. Instead, the old paintings were covered by a thin wash of white paint that acted as a primer and then repainted.
3. This term was coined to describe the practice of applying a thin, bright paint layer on a thick, lighter layer of the same paint for better optics.

Materials and Manufacturers

Embedding cross-sections: Methyl methacrylate resin; Ward's Bio-Plastic (Wards Natural Science Establishment)

Preparation of thin sections: Quality Thin Sections, 9835 E. Celester Drive, Tucson, Ariz., 85730

Mounting of dispersed pigment samples: Cargille Meltmount (n_D = 1.662 @ 25°C)

Fluorescent dyes: Triphenyltetrazolium chloride; fluorescamine; antimony pentachoride; all from Sigma Chemical Co. (now Sigma-Aldrich), St. Louis, Mo.

Reference standards: Indigo (Mannings Co., East Berlin, Pa.); Prussian blue (Weber); huntite, one sample from Nevada (National Museum of Natural History, Smithsonian Institution, sample NMNH 112519), second sample from a site in West Kimberly, Australia (courtesy Dr. I. MacLeod, Western Australian Maritime Museum); hydromagnesite, San Benito County, Calif. (National Museum of Natural History, Smithsonian Institution, sample NMNH 121511)

Instruments

Optical microscopy: Nikon LABOPHOT2-POL polarizing microscope (Nikon Corporation); Leitz Orthoplan polarizing microscope

Scanning electron microscopy: Leica 440 Serial scanning electron microscope

X-ray powder diffraction: Philips XRG 2600 x-ray generator with Gandolfi powder diffraction cameras

Energy dispersive x-ray microanalysis: JEOL JSM-840A

Fourier transform infrared spectroscopy: Analect RFX-65 Fourier transform infrared spectrometer with XAD microscope and Analect ATC 652-286 FTIR Data System; Polaris FTIR Spectrometer (Mattson Instruments) with WinFIRST Fourier Infrared Software system.

FTIR spectral libraries used: Gettens Collection of Aged Materials of the Artists: FTIR Spectral Library and Catalogue of the Raw Materials, compiled by Amy Snodgrass and Beth A. Price, 1994; Infrared User's Group, 1994 Spectra Collection compiled by the Getty Conservation Institute

Ultraviolet-visible absorption spectroscopy: Perkin-Elmer 552 Spectrophotometer

Viewing thin-layer chromatography plates: Chromato-vue viewing system (UVP, San Gabriel, Calif., 91778)

References

Bandaranayake, S. 1986. *Rock and wall paintings of Sri Lanka.* Colombo: Lake House Bookshop.

Catling, D., and J. Grayson. 1982. *Identification of vegetable fibres.* London: Chapman and Hall.

Charles, S. P. 1990. Painting (1200 A.D.–1400 A.D.). Section III. *Painting: Archaeological Department centenary (1890–1990) commemorative series,* ed. Nandadeva Wijesekera, vol. 5, 69–77. Colombo: Archaeological Department of Sri Lanka.

Coomaraswamy, A. K. 1956. *Medieval Sinhalese art.* 2nd ed. New York: Pantheon Books.

Davids, T. W. R. 1875. Sigiri, the lion rock, near Pulastipura, Ceylon, and the thirty-ninth chapter of the Mahavamsa. *Journal of the Royal Asiatic Society, Great Britain and Ireland* 7:191–211.

Davy, J. 1821. *An account of the interior of Ceylon and its inhabitants with travels in that island.* London: Longman.

de Silva, R. H. 1962. The evolution of the technique of Sinhalese wall painting: A study employing museum laboratory techniques. Ph.D. thesis, University of Oxford.

——. 1963. The problem of the binding medium particularly in wall painting. *Archaeometry* 6:56–64.

——. 1965. The technique of ancient Sinhalese wallpainting—Sigiri. *Paranavitana felicitation volume on art and architecture and Oriental studies,* ed. N. A. Jayawickrama, 89–121. Colombo: M. D. Gunasena and Co.

——. 1971. The evolution of the techniques of Sinhalese wall painting and comparison with Indian painting methods. *Ancient Ceylon* 1:90–104.

——. 1975. Rock painting in Sri Lanka. *Conservation in the archaeology and the applied arts.* Reprints of the contributions to the Stockholm Congress, 2–6 June 1975, 69–71. London: International Institute for Conservation.

——. 1990. Painting (early period 247 B.C. to 800 A.D.). *Painting: Archaeological Department centenary (1890–1990) commemorative series,* ed. B. D. Nandadeva and R. O. B. Wijesekera, vol. 5, 3–44. Colombo: Archaeological Department of Sri Lanka.

de Silva, L. B., S. Rodrigo, and R. O. B. Wijesekera. 1962. Extractives of Ceylonese *Dipterocarpaceae. Journal of science and industrial research* 21B (August).

Geevananda, Gunawardana, et al. 1980. Distribution of some triterpenes and phenolic compounds in the extractives of endemic *Dipterocarpceae* species of Sri Lanka. *Phytochemistry* 19:1099–102.

Gunasinghe, S. 1978. *An album of Buddhist paintings from Sri Lanka (Ceylon) (Kandyan period).* Colombo: National Museum.

Hofenk de Graaff, J. H. 1974. A simple method for the identification of indigo. *Studies in conservation* 19:54–55.

Isenberg, I. H. 1967. *Pulp and paper microscopy.* 3rd ed. Appleton, Wisc.: Institute of Paper Chemistry.

McCrone, W. C. 1982. The microscopical identification of artists' pigments. *Journal of the International Institute for Conservation: Canadian group* 7(1–2).

McCrone, W. C., R. G. Draftz, and J. G. Delly. 1967. *The particle atlas: A photomicrographic reference for the microscopical identification of particulate substances.* Ann Arbor, Mich.: Ann Arbor Science Publishers.

McCrone, W. C., L. B. McCrone, and J. G. Delly. 1984. *Polarized light microscopy.* Ann Arbor, Mich.: Ann Arbor Science Publishers, Inc.

Nandadeva, B. D. 1997. Materials and techniques of Kandyan and southern schools of mural painting of Sri Lanka: Mid-eighteenth to late nineteenth centuries. Ph.D. diss., Art Conservation Department, University of Delaware.

Ollis, W. D., et al. 1965. The constitution of gambogic acid. *Tetrahedron* 21:1455–70.

Stoves, J. L. 1958. *Fiber microscopy: Its techniques and application.* Princeton and New York: D. van Nostrand Co.

Winter, J. 1983. The characterization of pigments based on carbon. *Studies in conservation* 28:49–66.

——. 1987. Organic colors in East Asian paintings: The identification of gamboge. *ICOM Committee for Conservation, eighth triennial meeting, Sydney, Australia, 6–11 Sept. 1987 preprints.* Los Angeles: The Getty Conservation Institute.

Wolbers, R., and G. Landrey. 1987. The use of direct reactive fluorescent dyes for the characterization of binding media in cross-sectional examinations. *AIC preprints,* 168–202. American Institute for Conservation fifteenth annual meeting, Vancouver. Washington, D.C.: AIC.

Glass, Jade, and Ceramics

21 Compositions and Particles Embedded in Korean Ancient Glass Beads

Heisun Yu and Hyung Tae Kang

ABSTRACT Fifty-eight glass beads collected from sites of the Paekche period (ca. second to sixth century C.E.) were analyzed by scanning electron microscopy with energy dispersive spectroscopy (SEM-EDS) for nine oxides (SiO_2, Al_2O_3, Na_2O, K_2O, CaO, MgO, CuO, TiO_2 and Fe_2O_3) and the results classified with the statistical method of principal component analysis (PCA). Spherical or irregularly shaped particles in the glass were analyzed by EDS and micro-beam x-ray diffraction. These particles, which function as colorants and opacifiers, were identified as lead tin oxide ($PbSnO_3$), lead antimony oxide, copper sulfide (CuS or Cu_2S), and copper oxide (Cu_2O). For lead-containing glasses, lead isotope ratios were determined by thermal ionization mass spectroscopy (TIMS). Linear discriminant analysis (LDA) was applied to the lead-isotope data, and the lead in the glass was found to be similar to lead ore samples from southern China.

Introduction

Ancient glass can be grouped based on the modifiers and stabilizers used, allowing study of the characteristics of glass of different periods and geographic areas. Colorants are also important, and blue, dark green, and violet glass in particular contain information regarding import, distribution, and provenance of colorants during the manufacturing process of glass. Therefore, systematic study of this glass will make a great contribution to the understanding of cultural exchange.

In this study, fifty-eight glass bead samples from the Paekche period (ca. second to sixth century C.E.) were analyzed to determine their composition. Information about the colorants used in the beads was also obtained through analysis of the distribution, size, composition, and structure of the particles included in the glass. Lead-isotope analysis was used on a small number of the beads to obtain information about their provenance.

Analytical Methods

Samples

Fifty-eight glass bead samples excavated from Paekche-period tombs and temples were studied. The characteristics of the beads are summarized in table 1. Thirteen of the samples were from the Neungsan-ri temple in Buyeo (sixth century), and these included glass of many colors, such as yellow, green, orange, red, and blue. Particles that seem to be colorants were found in the yellow, green, orange, and red glass. There did not seem to be particles in the blue glass. Thirty-one samples from Muryeong's tomb in Gongju (sixth century) were also studied. Like the Neungsan-ri glass, particles that seem to be colorants were found in the yellow, green, orange, and red glass.

Finally, fourteen glass bead samples from the Poongnap tomb in Seoul (second to fourth century C.E.) were studied. Unlike the glass from the above two sites, few colors were found, and the surface of the glass was not homogeneous. Large yellow particles, approximately 0.1 mm in diameter, were found in the green glass from the Poongnap tomb.

Methods

SEM-EDS, micro-focus x-ray diffraction (micro-XRD), and TIMS were used to analyze the beads. The SEM work was carried out with a Hitachi S-3500N microscope using an accelerating voltage of 20 kV and 75 μA beam current at a working distance of 15 mm. Quantitative EDS analysis was carried out for nine major components—SiO_2, Al_2O_3, Na_2O, MgO, CaO, K_2O, CuO, Fe_2O_3 and TiO_2—using a Kevex Super Dry system. Measurements were made on an area 60 × 60 mm^2 for two hundred seconds. Both standard samples and excavated samples were analyzed under the same conditions, and a XPP/ASAP method was used for quantification. Principal component analysis (PCA) was conducted on the glass compositions using the amounts determined for all nine major oxides.

Micro-XRD was used to identify the crystal structure of particles embedded in the yellow, green, and orange glass samples that contained relatively large particles. A Bruker's GADDS system was used with an accelerating voltage of 40 kV and a current of 40 mA. Collection times ranged from two hundred to six hundred seconds, and the beam was focused with 0.1–0.5 mm collimators.

Lead-isotope data were obtained using TIMS for samples with relatively high lead contents. These included one

Table 1. *Descriptions and compositions of glass samples*[1]

Sample	Color	Particles	Na_2O	MgO	Al_2O_3	SiO_2	K_2O	CaO	Fe_2O_3	TiO_2	CuO	Total
1	yellow	yellow	17.34	0.25	11.60	63.98	0.62	2.89	0.45	0.50	ND	97.63
2	yellow	yellow	15.76	0.35	9.25	68.71	0.67	1.72	0.44	0.70	ND	97.6
3	green	yellow	16.59	0.35	10.89	64.04	0.58	3.30	0.67	0.37	0.91	97.7
4	green	yellow	16.37	0.18	11.20	61.52	0.59	2.71	0.59	0.48	0.62	94.26
5	red	black	15.92	0.17	10.48	64.24	0.82	2.96	0.45	0.78	1.28	97.1
6	red	ND	16.00	0.33	10.50	64.45	0.81	2.96	0.46	0.94	1.67	98.12
7	red	black	14.95	0.07	10.30	64.21	0.83	3.00	0.47	0.85	1.38	96.06
8	orange	red black	13.09	0.60	10.91	58.91	0.67	3.51	0.75	0.53	10.62	99.59
9	orange	red black	12.27	0.33	10.74	60.21	0.69	3.34	0.42	0.48	10.29	98.77
10	dk blue	ND	18.59	ND	11.35	64.19	0.68	3.38	0.42	0.17	1.18	99.96
11	dk blue	ND	15.44	0.33	6.22	74.75	0.41	1.99	0.47	0.62	ND	100.23
12	lt blue	ND	18.23	ND	9.35	68.87	0.67	2.95	0.39	0.43	0.93	101.82
13	lt blue	ND	18.15	ND	9.02	66.60	0.62	2.92	0.37	0.63	0.63	98.94
14	blue	ND	15.86	2.51	4.70	64.46	3.03	6.44	1.18	1.44	trace	99.62
15	yellow	yellow	16.76	1.46	6.37	69.77	0.38	1.04	1.70	1.92	trace	99.4
16	yellow	yellow	17.74	1.44	6.52	70.17	0.30	0.96	1.92	1.99	trace	101.04
17	orange	red black	11.97	1.75	11.36	57.02	2.12	3.45	2.97	1.86	6.86	99.36
18	orange	red black	10.04	1.95	11.28	59.84	2.29	3.63	2.70	2.15	6.77	100.65
19	red	black	14.20	1.40	9.77	63.70	2.08	3.15	1.90	1.87	0.36	98.43
20	red	black	14.07	1.32	10.0	64.34	2.11	3.26	1.98	2.18	0.36	99.62
21	lt green	yellow	15.81	1.53	5.83	70.46	0.38	1.60	1.73	2.25	0.33	99.92
22	lt green	yellow	15.70	1.56	5.82	70.00	0.39	1.63	1.72	2.16	0.28	99.26
23	lt green	yellow	15.71	1.57	5.84	71.57	0.39	1.64	1.74	2.17	0.29	100.92
24	bl green	ND	14.62	1.44	9.49	65.10	2.37	3.95	1.32	1.93	0.33	100.55
25	bl green	ND	14.01	1.41	9.38	64.82	2.39	4.10	1.58	1.91	0.30	99.9
26	bl green	ND	14.07	1.06	9.25	64.01	2.29	4.04	1.53	1.94	0.27	98.46
27	bl green	ND	13.73	1.05	9.17	64.16	2.30	4.13	1.61	2.04	0.29	98.48
28	gr blue	ND	16.41	1.12	7.70	65.48	1.29	3.12	1.47	1.64	0.35	98.58
29	gr blue	ND	16.54	1.12	7.52	65.62	1.31	3.18	1.30	1.67	0.34	98.6
30	blue	ND	17.04	1.13	7.60	63.50	2.39	3.53	2.08	1.79	trace	99.06
31	dk green	ND	15.85	2.30	4.43	62.94	3.07	6.61	1.88	1.52	trace	98.6
32	blue	ND	16.46	1.19	8.03	62.19	2.18	4.58	2.50	1.78	trace	98.91
33	blue	ND	15.76	2.66	4.61	62.39	2.18	6.70	1.98	1.48	trace	97.76
34	lt blue	ND	15.24	1.17	5.81	65.82	0.38	4.00	2.39	1.62	0.19	96.62
35	lt blue	ND	16.85	1.18	5.80	65.83	1.20	4.18	1.50	1.61	trace	98.15
36	lt blue	ND	18.20	1.17	6.09	68.77	1.12	3.99	1.62	1.45	trace	102.41
37	lt blue	ND	18.03	1.12	5.77	68.34	1.08	3.97	1.51	1.39	trace	101.21
38	blue	ND	17.71	1.41	7.68	66.91	0.54	3.03	1.21	1.48	0.35	100.32
39	blue	ND	17.89	1.56	7.56	64.93	0.53	1.70	1.75	2.14	0.41	98.47
40	violet	ND	2.42	1.18	4.05	72.51	11.99	2.48	1.15	1.48	trace	97.26
41	violet	ND	2.42	1.22	4.99	71.14	11.64	2.67	2.67	1.61	trace	98.36
42	gold	ND	18.37	2.17	4.51	60.37	3.24	5.40	2.94	1.26	trace	98.26
43	yel-red	ND	17.54	2.20	4.76	59.14	3.30	6.18	3.85	1.37	0.20	98.54
44	black	ND	16.63	1.14	7.82	68.38	2.11	2.15	1.37	2.20	trace	101.8
45	lt-blue	ND	11.93	1.44	5.23	70.15	0.42	4.15	1.80	1.77	0.13	97.02
46	lt-green	yellow	16.54	1.41	9.04	65.25	0.72	3.08	1.57	2.02	0.34	99.97
47	lt-green	yellow	15.00	1.09	5.64	62.07	2.41	2.67	1.73	2.21	1.27	94.09
48	dk-green	ND	2.81	2.87	10.99	52.78	1.33	18.9	6.33	2.72	ND	98.73
49	dk-green	ND	2.66	3.48	9.52	49.73	1.21	26.5	3.19	2.52	0.27	99.08
50[2]	dk green	brown	2.52	3.51	9.27	47.54	1.08	26.6	2.36	2.42	0.22	95.52
51[2]	dk green	brown	2.64	2.95	8.31	44.94	4.47	23.4	4.68	2.71	ND	94.1
52[2]	dk green	ND	12.84	3.15	8.84	63.85	0.59	2.54	3.54	2.80	ND	98.15
53[2]	dk green	ND	12.95	3.29	8.96	61.96	0.53	2.88	5.20	2.81	ND	98.58
54	dk green	yellow	ND	ND	3.49	53.17	0.08	0.50	0.50	ND	40.1	97.84
55	lt blue	ND	14.21	1.98	7.08	68.16	0.95	2.60	1.72	1.72	1.31	99.73
56	lt blue	ND	13.82	2.05	7.11	67.52	0.97	2.66	1.62	1.62	0.89	98.26
57	lt blue	ND	14.06	1.94	6.87	68.83	0.96	2.61	1.60	1.60	1.16	99.63
58	lt blue	ND	14.31	1.98	6.89	68.55	0.95	2.59	1.59	1.59	1.07	99.52

1. Samples 1–13 are from Neungsan-ri. Samples 14–44 are from Muryeong. Samples 45–58 are from Poongnap

2. Heterogeneous surface

ND = None detected

Neungsan-ri sample, two Muryeong samples, and one Poongnap sample. LDA was used to analyze the lead-isotope data.

Results and Discussion

Composition

The EDS compositional data determined for the samples are given in table 1. They were found to be soda glass, except for samples 40 and 41, which were potash glass. Figure 1 shows their distributions, as determined by PCA analysis.

Group I is composed of glass beads excavated from the Neungsan-ri temple. It is a soda glass where the Na_2O content is between 12.3 and 18.6 wt. %. Group I can be divided into two subgroups: group Ia, with low Na_2O content (about 13 wt. %) and high MgO (0.3–0.6 wt. %); and group Ib, with high Na_2O content and low MgO content.

Group II includes glass beads from the Muryeong tomb and blue glass from Poongnap tomb. The group can be divided into three subgroups: IIa, a group with less than 10 wt. % Na_2O and less than 2 wt. % MgO; IIb, a group with high Na_2O (16–18 wt. %) and over 2 wt. % MgO; and IIc, a group with high Na_2O and less than 2 wt. % MgO. Some of the glass from the Poongnap tomb belongs to the IIc group with the same tendency as the ones from the Muryeong tomb in terms of Na_2O and MgO.

Group III includes dark-green glass beads excavated from the Poongnap site and has a high MgO content (about 3 wt. %). In group III, there are two subgroups: IIIa, a high CaO (20 wt. %) potash glass with very low soda (2.5–2.8 wt. % Na_2O); and IIIb, a soda-glass group (about 13 wt. % Na_2O).

Particles Embedded in Glass Beads

Results of EDS mapping and XRD analysis of particles included in the glass are shown in table 2 and plates 34–39. It was possible to analyze particles embedded in yellow, green, and orange glass with micro-XRD to identify their crystal structures, but in the case of the red and dark-green glasses, it was impossible to identify the structure due to the very small size of the particles. In these cases, information on the particles was obtained using EDS mapping. The majority of the glasses with particles had only one type; however, the dark-green glass from the Poongnap tomb was found to have two types.

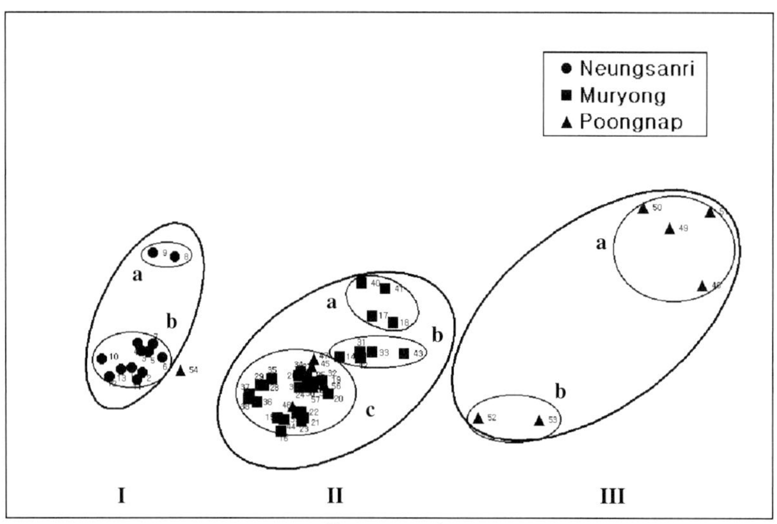

***Figure 1.** PCA mapping result of fifty-eight glass beads using nine compounds*

***Table 2.** Analytical results of particles in Korean ancient glass beads*

Color (particle)	Element or minerals	Shape of particles	Nungsan-ri temple Ratio	Nungsan-ri temple Size (μm)	Muryeong tomb Ratio	Muryeong tomb Size (μm)	Poongnap Ratio	Poongnap Size (μm)
Yellow (yellow)	Pb, Sn, O ($PbSnO_3$)	Irregular	Pb/Sn 5.4	10–30	Pb/Sn 5.1	15–30	—	—
Green (yellow)	Pb, Sn, O ($PbSnO_3$)?	Irregular	Pb/Sn 4.3	5–20	Pb/Sn 4.4	<30	Pb/Sn 4.8	0.1 mm
Orange (black)	Cu, O (Cu_2O)	Spherical		30–120		60-100	—	—
Red (black)	CuS or Cu_2S	Spherical	Cu/S 3.95	3–6	Cu/S 3.68	80	—	—
Dark green (black, yellow)	Fe, O Pb, Sb, O($Pb_2Sb_2O_7$?)	Spherical Irregular	— —	— —	— —	— —	 Pb/Sb 3.5	< 50 < 10

Yellow glass excavated from the Neungsanri temple and the Muryeong tomb contained yellow particles of irregular shape with a size of 10–30 μm. The particles are composed of Pb, Sn, and O, and the ratio of Pb to Sn is 5.4 in Neungsan-ri glass and 5.1 in Muryeong glass. Micro-XRD results showed that the yellow particles in the yellow glass from the Neungsan-ri temple were $PbSnO_3$.

Green beads were found in all three sites. The glass extracted from the Poongnap tomb was found to have especially small yellow particles 0.1 mm in diameter. The particles were composed of Pb, Sn, and O. The Pb-to-Sn ratio is 4.3 in glass from Neungsan-ri, 4.4 in glass from the Muryeong tomb, and 4.8 in glass from Poongnap. The small particles in the glass are irregular in shape, and those from Poongnap are oval.

Orange glass from the Neungsan-ri and Muryeong sites has round particles that are 30–120 μm in size. Plate 36 shows dark-red particles in the orange glass from the Muryeong tomb. Red color can be seen to spread outward from the particles, which were identified as Cu_2O using micro-XRD.

Very fine particles are visible in a back-scattered electron image of the red glass excavated from the Muryeong tomb. In the case of the Neungsan-ri glass, it was impossible to identify the structure using micro-XRD because the particle size is too small. Although the round particles in Muryeong tomb glass are relatively large (80 μm), it was still hard to identify the particle structure, as the XRD pattern was unclear. However, using EDS mapping, the particles were found to be composed of Cu and S, possibly CuS or Cu_2S. The Cu-to-S ratio for the particles is 3.95 and that for Cu_2S is 3.96, while that for CuS is 1.98, making it most likely that they are Cu_2S.

There are two types of particles in the green glass extracted from the Poongnap site. One is reddish-brown in color, composed of Fe and O. The other appears yellow and is composed of Pb, Sb, and O. Accurate analysis was impossible using micro-XRD, but the structure was estimated from EDS results. Bright yellow particles with irregular shape embedded in the dark-green glass (no. 54) were different in composition (Pb, Sb, O) from yellow particles found in the other yellow and green glasses (Pb, Sn, O). With EDS mapping, we found that there were high amounts of calcium in the rough glass surface, and the results were confirmed with SEM/EDS analysis, where the CaO content was determined to be as high as 26.6 wt. %.

Table 3. Lead-isotope data and discriminate scores

No.	206/204	207/204	208/204	207/206	208/206	DS_1[1]	DS_2
1	18.254	15.731	38.404	0.8618	2.1039	0.198	0.385
14	18.218	15.699	38.332	0.8617	2.1041	0.166	0.279
16	18.213	15.676	38.263	0.8607	2.1008	0.076	0.216
46	18.207	15.698	38.307	0.8622	2.1039	0.174	0.280

1. $DS_1 = —0.571X_1 + 1.916X_2 — 0.091X_3 + 8.292X_4 + 1424X_5 — 53.13$
$DS_2 = 1.025X_1 + 3.231X_2 — 0.487X_3 + 7.280X_4 + 3.140X_5 — 63.33$

Lead-Isotope Ratio

Discriminate scores, DS_1 and DS_2, were produced by putting the lead isotope ratio data of the unknown Paekche samples into the linear discriminate function given in table 3, where X1, X2, X3, X4, and X5 are the values of 206Pb/204Pb, 207Pb/204Pb, 208Pb/204Pb, 207Pb/206Pb, and 208Pb/206Pb, respectively. The scores are plotted (DS_1 vs. DS_2) along with results of the analyses of lead ore samples of known provenance from Korea, China, and Japan in figure 2. The lead used for all four Paekche samples gave similar results to those of lead from the southern part of China.

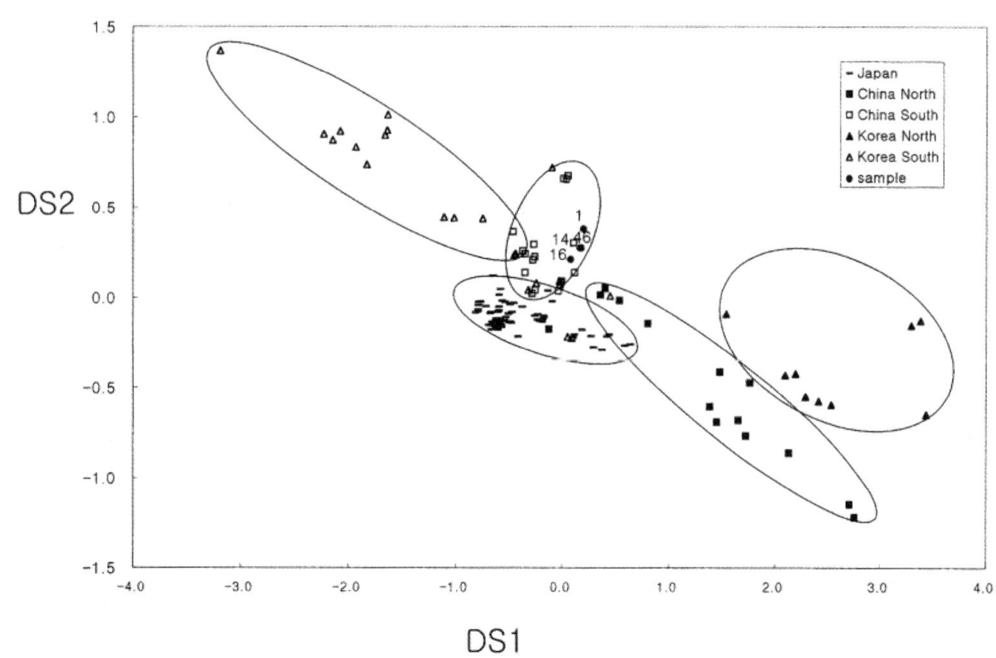

Figure 2. *Distribution of lead ore in Korea, China, and Japan by LDA*

Conclusions

Although most of the glass beads excavated from Neungsan-ri, Muryeong, and Poongnap are soda glass, they can be divided depending on tomb site. Subgroups of glass from individual sites can be formed, and these depend particularly on Na_2O and MgO.

Yellow particles in yellow and green glass were found to be $PbSnO_3$ in glass from all three tombs. The colorant in the orange glass was Cu_2O and EDS mapping suggests that the colorant for the red glass was Cu_2S. Dark-red particles in dark-green glass from the Poongnap tomb are a combination of Fe and O, and yellow particles are a combination of Pb, Sb, and O.

Lead isotope analysis was conducted for four glass beads, including samples excavated from all three sites studied. Comparison with lead ores of known provenance indicates that lead ores from southern China were used in the production of these four samples.

Selected Reading

Jo, K.M., H. S. Yu, and H. T. Kang. 1999. Scientific analysis of the glass from Hwangnam-daech'ong tomb no. 98. *Conservation science in museums* 1.

Kim, Y. S. 1979. Analysis of major chemical components of cultural samples excavated at Hwang Ryong-sa temple site. *Kobunkazai no kagaku* 24:44–50.

Kuleff, I., R. Djingova, and I. Penev. 1984. Analysis of ancient and medieval glasses by INAA. *Journal of radio analytical and nuclear chemistry* 83:333–43.

Lee, I. S. 1990. Archaeological study of ancient glass in Korea. Ph.D. thesis, Hanyang University.

Mabuchi, Hisao, and Yoshimichi Emoto. 1980. Tokyo: Kodai garasu no kagaku bunseki. In *Tokyo no kodai garasu*, 162–67. Tokyo: Tokyo Kokuritsu Hakubutsukan.

22 Scientific Study on the Glass Beads Found in the Yayoi Period of Japan

Takayasu Koezuka and Kazuo Yamasaki

ABSTRACT The early cultural history of Japan is divided into the Jomon and Yayoi periods. Glasses have never been found from the Jomon period. However, the earliest glasses are found in northern Kyushu in the Yayoi period, from circa the end of the third century B.C.E. to the early second century B.C.E.

In the Yayoi period, the glass systems (chemical classifications) mainly in the form of beads are potash glass (K_2O-SiO_2 system) and lead-barium glass (PbO-BaO-SiO_2 system), and both of these glass systems appeared almost at the same period in Japan. The lead-barium glasses circulated only in the Yayoi period, though potash glasses circulated until the late Kofun (Tumulus) period (sixth century). It is not well understood where the potash glasses were made, although the lead-barium glasses found in Japan were made in China (Yamasaki 1974; Brill et al. 1979, 87; Yamasaki and Murozumi 1991).

Chemical compositions of these potash glasses and lead-barium glasses are quite similar to those found in China and Korea, and these glasses were probably brought to Japan from China or through the Korean Peninsula.

A small quantity of soda-lime glasses appeared from the first century C.E. to the second century C.E., and they had two different systems (Na_2O-CaO-SiO_2 and Na_2O- Al_2O_3-CaO-SiO_2).

The opaque red beads known as "Indo-pacific monochrome drawn glass beads," which are called *mutisalah* (Glover and Henderson 1994), have been found from sites of the late Yayoi period (second century C.E.) and were identified to be soda-lime glass. The *mutisalah* found in Japan were similar to the opaque red beads found from Arikamedu in India. It is thought that these beads were manufactured in south and southeast Asia and brought to Japan via China and the Korean Peninsula.

Introduction

In recent years archaeological excavations at more than ten thousand sites were carried out in Japan, and large amounts of relics, including glasses, have been found. We have carried out quantitative analyses of about twelve hundred ancient glasses (including 450 samples of the Yayoi period) excavated in Japan and found that the chemical compositions of glasses changed with the historical ages.

In this report, summaries of the chemical compositions and structures of glass beads in the Yayoi period will be presented and discussed.

Potash Glasses (K_2O-SiO_2 System)

In the first half of the late Yayoi period (first century C.E.) (fig. 1), large quantities of potash glasses have been found in the areas ranging from the Kyushu to the Kanto district (fig. 2) but rarely found in Tohoku and Hokkaido.

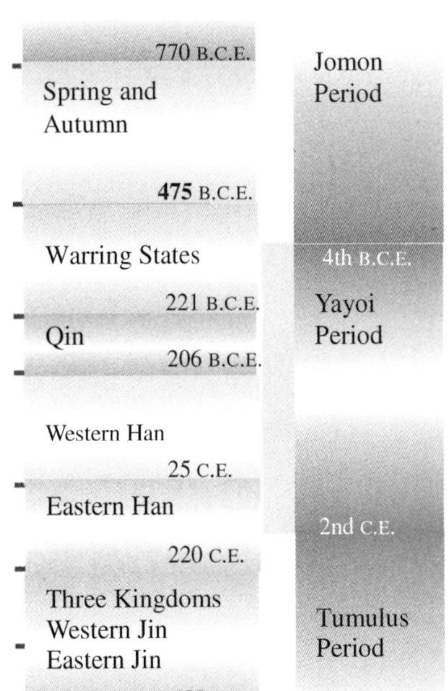

Figure 1. *Chronology of Japan and China*[1]

Type of Glasses

Almost all of the objects are small, monochrome drawn beads usually less than 6 mm in diameter (fig. 3) and monochrome drawn micro-beads less than 2 mm in diameter (fig. 4). Tubular monochrome beads (fig. 5) usually less than 30 mm in length are also found with a few comma-shaped beads (fig. 6). These small beads and micro-beads belong to the "Indo-Pacific beads" (Francis 1990).

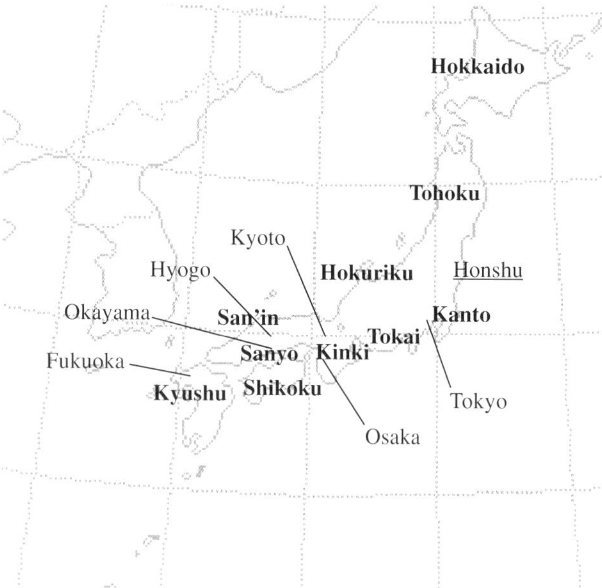

Figure 2. *Map of Japan showing the districts and several prefectures*

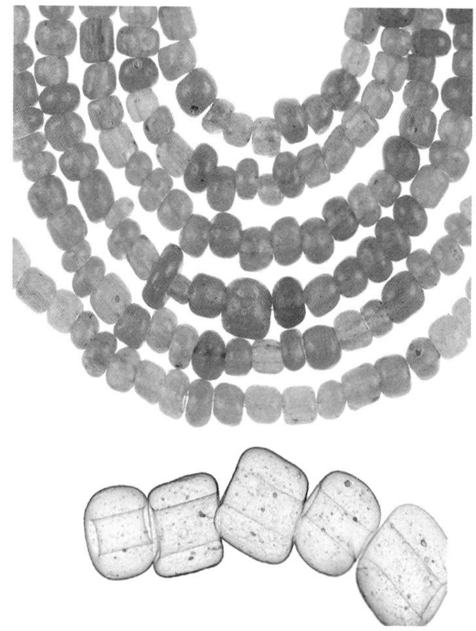

Figure 3. *Small beads (K_2O-SiO_2 system), monochrome, drawn, 2.0–3.0 mm in length, 2.5–3.2 mm in diameter, Sasaka tombs (1st century C.E.) in Kyoto Prefecture.* Below: *CR image*

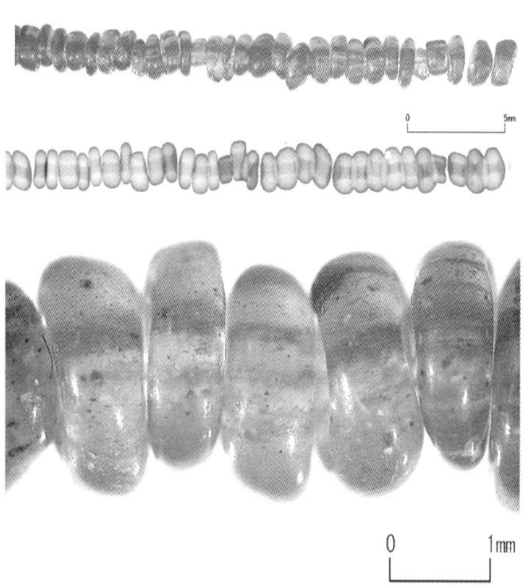

Figure 4. Above: *Dark blue micro beads, transparent, from the Higashi-Umeda tombs site in Hyogo Prefecture (2nd century C.E.), about 1.5 mm in diameter, 1.0 mm in length.* Center: *CR image.* Below: *close-up image of bead, showing drawn technique*

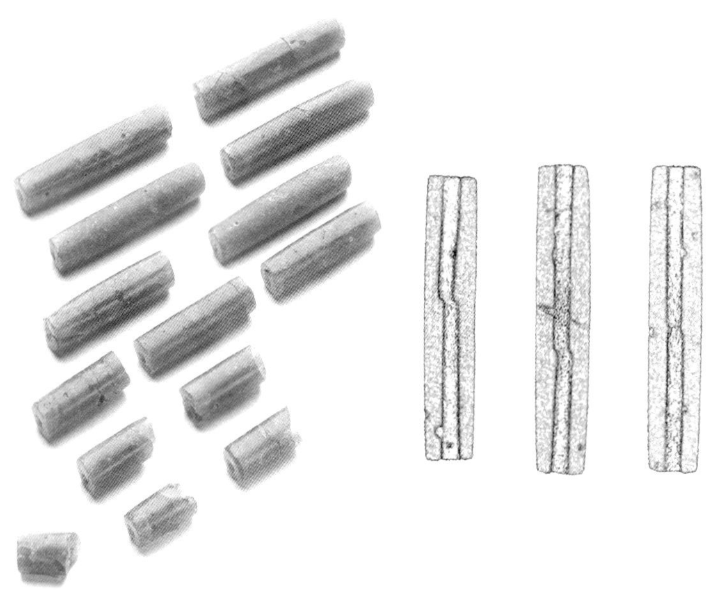

Figure 5. *Tubular beads from Misaka-Jinja tombs site in Kyoto Prefecture. Light blue, translucent, about 3 cm in length.* Right: *X-CT image shows that these tubular beads are made by the boring technique*

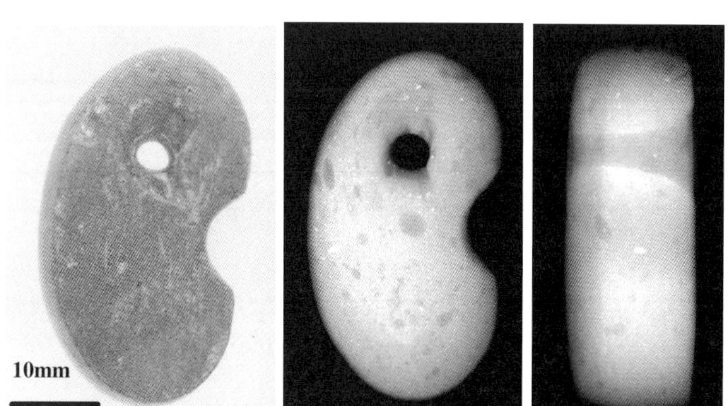

Figure 6. Left: *Comma-shaped bead (K_2O-SiO_2 system) from Sasaka tombs site in Kyoto Prefecture.* Center and right: *CR image. The bubbles were arranged in a concentric-ring structure. This shows that after molding, the hole was made by using a revolving rod*

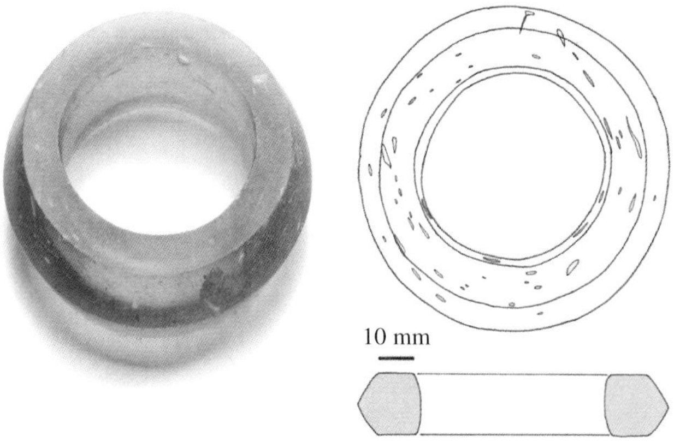

Figure 7. Large pale blue bracelet (K_2O-SiO_2 system), outer diameter: 9.7 cm, inner diameter: 5.8 cm, thickness: 1.8 cm, transparent. The bubbles in the glass are arranged as shown in the drawing

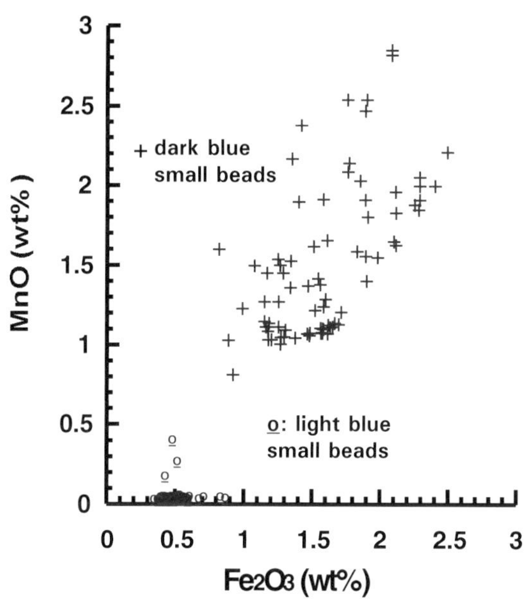

Figure 8. Plot of iron oxide vs. manganese oxide in light blue and dark blue small beads

Recently, an unusually large bracelet (outer diameter: 9.7 cm; inner diameter: 5.8 cm; thickness: 1.8 cm; pale blue, similar color to aquamarine crystal, transparent) made of potash glass was found for the first time together with thirteen bronze bracelets, eleven iron daggers, some iron fish spears, and 272 stone beads in the Ofuro-Minami tombs (ca. second century C.E.) in northern Kyoto Prefecture (fig. 7) (Koezuka 2000).

Most potash glasses are light blue or dark blue, and a few are greenish in color. Observation of more than fifty thousand small blue beads reveals that the ratio of light blue and dark blue colored beads varies from site to site.

In the Kyushu district, dark blue and light blue beads are roughly equal in proportion, while in the east region (San'in, Kinki, Hokuriku, Tokai, and Kanto districts), the ratio is about one to four. Therefore, the proportion of dark blue and light blue beads in these two regions may reflect different trade routes and distribution. In both regions, all of the small light blue beads are made with the drawing technique (cut from a tube) and are of similar size. All of the small dark blue beads are likewise made by the drawing technique, and some variation in coloration (dark blue to purplish-blue) is also observed in both regions.

Among tubular beads of potash glass, which are small in number and not as well understood, a distinctive type has emerged. They were light blue in color with holes bored from both sides by a drill. It is a peculiar type of construction that has never been found in the tubular beads made of lead-barium glass. This type of tubular bead has been found in Hokuriku, Kinki, San'in, and Sanyo districts, but not in Kyushu, suggesting limited distribution.

Chemical Compositions

The chemical compositions of the potash glass including small beads, tubular beads, and comma-shaped beads, show no significant variation. However, the potassium contents on the surface of the weathered portion decrease remarkably.

The main constituents of the potash glass are potassium oxide (K_2O), 17.9 ± 2.5%; silicon dioxide (SiO_2), 75.4 ± 2.8%; aluminum oxide (Al_2O_3), 2.7 ± 0.8%; and sodium oxide (Na_2O), magnesium oxide (MgO), and calcium oxide (CaO), each less than 1.5%. Metal oxide contents vary with the glass colors.

Dark blue beads, some of which are purplish, contain about 0.04–0.1% cobalt oxide (CoO) as well as iron oxide (Fe_2O_3), 1.0-2.5%; manganese oxide (MnO), 1.0–2.5%; copper oxide (CuO), 0.02 ± 0.01%; and, scarcely detected, lead oxide (PbO < 0.01%) (fig. 8).

The dark blue potash glasses colored with cobalt are richer in manganese oxide compared to glasses of other colors (table 1), indicating the cobalt colorant has a high manganese content. The chemical compositions of these dark blue glasses found in Japan are quite similar to those found in China (Fukang 1991) and Korea (Lee 1993).

Light blue glasses (some are greenish-blue in color) contain iron oxide 0.48 ± 0.07%, manganese oxide 0.01% (virtually at the limit of detection), copper oxide 1.14 ± 0.2%, and lead oxide about 0.1%. A trace of tin oxide (SnO_2) was detected. Copper was apparently used in the coloring of the light blue glass, and the presence of lead and tin together with copper suggests that the colorant was either copper metal or a copper alloy.

An unusual find is a pale blue transparent glass bracelet resembling aquamarine. Analysis shows that the iron oxide content is only 0.4%, with less than 0.05% each of copper oxide, manganese oxide, and lead oxide. This result suggests that the potash-glass bracelet was intended to be transparent and colorless and was therefore fabricated without the addition of colorants, but iron that was present as impurities in the raw materials was reduced, yielding a pale blue color.

Lead-isotope ratios were measured for these beads (table 2), but there is very little data available to ascertain the origin of lead ores. However, the ratios differ markedly from those obtained by R. H. Brill for yellow opaque potash glass, dark blue glass, and red opaque glass unearthed in Vietnam

Table 1. Chemical compositions of some potash glasses (% wt.). These samples were analyzed by using EDXA. (Before analysis, the weathered portions were removed.) The pale blue bracelet was analyzed by nondestructive EDXA technique. (A fresh small broken surface was measured.)

	Small bead, light blue; Sasaka tombs site	Small bead, dark blue; Sasaka tombs site	Tubular bead, light blue; Sasaka tombs site	Comma-shaped bead, light blue; Sasaka tombs site	Micro bead, dark blue; Higashi-Umeda tombs site	Bracelet, pale blue; Ofuro-Minami tombs site
SiO_2	76.8	74.8	77.7	76.3	76.5	82.8
Al_2O_3	2.3	1.4	2.4	2.5	1.6	2.7
Fe_2O_3	0.52	1.88	0.40	0.5	2.1	0.4
TiO_2	0.06	0.08	0.05	0.05	0.07	0.01
CaO	0.8	1.12	0.7	0.7	0.7	0.7
MgO	0.3	0.4	0.4	0.3	0.3	0.2
Na_2O	0.03	0.4	0.3	0.3	0.3	0.4
K_2O	17.7	18.3	16.9	18.3	17.1	13.0
PbO	0.09	<0.02	0.08	0.07	<0.02	—
MnO	0.02	1.58	0.02	0.02	1.9	0.04
CuO	1.40	0.02	1.09	1.00	0.04	0.01
CoO	—	0.06	—	—	0.06	—

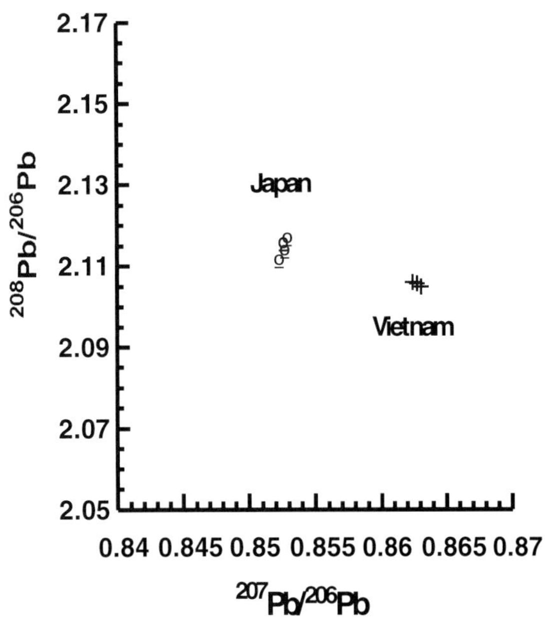

Figure 9. Lead-isotope ratios of potash-glass beads. Sample from the Sasaka and Misaka-Jinja tombs site (1st century C.E.) in Kyoto Prefecture. Vietnamese glass data are obtained from Brill

(Brill and Shirahata 1995). The lead-isotope ratios of light blue potash glasses are not in the same area with those of Vietnamese glasses, although they are distributed near the area of Chinese Han bronzes (fig. 9).

Lead-Barium Glass (PbO-BaO-SiO_2)

The earliest lead-barium glasses have been found from the end of the early Yayoi period to the beginning of the middle Yayoi period in the archaeological sites of northern Kyushu (Yamasaki 1973). In other words, lead-barium glasses and potash glasses appeared almost at the same period and in the same region. In the subsequent latter half of the middle Yayoi to the end of the Yayoi period, large quantities of lead-barium glasses have been found in areas

Table 2. Lead-isotope ratios of potash glasses 175. Sample nos. 1–4: small bead, 3–5 mm in diameter, light blue, translucent, from Sasaka tombs. Sample no. 5: Misaka-Jinja tombs in Kyoto Prefecture

No.	$^{207}Pb/^{206}Pb$	$^{208}Pb/^{206}Pb$	$^{206}Pb/^{204}Pb$
1	0.8522	2.1101	18.405
2	0.8526	2.1126	18.417
3	0.8525	2.1143	18.389
4	0.8525	2.1144	18.409
5	0.8528	2.1155	18.394

ranging from the Kyushu to the Kinki districts, but this glass system has hardly ever been found after the Yayoi period.

Type of Glasses

Among lead-barium glass objects, tubular beads are the most abundant, followed by small beads and small amounts of comma-shaped beads (fig. 10).

It is believed that almost all objects were imported as products, such as tubular beads, small beads, or *pi*-disks, although some of the comma-shaped beads were manufactured in Japan with molds (fig. 11) by using wastes or cullets brought from China or Korea (Fujita 1994).

Almost all the tubular beads were fabricated by the winding (fig. 12) or the drawing technique. The winding

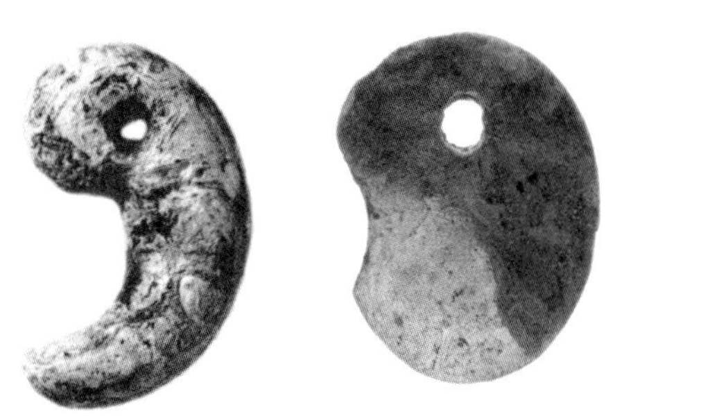

Figure 10. Comma-shaped beads. Left*: Suku-Okamoto site, heavily weathered, 3.1 cm in length;* right*: a variety of comma-shaped bead, Sasaka tombs, green transparent, 1.7 cm in length*

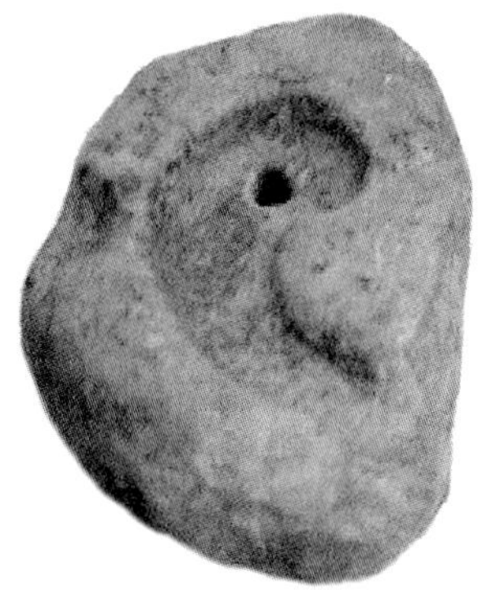

Figure 11. *Mold for comma-shaped bead from Suku-Akaide site in Fukuoka Prefecture; about 6 cm in length*

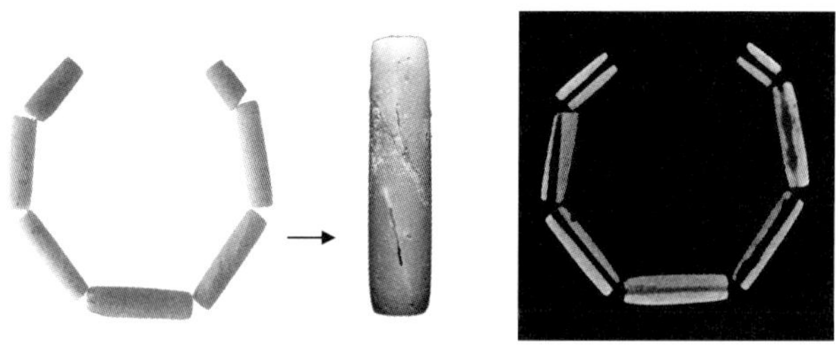

Figure 12. *Light green translucent tubular bead from Naibayama site in the Hyogo Prefecture. 2.2–0.7 cm in length.* Right*: X-CT image. These tubular beads were made by the core-winding technique*

technique involves winding molten glass directly around a metal core, around which substances such as clay are used to make removal of the core easy.

More than fifteen hundred light blue tubular beads made by drawing and twisting (plates 40 and 41) have been unearthed from sites from about the first century B.C.E. to the end of the second century C.E. These tubular beads are supposed to have been made by twisting blue glass tubes, which for some reason produced bands that weathered more easily than the surrounding portions. The weathered parts turned white, giving a pattern of spiral bands. Microcrystalline bright blue particles were observed in the weathered parts.

The blue particles were identified as "Han-blue ($BaCuSi_4O_{10}$)" by the powder diffraction method (fig. 13). Han blue has been known only as a pigment in the ancient Chinese palette and has never been reported in lead-barium glasses in either Japan, Korea, or China. The crystal system of Han blue is tetragonal, and the lattice constants are a-axis 7.43-7.45 Å and c-axis 16.13-16.21 Å, as calculated by the Willson and Pike method (Koezuka and Yamasaki 2001). These data are similar to those for Han blue hitherto reported for the Chinese faience beads (table 3) (Yamasaki, Koezuka, and Shirahata 1995).

The powdered Han blue might have been added into molten glass as a colorant. In transparent or translucent green lead-barium glasses, Han-blue or other crystalline colorants have not been detected.

Chemical Compositions

It is difficult to determine the chemical composition of archaeological finds of lead-barium glass found in Japan, because most are heavily weathered without good preservation (Brill et al. 1979).

Generally, cerussite ($PbCO_3$) or pyromorphite ($Pb_5(PO_4)_3Cl$) are formed as weathering products on the surface of glasses.

More than forty glass samples were polished to disclose fresh parts, and their analyses showed variable chemical compositions as shown in the following: 1–6% sodium oxide (Na_2O), 36–58% silicon dioxide (SiO_2), 7–17% barium oxide (BaO), and 28–46% lead oxide (PbO). Green lead-barium glasses have 1–2% copper oxide (CuO) contents, and light blue glasses have about 0.5% (table 4.).

There is no clear correlation between lead oxide and barium oxide contents (fig. 14). These chemical composi-

Table 3. *The lattice constants of Han blue in the lead barium-tubular bead. The Miyauchi sample could not be determined exactly due to the small amount of the blue particles*

Sample	a-axis (Å)	c-axis (Å)	Ratio (c/a)
Arimoto site in Okayama Prefecture.	7.45 ± 0.004	16.21 ± 0.06	2.18
Akasaka-Imai site in Kyoto Prefecture	7.44 ± 0.004	16.13 ± 0.01	2.17
Miyauchi site in Tottori Prefecture	7.39 ± 0.06	16.11 ± 0.14	2.18
Faience bead (China)	7.43 ± 0.01	16.15 ± 0.07	2.17
PDF 12-0510	7.44	16.11	2.17

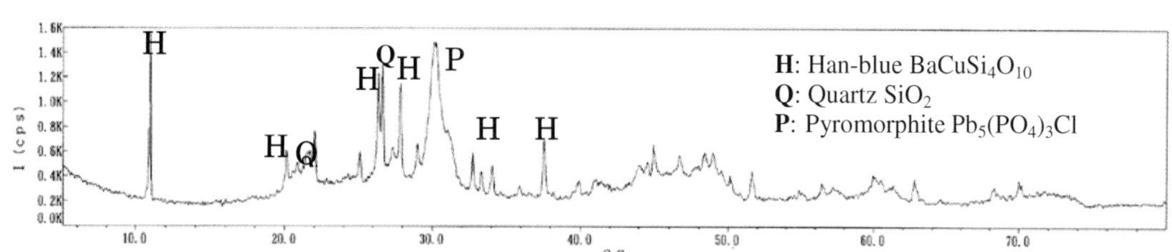

Figure 13. *X-ray diffraction spectra obtained from the Akasaka-Imai tumulus bead. Han blue is the colorant for the light blue lead-barium glass. Quartz is an unmelted ingredient in the lead-barium glass. Pyromorphite is regarded as a weathering product formed by the presence of a buried human body*

***Table 4.** Chemical compositions of some lead-barium glasses (wt. %). These samples were analyzed by using EDXA. (Before analysis, the weathered parts were removed.) The tubular bead from Sugu-Okamoto sample was analyzed by the wet-chemical method (K. Yamasaki 1974)*

	Comma-shaped bead (3 pieces), green transparent; Sasaka tombs site, Kyoto Prefecture	Small quadrilateral bead (3 pieces), green, transparent; Karako site, Nara Prefecture	Tubular bead, green, transparent; Sugu-Okamoto site, Fukuoka Prefecture	Tubular bead (7 pieces), light green; Naebayama site, Hyogo Prefecture	Tubular bead, (3 pieces), light blue, translucent; Arimoto site, Okayama Prefecture
SiO_2	39.1–41.6	30.7–41.0	38.0	38.8–57.7	43.3–45.9
Al_2O_3	0.5–0.7	0.6–1.0	0.35	0.3–0.5	0.6–0.7
Fe_2O_3	0.2–0.3	0.1	0.29	0.1–0.2	0.1
CaO	0.8–0.9	0.3	1.1	0.5–1.3	3.7–4.2
MgO	0.3–0.6	0.3–0.4	0.15	0.1	0.3–2.5
Na_2O	1.9–2.3	1.0–1.2	3.9	3.5–5.8	0.1–1.1
K_2O	0.1	0.1–0.2	0.19	0.1–0.4	0.1
PbO	39.3–43.1	40.8–50.7	38.5	28.4–43.0	30.5–33.4
BaO	12.9–14.6	10.2–14.7	14.0	7.1–16.5	14.8–16.2
CuO	0.3–0.7	0.5–0.9	0.78	0.2–0.7	0.3–0.4
MnO	0.1–0.6	0.5		0.1	0.1

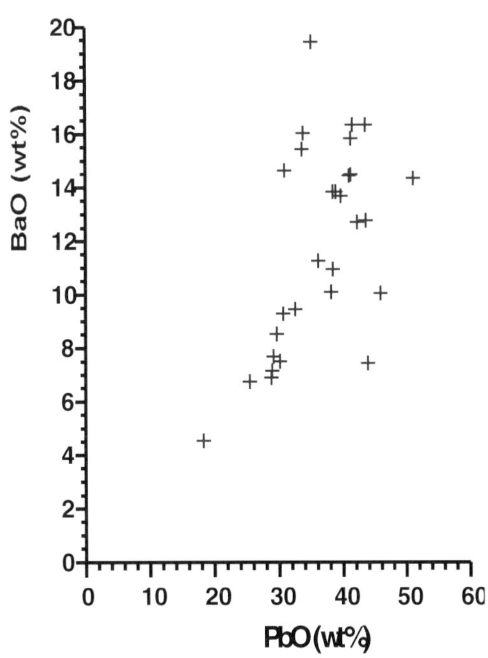

***Figure 14.** The relationship between lead and barium contents in the lead-barium glasses*

tions of lead-barium glass found in Japan are quite similar to those found in China and Korea, and furthermore, the distribution of lead-isotope ratios are considered to be the same (table 5).

Lead glasses without barium were found by Yamasaki (Oga and Gupta 2000), but objects of this glass system were very small in number. The lead oxide contents of these glasses are about 70%.

Soda-Lime Glass (Na_2O-Al_2O_3-CaO-SiO_2 and Na_2O-CaO-SiO_2 System)

Soda-lime glass first appeared as a few small beads and tubular beads in Japan considerably later than lead-barium and potash glasses (i.e., in the late Yayoi period [first century C.E.]). Later, large numbers of soda-lime glasses were distributed in the Kofun period (third to sixth century C.E.) (Koezuka and Yamasaki 1995).

Type of Glasses and Chemical Compositions

Soda-lime glasses are divided by their chemical compositions into two main types. One has high alumina (Al_2O_3 > 4%) and relatively low lime contents (CaO > 4–5%) (Na_2O-Al_2O_3-CaO-SiO_2 system); the other has low alumina and high lime contents (Na_2O-CaO-SiO_2 system). Most soda-lime glass beads with opaque red, yellow green, light blue, and dark blue colorations in the Yayoi period had high alumina content, and only a few dark blue beads had low lime content. All these high alumina soda-lime glass beads had the characteristics of "Indo-Pacific beads" (usually less than 6 mm in diameter, monochrome, and made by the drawing method).

Some small opaque red beads, typical Indo-Pacific beads, which are called *mutisalah*, have been found in northern Kyushu. About three hundred beads were found in the Hirabaru site (late second century C.E.) (fig. 15) (Watanabe 1991), and only a few were found in other sites (Oga and Guputa 2000). Drawn opaque red glass tubes, which are similar to the unfinished beads found in Arikamedu, India (Glover and Henderson 1994), were found in the Minami-Hachiman site and the Imajuku-Goroe site in Fukuoka Prefecture, northern Kyushu (fig. 16).

Out of the ten *mutisalah* beads from the Hirabaru site, the eight beads analyzed were of the high-alumina and low-lime type with the following compositions: SiO_2 60–70%, Al_2O_3 4.7–11%, Na_2O 16–20%, K_2O 1.1–2.5%, MgO less than about 1%, CaO 2.0–3.0%, TiO_2 0.27–0.33%, Fe_2O_3 0.8–1.3%, CuO 1.0–2.5%, MnO 0.1–0.2%, PbO 0.3–0.5%. The remaining two beads had lower aluminum oxide (3.0–3.2%) and somewhat higher calcium oxide (3.7–4.2%), with similar values for other oxides. The Minami-Hachiman and Imajuku-Goroe sites have each yielded one tubular bead, which appeared to be the intermediate (unfinished bead) for *mutisalah* beads, because they have characteristics similar to the drawn glass tubes of Arikamedu in India. By applying the nondestructive energy dispersive x-ray analysis

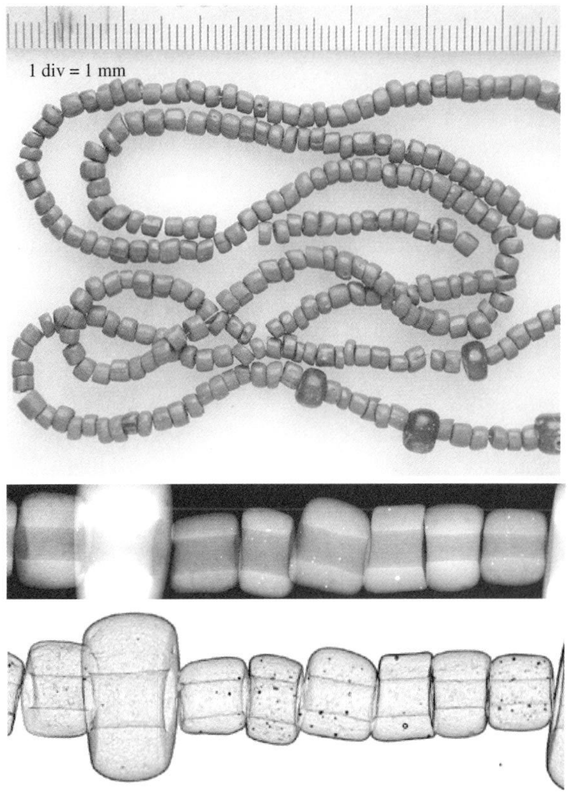

Figure 15. Above: Mutisalah *(opaque red drawn beads) from Hirabaru site (late 2nd century C.E.).* Center: *CR image.* Below: *Enhanced CR image of* mutisalah. *Small beads are dark blue potash-glass beads*

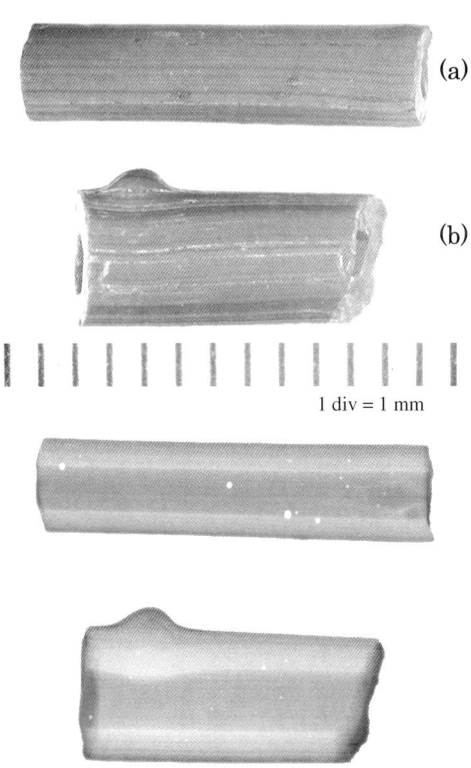

Figure 16. Above: *Opaque red drawn glass tube and unfinished bead from (a) the Minami-Hachiman site (2nd century C.E.) and (b) the Imajuku-Goroe site (2nd century C.E.) in Fukuoka Prefecture.* Below: *CR image. A small white round spot shows the existence of the material with a high absorption of x-rays*

(EDXA) method, they were identified to be a high-alumina, low-lime type, with lower (< 0.1%) lead oxide content.

Mutisalah beads unearthed in Japan include a few from the first and second century C.E. (Yayoi period), none from the third to fourth century C.E., and many from the fifth and sixth century C.E. (middle and late Kofun period). The *mutisalah* beads of the Kofun period are fundamentally similar to the Yayoi *mutisalah* beads, but some irregular beads have less than 4% aluminum oxide. Others have less than 0.1% lead oxide, and yet others have about 0.4% nickel oxide. Recently, *mutisalah* beads made of potash glass (opaque red, about 6 mm in diameter) have been found in the Rakuro-Jo site (first century B.C.E.–first century C.E.) in D.P.R. Korea (North Korea). *Mutisalah* beads are presumed to have been made in the various areas of south to southeast Asia.

One soda-lime bead of the Yayoi period was found at the Sasaka tombs (first century C.E.) in northern Kyoto Prefecture, which have yielded about seven thousand small beads and tubular beads of potash glasses with light blue and dark blue coloration. More than two hundred beads from the same site were analyzed, and this soda-lime glass of the greenish-blue color was proved to be of the high-alumina low-lime type (table 6).

Table 5. *Lead-isotope ratios for lead-barium tubular beads. Lead-isotope ratios of Arimoto and Akasaka tubular beads are different, although these beads have the same size, shape, and date*

Sample	$^{207}Pb/^{206}Pb$	$^{208}Pb/^{206}Pb$	$^{206}Pb/^{204}Pb$
Tubular bead, Arimoto site	0.85731	2.11833	18.2910
Tubular bead, Akasaka-Imai	0.91206	2.25081	17.1170
Tubular bead, Kyunai site	0.84947	2.10776	18.5840

Another remarkable find, segmented tiny beads from the Hirabaru site in Fukuoka Prefecture, shows an unusual forming technique for small beads. This technique is to cover a glass tube with a different kind of glass, resulting in a structure similar to gold-foil glass artifacts that have been found at Kofun-period sites. These beads are about 2 mm thick with outer diameters of 4~5 mm, dark blue on the outside and colorless and transparent with many bubbles on the inside (fig. 17). They are the high-alumina type of glass: SiO_2 67%, Al_2O_3 4.1%, Na_2O 15.4%, K_2O 0.5%, MgO about 3%, CaO 4.3%, TiO_2 0.1%, Fe_2O_3 2.3%, CuO 0.3%, MnO 1.3%, CoO 0.1%, PbO < 0.1%.

These segmented dark blue beads are colored with cobalt and have a high manganese content, a characteristic similar to the dark blue potash glass described above (table 6).

Hence, the cobalt ore in both cases probably came from the same source. On the contrary, many dark blue beads found in the sites of the Kofun period, which are colored with cobalt and are of the low-alumina, high-lime type (Na_2O-CaO-SiO_2 system), have an extremely low manganese content (MnO < 0.4%), indicating that the source of cobalt ores differs from that of the potash glass.

If the Chinese cobalt ores have a high manganese content, dark blue soda-lime glasses with a high-alumina content might have been made in south to southeast Asia with

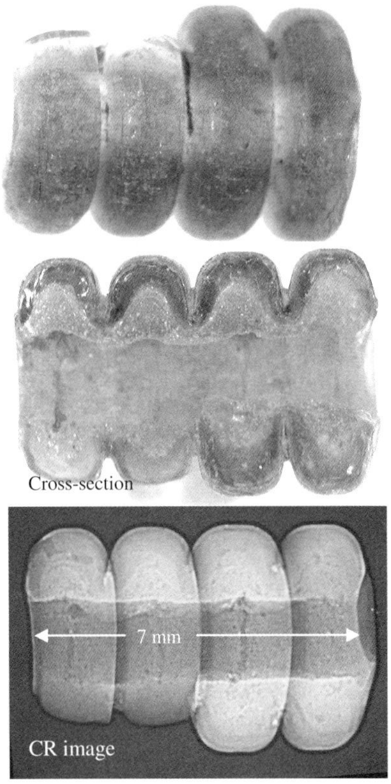

***Figure 17**. Segmented bead, from Hirabaru site in Fukuoka Prefecture (2nd century C.E.)*

***Table 6.** Chemical compositions (wt. %). Sasaka tombs site, Kyoto Prefecture, first century C.E.; 3–4 mm in diameter*

	Bluish-green small bead
SiO_2	62.5
Al_2O_3	8.0
Fe_2O_3	1.1
TiO_2	0.3
CaO	3.7
MgO	0.2
Na_2O	20.9
K_2O	2.1
PbO	0.13
MnO	0.12
CuO	0.68
CoO	–

such ores. These data are rather limited, due to the small number of glass beads available for analysis.

Conclusions

All of the small potash and soda-lime glass beads found in the sites of the Yayoi period in Japan belong to the Indo-Pacific bead type, whereas round small lead-barium beads are made by the winding technique and do not belong to the Indo-Pacific bead type.

Since the Indo-Pacific beads are widely distributed from south Asia to southeast Asia from around the third century B.C.E. to the tenth century C.E., it is presumed that these beads were brought into Japan via China or Korea, and some green comma-shaped lead-barium glass beads and light blue tubular potash-glass beads were made in Japan by remelting broken pieces or waste glass.

The potash glasses found in Japan in the Yayoi period show no variation in chemical compositions between small beads, tubular beads, and comma-shaped beads, suggesting similar batch materials for most potash-glass beads.

There is a possibility that these potash glasses were made in India and not in China, because unfinished dark blue potash glasses were found from Arikamedu in India. The dark blue potash glasses, including micro-beads colored with cobalt, show high contents of manganese without exception. This indicates that the colorant might have been a high manganese cobalt ore (i.e., Chinese cobalt ores). This is quite different from the dark blue low-alumina soda-lime glasses of the Kofun period, colored with low-manganese cobalt ore. Therefore, it is provable that these dark blue potash glasses using presumed-to-be Chinese cobalt ores were made in south to southeast Asia.

There is also a possibility that the potash glasses were made in south and southeast Asia because some *mutisalah* made of potash glass were found in which the colloidal technique (copper-red) not found in China at that period is used (Fukang 1991).

The Ofuro-Minami bracelet, the largest glass ornament yet to be found, may have been made in southeast Asia, because no bead with similar coloring has been found. It is very interesting that a similar glass bracelet has been found in Ban Don Ta Phet in the Kanchanaburi Province of west central Thailand (fig. 18) (Glover and Henderson 1994).

The lead-barium glass unearthed in Japan varies widely in chemical composition, and some distinctive light blue tubular beads are colored with Han blue (barium copper tetra-silicate), indicating that glasses made in different areas of China were brought to Japan.

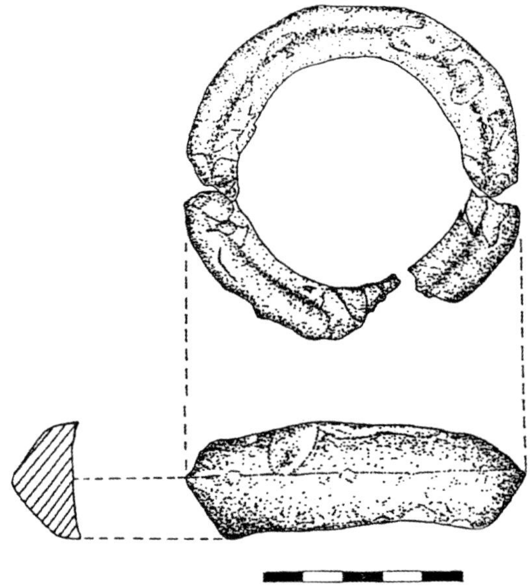

***Figure 18.** Glass bracelet from Ban Don Ta Phet in Thailand. Drawing by L. Ryan (Glover and Henderson 1994)*

The soda-lime glass, which appeared in the late Yayoi period, has two compositional types. One is the low-alumina high-lime type (Na_2O-CaO-SiO_2 system), including many dark blue small beads found at the Kofun-period sites and a few at the Yayoi-period sites. The other is the high-alumina low-lime type (Na_2O-Al_2O_3-CaO-SiO_2 system), which includes small red opaque beads *(mutisalah)* and small yellow-green beads, which are clearly Indo-Pacific beads. The dark blue segmented beads of two-layer construction from the Hirabaru site in Fukuoka Prefecture are included in this type that have Indo-Pacific elements, but they have a Chinese element also in that the cobalt ore used also has a high manganese content.

Note

1. The Yayoi period is divided into three phases: early, middle, and late. The early phase began in around the fourth century B.C.E. in northern Kyushu. Early Yayoi culture then gradually spread from Kyushu along the western and central Honshu. During the middle period (second to first B.C.E.), it extended to the central highlands and eastern Honshu.

References

Brill, R. H., and H. Shirahata. 1995. Lead isotope analyses of some Asian glasses. In *Proceedings of the seventeenth international congress on glass,* ed. Fangtian Gong, vol. 6, 491–96. Beijing: International Academic Publishers.

Brill, R. H., K. Yamasaki, I. L. Barnes, et al. 1979. Lead isotopes in some Japanese glasses. *Ars Orientalis* 11:87–109.

Francis, Peter, Jr. 1990. Glass beads in Asia part two: Indo-Pacific beads. *Asian perspectives* 29(1):1–23.

Fujita, H. 1994. *Yayoijidai no garasu no Kenkyu.* Tokyo: Meicho Shuppan, 227–51.

Glover, Ian, and Julian Henderson. 1994. Early glass in south and south east Asia and China. In *Colloquies on art and archaeology in Asia,* no. 17. London: Percival David Foundation on Chinese Art. 148.

Koezuka, T. 2000. *Ofuro-Minami Funbogun.* Kyoto: Iwataki-Cho Kyoiku Iinkai. 47–51.

Koezuka, T., and K. Yamasaki. 1995. Chemical compositions of ancient glasses found in Japan: Historical survey. In *Proceedings of the seventeenth international congress on glass,* ed. Fangtian Gong, vol. 6, 469–74. Beijing: International Academic Publishers.

——. 2001. Detection of Han blue pigment in the blue lead-barium glass beads excavated in Japan. In *Proceedings of the nineteenth international congress on glass.*

Lee, In-Sook, in collaboration with R. H. Brill and P. M. Fenn. 1993. Chemical analyses of some ancient glasses from Korea. In *Annales du re congres de Association International pour l'Histoire du Verre.* 163–71.

Oga, K., and S. Gupta. 2000. The Far East, southeast and south Asia: Indo-Pacific beads from Yayoi tombs as indicator of early maritime exchange. *South Asian studies* 16:73–88.

Watanabe, M. 1991. Hirabaru yayoi kofun no Tamarui ni tuite. In *Hirabaru Yayoi kofun,* ed. Dairoku Harada, 235–45. Kyushu: Kankyo Kanrikyokai.

Yamasaki, K. 1973. Recent technical studies of works of art in Japan. *Application of science in examination of works of art.* Boston: Museum of Fine Arts. 229–34.

——. 1974. Chemical compositions of ancient glasses found in Japan. In *Proceedings of the tenth international congress on glass in Kyoto* 9:15–20.

Yamasaki, K., T. Koezuka, and H. Shirahata. 1995. Scientific studies on several ancient Chinese glasses. In *Proceedings of the seventeenth international congress on glass,* ed. Fangtian Gong, vol. 6, 497–502. Beijing: International Academic Publishers.

Yamasaki, K., and M. Murozumi. 1991. Similarities between ancient Chinese glasses and glass excavated in Japanese tombs. In *Scientific research in early Chinese glass,* 91–98. Corning Museum of Glass.

Zhang, Fukang. 1991. Scientific studies of early glasses excavated in China. In *Scientific research in early Chinese glass,* ed. R. H. Brill, 157–65. Corning, N.Y.: Corning Museum of Glass.

23 Exploring Issues of Geological Source for Jade Worked by Ancient Chinese Cultures with the Aid of X-Ray Fluorescence Spectroscopy

Janet G. Douglas

ABSTRACT Most of the earliest jades from China are known to be composed of the compact variety of tremolite-actinolite called "nephrite." A challenging question for art historians, archaeologists, and other scholars working with ancient Chinese jade is the geological source of raw nephrite, because it is not known which nephrite deposits were worked in antiquity. In total, 145 jade objects composed of nephrite from the Freer and Sackler collections that date from the Neolithic (4000–1700 B.C.E.) to the Han dynasty (206 B.C.E.–220 C.E.) were chosen to investigate issues related to the geological source of these materials with the aid of x-ray fluorescence spectroscopy (XRF).

Two major types of geologic occurrences of Chinese nephrite are known: nephrite associated with metamorphosed dolomitic marbles and nephrite associated with serpentinites. Geological sources of nephrite for the jades in this study are most likely associated with dolomitic marbles because they were consistently low in Cr_2O_3 and NiO, as are nephrites associated with dolomitic marbles. Minor elemental oxide compositions of jades from the Hongshan, Liangzhu, and Longshan cultures of the Neolithic period suggest that each has a simple source pattern for its raw nephrite. Longshan jades were found to have particularly high FeO and MnO contents, suggesting an unusual source rich in iron and manganese. Jades of the Shang and Western Zhou dynasties show a wide range of compositions, suggesting multiple nephrite sources for these objects. The small group of Eastern Zhou jades from Jincun in Luoyang, Henan Province, were found to be visually and compositionally similar, suggesting one deposit or multiple related deposits for this material.

The use of minor elemental oxides determined by XRF is rapid, noninvasive, and extremely useful when used in conjunction with the art-historical considerations of shape and form and other visual evidence, such as workmanship, in developing the cultural context of jade objects and in answering questions of authentication.

Introduction

It is well known that most of the earliest jades from China are composed of the compact variety of tremolite-actinolite called "nephrite." A challenging question for art historians, archaeologists, and other scholars working with ancient Chinese jade is the geological source of raw nephrite, because it is not known which nephrite deposits were worked in antiquity. Very little scientific work has been done to investigate jade sources. In this study, 145 jade objects composed of nephrite from the Freer and Sackler collections were studied using x-ray fluorescence spectroscopy (XRF) to investigate issues related to the geological source of these materials. Some of these jades are shown in plate 42.

XRF is a useful tool for this study, because it allows examination of jades beyond basic mineral composition directly from the surface without the need to remove a sample. Two particular issues are addressed: (1) What can be learned about geological sources by comparing the composition of modern nephrite specimens from various locations in Asia with ancient Chinese jades?; and (2) How much compositional variability is there in the nephrite materials used by ancient cultures in China indicating one or more geological sources for their raw material? The approach outlined in this study will aid with authenticity questions about Chinese jades, which is important in light of the current prolific market in jade forgeries.

The Chinese word for jade, *yu,* refers to the beautiful, polished appearance of a stone rather than its mineral content. Although various stone materials were used by early cultures, nephrite *yu* was clearly the material of choice. "Nephrite" is a term that refers to the massive, fine-grained variety of amphibole (actinolite-tremolite series). Its general composition can be defined as $Ca_2(Mg,Fe)_5(OH)_2(Si_4O_{11})_2$, where actinolite refers to the iron-rich member of the series and tremolite refers to the iron-poor member. Another jade mineral, jadeite, $NaAl(SiO_3)_2$, was not worked in the early periods of Chinese history and in fact was not imported from Burma until the late seventeenth century. Other jadelike materials were used in antiquity in China, such as serpentine (antigorite and lizardite), diopside, quartz (chalcedony), marble, talc, and various other rocks and minerals.

Amphibole minerals typically have some ionic replacement within their structure, so a wide range of chemical compositions can occur in nature. In the case of tremolite-actinolite, ionic replacements of Fe with minor amounts of Mn, Cr, and Ni are of particular interest in terms of the geological source for these materials.

Nephrite Sources in China and Neighboring Regions

Our current state of knowledge of jade quarrying and manufacturing techniques in ancient China is limited. The remains of one jade workshop were discovered in 1997 in the Baohua mountain range north of Dingshadi in Jiangsu Province, in the proximity of remains of the Neolithic cultures of Majiaban, Songze, and Liangzhu (Lu and Tao 2001). The Nanjing Museum Institute of Archaeology and the Institute of Geological Research of Huadong are currently excavating this area. Raw jade pebbles can still be found along the river's bank, suggesting that the geological source for the material is close by.

At the Dingshadi workshop, hammering, chiseling, and burning to crack the jade were techniques used to expose the inner matrix of a pebble to determine if it was suitable for use. The workshop has also yielded a variety of stone tools that may have been used to work jade through cutting, drilling, surface abrading, polishing, and incising. Numerous jade remains with evidence of working were also found. Two types of nephrite were found in the area, including *toushanshi,* or "tremolitic" white jade, and *yangchishi,* or "actinolitic" green jade. No scientific analysis has yet been undertaken on the remains from the Dingshadi jade workshop; therefore, this newly discovered site presents an excellent opportunity to characterize local raw jade, worked jade remains, and excavated cultural jades. Researchers also plan to investigate the areas surrounding the jade workshop to find the in situ nephrite deposits worked to produce jade at this site.

Nephrite usually occurs in small pods or lenses, typically meters in width, that form through metamorphic processes involving relatively high temperature and pressure. Two major types of geologic occurrences of Chinese nephrite are known: nephrite associated with metamorphosed dolomitic marbles and nephrite associated with serpentinites (Wen and Jing 1992). Dolomitic marbles contain a predominance of dolomite, $CaMg(CO_3)_2$, and the associated nephrite tends to have a low iron content with no significant ionic substitution by nickel and chromium. Serpentinites contain a predominance of serpentine, $Mg_3(Si_2O_5)(OH)_4$, and the associated nephrite tends to have a higher iron content with some ionic substitution by chromium and nickel, which are elements typically present in serpentinites. Minor elemental composition of nephrite based on these ionic substitutions often indicates from which type of geological deposit the raw material has originated.

No modern nephrite deposits in China are known to have been worked in antiquity. As nephrite tends to occur in small, localized deposits, it is possible that some in situ nephrite deposits used by ancient people have been depleted. Archaeologists and other jade researchers have considered the possible sources of jade for the early Chinese cultures from modern-day nephrite deposits as summarized in table 1. These ideas have formed through a combination of proximity to archaeological localities yielding jades and visual similarities in material to these jades, and they remain to be systematically explored. Figure 1a shows the proximity of the locations of these modern-day nephrite deposits to the regions of the Neolithic cultures of Liangzhu, Hongshan and Longshan. Figure 1b shows the modern-day nephrite deposits that may have been used by the post-Neolithic cultures of Eastern Zhou, Western Zhou, and the Shang dynasties.

Little scientific work has been done to investigate these possible sources for jade, although the hydrogen and oxygen isotopic analysis (δD and $\delta^{18}O$) has been explored as a technique to resolve some of these issues (Wen and Jing 1996). The results of this work show a relatively limited distribution in isotope compositions of jades from the early periods, suggesting a simple pattern of local nephrite sources. The isotopic compositions become more widely distributed starting in the Western Zhou dynasty, suggesting diverse sources of nephrite at that time. With the increase in trade starting in the late Shang and Western Zhou dynasties, jade would have been more likely to be traded over wide regions.

Table 2 shows XRF data of four minor elemental oxides for all modern-day nephrite mineral specimens from China and neighboring regions currently in the reference collection in the Freer and Sackler's Department of Conservation and Scientific Research. Classification of these mineral specimens into those associated with dolomitic marbles and those associated with serpentinites is based on the geological occurrences of nephrite reported in the literature (Babkin et al. 1973; Kim, Lee, and Chang 1986; Kolesnik 1966; Sekerina 1988; Sekerin and Sekerina 1986; Tan et al. 1978; Harlow and Sorensen 2001; Wen, pers. comm.).

Table 1. *Possible geological sources for jades produced by some early Chinese cultures. These connections remain to be explored*

Province	Regional location	Period
Nephrite from deposits associated with dolomitic marbles		
Liaoning	Kuandian	Neolithic, Hongshan culture
Jiansu	Liyang, Xiaomeiling	Neolithic, Liangzhu culture
Gansu	Lintao, Tao River Valley	Late Neolithic, Longshan culture to early Shang
Sichuan	Wenchuan	Eastern Zhou (Spring and Autumn)
Xinjiang	Kunlun Mts.	Han
Nephrite from deposits associated with serpentinites		
Xinjiang	Tianshan Mts., Manasi	Shang and Western Zhou
Henan	Xichuan	Eastern Zhou (Spring and Autumn)

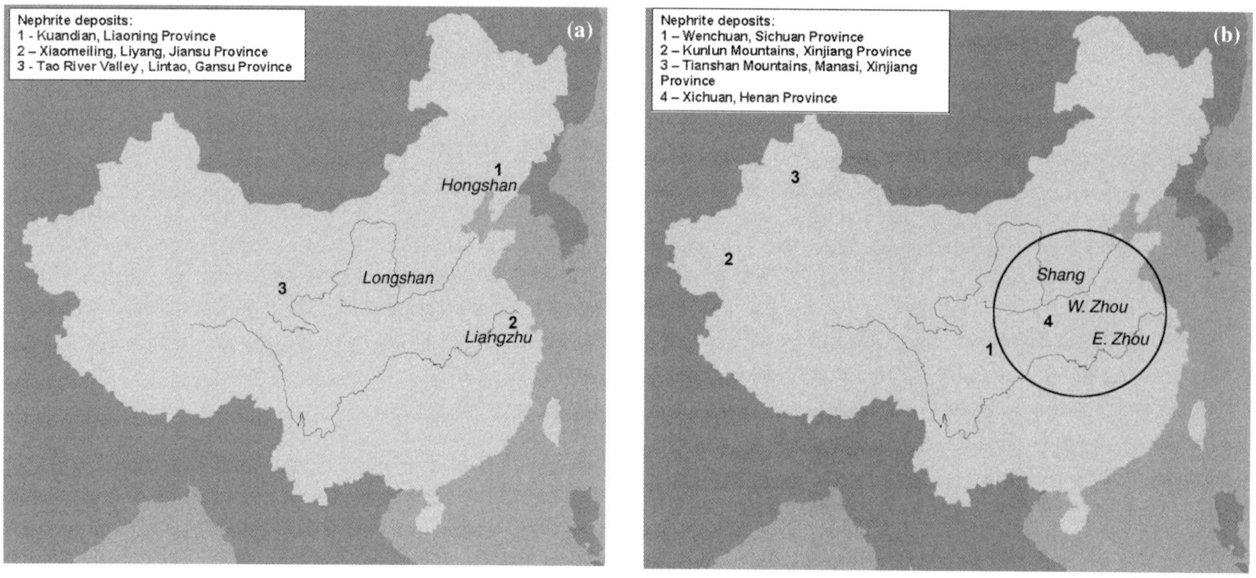

Figure 1. *Location of ancient cultures in China and modern-day nephrite deposits that may have been exploited as jade sources in antiquity*

Table 2. *Modern nephrite samples from China and neighboring regions*

	Cr_2O_3	MnO	FeO	NiO
China				
Nephrite associated with dolomitic marbles				
Xinjiang, Kunlun Mts., USNM-104669 (1)	0.03	0.07	0.34	0.00
Xinjiang, Kunlun Mts., USNM-104669 (2)	0.02	0.06	0.43	0.00
Xianjiang, Kunlun Mts., USNM	0.03	0.03	0.20	0.00
Gansu, Lintao, LY02[a]	0.05	0.04	0.88	0.00
Gansu, Lintao, LY12[a]	0.07	0.05	1.66	0.00
Gansu, Lintao, LY14[a]	0.07	0.06	1.52	0.00
Gansu, Lintao, LY21[a]	0.05	0.05	0.67	0.00
Gansu, Lintao, LY22[a]	0.07	0.05	0.36	0.00
Gansu, Lintao, LY23[a]	0.07	0.22	1.89	0.00
Gansu, Lintao, LY24[a]	0.06	0.09	1.71	0.00
Jiansu, Liyang, Xiaomeiling[a]	0.01	0.01	0.05	0.00
Nephrite associated with serpentinites				
Xinjiang, Tianshan Mts., Manasi S002[a]	0.23	0.05	3.89	0.21
Neighboring regions				
Nephrite associated with dolomitic marbles				
South Korea, Chuncheon, USNM-137417	0.01	0.02	0.15	0.00
Siberia, Vitim[b]	0.05	0.04	0.27	0.00
Nephrite associated with serpentinites				
Taiwan, Hualien, USNM-119446	0.43	0.07	3.28	0.36
Taiwan, Hualien, USNM-119446	0.42	0.07	3.22	0.35
Taiwan, Haulien, USNM-119446	0.30	0.06	3.56	0.30
Taiwan, Hualien, USNM-122843	0.07	0.12	4.80	0.08
Taiwan, Haulien, FGA	0.50	0.07	3.24	0.23
Taiwan, Haulien, FGA	0.16	0.09	3.67	0.12
Taiwan, Hualien, FGA	0.18	0.12	4.74	0.12
Siberia, Chara Jelga, USNM-84504 (1)	0.12	0.15	4.19	0.26
Siberia, Chara Jelga, USNM-84504 (2)	0.11	0.06	6.88	0.20
Siberia, Chara Jelga, USNM-84504 (3)	0.09	0.06	7.18	0.24
Siberia, East Sayan Mts., Onot River, USNM-8459	0.14	0.08	2.59	0.18
Siberia, East Sayan Mts., Onot River, AMNH-G44070 (1)	0.02	0.19	3.05	0.01
Siberia, East Sayan Mts., Onot River, AMNH-G44070 (2)	0.04	0.19	3.46	0.00
Siberia, USNM-R6773 (gray nephrite)	0.03	0.04	0.24	0.00
Siberia, USNM-R6775	0.11	0.07	5.23	0.10

a. Gift of Wen Guang b. Gift of Robert Frey

All of the nephrite specimens from deposits associated with dolomitic marbles from China and neighboring regions have especially low Cr_2O_3 content of 0.02–0.07% (by wt.), and NiO was below the detection limits of the XRF (<0.01% by wt.). Nephrite specimens from deposits associated with serpentinites, on the other hand, show a different compositional trend. With the exception of the Siberian nephrite samples from the East Sayan Mountains, AMNH-G44070(1) + (2), and the gray nephrite, R6773, the nephrites associated with serpentinites contain appreciable Cr_2O_3, 0.07–0.43% (by wt.) and NiO, 0.08–0.36% (by wt.).

The XRF data for nephrite specimens as well as data in the geological literature cited above show that nephrite associated with dolomitic marbles is typically distinguishable from nephrite associated with serpentinite based on Cr_2O_3 and NiO contents. Nephrites from dolomitic marble sources rarely have appreciable amounts of Cr_2O_3 and NiO. Nephrites associated with serpentinites, however, usually have appreciable minor amounts of Cr_2O_3 and NiO, although not in every case. Therefore, higher amounts of Cr_2O_3 and NiO are strongly indicative of geological association of serpentinites, but it is not possible to conclusively associate a nephrite with low Cr_2O_3 and NiO contents with either type of deposit.

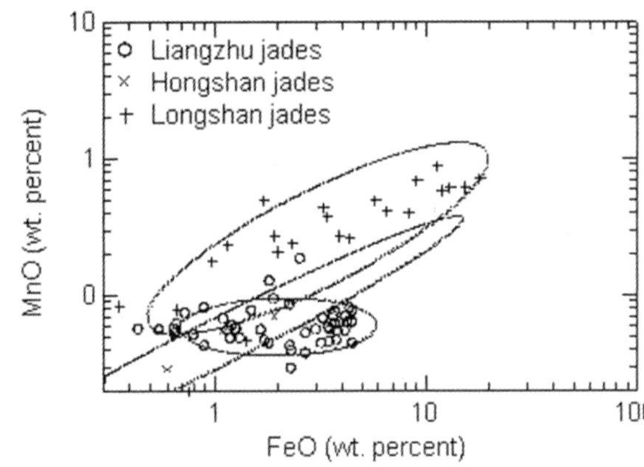

Figure 2. FeO vs. MnO plots derived from jades of Neolithic cultures in China

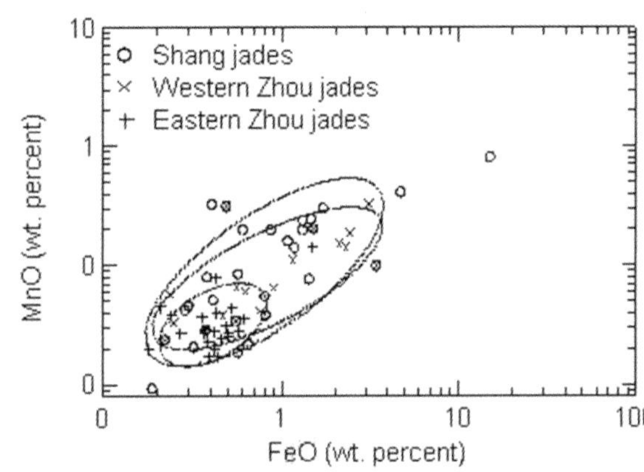

Figure 3. FeO vs. MnO plots derived from jades of the Shang, Western Zhou, and Eastern Zhou dynasties

Jade Materials Used by Ancient Chinese Cultures

The compositional ranges of the four oxides determined by XRF on 145 ancient Chinese jades from the Freer and Sackler collections are given in table 3. All jades in the table have a notably low Cr_2O_3 content of 0.01–0.07% (by wt.), and NiO content of most jades is below the detection limits of the XRF. The XRF data for the entire group suggest that these jades are from geological deposits associated with dolomitic marbles.

Having established that Cr_2O_3 and NiO contents are relatively low in the entire body of jades in this study, it is informative to examine the XRF data for FeO and MnO. For purposes of discussion, FeO can be divided into the categories of low (≤1.00%), medium (>1.00 to ≤5.00%) and high (>5.00%). MnO can be divided into the categories of low (≤0.10%) and high (>0.10%). Although FeO and MnO do not provide evidence of the two types of geological deposit from which the nephrite originated, these oxides do show some interesting trends within the early cultural groups. Figure 2 shows FeO versus MnO plots for jades from the Neolithic cultures of Liangzhu, Hongshan, and Longshan; figure 3 shows similar plots for the Shang, Western Zhou, and Eastern Zhou dynasties. These plots, discussed in more detail below, are restricted to those objects that have well-accepted analogues from excavated sites in China.

Munsell color notations and standard color names (Kelly and Judd 1955) were determined for the jades in this study.[1] The color of unaltered jades correlates fairly well with minor oxide compositions, as will be discussed below.[2] Those jades with weathered or otherwise discolored surfaces show less correlation.

Table 3. Compositional ranges of Cr_2O_3, FeO, MnO, and NiO for jades from various Chinese cultures[a]

Period	No. of jades	Cr_2O_3	MnO	FeO	NiO
Neolithic, Hongshan culture	3	0.02–0.05	0.03–0.07	0.60–1.94	0.00–0.00
Neolithic, Liangzhu culture	49	0.00–0.05	0.03–0.18	0.44–4.52	0.00–0.01
Neolithic, Longshan culture to Late Shang	24	0.01–0.06	0.02–0.89	0.35–17.95	0.00–0.00
Shang and Western Zhou	47	0.01–0.07	0.01–0.78	0.19–15.26	0.00–0.01
Eastern Zhou	22	0.01–0.06	0.02–0.14	0.18–1.48	0.00–0.01

a. Two jades composed of nephrite examined by XRF are not included in this table. The first is a ceremonial implement (F1917.49) with slightly elevated Cr_2O_3 content of 0.15% (by wt.) and NiO of 0.03% (by wt.). The attribution of this jade was questioned because it is devoid of ornamentation and detail. It is similarly devoid of tool marks that might help place its origin and age. The material is a rich green, which differentiates it from other jades within the Freer and Sackler collections. No parallels have been found in archaeological reports.

The second is a Shang *bing* (51987.838), with Cr_2O_3 content of 0.01% (by wt.) and NiO content of 0.03% (by wt.). This nephrite has a slight layering similar to jade from Fengtien, Taiwan. Its attribution should be questioned, as no similar layering has been observed on any other ancient jades.

Hongshan Jades

One of the earliest well-developed jade working cultures is the Hongshan culture that existed along the western Liao River valley in northeastern China from about 4000 to 2000 B.C.E. Only three well-established Hongshan jades are in the Freer and Sackler collections, including a toothed plaque (F1991.52, plate 42a) and two tubular headpieces (S1987.842 and S1987.626). The Hongshan jades are very similar in color and have textures that range from extremely homogenous to mottled and slightly layered. All three were found to have low FeO and MnO contents by XRF, as shown in table 4. It is not possible to adequately characterize the nephrite materials used by people of the Hongshan culture given the small number of objects this group comprises. Difficult access to archaeologically attested material and the

Table 4. *Early Chinese jades arranged by increasing iron content (by % wt.), along with jade color (Kelly and Judd 1955)*

Accession no.	Object type	FeO	MnO	Color (Munsell notation)
Hongshan jades				
F1991.52	toothed mask plaque	1.94	0.07	Grayish yellow green (5GY 5/2)
S1987.626	tubular headpiece	1.10	0.05	Moderate olive green (5GY 4/4)
S1987.842	tubular headpiece	0.61	0.03	Moderate olive green (5GY 4/4)
Liangzhu jades				
F1919.46	bracelet	0.44	0.06	Light yellow green (2.5GY 8/6)
F1917.384	*cong*	0.65	0.05	Light yellow green (5GY 8/2)
F1919.47	*cong*	0.90	0.04	Light yellow green (5GY 7/4)
S1987.734	*shi*	1.21	0.06	Gray yellow green (10Y 8/4)
F1916.512	ax fitting	1.26	0.06	Gray yellow green (10Y 8/4)
F1916.380	*bi*	1.82	0.04	Moderate olive green (5GY 3/4)
F1916.628	ornament	1.91	0.09	Light olive (5Y 5/4)
F1917.83	*bi*	3.48	0.05	Moderate olive green (7.5GY 3/4)
S1987.929	*bi*	3.71	0.08	Moderate yellow green (2.5GY 5/4)
F1968.30	*cong*	3.73	0.06	Moderate yellow green (5GY 5/4)
F1919.58	*bi*	3.92	0.06	Grayish olive (5G 3/2)
F1915.51	*bi*	4.16	0.07	Moderate yellow green (2.5GY 6/4)
F1917.78	ax	4.36	0.08	Moderate yellow green (5GY 3/4)
F1917.19	*bi*	4.51	0.06	Moderate yellow green (5GY 6/4)
F1918.14	*cong*	4.52	0.08	Moderate yellow green (5GY 7/4)
Longshan jades				
F1917.396	*ge*	0.50	0.23	Light olive gray (10Y 5/1)
F1970.40	*dao*	0.87	0.15	Light olive gray (5Y 5/1)
S1987.450	*dao*	1.19	0.14	Greenish gray (5GY 5/1)
F1916.416	knife	1.33	0.23	Light olive gray (7.5Y 5/2)
F1916.162	*zhang*	1.42	0.23	Olive gray (10Y 4/1)
F1918.1	*dao*	1.61	0.11	Grayish olive green (5GY 4/2)
F1916.490	*dao*	1.71	0.50	Light olive gray (7.5Y 5/2)
F1983.3	*dao*	1.74	0.14	Grayish olive (5Y 4/2)
F1916.417	*dao*	1.90	0.28	Grayish olive (10Y 4/2)
F1917.24	*dao*	2.32	0.24	Grayish olive (5Y 4/2)
F1917.23	*gui*	2.36	0.48	Grayish olive (5Y 4/2)
F1917.30	*dao*	3.36	0.38	Grayish olive (5Y 4/2)
F1916.161	*dao*	4.33	0.26	Grayish olive green (2.5GY 2/2)
F1916.165	*zhang*	5.75	0.50	Dark grayish olive (5Y 2/1)
F1916.166	*zhang*	6.54	0.41	Dark grayish olive (5Y 2/1)
F1916.492	*zhang*	8.28	0.41	Dark grayish olive (5Y 2/1)
F1916.495	*zhang*	9.07	0.70	Olive black (10Y 2/1)
F1916.491	*zhang*	10.57	0.76	Dark grayish olive (5Y 2/1)
F1916.168	*zhang*	11.50	0.73	Dark grayish olive (7.5Y 2/2)
F1916.494	*zhang*	11.83	0.59	Dark grayish olive (5Y 2/1)
F1916.164	*zhang*	13.02	0.62	Dark grayish olive (5Y 2/1)
F1919.60	*dao*	13.59	0.59	Grayish olive (5Y 3/2)
S1987.449	*dao*	15.26	0.78	Grayish olive (10Y 3/2)
Eastern Zhou jades				
F1932.39	pendant	0.37	0.03	Pale yellow green (10Y 9/2)
F1932.40	pendant	0.37	0.03	Pale greenish yellow (10Y 9/4)
F1931.19	cup	0.39	0.03	Pale greenish yellow (10Y 8/4)
F1930.28	comb	0.42	0.02	Light yellow (2.5GY 8/6)
F1930.31	cup	0.43	0.08	Pale greenish yellow (10Y 8/4)
F1932.43	dancing figure	0.44	0.02	Grayish yellow (5Y 8.5/6)
F1930.27	bead (1)	0.49	0.03	Light greenish yellow (10Y 8/2)
F1932.38	ornament	0.61	0.04	Pale greenish yellow (7.5Y 8.5/4)

flood of modern forgeries further complicate our knowledge of Hongshan jade materials, but strides are being made to understand and identify Hongshan jades based on their materials and manufacturing methods (So and Douglas 1998).

Liangzhu Jades

The well-known Liangzhu culture existed along the lower Yangzi River from about 3500 to 2000 B.C.E. in the region near modern-day Shanghai. The Liangzhu culture produced some of the finest early jades. Some variations in color and translucency appear to be due to processes such as natural weathering, burial alteration, and surface accretions. The forty-nine Liangzhu in this study include *bi* and *cong* and other ornamental objects, and these materials were typically found to be similar in texture and color. Some of the Liangzhu jades in the Freer and Sackler collections consist of fine-grained, homogeneous nephrite with a relict texture from the original rock, which the nephrite replaced through metasomatic processes during its formation. One jade in particular, as shown in figure 4, has a relict texture of vein calcite, as evidenced by large calcite twins in the nephrite structure (Tsien, Tan, and Douglas 1996). This texture is evidence that the nephrite formed through a metasomatic replacement of preexisting vein calcite. To find the geological source of this material, one must look in areas containing coarse-grained calcite veins or their metamorphosed equivalents.

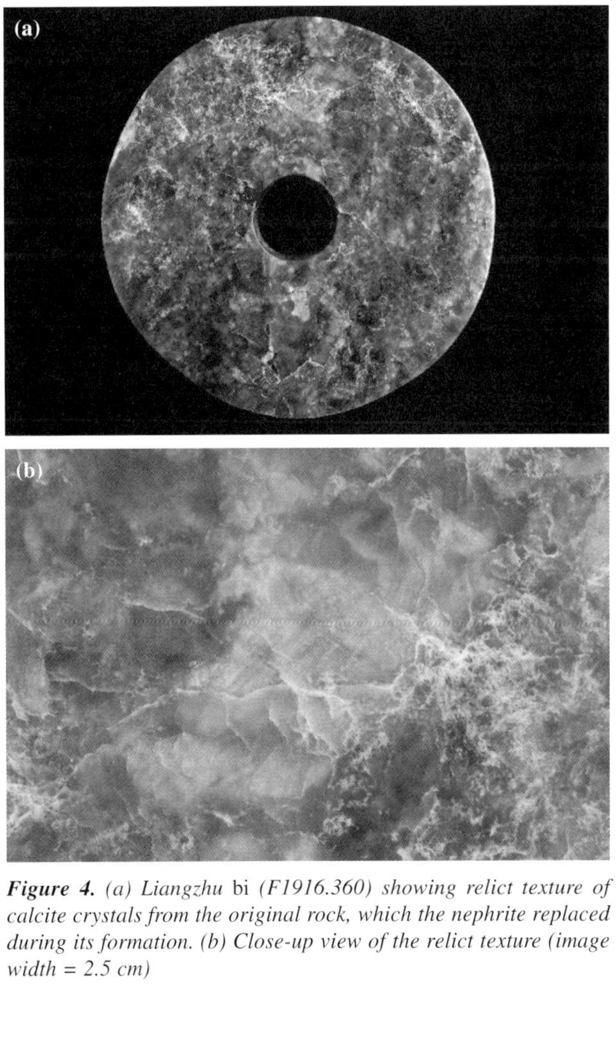

Figure 4. *(a) Liangzhu* bi *(F1916.360) showing relict texture of calcite crystals from the original rock, which the nephrite replaced during its formation. (b) Close-up view of the relict texture (image width = 2.5 cm)*

As shown in figure 2, the Liangzhu jades show a fairly narrow compositional range of FeO and MnO. All have MnO contents less than 0.10% (by wt.) and FeO content of less than 5.00% (by wt.). Although most of these jades are mottled, the overall color ranges from light yellow green (2.5GY 8/6) for the lowest iron nephrite to moderate yellow green (5GY 7/4) for the highest iron content nephrite (table 4). The composition and textures of these materials suggest the same or related nephrite sources.

Longshan Jades

The Longshan culture existed from about 2300 to 1800 B.C.E. over a vast area of northern China and includes the upper, middle, and lower Yellow River valleys. These objects typically consist of *dao* (knife), *ge* (blade), *gui* (blade), *zhang* (scepter), and, to a lesser extent, plain undecorated *cong* and *bi,* along with other ornaments. Because these jades have been found and may have been made over a wide area, it becomes difficult to assign them to any specific cultural group within the Longshan culture. However, this study shows that most of the thirty-three jades in the Longshan group typically have a higher FeO and MnO content in contrast to the jade objects from the Liangzhu and Hongshan cultures of the Neolithic (see fig. 2). The objects consist mostly of *dao, zhang,* and various blades, and some of these jades are listed in table 4. The dark grayish olive *zhang* are particularly rich in FeO and MnO. These relatively high FeO and MnO contents are probably responsible for the dark grayish olive color, suggesting that these jades are composed of actinolitic rather than tremolitic nephrite. Several Longshan blades have lower iron contents of less than 2% (by wt.) but still show elevated amounts of MnO. This group includes *dao* (F1916.163, F1916.417, F1916.490, F1918.1, F1970.40, F1983.3) and *zhang* (F1916.162, S1987.630). The Longshan people were using a geological source or sources of nephrite particularly rich in MnO, a potentially diagnostic feature of the nephrite that may help pinpoint its source.

Some of these jades with relatively high FeO and MnO contain a subnodular structure, suggesting that they may have originated from a parent rock consisting of a sedimentary conglomerate (see plate 43). Table 4 also lists a series of these jades arranged in increasing color value with increasing FeO and MnO. These particular jades are attributed to the Longshan-Qijia culture of northern China. Some of these jades have lightened surfaces due to natural alteration, but their color generally becomes darker with increasing FeO and MnO.

Several of the Longshan jades are low in FeO and MnO. Most notable are four mask plaques (LTS1985.1.276.1, LTS1985.1.276.2, LTS1985.1.276.3, F1953.9) that are lighter in color, ranging from light to moderate yellow green (5GY 8/4 to 2.5G 6/6). The low FeO and MnO compositions may indicate that a nephrite material from a different geological source from the blades discussed above was used for the small plaques.

Shang and Western Zhou Jades

The cultures of the Shang (1700–1050 B.C.E.) and Western Zhou (1050–771 B.C.E.) dynasties each occupied vast regions generally ranging from Mongolia to the north, Gansu and the upper middle Yellow River to the west, and the Yangzi River valley to the south. Small jade objects, often in the shape of animals, became a tradition during the Shang dynasty, and this tradition continued well into the Western Zhou dynasty. One notable difference is that jade shapes of the Shang dynasty are often three-dimensional, whereas those from the Western Zhou tend to be simple, flat plaques.

Shang and Western Zhou jades show a wider range of FeO and MnO contents than those for the jades from Neolithic cultures described earlier. Jades of high FeO and MnO content, such as the *dao* (S1987.449), suggest that a jade source or sources similar to those used by the earlier Longshan cultures were still being used. Other jades with low FeO (<1%) but significant MnO contents (>0.10%), such as the collared disk, *yuan* (S1987.458), with FeO of 0.88% and MnO of 0.19% (by wt.), suggest a related geological source.

Geologic sourcing of jades from the Shang and Western Zhou dynasties without known provenance is more complex, as these jades tend to exhibit a wide variety of shapes and are worked generally in the same way. Scholars currently debate whether nephrite from Xinjiang Province was imported to the Shang region in eastern China. If so, this probably occurred relatively late in the Shang dynasty (Kao 1985). Clearly, a wide variety of jade, probably from several sources, was used during these periods, and these materials need to be more fully characterized to understand their use and distribution. Future compositional characterization of jades from the recent excavations of the Sanxingdui culture in Sichuan Province and the Doyangzhou culture in Jiangxi Province will shed light on the jade and jade sources during these periods.

Eastern Zhou Jades

The Eastern Zhou jades in this study are from Jincun, in Luoyang, Henan Province, and therefore form a tight group of well-provenanced material from one site. Five XRF analyses are from ten beads strung on a gold chain (F1930.27). Together, the jades of this group are pale yellow, very fine-grained, homogeneous, and free of mineral inclusions. With the exception of one, these jades contain FeO less than 1.00% (by wt.) and MnO less than 0.10% (by wt.), indicating they are composed of a high-quality tremolitic nephrite. Some of these jades are listed along with their color in table 4. These analyses are consistent with those for nephrites associated with dolomitic marbles, although they are generally lower in FeO than the Lintao nephrite specimens. Xinjiang and Jiangsu nephrite or some other unknown local deposit might be considered as sources.

Conclusions

The XRF data on the jades in this study support geological sources associated with dolomitic marbles rather than serpentinites. Few Hongshan jades were analyzed because there are few well-attested jades in the Freer and Sackler collections. XRF data on these Longshan jades, however, show that they are compositionally similar to one another. Understanding Hongshan jade materials is extremely important given their rarity and value along with the large quantities of forgeries on the market in modern times. The Liangzhu jade materials appear to be remarkably consistent with some variation in FeO content, suggesting one or more related geological sources. Longshan jade materials may have had more than one geological source, but the dark, almost black jade, along with those with the subnodular structure, probably originated from one source or related sources. They are unusually high in FeO and MnO contents, which may help pinpoint potential geological sources. Jades from the Shang and Western Zhou dynasties show a wide range in compositions and may have originated from several geological sources. Issues relating to Shang and Western Zhou jades can only be resolved with the study of excavated or well-provenanced materials from these cultures. The Eastern Zhou jades in our collections from Jincun in Luoyang, Henan Province, form a tight group of homogeneous, tremolitic nephrite of similar composition. More variations of jade materials may be found as jades from other Eastern Zhou sites are analyzed. No evidence for the use of Siberian nephrite or other nephrites associated with serpentinites was found in this study.

The use of minor elemental oxides determined by XRF is quick, noninvasive, and extremely useful in developing the cultural context of jade objects, when used in conjunction with the art-historical considerations of shape and form and other visual evidence, such as tool marks and methods of manufacture. This study can serve as a basis from which to start understanding jade materials in relation to their geological sources.

Acknowledgments

I would like to thank Dr. Jenny So, Chinese University of Hong Kong; Mr. Wen Guang, Institute of Geological Sciences in Beijing, China; and Dr. Su Rongyu, Chinese Academy of Sciences, Institute for the History of Natural Science, for many helpful discussions on the art-historical and archaeological aspects of early Chinese jades. I also greatly appreciate the generous help of the staff of the Department of Conservation and Scientific Research at the Freer Gallery of Art and Arthur M. Sackler Gallery. The Freer and Sackler's Department of Imaging and Photographic Services supplied most of the photographs for this paper.

Appendix 1: Analytical Procedure

X-ray fluorescence analysis of Chinese jades was carried out using an OMEGA FIVE, a modified Spectrace 6000 with an open-beam architecture, which allows XRF analysis directly from the surface of an object using a helium chamber (Douglas 1996). The semiquantitative XRF procedure uses a 15 kV tube voltage, 0.99 mA tube current, no beam filter, 200 seconds live time. The Fundamental Parameters Program (Version 1.34a) provided by Spectrace, Inc., was used for this analysis.

XRF is a noninvasive analytical technique that allows elemental analysis directly from the surface of an object, but some limitations need to be considered. A flat, polished, and clean surface always gives the most reliable XRF analysis. Fortunately, this is the case with many jades, especially *bi, cong,* and large blades. Because alteration or contamination on the surface of jades will be included in the analysis, it is important to choose areas that are unaltered and free of discoloration. In the present study, at least two spots were chosen for the XRF analysis if possible, to ensure that the analysis is representative. In practice, the largest beam aperture with a diameter of 3.3 cm is used, which provides the most x-ray signal at the detector.

Precision and accuracy of the XRF analytical technique can be estimated by comparing multiple analyses of nephrite reference materials. Twenty replicate analyses of nephrite (USNM R6778) gave the following average compositions (% by weight): $SiO_2 = 59.6 + 1.6$, $K_2O = 0.00 \pm 0.00$, $CaO = 14.02 \pm 0.67$, $TiO_2 = 0.10 \pm 0.02$, $Cr_2O_3 = 0.04 \pm 0.01$, $MnO = 0.03 \pm 0.01$, $FeO = 0.18 \pm 0.01$, $NiO = 0.00 \pm 0.00$, $MgO+Al_2O_3$ (by difference) $= 26.1 \pm 2.0$.

Notes

1. The Munsell system of color notation was developed by A. H. Munsell and identifies three color attributes: hue, value, and chroma. These three attributes are arranged into an orderly scale that can be used to accurately describe color. The color designation is written as hue value/chroma, as in the example 10Y 8/4. The Munsell color notations are determined by comparing the sample with small color chips from the *Munsell book of color* under the conditions of average daylight, 45° illumination, and normal viewing. The color of nephrite is often mottled so that the Munsell color notation is not always exact. The most common color in the nephrite is recorded.

   The color names reported in this study are derived from a standard dictionary of color names based on the Munsell system of color notation (Kelly and Judd 1955). The color names, also based on hue, value, and chroma, are designed to be simple and easily understood by most people.
2. Many ancient Chinese jades have colors ranging in the yellow-green to green hues, with minor variations in value and chroma. The color of nephrite can depend on a number of factors, such as the chemical composition, the presence of accessory minerals, and surface alterations, such as discoloration, burial alteration, or surface treatments. However, in the absence of chromium, the color of nephrite is predominantly attributed to its iron content. A tremolitic nephrite low in iron will be very pale yellowish-green in color. Actinolitic nephrite, which has a higher content of Fe^{+3}, will be green. Yellowish-brown and brown hues may be due to the presence of oxidized iron, Fe^{+2}, in the nephrite mineral structure. Similarly, weathering of the surface can also cause a brown coloration due to the oxidation of iron. Small amounts of chromium will cause a rich green color in nephrite.

References

Babkin, A., S. Gurulev, K. Sambuyev, et al. 1973. A new nephrite deposit in eastern Siberia. *Acad. sci. USSR, Doklady Akad. Nauk SSR* 204:137–39.

Douglas, J. 1996. The study of Chinese archaic jades using non-destructive x-ray fluorescence spectroscopy. *Acta geologica Taiwanica: Mineralogical studies of archaic jades* 32:43–54.

Grudinin, M., A. Sekerin, and N. Serkerina. 1979. Nature of nephrites of various color. *Soviet geology and geophysics* 20(2):122–24.

Harlow, G., and S. Sorensen. 2001. Jade: Occurrence and metasomatic origin. *Australian gemmologist* 21:7–10.

Kao, J. 1985. The archaeology of ancient Chinese jades: A case study from the late Shang period site of Yinxu. Ph.D. diss., Department of Anthropology, Harvard University.

Kelly, K., and D. Judd. 1955. *The ISCC-NBS method of designating colors and a dictionary of color names*. National Bureau of Standards Circular no. 553. Washington, D.C.: U.S. Government Printing Office.

Kim, S., D. Lee, and S. Chang. 1986. A mineralogical and gemological characterization of the Korean jade from Chuncheon, Korea. *Journal of the Geological Society of Korea* 22(3):278–88.

Kolesnik, Y. 1966. *Nefrity Sibiri*. Akad. Nauk SSSR, Sib. Otdeleniye, Ins. Geologiya i Geofiziki, Novosibirsk, 150. (English translation: Sofonov, A. 1970. *Nephrites of Siberia. International geology review* 12(10):1–107.)

Kollmorgen Corporation. 1966. *Munsell book of color.* Vols. 1 and 2. Baltimore, Md.: Kollmorgen Corporation.

Lu, J., and H. Tao. 2001. Prehistoric jade working based on remains at the site of Dingshadi. In *Enduring art of Jade Age China,* vol. 2, E. Childs-Johnson, 31–42. New York: Throckmorton Fine Art.

Sekerin, A., and N. Sekerina 1986. Petrology of granites in deposits of light-colored nephrite in the central Vitim highland country. *Soviet geology and geophysics* 27(12):34–40.

Sekerina, N. 1988. Conditions of localization of deposits of apocarbonate nephrite in the middle-Vitim mountain country. *Soviet geology and geophysics* 29(11):96–101.

So, J., and J. Douglas. 1998. Understanding and identifying jades from the Hongshan culture. In *East Asian jade: Symbol of excellence,* vol. 1, ed. T. Chung, 148–63. Hong Kong: University of Hong Kong.

Tan, L., C. Lee, C. Chen, et al. 1978. *A mineralogical study of the Fengtien nephrite deposits of Hualien, Taiwan*. NSC-Special Publication no. 1:81. National Science Council, Taipei.

Tsien, H., L. Tan, and J. Douglas. 1996. Geology of tremolite rock and petrofabrics of archaic Chinese Yu. *Acta geologica Taiwanica: Mineralogical studies of archaic jades* (32):85–101.

Wen, G., and Z. Jing. 1992. Chinese Neolithic jade: A preliminary geoarchaeological study. *Geoarchaeology* 7(3):251–75.

——. 1996. Mineralogical studies of Chinese archaic jade. *Acta geologica Taiwanica: Mineralogical studies of archaic jades* 32:55–83.

24 Evolution of the Tilework of Friday Mosque in Isfahan, Based on Glaze Analysis

S. Durali, P. Oliaiy, F. Shokouhi, M. Lamehi, and J. Rahighi

ABSTRACT Friday Mosque of Isfahan (Masjid-i juma) is a unique complex, the result of a millennium of architectural experience and Islamic tilework. The mosque contains masterpieces from the time of the Saljuqs (1038–1221), Mozafarids (1314–93), Safavids (1502–1732), and Qajars (1779–1924) as well as contemporary tilework. Even though there have been architectural changes in the building during each period, the decoration of each era has kept its originality and integrity. Therefore, the evolution of color and glaze can be studied by analysis of their characteristic elements. More than sixty glaze samples of different colors, such as turquoise, azure, black, white, green, and yellow, have been taken from Safavid masterpieces: the western *iwan* and the southern *iwan;* from the Mozafarid mihrab, the Saljuq decorations, and the Qajar restorations. Elemental analysis was carried out using proton-induced x-ray emission spectroscopy (PIXE). The quantities of colorants and the lead content of the glazes have been determined. The correlation between color, appearance, and elemental content were investigated, and preparation methods were discussed.

The most important result of the present research is that it can be used as a model for studying the history of tilework and restoration in other buildings of the same periods in the city of Isfahan.

Introduction

The evolution of tilework throughout the centuries clearly reveals a tradition of unquenchable vitality with a constant practice of renewal. Tilework as a form of architectural decoration offers endless possibilities of style and theme. The chronological record for the use of tile is impressive, ranging from Elamite times (ca. 1250 B.C.E.) through the polychrome glazed bricks of Achaemenian palaces to the present day. The validity and variation in style—from the tentative experiments of the fifth to eleventh century C.E. in enlivening a matte brick surface with turquoise glaze to the glittering panoramas of Safavid (1502–1732) and Qajar (1779–1924) buildings—is apparent (Scarce 1989).

In general the impact of tilework depends on three factors, which were admirably fulfilled in Iran: the appropriate position in an architectural scheme, the total compositional effect, and the quality of each individual tile (Porter 1995).

It would seem logical that as there was a great demand for tilework, as such it would require the support of a considerable industry, but there was never a large-scale industry. Workshops were modest with a few craftsmen and apprentices and were operated on a family basis, suggesting that they were continuing a well-established tradition of small units using locally available materials.

Despite the lack of written documents, much information can be gained from the tiles themselves, which adorn buildings in Iran and within its sphere of influence. The basic construction material in Iran is brick. A weather-resistant surface was achieved by covering the brick with an impermeable layer such as that provided by a glaze (Tite et al. 1998). Functional and aesthetic reasons explain the importance and continuity of tilework as an architectural ornament. Tilework was used to emphasize rather than to conceal by being used in a pattern suitable to the shape of the structure to be decorated or by taking architectural units and using them as a design component. The most striking effect of tilework is of brilliant and intense color, and the extensive use of tilework is indicative of how the delight in color was deeply rooted in Persian culture (Wulff 1976).

The study of Persian tilework abounds with problems, because many of the tiles have been removed from their original monuments. However, in the Friday Mosque of Isfahan, where the building can be dated, it is possible to attempt some chronological sequence of the development of tilework.

Historical Background

The Great Mosque or Friday Mosque is equivalent in importance to a Christian cathedral. The Friday Mosque in Isfahan is a dynamic expression of some one thousand years of Iranian architecture encapsulated into the small area it occupies. The origin of this mosque dates back to the early Islamic period, some fourteen centuries ago. The mosque underwent alterations and additions under successive imperial rulers. It therefore brings together many of the themes and decorative ideas that achieved individual expression in other monuments throughout the city.

Friday Mosque, which has four *iwan* (Galdieri 1972, 1973, 1984) surrounding the courtyard of the mosque, is an example of the Iranian four-*iwan* type of mosque building

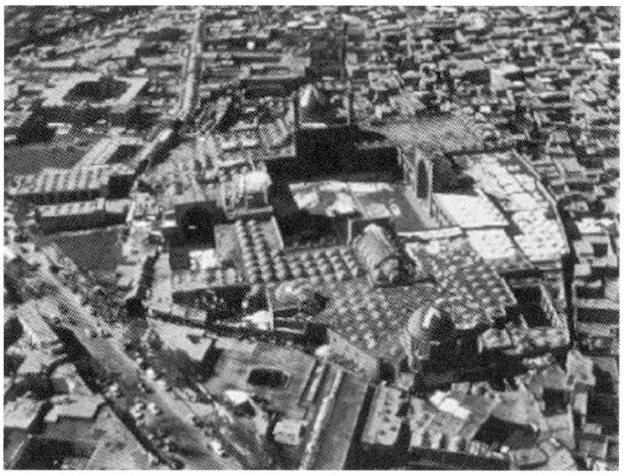

Figure 1. *Masgid-I Guma in Isfahan (after O'Kane, B. 1994)*

(fig. 1). It was founded in the seventh century on the ruins of a building constructed during the Sassanian dynasty (224–642 C.E.) (Laleh 2000). The building layout was originally on a northeast-to-southwest axis as opposed to the present north-to-south axis of the mosque. In the eighth century, during the Abbasid rule when the city of Isfahan developed, the mosque was expanded and modified to face south. During the tenth-century Buyid rule (932–1062), a few annexes were added to the building, and some changes were made to the appearance of the building without changing its original design.

In the eleventh to thirteenth century—the Saljuq period—some very substantial modifications were made to the building, and the original layout was vastly altered. After the attack of the Mongols (1221) and up until the Il-Khanids, no important changes occured to the building. One of the most important parts added to Friday Mosque during the Mozafarid period (fifteenth century) is the Mozafarid *madrasa* that is next to the eastern *iwan*. The *madrasa*, with an area of 65 × 25 m^2, includes a big yard with a large *iwan* located in the south of the building and a hall facing east–west. There is also a mihrab in this hall (fig. 2) (Golombek and Wilber 1988).

There was no major alteration in the mosque during the Safavid period (sixteenth to eighteenth century). Architectural development in the city of Isfahan initiated a number of changes in its frontispiece and internal decoration, which have mainly remained intact since that time. Some of the inscriptions from the Safavid period talk about maintenance and decoration of the mosque in that period with no clear mention of the details. These inscriptions are at the entrance on the inside wall of the southern *iwan* and on that of the eastern *iwan* (fig. 3).

During Qajar rule (eighteenth to nineteenth century), some maintenance and restoration were carried out (fig. 4). On the present-day front side of the mosque, to the south-east, there is a ceramic inscription written in Thuluth script describing the various maintenance works carried out during the rule of Fathali Shah in the Qajar period.

Experimental Procedure

The Van de Graaff accelerator at the Department of Nuclear Physics, Atomic Energy Organization of Iran was used to measure the elemental concentrations in the sample. The glaze samples were set on a sample holder and placed on the beam line in a chamber and bombarded by an ion beam under vacuum. A 2 MeV proton beam and Si(Li) detector were used for PIXE (Johansoon and Campbell 1988). The detector-target distance was 9 cm. A 175 μm Mylar absorber was placed in front of the Si(Li) detector to absorb the low-energy x-rays originating from the matrix elements in the sample. The PIXE spectra were analyzed by means of the

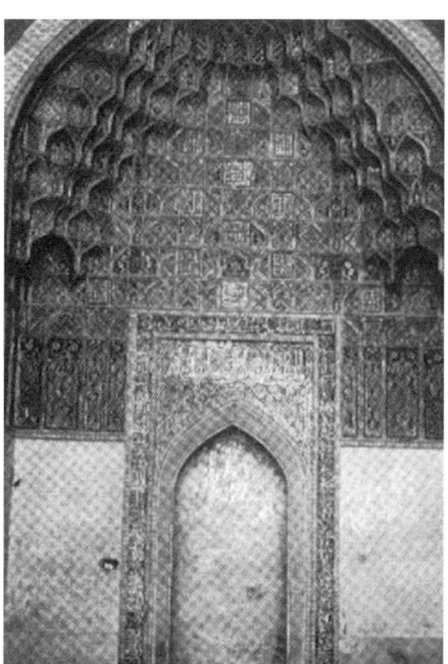

Figure 2. *Mozafarid mihrab (after O'Kane, B. 1994)*

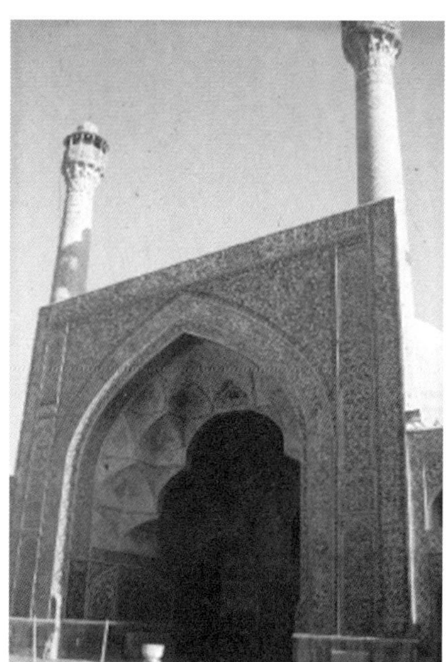

Figure 3. *Southern iwan*

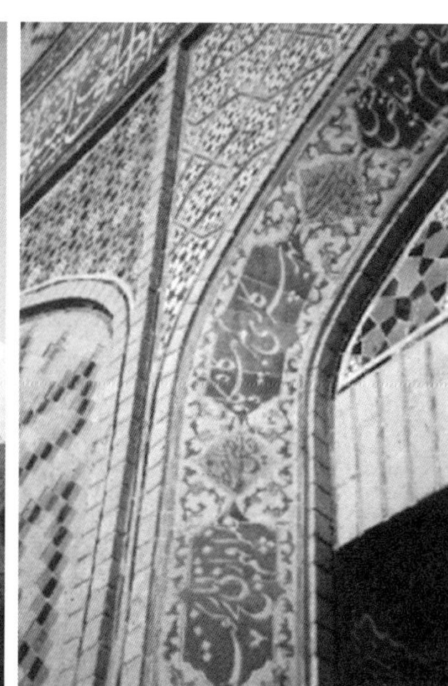

Figure 4. *Qajar restoration*

Table 1. *Sample descriptions*

Sample nos.	Dynasty	Date[a]	Color	Opacity (binocular microscope observation)
1, 2	Safavid	1502–1732	Dark yellow	Very fine opaque inclusions
3, 4	Qajar	1779–1924	Pale yellow	Gas bubbles, undissolved white crystals
5	Safavid	1502–1732	Green	Very fine opaque inclusions
6, 7	Mozafarid	1314–93	Dark green	Gas bubbles, fine green crystals
8, 9	Safavid	1502–1732	White	Very fine opaque inclusions
10	Mozafarid	1314–93	White	Very fine opaque inclusions
11, 12	Qajar	1779–1924	White	Gas bubbles, undissolved white crystals
13–19	Safavid	1502–1732	Deep blue (azure)	Very fine opaque inclusions
20, 21	Mozafarid	1314–93	Deep blue (azure)	Very fine opaque inclusions
22	Qajar	1779–1924	Deep blue (azure)	Gas bubbles, undissolved white crystals
23	Seljuq	1038–1221	Dark blue (azure)	Very fine opaque inclusions
24–33	Safavid	1502–1732	Blue	Very fine opaque inclusions
34–37	Mozafarid	1314–93	Light blue	Very fine opaque inclusions
38	Qajar	1779–1924	Pale blue	Gas bubbles, undissolved white crystals
39	Seljuq	1038–1221	Bright blue	Very fine opaque inclusions
40	Safavid	1502–1732	Black	Very fine opaque inclusions
41	Mozafarid	1314–93	Brown-black	Fine white crystals
42	Qajar	1779–1924	Red-black	Gas bubbles, fine red crystals

a. All dates C.E.

linear least squares fitting program, Gupix (Maxwell, Campbell, and Teesdale 1988). The beam spot on the target was 4 mm^2 (2 × 2). Beam current on the target in this study was about 10 nA of protons.

More than sixty samples of different colors, such as turquoise, azure, black, white, green, and yellow, were taken from various parts of the mosque: western *iwan* (fig. 5) and southern *iwan* (fig. 6), Mozafarid mihrab (fig. 2), Saljuq decorations, and Qajar restorations (fig. 4) were all analyzed. The sample size was about 3 × 3 mm^2. The sample surface was cleaned with acetone prior to measurement. The samples were taken from different tileworks of each part, and normally a minimum of two analyses were done on each glaze sample, and the results were then averaged (table 2).

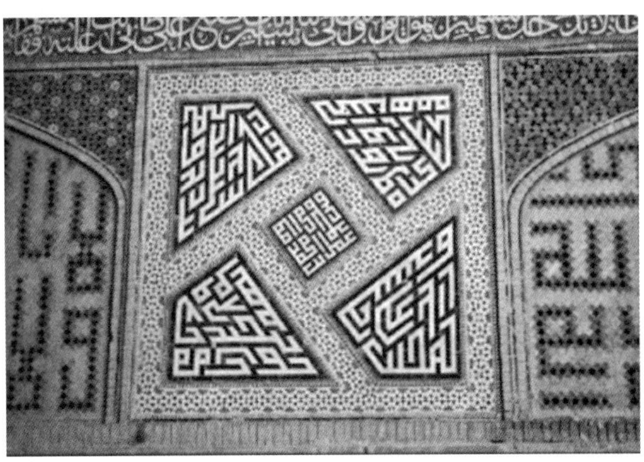

Figure 5. *Safavid tilework, Western iwan*

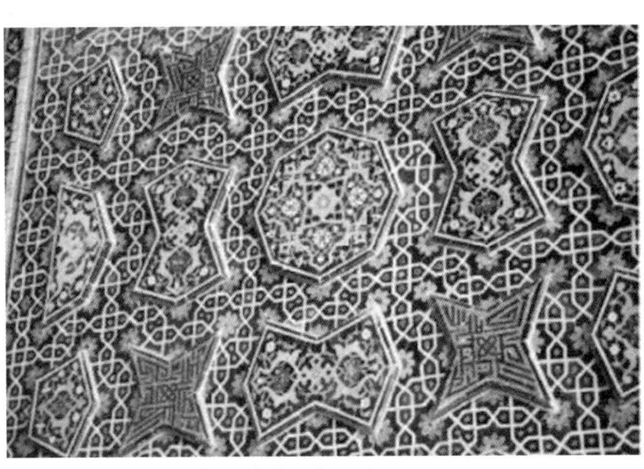

Figure 6. *Safavid tilework, Southern iwan*

Results and Discussions

In Figure 7, the elemental constituents of yellow glazes from the Safavid and Qajar dynasties are shown. The yellow color in the Qajar sample is perplexing, because lead is the only significant nonmatrix element. Normally such a color results from lead antimonate, $Pb_2Sb_2O_7$, but powder x-ray studies on this sample did not show such a composition. Abu'l-Qasim suggested the yellow color is a result of mixing lead and tin (Allen 1973).

The spectra for the green-colored samples are presented in figure 8. It shows that copper is the highest nonmatrix constituent in both the Safavid and Mozafarid samples. The color is probably caused by copper in the presence of iron and lead. The only difference between the two samples seems to be the higher content of lead in the Safavid sample, although they do not look visually the same.

Figure 9 shows the comparison between white-colored samples from the Safavid, Mozafarid, and Qajar periods. The major coloring element is tin. Opaque white glaze is produced by adding tin oxide or arsenic to the batch (Mason and Tite 1997). Tin oxide is not actually a colorant but produces a white opacity in the glaze due to its presence as a colloid.

In figure 10 the spectra of azure samples from the Safavid, Mozafarid, Qajar, and Saljuq periods are presented. The major coloring elements are cobalt and copper. Lead is almost entirely absent in the Mozafarid sample. The percentage of copper in the Saljuq sample is higher than in the

Table 2. *PIXE elemental analyses for tile samples*

Sample no.	Color	Dynasty	Location	Elements (% wt.)										
				Ti	Cr	Mn	Fe	Co	Cu	Zn	As	Sn	Ba	Pb
1	Yellow	Safavid	Southern *iwan*	np	np	np	0.4	np	np	0.1	np	4.4	np	56.9
2	Yellow	Safavid	Western *iwan*	np	np	np	0.1	np	np	bd	np	3.8	np	47.2
3	Yellow	Qajar	Western *iwan*, restoration	np	np	np	0.2	np	np	np	np	2.4	np	44.9
4	Yellow	Qajar	Courtyard, restoration	np	np	np	0.1	np	np	np	np	3.2	np	37.7
5	Green	Safavid	Southern *iwan*	bd	np	np	0.6	np	1.0	np	np	np	np	45.1
6	Green	Mozafarid	Courtyard, restoration	bd	np	np	1.4	np	2.3	np	np	np	np	np
7	Green	Mozafarid	Courtyard, restoration	bd	np	np	2.4	np	1.4	np	np	np	np	np
8	White	Safavid	Southern *iwan*	bd	np	0.1	0.4	np	0.1	np	np	4.9	np	8.4
9	White	Safavid	Western *iwan*	np	np	np	0.4	np	0.1	np	np	6.7	np	13.9
10	White	Mozafarid	Mozafarid mihrab	np	np	np	0.2	np	np	np	np	1.4	np	11.2
11	White	Qajar	Western *iwan*, restoration	np	np	np	0.3	np	0.1	np	np	2.1	np	6.9
12	White	Qajar	Courtyard, restoration	np	np	np	0.4	np	0.1	np	np	3.8	np	10.4
13	Azure	Safavid	Southern *iwan* 1	bd	np	bd	1.3	0.3	0.1	np	0.3	np	np	np
14	Azure	Safavid	Southern *iwan* 2	0.1	np	0.1	1.1	0.4	0.2	np	0.4	np	np	np
15	Azure	Safavid	Southern *iwan* 3	0.1	np	0.1	0.7	0.4	0.3	np	np	np	np	3.2
16	Azure	Safavid	Southern *iwan* 4	bd	np	bd	1.0	0.2	0.2	np	0.3	np	np	np
17	Azure	Safavid	Western *iwan* 1	bd	np	0.1	0.9	0.2	0.1	np	np	np	np	2.0
18	Azure	Safavid	Western *iwan* 2	bd	np	0.1	0.8	0.2	0.1	np	np	np	np	3.0
19	Azure	Safavid	Western *iwan* 3	np	np	0.10	0.9	0.2	0.1	np	np	np	np	2.1
20	Azure	Mozafarid	Mozafarid mihrab 1	np	np	bd	1.5	0.9	bd	np	np	np	np	np
21	Azure	Mozafarid	Mozafarid mihrab 2	bd	np	bd	1.2	0.5	bd	np	np	np	np	0.7
22	Azure	Qajar	Courtyard, restoration	0.1	np	0.1	1.6	0.3	0.2	np	np	np	np	np
23	Azure	Seljuq	Eastern *iwan*	np	np	np	0.6	0.2	0.9	np	np	np	np	2.1
24	Turquoise	Safavid	Southern *iwan* 1	np	np	np	0.4	np	2.4	np	np	3.9	np	10.9
25	Turquoise	Safavid	Southern *iwan* 2	np	np	np	0.5	np	2.9	np	np	4.9	np	11.7
26	Turquoise	Safavid	Southern *iwan* 3	np	np	np	0.5	np	2.7	np	np	4.0	np	12.8
27	Turquoise	Safavid	Southern *iwan* 4	np	np	np	0.4	np	2.5	np	np	3.7	np	12.3
28	Turquoise	Safavid	Southern *iwan* 5	np	np	np	0.5	np	2.6	np	np	4.7	np	13.9
29	Turquoise	Safavid	Western *iwan* 1	bd	np	0.1	0.4	np	2.1	np	np	2.5	np	5.7
30	Turquoise	Safavid	Western *iwan* 2	np	np	0.1	0.3	np	2.3	np	np	2.5	np	10.2
31	Turquoise	Safavid	Western *iwan* 3	bd	np	0.1	0.4	np	2.2	np	np	2.1	np	5.1
32	Turquoise	Safavid	Western *iwan* 4	np	np	np	0.3	np	2.7	np	np	2.5	np	9.7
33	Turquoise	Safavid	Western *iwan* 5	np	np	bd	0.3	np	2.1	np	np	2.3	np	9.7
34	Turquoise	Mozafarid	Mozafarid mihrab 1	np	np	np	0.2	np	0.7	np	np	2.4	np	9.1
35	Turquoise	Mozafarid	Mozafarid mihrab 2	np	np	np	0.3	np	0.5	np	np	1.4	np	5.8
36	Turquoise	Mozafarid	Mozafarid mihrab 3	np	np	bd	0.2	np	1.0	np	np	2.4	np	10.0
37	Turquoise	Mozafarid	Mozafarid mihrab 4	np	np	np	0.3	np	0.9	np	np	2.5	np	14.0
38	Turquoise	Qajar	Courtyard, restoration	bd	np	np	0.3	np	1.1	np	np	1.7	np	18.2
39	Turquoise	Seljuq	Eastern *iwan*	bd	np	np	0.2	np	1.6	np	np	0.3	np	9.7
40	Black	Safavid	Southern *iwan*	0.1	np	4.5	0.9	np	0.2	np	np	np	0.5	4.5
41	Black	Mozafarid	Mozafarid mihrab	0.1	np	2.5	0.6	np	0.3	np	np	np	0.4	6.0
42	Black	Qajar	Courtyard, restoration	np	0.2	1.6	0.7	np	np	np	np	np	np	0.4

Key: bd = below detection; np = not present

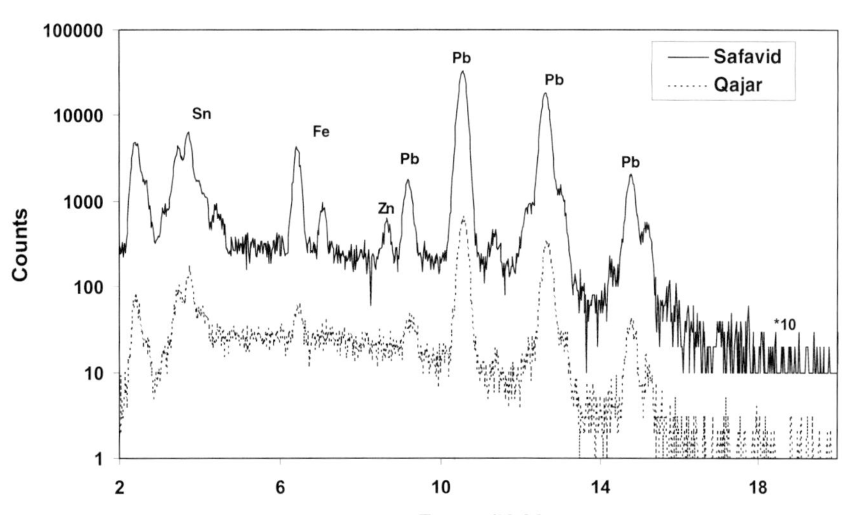

Figure 7. *X-ray spectrum of yellow glaze from the Safavid and Qajar dynasties*

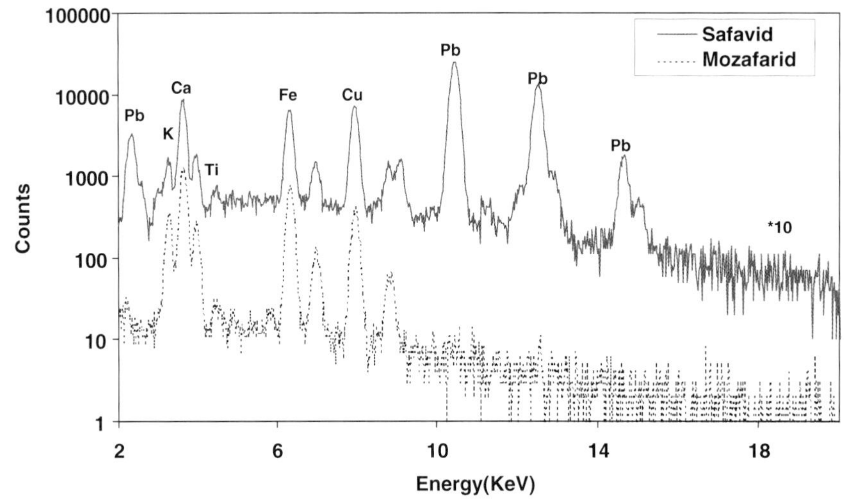

Figure 8. *X-ray spectrum of green glaze from the Safavid and Mozafarid dynasties*

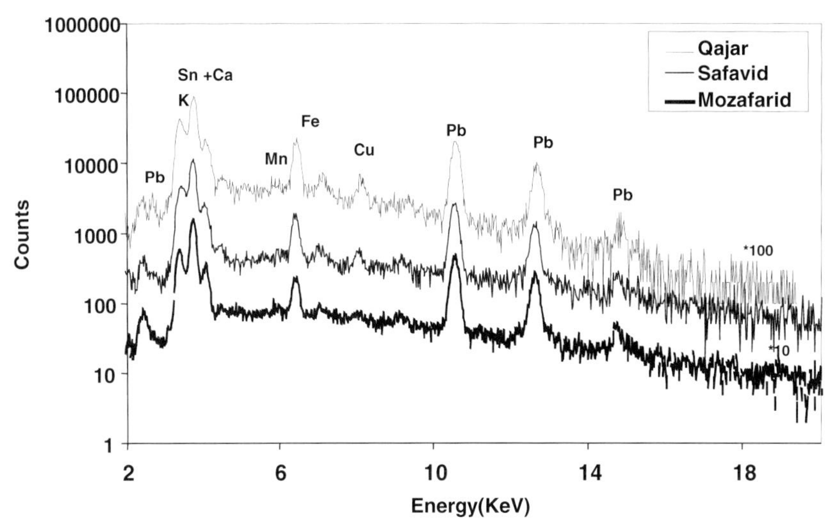

Figure 9. *X-ray spectrum of white glaze from the Safavid, Mozafarid, and Qajar dynasties*

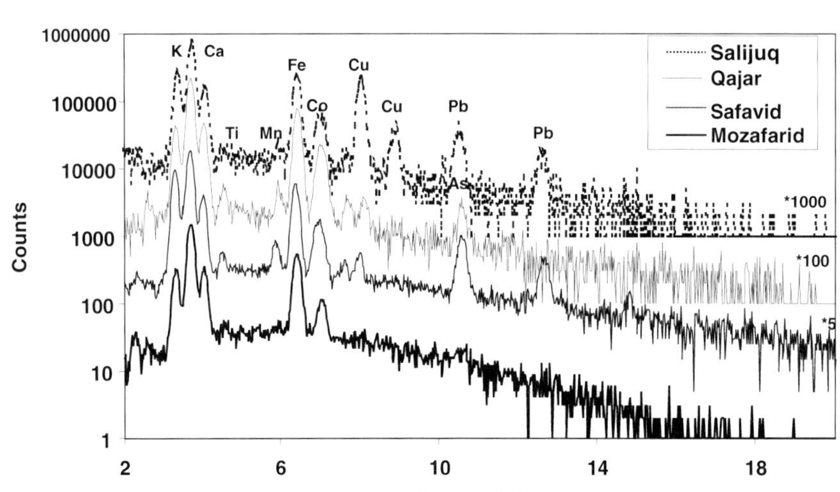

Figure 10. *X-ray spectrum of azure glaze from the Safavid, Mozafarid, Qajar, and Saljuq dynasties*

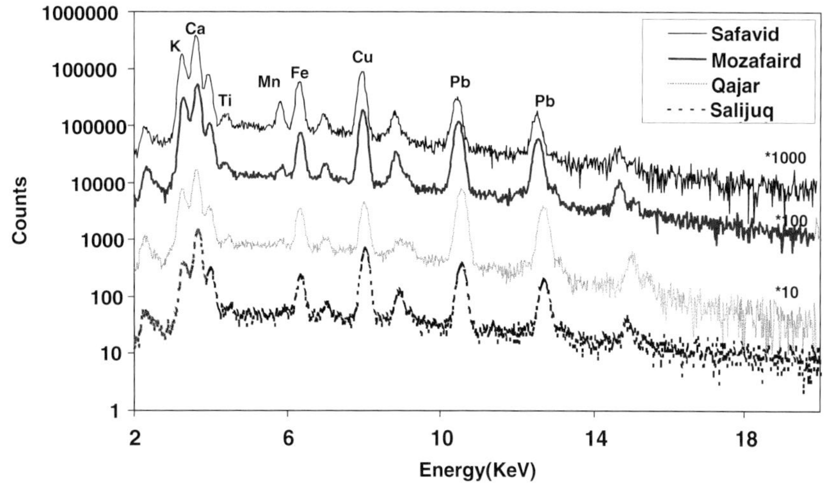

Figure 11. *X-ray spectrum of turquoise glaze from the Safavid, Mozafarid, Qajar, and Saljuq dynasties*

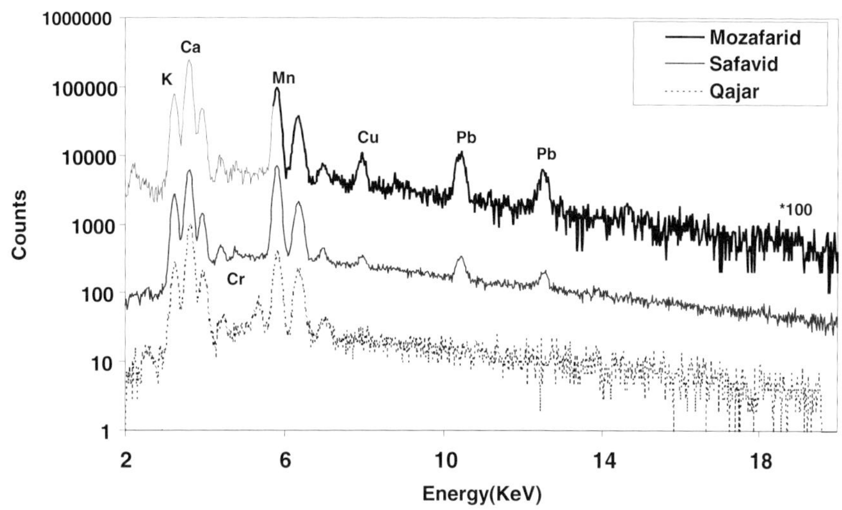

Figure 12. *X-ray spectrum of black glaze from the Safavid, Mozafarid, and Qajar dynasties*

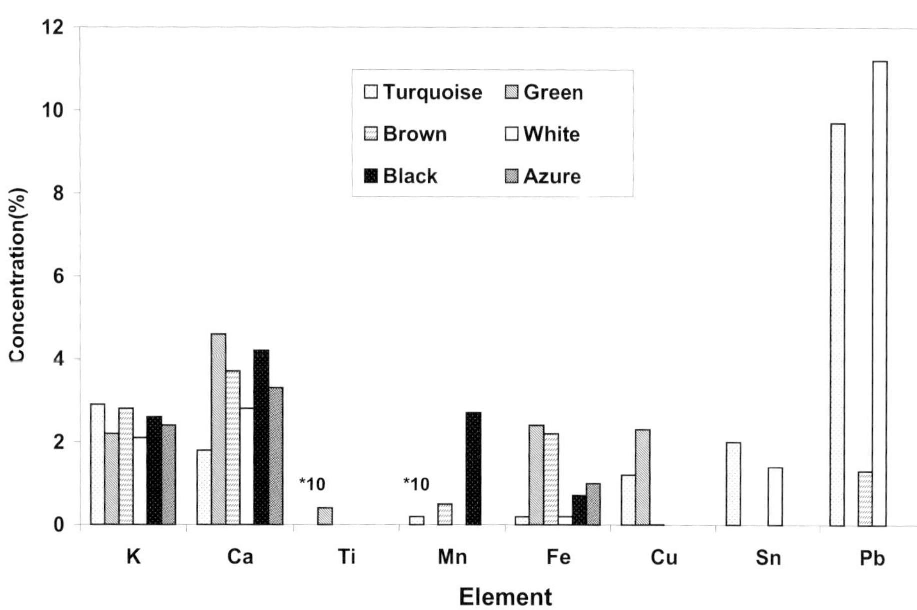

Figure 13. *Histogram of the percent concentration of elements in green-, azure-, brown-, white-, turquoise-, and black-glazed tile samples from Mozafarid mihrab*

others, but the relationship between colors and copper contents is not clear (table 1).

In figure 11 the spectra of turquoise glazes from the Safavid, Mozafarid, Qajar, and Saljuq periods are shown. Copper is the major coloring element, in the presence of iron. Compared to the azure-colored glazes, these glazes have higher lead contents. The copper content of the Mozafarid sample is lower than the others and, as shown in table 1, the visual color is light blue.

Figure 12 displays a comparison between black samples from the Safavid, Mozafarid, and Qajar periods. The coloring elements of these glazes are manganese and iron. The combination of iron, manganese, and sulfur in the glaze coupled with a reducing atmosphere in the furnace produced a black glaze. The visual color of these three samples is not the same. The sample from the Qajar period has a red tint, there are too many gas bubbles and impurities in the glaze, and although the Mozafarid black glaze seems homogenous, it is brown-black in color.

Figure 13 shows the percent concentration of elements in green-, azure-, white-, turquoise-, and black-glazed tile samples from the Mozafarid mihrab. As can be seen, iron is present in all of the samples. Actually, iron is the most important impurity that comes from the raw materials. It can give a range of colors from pale yellow through to blues and greens, depending on the oxidation state of the ion, the presence of transition metal ions, and the composition of the base glaze (Pollard and Heron 1996). Under oxidizing furnace conditions, iron forms the Fe^{III} species, which usually imparts darker green or brown colors, whereas under reducing conditions, Fe^{II} predominates, giving pale blues and greens. Indeed, an understanding of the principles underlying the color of glazes is complex and requires a discussion of such factors as the redox conditions in the furnace and the nature of the coordination sphere around the coloring ion.

Conclusions

Various coloring agents were added to the colorless lead and alkaline glazes. The question of which minerals and pigments were used is complex because they varied according to what was locally available, which impurities were present, their reactions in alkaline or lead media, and the conditions of firing.

Despite the apparent differences of color, the similarity in the composition of the glazes in the use of elements like copper for green, cobalt for blue, and tin oxide for white suggests that availability of resources imposed limits on the variety of production techniques. Due to the fact that lead contents vary in glazes from the same period, comparison of lead contents between samples of different periods does not provide a good basis to study the evolution of tileworking. The appearance of the glazes is dependent on temperature, and it could have been optimized by the selection of an appropriate firing temperature. The inhomogeneity of the glazes and the presence of gas bubbles, especially in the Qajar samples, indicate that the glaze layer did not achieve a suitable maturation temperature.

Sometimes the opacity is due to colorless crystals, gas bubbles, and undissolved quartz and feldspars, which have been seen in the glazes of Qajar period. The numerous quartz grains in the glazes could have served to make the glazes more resistant (Kleinmann 1984), but the quartz grains in these samples are angular and not rounded; therefore, they could be undissolved quartz due to the low firing temperature. The concentration of crystals depends also on other factors—in particular, glaze composition, firing duration, and cooling rate. Differentiation between the effects of all of these different variables is likely to prove very difficult (Tite, Freestone, Mason, et al. 1998). Glaze transparency depends on the absence of crystalline or other phases and the color range associated with the added metal ions, which depends on both the glaze composition and redox conditions during firing. The dark blue (azure) glaze of the Qajar period is very diffuse and mottled with a pitted glaze surface. This is the same for black and turquoise glaze of this period and does not seem to represent a special technology. However, the colors, such as azure and turquoise, are the most commonly found in tiling decorations in Iran, and information can be gathered from the systematic analysis of samples taken from different periods.

For further study, analysis of the samples from local kilns, analysis of the samples from different buildings in the same period, and study of the effect of natural agents on the color and glaze should be investigated.

The accomplishment of these tileworks shows that the technique, although limited to the range of pigments that would withstand the heat required to fire ceramic and glaze in one firing, was well understood by the potters.

Acknowledgments

We would like to thank the technical staff of the Van de Graaff laboratory, Mr. A. Malekzadeh and Mr. B. Movahed, for their help during the experiments; Dr. Haideh Laleh for her useful comments; the head of the Cultural Heritage Organization of Isfahan; and also Mr. Hossein Abadi and Mr. Pour Shiravi for their kind help.

References

Allen, J. W. 1973. Abu'l-Qsim's treatise on ceramics. *Iran* II:111–20.

Galdieri, E. 1972. *Isfahan: Masjid-i Gumᶜa*. Vol. I. Rome: IsMEO.

——. 1973. *Isfahan: Masjid-i Gumᶜa*. Vol. II. Rome: IsMEO.

——. 1984. *Isfahan: Masjid-i Gumᶜa*. Vol. III. Rome: IsMEO.

Golombek, L., and D. Wilber. 1988. *The Timurid architecture of Iran and Turan*. Princeton: Princeton University Press.

Johansoon, S. A., and J. L. Campbell. 1988. *"PIXE": A novel technique for elemental analysis*. New York: John Wiley and Sons.

Kleinmann, B. 1984. History and development of early Islamic pottery glazes. Paper presented at twenty-fourth international archaeometry symposium, 14–18 May, at Smithsonian Institution, Washington, D.C.

Laleh, H. 2000. Masaid-i aumᶜa in Isfahan. In *Great Islamic encyclopedia,* ed. K. Mosavi Bojnoordi. Tehran: Islamic Culture and Guidance.

Mason, R. B., and M. S. Tite. 1997. The beginning of tin-opacification of pottery glazes. *Archaeometry* 39(1):41–58.

Maxwell, J. A., J. L. Campbell, and W. J. Teesdale. 1988. The Guelph PIXE software package. *Nuclear instruments and methods in physics research* B43:218–30.

O'Kane, B. 1994. Iran and Central Asia. In *The mosque,* ed. M. Frishman and H. Uddin Khan. London: Thomas and Hudson.

Pollard, A. M., and C. Heron. 1996. *Archaeological chemistry.* Washington, D.C.: Royal Society of Chemistry.

Porter, V. 1995. *Islamic tile.* London: British Museum Press.

Scarce, J. 1989. Tilework. In *The art of Persia,* ed. R. W. Ferrier, 271–94. London: New Haven and London Publications.

Tite, M. S., I. Freestone, R. Mason, et al. 1998. Review article lead glazes in antiquity-methods of production and reasons for use. *Archaeometry* 40(2):241–60.

Wulff, H. E. 1976. *The traditional crafts of Persia.* Cambridge, Mass.: MIT Press.